W0231972

PATTERN RECOGNITION AND SIGNAL PROCESSING

NATO ADVANCED STUDY INSTITUTES SERIES

Proceedings of the Advanced Study Institute Programme, which aims at the dissemination of advanced knowledge and the formation of contacts among scientists from different countries.

The series is published by an international board of publishers in conjunction with NATO Scientific Affairs Division

A	Life Sciences	Plenum Publishing Corporation
B	Physics	London and New York
C	Mathematical and Physical Sciences	D. Reidel Publishing Company Dordrecht and Boston
D	Behavioural and Social Sciences	Sijthoff & Noordhoff International Publishers B.V.
E	Applied Science	Alphen aan den Rijn, The Netherlands and Winchester, Mass., USA

Series E: Applied Science — No. 29

PATTERN RECOGNITION AND SIGNAL PROCESSING

edited by

C. H. CHEN

Professor of Electrical Engineering
Southeastern Massachusetts University
North Dartmouth, Mass., USA

SIJTHOFF & NOORDHOFF 1978
Alphen aan den Rijn — The Netherlands

Proceedings of the NATO Advanced Study Institute on
Pattern Recognition and Signal Processing
E.N.S.T., Paris, France
June 25-July 4, 1978

ISBN-13: 978-94-009-9943-5 **e-ISBN-13: 978-94-009-9941-1**
DOI: 10.1007/978-94-009-9941-1

Softcover reprint of the hardcover 1st edition 1978

TABLE OF CONTENTS

LIST OF PARTICIPANTS

ABOUTAJDINE, Driss
MOROCCO

EL ABIDINE AMRI, Zine
MOROCCO

ARENA, Luigi
ITALY

BAKKE, Knut I.
NORWAY

BANERJI, Ranan
U.S.A.

BASAR, Tangül
TURKEY

BEATTY, Jackson
U.S.A.

BERG, Øivind
NORWAY

BÖHME, Johann F.
GERMANY

BROCK-NANNESTAD, L.
DENMARK

BROLLEY, John E.
U.S.A.

CARAYANNIS, George
BELGIUM

CARDOSO, Julio A.
PORTUGAL

CEROFOLINI, Luigi
ITALY

CETINCELIK, Muammer
TURKEY

CHEN, C. H.
U.S.A.

CHEN, W. W.
U.S.A.

COCKETT, J. R. B.
U.K.

CORCI, Patrick
FRANCE

DALE, Christopher
CANADA

DEHNE, John S.
U.S.A.

DEMORI, Renato
ITALY

DEVIJVER, Pierre A.
BELGIUM

DUARTE, Vasco
ITALY

DURRANI, T. S.
U.K.

ENDER, J.
GERMANY

FARANTATOS, Christos S.
GREECE

FAURE, C.
FRANCE

FLEURET, J.
FRANCE

FU, K. S.
U.S.A.

GARIBOTTO, Giovanni
ITALY

GECKINLI, Nezih C.
TURKEY

GOUTIS, C. E.
U.K.

GREEN, Roger
U.K.

GUEGUEN, C.
FRANCE

HACCOUN, David
CANADA

HALEY, William L.
U.S.A.

HAND, D. J.
U.K.

HAYKIN, Simon
CANADA

HEIER, Halvor
NORWAY

HELME, B.
ITALY

HICKS, P. J.
U.K.

HILL, W. J.
U.K.

HODES, Isidore
U.S.A.

HODGE, David C.
U.S.A.

HODGKISS, William
U.S.A.

JAIN, A.K.
U.S.A.

JENSEN, Gert Hvedstrup
NORWAY

JOHNSEN, Jarl K.
NORWAY

KAHN, Edmond
FRANCE

KAMMENOS, Panayotis
GREECE

KITTLER, Josef
U.K.

KOTSIS, Demetres
GREECE

KRUGER, R.P.
U.S.A.

KURSS, Herbert
U.S.A.

LAI, David C.
U.S.A.

LECROART, J. L.
FRANCE

LEMKE, Heinz
GERMANY

LONGO, Maurizio
ITALY

LU, Shin-yee
U.S.A.

MAHMOUD, M.
FRANCE

MAITRE, H.
FRANCE

MAKOV, U. E.
U.K.

MCNARY, Charles Allen
U.S.A.

MICLET, L.
FRANCE

MINDEN, Gary J.
U.S.A.

MODESTINO, J. W.
U.S.A.

MOLIN, Carrado
ITALY

MONDT, J. C.
NETHERLANDS

MONTEIRO, Isabel M.
PORTUGAL

NADLER, Morton
FRANCE

NAVARRO, A. M.
NETHERLANDS

OLSON, R. Craig
U.S.A.

OOSTERLINCK, A.
BELGIUM

OZAKI, T.
U.K.

PADILHA, Jorge
PORTUGAL

PAU, L. F.
FRANCE

PEKMESTZI, Kiamal
GREECE

REMOND, Antoine
FRANCE

REPPERGER, Daniel W.
U.S.A.

RETTER, M. L.
U. K.

SÄDTLER, E.
GERMANY

SCHOONEVELD, C. van
NETHERLANDS

SCLOVE, Stanley L.
U.S.A.

STARK, Henry
U.S.A.

STEPANISHEN, Peter R.
U.S.A.

SUEN, Ching Y.
CANADA

TACCONI, Giorgio
ITALY

TAM, W.
CANADA

TASTO, M.
GERMANY

TONG, Howell
U.K.

TRULSEN, Jan
NORWAY

VASCONCELOS, H.
PORTUGAL

VISWANATHAN, R.
U.S.A.

VOLES, Roger
U.K.

YAVUZ, Davras
TURKEY

ZISKIND, Ilan
ISRAEL

PREFACE

Both pattern recognition and signal processing are rapidly growing areas. Organized with emphasis on many inter-relations between the two areas, a NATO Advanced Study Institute on Pattern Recognition and Signal Processing was held June 25th - July 4, 1978 at the E.N.S.T. (Department of Electronics) in Paris, France. This volume is the Proceedings of the Institute. It contains what I believed to be a truly outstanding collection of papers which cover all major activities in both pattern recognition and signal processing. The papers are grouped by topics as follows:

I. Syntactic Methods: paper numbers 1, 2.
II. Statistical Methods: paper numbers 3, 4, 5, 6.
III. Detection and Estimation: paper numbers 7, 8.
IV. Image Processing, Modelling, and Analysis: paper numbers 9, 10, 11, 12.
V. Speech Application: paper numbers 13, 14.
VI. Radar Application: paper number 15.
VII. Seismic Application: paper number 16.
VIII. Biomedical Application: paper numbers 17, 18, 19.
IX. Reconstruction From Projections: paper numbers 20, 21.
X. Signal Modelling and Application: paper numbers 22, 23, 24.
XI. NATO Pattern Recognition Research Study Group Report: paper number 25.

It is my strong belief that there is a need for continuing interaction between pattern recognition and signal processing. The book will serve as a useful text and reference for such a need, and for both areas.

Finally on behalf of all participants of the Institute, I would like to thank Drs. T. Kester and M. N. Özdas of NATO for their support.

No. Dartmouth, Massachusetts
August 1978

C. H. Chen

SYNTACTIC PATTERN RECOGNITION AND ITS APPLICATIONS TO SIGNAL PROCESSING*

K. S. Fu

School of Electrical Engineering
Purdue University
W. Lafayette, Indiana 47907 U.S.A.

1. INTRODUCTION

The many different mathematical techniques used to solve pattern recognition problems may be grouped into two general approaches [1,2]. They are the decision-theoretic (or discriminant) approach and the syntactic (or structural) approach [3]. In the decision-theoretic approach, a set of characteristic measurements, called features, are extracted from the patterns. Each pattern is represented by a feature vector, and the recognition of each pattern is usually made by partitioning the feature space. On the other hand, in the syntactic approach, each pattern is expressed as a composition of its components, called subpatterns and pattern primitives. This approach draws an analogy between the structure of patterns and the syntax of a language. The recognition of each pattern is usually made by parsing the pattern structure according to a given set of syntax rules. In this paper, we briefly review the recent progress in syntactic pattern recognition and some of its applications.

A block diagram of a syntactic pattern recognition system is shown in Fig. 1. We divide the block diagram into the recognition part and the analysis part, where the recognition part consists of preprocessing, primitive extraction (including relations among primitives and subpatterns), and syntax (or structural) analysis, and the analysis part includes primitive selection and grammatical (or structural) inference.

*This work was supported by the National Science Foundation Grant ENG 76-18567, and the U. S. - Italy Cooperative Science Program.

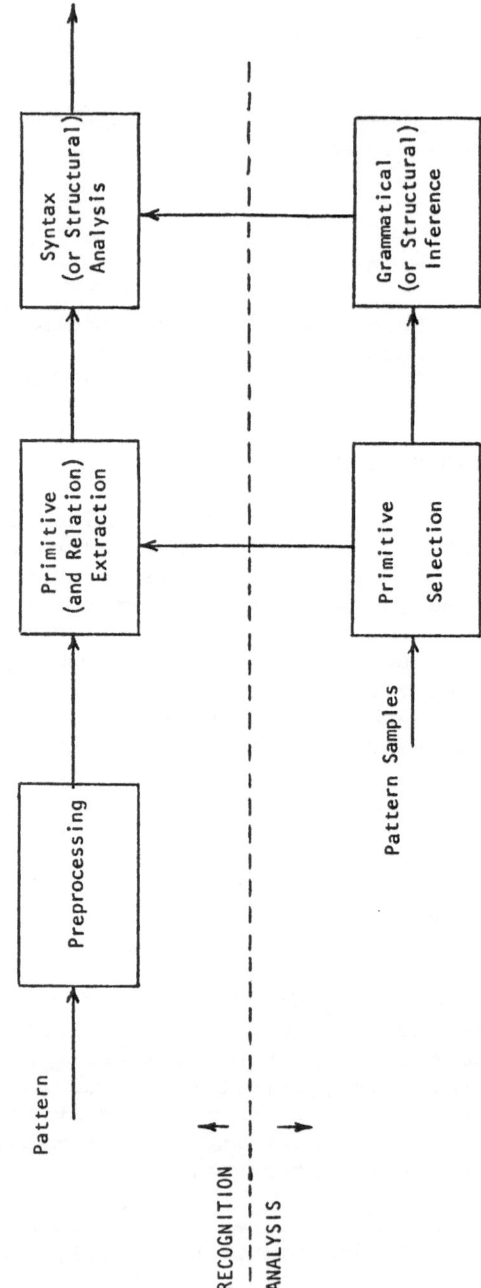

Figure 1 Block Diagram of a Syntactic Pattern Recognition System

In syntactic methods, a pattern is represented by a sentence in a language which is specified by a grammar. The language which provides the structural description of patterns, in terms of a set of pattern primitives and their composition relations, is sometimes called the "pattern description language". The rules governing the composition of primitives into patterns are specified by the so-called "pattern grammar". An alternative representation of the structural information of a pattern is to use a "relational graph", of which the nodes represent the subpatterns and the branches represent the relations between subpatterns.

Figure 2 gives an illustrative example for the description of the boundary of a submedian chromosome. The hierarchical structural description is shown in Fig. 2(a), and the grammar generating submedian chromosome boundaries is given in Fig. 2(b).

2. PRIMITIVE SELECTION AND PATTERN GRAMMARS

Since pattern primitives are the basic components of a pattern, presumably they are easy to recognize. Unfortunately, this is not necessarily the case in some practical applications. For example, strokes are considered good primitives for script handwriting, and so are phonemes for continuous speech, however, neither strokes nor phonemes can easily be extracted by machine. A compromise between its use as a basic part of the pattern and its easiness for recognition is often required in the process of selecting pattern primitives.

There is no general solution for the primitive selection problem at this time [3-5]. For line patterns or patterns described by boundaries or skeletons, line segments are often suggested as primitives. A straight line segment could be characterized by the locations of its beginning (tail) and end (head), its length, and/or slope. Similarly, a curve segment might be described in terms of its head and tail and its curvature. The information characterizing the primitives can be considered as their associated semantic information or as features used for primitive recognition. Through the structual description and the semantic specification of a pattern, the semantic information associated with its subpatterns or the pattern itself can then be determined. For pattern description in terms of regions, half-planes have been proposed as primitives. Shape and texture measurements are often used for the description of regions.

After pattern primitives are selected, the next step is the construction of a grammar (or grammars) which will generate a language (or languages) to describe the patterns under study. It is known that increased descriptive power of a language is paid for in terms of increased complexity of the syntax analysis system

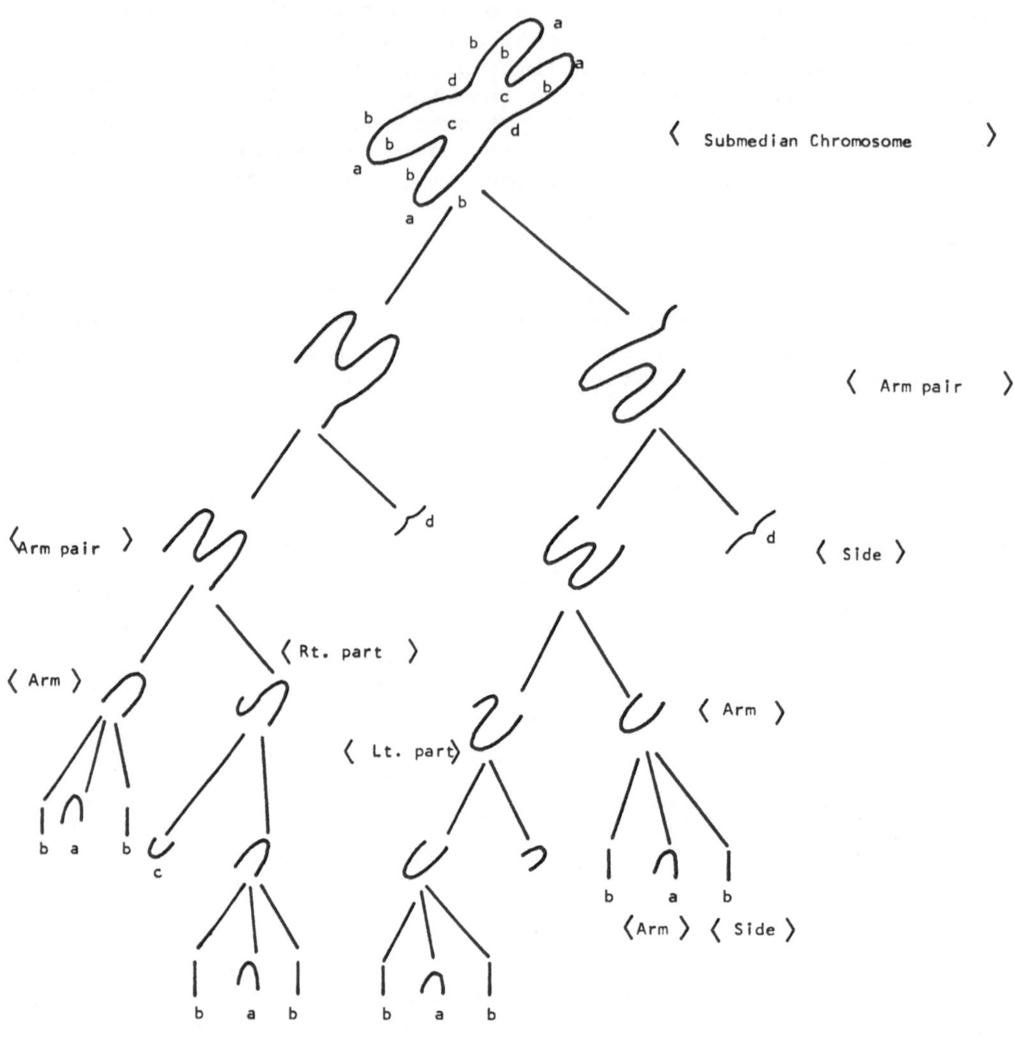

babcbabdbabcbabd

(a)

G = (V_N, TV_T, P, <Submedian>)

where V_N = {<Submedian>, <Arm pair>, <Lt. part>,
<Rt. part>, <Arm>, <Side>}

$$V_T = \left\{ \overset{\cap}{a}, \ \overset{|}{b}, \ \overset{\cup}{c}, \ \overset{\langle}{d} \right\}$$

and P: <Submedian> → <Arm pair> <Arm pair>

<Arm pair> → <Arm pair><Side>

<Arm pair> → <Arm><Rt. part>

<Arm pair> → <Lt. part><Arm>

<Rt. part> → c<Arm>

<Lt. part> → <Arm>c

<Arm> → b<Arm>

<Arm> → <Arm>b

<Side> → b<Side>

<Side> → <Side>b

<Arm> → a

<Side> → b

<Side> → d

(b)

Figure 2 Syntactic Representation of Submedian Chromosome

(recognizer or acceptor). Finite-state automata are capable of recognizing finite-state languages although the descriptive power of finite-state languages is also known to be weaker than that of context-free and context-sensitive languages. On the other hand, nonfinite, nondeterministic procedures are required, in general, to recognize languages generated by context-free and context-sensitive grammars. The selection of a particular grammar for pattern description is affected by the primitives selected, and by the tradeoff between the grammar's descriptive power and analysis efficiency.

If the primitives selected are very simple, more complex grammars may have to be used for pattern description. On the other hand, a use of sophisticated primitives may result in rather simple grammars for pattern description, which in turn will result in fast recognition algorithms. The interplay between the complexities of primitives and of pattern grammars is certainly very important in the design of a syntactic pattern recognition system. Context-free programmed grammars, which maintain the simplicity of context-free grammars but can generate context-sensitive languages, have recently been suggested for pattern description [3].

A number of special languages have been proposed for the description of patterns such as English and Chinese characters, chromosome images, spark chamber pictures, two-dimensional mathematics, chemical structures, spoken words, and fingerprint patterns [3,6]. For the purpose of effectively describing high dimensional patterns, high dimensional grammars such as web grammars, graph grammars, tree grammars, and shape grammars have been used for syntactic pattern recognition [3,7,8].

Ideally speaking, it would be nice to have a grammatical (or structural) inference machine which would infer a grammar from a given set of patterns. Unfortunately, not many convenient grammatical inference algorithms are presently available for this purpose [9-15]. Nevertheless, recent literatures have indicated that some simple grammatical inference algorithms have already been applied to syntactic pattern recognition, particularly through man-machine interaction [16-19].

3. SYNTACTIC PATTERN RECOGNITION USING STOCHASTIC LANGUAGES

In some practical applications, a certain amount of uncertainty exists in the process under study. For example, due to the presence of noise and variation in the pattern measurements, segmentation error and primitive extraction error may occur, causing ambiguities in the pattern description languages. In order to describe noisy and distorted patterns under ambiguous situations, the use of stochastic languages has been suggested [3].

With probabilities associated with grammar rules, a stochastic grammar generates sentences with a probability distribution. The probability distribution of the sentences can be used to model the noisy situations.

A stochastic grammar is a four-tuple $G_s = (V_N, V_T, P_s, S)$ where P_s is a finite set of stochastic productions. For a stochastic context-free grammar, a production in P_s is of the form

$$A_i \xrightarrow{P_{ij}} \alpha_j, A_i \in V_N, \alpha_j \in (V_N \cup V_T)^*$$

where P_{ij} is called the production probability. The probability of generating a string x, called the string probability p(x), is the product of all production probabilities associated with the productions used in the generation of x. The language generated by a stochastic grammar consists of the strings generated by the grammar and their associated string probabilities.

By associating probabilities with the strings, we can impose a probabilistic structure on the language to describe noisy patterns. The probability distribution characterizing the patterns in a class can be interpreted as the probability distribution associated with the strings in a language. Thus, statistical decision rules can be applied to the classification of a pattern under ambiguous situations (for example, use the maximum-likelihood or Bayes decision rule). A block diagram of such a recognition system using maximum-likelihood decision rule is shown in Figure 3. Furthermore, because of the availability of the information about production probabilities, the speed of syntactic analysis can be improved through the use of this information [3]. Of course, in practice, the production probabilities will have to be inferred from the observation of relatively large numbers of pattern samples [3]. When the imprecision and uncertainty involving in the pattern description can be modeled by using the fuzzy set theory, the use of fuzzy languages for syntactic pattern recognition has recently been suggested [53].

Other approaches for the recognition of distorted or noisy patterns using syntactic methods include the use of transformational grammar [26,27] and approximation [25], and the application of error-correcting parsing techniques [24,28-31].

4. SYNTACTIC RECOGNITION AND ERROR-CORRECTING PARSING

Conceptually, the simplest form of recognition is probably "template-matching". The sentence describing an input pattern is matched against sentences representing each prototype or reference pattern. Based on a selected "matching" or "similarity" criterion,

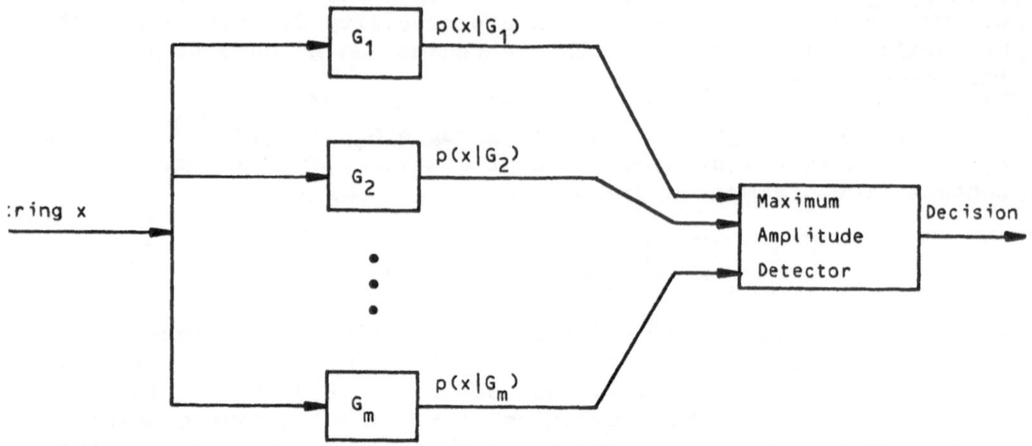

Figure 3 Maximum-Likelihood Syntactic Recognition System

the input pattern is classified in the same class as the prototype pattern which is the "best" to match the input. The structural information is not recovered. If a complete pattern description is required for recognition, a parsing or syntax analysis is necessary. In between the two extreme situations, there are a number of intermediate approaches. For example, a series of tests can be designed to test the occurrence or nonoccurrence of certain subpatterns (or primitives) or certain combinations of them. The result of the tests, through a table lookup, a decision tree, or a logical operation, is used for a classification decision. Recently, the use of discriminant grammars has been proposed for the classification of syntactic patterns [20].

A parsing procedure for recognition is, in general, nondeterministic and, hence, is regarded as computationally inefficient. Efficient parsing could be achieved by using special classes of languages such as finite state and deterministic languages for pattern description. The tradeoff here between the descriptive power of the pattern grammar and its parsing efficiency is very much like that between the feature space selected and the classifier's discrimination power in a decision-theoretic recognition system. Special parsers using sequential procedures or other heuristic means for efficiency improvement in syntactic pattern recognition have recently been constructed [21-24].

In practical applications, pattern distortion and measurement noise often exist. Pattern segmentation errors and misrecognitions of primitives (and relations) and/or subpatterns will lead to erroneous or noisy sentences rejected by the grammar characterizing its class. Recently, the use of an error-correcting parser as a recognizer of noisy and distorted patterns has been proposed [24, 28-31]. In the use of an error-correcting parser as a recognizer, the pattern grammar is first expanded to include all the possible errors into its productions. The original grammar is transformed into a covering grammar that generates not only the correct sentences, but also all the possible erroneous sentences. For string grammars, three types of error - substitution, deletion and insertion - are considered. Misrecognition of primitives (and relations) are regarded as substitution errors, and segmentation errors as deletion and insertion errors. A distance between two strings is defined as the minimum number of error transformations required to derive one from the other. If the pattern grammar is nonstochastic, the minimum-distance criterion [30] for error-correcting parsing can be applied. On the other hand, if the grammar is stochastic, the maximum-liklihood and Bayes criteria can be used [24,30]. One difficulty about this approach is the parsing time, in particular, when all the three types of error are considered.

The sequential parsing procedure suggested by Persoon and Fu [23] has been applied to error-correcting parser to reduce the parsing time [24]. By sacrificing a small amount of error-correcting power (that is, allowing a small error in parsing), a parsing could be terminated much earlier before a complete sentence is scanned. The trade-off between the parsing time and the error committed can be easily demonstrated. In addition, error-correcting parsing for transition network grammars [30] and tree grammars [31] has also been studied. For tree grammars, five types of error - substitution, deletion, stretch, branch, and split - are considered. The original tree pattern grammar is expanded by including the five types of error transformation rule. The tree automaton constructed according to the expanded tree grammar and the minimum-distance criterion is called an error-correcting tree automaton (ECTA). When only substitution errors are considered, the structure of the tree to be analyzed remains unchanged. Such an error-correcting tree automaton is called a "structure-preserved error-correcting tree automaton" (SPECTA)* [32]. Another approach to reduce the parsing time is the use of parallel processing [33].

5. SYNTACTIC CLUSTERING

Fu and Lu [34,35] have recently suggested to use the distance or the weighted distance between two sentences (two strings or two trees) and between a sentence and a language as a similarity measure between two syntactic patterns and between a syntactic pattern and a class of syntactic patterns. With such similarity measures, cluster analysis can be performed on syntactic patterns using any existing cluster-seeking algorithm [2,36]. As in the decision-theoretic approach, the clustering procedure is nonsupervised. After the cluster analysis is completed, a grammatical inference procedure can be applied to infer a grammar for each cluster, a conventional (non-error-correcting) parser can be easily constructed for recognition. Consequently, the recognition efficiency is also improved.

When the distance (or weighted distance) is computed on a sentence-to-sentence basis, the use of error-correcting parser can be avoided in the cluster analysis [35]. Such a direct measure of similarity between two syntactic patterns will result in significant reduction of computations in cluster analysis although it is often not as flexible as that using error-correcting parsers. Also, when the cluster analysis is performed on a sentence-to-sentence basis, the clustering result is usually very sensitive with respect to the variations of pattern size and orientation. Normalizations with respect to pattern size and orientation are required in order to obtain reliable clustering result.

*One major application of SPECTA is texture discrimination [54].

6. SYNTACTIC APPROACH TO SHAPE ANALYSIS

Recently, syntactic methods have been applied to both shape description and recognition. Pavlidis and Ali [45] have proposed a general model of syntactic shape analyzer. The first major component of the model is a curve-fitting algorithm which achieves the noise elimination and data reduction. The split-and-merge algorithm is used to obtain a polygonal approximation of the boundary of the original picture or object. It is assumed that the boundaries of the objects of interest consist of concatenations of the following subpatterns or nonterminals: QUAD (arcs approximated by a quadric curve), TRUS (sharp protrusions or intrusions), LINE (long line segments), and BREAK (short segments with no regular shape). Each of the nonterminals has a set of attributes as its semantic information. The production rules of the proposed general shape grammar consist of both syntactic and semantic rules. Stochastic finite automata are used as parsers for shape recognition.

Another method recently proposed for syntactic shape description and recognition is the use of attributed grammars [18]. Two types of primitive with attributes are proposed. The first type is a curve segment with its direction (the vector from the starting point to the end point), total length, total angular change, and a measure of its symmetry as the four attributes. The second type of primitive is an angle primitive with its attribute specified by the angular change at the concatenating point of two consecutive curve segments. Finite-state and context-free attributed grammars are used for shape description and recognition. Each production rule of the attributed grammar has a symbolic part like the conventional grammar rule and a semantic part for processing the attributes of the terminals and nonterminals in the symbolic part. The primitive extraction process is embedded in the parsing of the strings describing the boundaries of objects. Modified Earley parser and finite automata are used as shape recognizers.

7. APPLICATIONS TO WAVEFORM AND SIGNAL PROCESSING

Syntactic pattern recognition has been applied to waveform analysis, ECG interpretation, speech recognition and understanding,* character recognition, fingerprint classification, recognition of two-dimensional mathematical notation, modelling of Earth Resources Satellite data, machine parts recognition and automatic visual inspection [6,8,32,37-44]. In this paper, we briefly review

*For a detailed review of the recent progress in speech recognition, and ECG processing, refer to the paper by R. DeMori in this Proceedings.

some of the recent applications of syntactic methods to waveform and signal processing.

Waveforms are basically one-dimensional signals, which appear to be naturally suitable for the application of syntactic methods. A waveform could be represented by a concatenation of waveform segments. However, the selection of primitives (basic waveform segments) and subpatterns could be quite different from different application points of view. Linear and/or quadric segments through functional approximation have been proposed as waveform primitives [6]. Ehrich and Foith [45] have proposed the use of a relational tree to describe a waveform in terms of its peaks and valleys. Sankar and Rosenfeld [46] have recently proposed an alternative method of using a peak relational tree in terms of fuzzy connectivity to describe waveforms.

Horowitz [6] has proposed a deterministic context-free grammar that can be used to recognize positive and negative peaks in a waveform represented by a string of "positive slope", "negative slope" and "zero slope" primitives. The grammar is

$$G = (V_N, V_T, P, W)$$

where $V_N = \{W, <w>, <p^{t}>, <P^{-}>, <p_1>, <n_1>,$

$<p_2>, <n_2>, <z>\}$

$V_T = \{p/, n\backslash, \underline{0}\}$

and P:

$$W \rightarrow <z><w><z>$$

$$W \rightarrow <z><w>$$

$$W \rightarrow <w><z>$$

$$W \rightarrow <w>$$

$$W \rightarrow <z>$$

$$<w> \rightarrow <p^{+}>$$

$$<w> \rightarrow <p^{-}>$$

$$<w> \rightarrow <p_1>$$

$$<w> \rightarrow <n_1>$$

$$<P^{+}> \rightarrow ><P^{-}><z><n_1>$$

$$<P^{+}> \rightarrow <P^{-}><n_1>$$

$\langle P^+ \rangle \rightarrow \langle p_1 \rangle \langle z \rangle \langle n_1 \rangle$

$\langle P^+ \rangle \rightarrow \langle p_1 \rangle \langle n_1 \rangle$

$\langle P^- \rangle \rightarrow \langle P^+ \rangle \langle z \rangle \langle p_1 \rangle$

$\langle P^- \rangle \rightarrow \langle P^+ \rangle \langle p_2 \rangle$

$\langle P^- \rangle \rightarrow \langle n_1 \rangle \langle z \rangle \langle p_1 \rangle$

$\langle P^- \rangle \rightarrow \langle n_1 \rangle \langle p_1 \rangle$

$\langle p_1 \rangle \rightarrow \langle p_1 \rangle\ p$

$\langle p_1 \rangle \rightarrow \langle p_2 \rangle\ p$

$\langle p_1 \rangle \rightarrow p$

$\langle n_1 \rangle \rightarrow \langle n_1 \rangle\ n$

$\langle n_1 \rangle \rightarrow \langle n_2 \rangle\ n$

$\langle n_1 \rangle \rightarrow n$

$\langle p_2 \rangle \rightarrow \langle p_1 \rangle\ 0$

$\langle p_2 \rangle \rightarrow \langle p_2 \rangle\ 0$

$\langle n_2 \rangle \rightarrow \langle n_1 \rangle\ 0$

$\langle n_2 \rangle \rightarrow \langle n_2 \rangle\ 0$

$\langle z \rangle \rightarrow \langle z \rangle\ 0$

$\langle z \rangle \rightarrow 0$

When parsing a specific string describing a waveform, a positive peak is recognized if and only if a section of the string is completely reduced by some production to the nonterminal $\langle P^+ \rangle$. Similarly, a negative peak is recognized if and only if a section of the input string is reduced to the nonterminal $\langle p^- \rangle$.

Stockman et. al. have suggested the use of a waveform parsing system to analyze carotid pulse waves [22]. A typical carotid pulse wave and its hierarchical structure representation are shown in Fig. 4(a) and Fig. 4(b) respectively. The context-free grammar generating various types (structures) of carotid pulse waves is given below.

$G = (V_N, V_T, P, \langle \text{CAROTID PULSE WAVE} \rangle)$

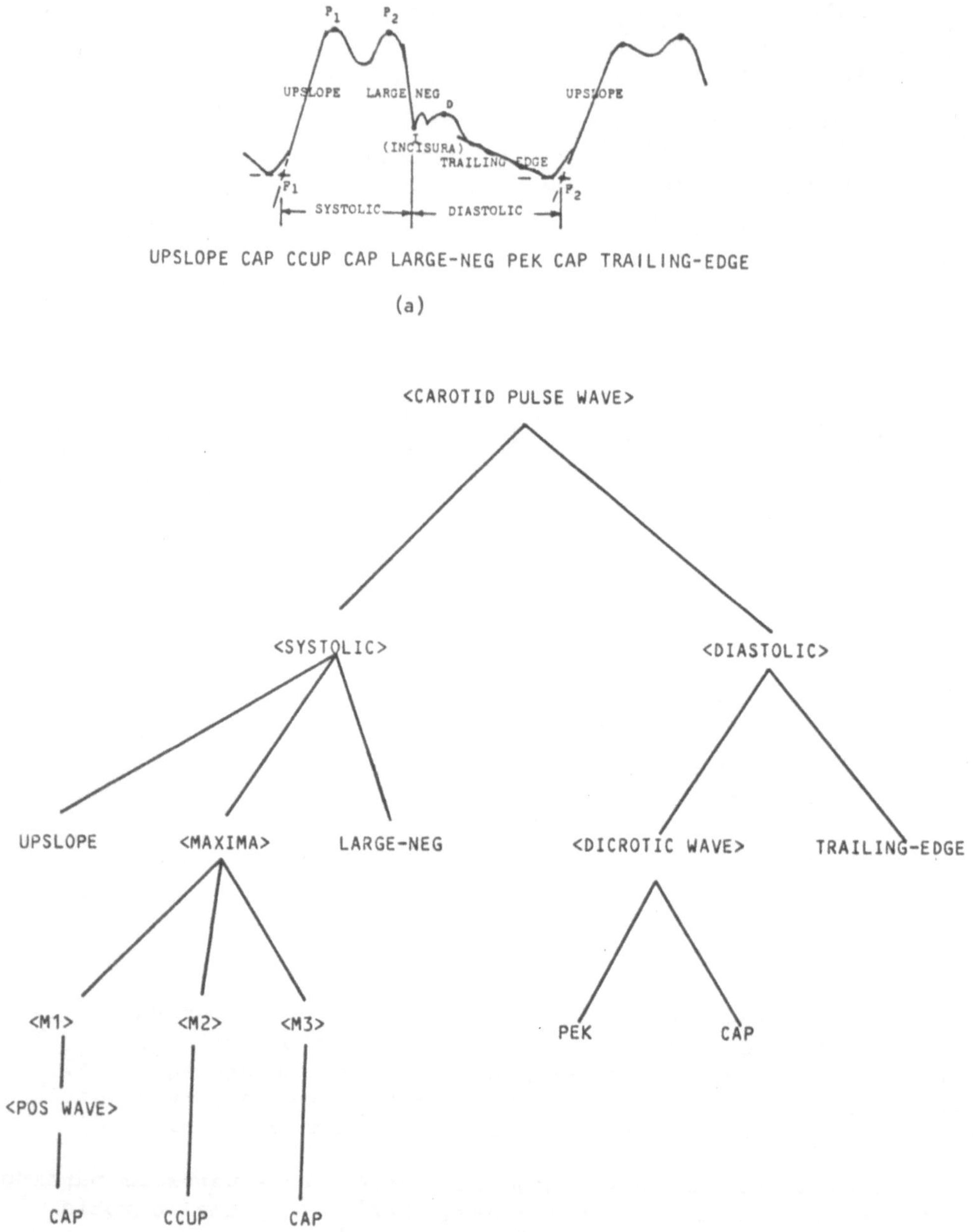

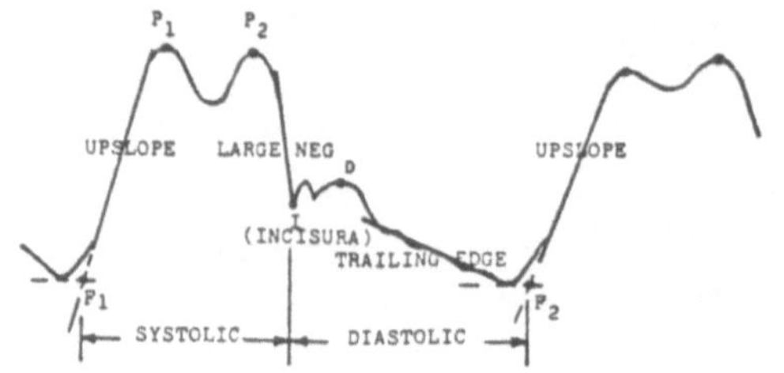

P1
P2
UPSLOPE
LARGE NEG
D
(INCISURA)
TRAILING EDGE
UPSLOPE
P1
P2
SYSTOLIC
DIASTOLIC
UPSLOPE CAP CCUP CAP LARGE-NEG PEK CAP TRAILING-EDGE
(a)
<CAROTID PULSE WAVE>
<SYSTOLIC>
<DIASTOLIC>
UPSLOPE
<MAXIMA>
LARGE-NEG
<DICROTIC WAVE>
TRAILING-EDGE
<M1>
<M2>
<M3>
PEK
CAP
<POS WAVE>
CAP
CCUP
CAP

(b) Structural representation of a Carotid pulse wave

Figure 4 A Typical Carotid Wave and Its Hierarchical Structure Representation

Primitives:

UPSLOPE --------- Line: long, large positive slope

LARGE-POS ------- Line: medium length, large positive slope

LARGE-NEG ------- Line: medium length, large negative slope

MED-POS --------- Line: medium length and positive slope

MED-NEG --------- Line: medium length and negative slope

TRAILING-EDGE --- Line: Long, medium negative slope

HOR ------------- Line: short, near 0 slope

CAP ------------- Parabola

PEK ------------- Parabola

CCUP ------------ Parabola

VCUP ------------ Parabola

RSHOLD ---------- Parabola: right half of parabolic maxima

LSHOLD ---------- Parabola: left half of parabolic maxima

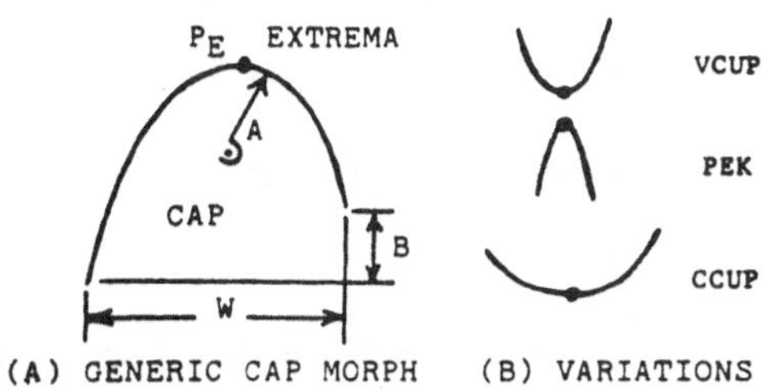

(A) GENERIC CAP MORPH (B) VARIATIONS

where V_N = {<CAROTID PULSE WAVE>, <SYSTOLIC>,
<DIASTOLIC>, <MAXIMA>, <M1>, <M1>, <M2>,
<M3>, <DICROTIC WAVE>, <POS WAVE>,
<NEG WAVE>}

V_T = {UPSLOPE, LARGE-POS, LARGE-NEG,
MED-POS, MED-NEG, TRAILING-EDGE
HOR, CAP, PEK, CCUP, VCUP
RSHOLD, LSHOLD}

and P:

<CAROTID PULSE WAVE> → <SYSTOLIC><DIASTOLIC>

<SYSTOLIC> → UPSLOPE <MAXIMA> LARGE-NEG

<MAXIMA> → <M1><M2><M3>

<MAXIMA> → MED-POS <M3>

<MAXIMA> → <M1> MED-NEG

<DIASTOLIC> → TRAILING-EDGE

<DIASTOLIC> → <DICROTIC WAVE> TRAILING-EDGE

<DICROTIC WAVE> → CAP, <DICROTIC WAVE> → HOR

<DICROTIC WAVE> → PEK, <DICROTIC WAVE> → PEK CAP

<M1> → LSHOLD, <M1> → <POS WAVE>

<M2> → CCUP, <M2> → <NEG WAVE>

<M3> → RSHOLD, <M3> → CAP

<POS WAVE> → CAP, <POS WAVE> → CAP LARGE-NEG

<NEG WAVE> → VCUP, <NEG WAVE> → VCUP LARGE-POS

Le Chevalier et. al. [47] have recently proposed the use of syntactic decoding method (error-correcting string parser with substitution error only) for syntactic signal processing. Waveform segments (e.g., segments of sinusoids) are selected as primitives. A signal is described as a sequence (concatenation) of primitives. In addition to the distance suggested in error-correcting parsing, the correction is also used as a similarity measure between two strings. Practical applications include signal detection, radar target identification and adaptive antenna processing

7. CONCLUDING REMARKS

We have briefly reviewed some recent advances in the area of syntactic pattern recognition and its application to signal processing. Due to noise and distortions in real world patterns, syntactic approach to pattern recognition was regarded earlier as only effective in handling abstract and artificial patterns. However, with the recent development of distance or similarity measures between syntactic patterns and error-correcting parsing procedures, the flexibility of syntactic methods has been greatly expanded. Errors occurring at the lower-level processing of a pattern (segmentation and primitive recognition) could be compensated at the higher level using structural information. Using a distance or similarity measure, nearest-neighbor and k-nearest-neighbor classification rules can be easily applied to syntactic patterns. Furthermore, with a distance or similarity measure, a clustering procedure can be applied to syntactic patterns. Such a nonsupervised learning procedure can also be very useful for grammatical inference in syntactic pattern recognition [19,34].

It has been noticed from the recent advances that semantic information has been used more and more with the syntax rules in characterizing patterns. Quite often, semantic information involving spatial information can be expressed syntactically such as attributed grammars, and relational trees and graphs [3,10,13, 18,43,45]. Parsing efficiency has become a concern in syntactic recognition. Special grammars and parallel parsing algorithms have been suggested for speeding up the parsing time. Structural information of an image can also be used as a guide in the segmentation process through the syntactic approach [40,48]. Syntactic representation of patterns such as hierarchical trees and relational graphs should also be very useful for database organization. Several recent publications have already shown such a trend [49-52].

In some applications, both the decision-theoretic and the syntactic methods may be used. One possibility is to use a decision-theoretic method for the recognition of primitives, and then to use a syntactic method for the recognition of subpattern or pattern itself. The second possibility is the use of stochastic grammars with which the syntactic recognition is made in the decision-theoretic sense (maximum-likelihood or Bayes). The term "mixed or combined approach" is often used to denote such an approach.

REFERENCES

1. K. S. Fu and A. Rosenfeld, "Pattern Recognition and Image Processing", _IEEE Trans. on Computers_, Vol. C-25, No. 12, 1976.
2. K. S. Fu, _Digital Pattern Recognition_, Springer-Verlag, 1976.
3. K. S. Fu, _Syntactic Methods in Pattern Recognition_, Academic Press, 1974.
4. K. Hanakata, "Feature Selection and Extraction for Decision Theoretic Approach and Structural Approach" in _Pattern Recognition-Theory and Application_, ed. by K. S. Fu and A. B. Whinston, Noordhoff International Publishing Co. Leyden, The Netherlands, 1977.
5. C. H. Chen, "On Statistical and Structural Feature Extraction" in _Pattern Recognition and Artificial Intelligence_, ed. by C. H. Chen, Academic Press, 1976.
6. K. S. Fu, _Syntactic Pattern Recognition Applications_, Springer-Verlag, 1977.
7. J. Gips, _Shape Grammars and Their Use_, Birkhauser Verlag, Basel and Stuttgart, 1975.
8. K. S. Fu, "Tree Languages and Syntactic Pattern Recognition" in _Pattern Recognition and Artificial Intelligence_, ed. by C. H. Chen, Academic Press, 1976.
9. K. S. Fu and T. L. Booth, "Grammatical Inference-Introduction and Survey", _IEEE Trans. on Systems, Man and Cybernetics_, Vol. SMC-5, Jan. and July 1975.
10. S. M. Chou and K. S. Fu, "inference for Transition Network Grammars", Proc. Third International Joint Conference on Pattern Recognition, Nov. 8-11, 1976, Coronado, Calif., U.S.A.
11. G. B. Porter, "Grammatical Inference Based on Pattern Recognition", Proc. Third International Joint Conference on Pattern Recognition, Nov. 8-11, 1976, Coronado, Calif., U.S.A.
12. L. Miclet, "Inference of Regular Expressions", Proc. Third International Joint Conference on Pattern Recognition, Nov. 8-11, 1976, Coronado, Calif., U.S.A.
13. J. M. Brayer and K. S. Fu, "Some Multidimensional Grammar Inference Methods" in _Pattern Recognition and Artificial Intelligence_, ed. by C. H. Chen, Academic Press, 1976.
14. J. M. Brayer and K. S. Fu, "A Note on the k-tail Method of Tree Grammar Inference", _IEEE Trans. on Systems, Man and Cybernetics_, Vol. SMC-7, No. 4, April 1977, pp. 293-299.
15. A. Barrero and R. C. Gonzalez, "A Tree Traversal Algorithm for the Inference of Tree Grammars", Proc. 1977 IEEE Computer Society Conference on Pattern Recognition and Image Processing, June 6-8, Troy, N. Y.

16. H. C. Lee and K. S. Fu, "A Syntactic Pattern Recognition System with Learning Capability", Proc. Fourth International Symposium on Computer and Information Sciences (COINS-72), Dec. 14-16, 1972, Bal Harbour, Florida.
17. J. Keng and K. S. Fu, "A System of Computerized Automatic Pattern Recognition for Remote Sensing", Proc. 1977 International Computer Symposium, Dec. 27-29, Taipei, Taiwan.
18. K. C. You and K. S. Fu, "Syntactic Shape Recognition Using Attributed Grammars", Proc. 8th EIA Symposium on Automatic Imagery Pattern Recognition, April 3-4, 1978, Gaithersburg, Md.
19. S. Y. Lu and K. S. Fu, "Stochastic Tree Grammar Inference for Texture Synthesis and Discrimination", Computer Graphics and Image Processing, Vol. 7, No. 5, Oct. 1978.
20. C. Page and A. Filipski, "Discriminant Grammars, An Alternative to Parsing for Pattern Classification", Proc. 1977 IEEE Workshop on Picture Data Description and Management, April 20-22, Chicago, Ill.
21. T. Pavlidis, "Syntactic Feature Extraction for Shape Recognition", Proc. 3rd International Joint Conf. on Pattern Recognition, Nov. 8-11, 1976, Coronado, Calif., pp. 95-99.
22. G. Stockman, L. N. Kanal and M. C. Kyle, "Structural Pattern Recognition of Carotid Pulse Waves Using a General Waveform Parsing System", Comm. ACM, Vol. 19, No. 12, Dec. 1976, pp. 688-695.
23. E. Persoon and K. S. Fu, "Sequential Classification of Strings Generated by SCFG's", International Journal of Computers and Information Sciences, Vol. 4, Sept. 1975.
24. S. Y. Lu and K. S. Fu, "Stochastic Error-Correcting Syntax Analysis for Recognition of Noisy Patterns", IEEE Trans. on Computers, Vol. C-26, No. 12, Dec. 1977, pp. 1268-1276.
25. T. Pavlidis, "Syntactic Pattern Recognition on the Basis of Functional Approximation" in Pattern Recognition and Artificial Intelligence, ed. by C. H. Chen, Academic Press, 1976.
26. A. K. Joshi, "Remarks on Some Aspects of Language Structure and Their Relevance to Pattern Analysis", Pattern Recognition, Vol. 5, No. 4, 1973.
27. B. K. Bhargava and K. S. Fu, "Transformation and Inference of Tree Grammars for Syntactic Pattern Recognition", Proc. 1974 IEEE International Conference on Cybernetics and Society, Oct., Dallas, Texas.
28. L. W. Fung and K. S. Fu, "Stochastic Syntactic Decoding for Pattern Classification", IEEE Trans. on Computers, Vol. C-24, July 1975.
29. M. G. Thomason and R. C. Gonzalez, "Error Detection and Classification in Syntactic Pattern Structures", IEEE Trans. on Computers, Vol. C-24, 1975.
30. K. S. Fu, "Error-Correcting Parsing for Syntactic Pattern Recognition", in Data Structure, Computer Graphics and Pattern Recognition, ed. by A. Klinger, et. al., Academic Press, 1977.

31. S. Y. Lu and K. S. Fu, "Error-Correcting Tree Automata for Syntactic Pattern Recognition", Proc. 1977 IEEE Conf. on Pattern Recognition and Image Processing, June 6-8, Troy, N. Y.
32. S. Y. Lu and K. S. Fu, "Structure-Preserved Error-Correcting Tree Automata for Syntactic Pattern Recognition", Proc. 1976 IEEE Conference on Decision and Control, Dec. 1-3, Clearwater, Florida.
33. N. S. Chang and K. S. Fu, "Parallel Parsing of Tree Languages", Proc. 1978 IEEE Computer Society Conference on Pattern Recognition and Image Processing, May 31 - June 2, Chicago, Ill.
34. K. S. Fu and S. Y. Lu, "A Clustering Procedure for Syntactic Patterns", IEEE Trans. on Systems, Man and Cybernetics, Vol. SMC-7, No. 10, Oct. 1977, pp. 734-742.
35. S. Y. Lu and K. S. Fu, "A sentence-to-Sentence Clustering Procedure for Pattern Analysis", IEEE Trans. on Systems, Man and Cybernetics, Vol. SMC-8, No. 5, May 1978.
36. R. O. Duda and P. E. Hart, Pattern Classification and Scene Analysis, Wiley, 1972.
37. F. Ali and T. Pavlidis, "Syntactic Recognition of Handwritten Numerals", IEEE Trans. on Systems, Man and Cybernetics, Vol. SMC-7, No. 7, July 1977, pp. 537-541.
38. J. M. Brayer and K. S. Fu, "Application of a Web Grammar Model to an ERTS Picture", Proc. Third International Joint Conference on Pattern Recognition, Nov. 8-11, 1976, Coronado, Calif., U.S.A.
39. R. Y. Li and K. S. Fu, "Tree System Approach to LANDSAT Data Interpretation", Proc. Symposium on Machine Processing of Remotely Sensed Data, June 29-July 1, 1976, Lafayette, Indiana.
40. J. Keng and K. S. Fu, "A Syntax-Directed Method for Land-Use Classification of LANDSAT Images", Proc. Symposium on Current Mathematical Problems in Image Science, Nov. 10-12, 1976, Monterey, Calif.
41. R. DeMori, "On Speech Recognition and Understanding" in Pattern Recognition: Theory and Application ed. by K. S. Fu and A. B. Whinston, Noordhoff Publ. Co., Leyden, Netherlands, 1977.
42. R. Jakubowski and A. Kasprzak, "A Syntactic Description and Recognition of Rotary Machine Elements", IEEE Trans. on Computers, Vol. C-26, No. 10, Oct. 1977, pp. 1039-1042.
43. J. F. Jarvis, "Regular Expressions as a Feature Selection Language for Pattern Recognition", Proc. Third International Joint Conference on Pattern Recognition, Nov. 8-11, 1976, Coronado, Calif., pp. 189-192.
44. J. L. Mundy and R. E. Joynson, "Automatic Visual Inspection Using Syntactic Analysis", Proc. 1977 IEEE Computer Society Conference on Pattern Recognition and Image Processing, June 6-8, Troy, N.Y.
45. R. W. Ehrich and J. P. Foith, "Representation of Random Waveforms by Relational Trees", IEEE Trans. on Computers, Vol. C-25, July 1976, pp. 725-736.

46. P. V. Sankar and A. Rosenfeld, "Hierarchical Representation of Waveforms", TR-615, Computer Science Center, University of Maryland, College Park, Md. 20742, Dec. 1977.

47. F. LeChevalier, G. Bobillot and C. Fugier-Garrel, "Syntactic Signal Processing", 1978 International Symposium on Information Theory, Oct. 10-14, Ithaca, N. Y.

48. S. Tsuji and R. Fugiwana, "Linguistic Segmentation of Scenes into Regions", Proc. Second International Joint Conference on Pattern Recognition, August 13-15, 1974, Copenhagen, Denmark.

49. T. Kunji, S. Weyle and J. M. Tenenbaum, "A Relational Data Base Scheme for Describing Complex Pictures with Color and Texture", Proc. Second International Joint Conference on Pattern Recognition, August 13-15, 1974, Copenhagen, Denmark.

50. R. H. Bonczek and A. B. Whinston, "Picture Processing and Automatic Data Base Design", <u>Computer Graphics and Image Processing</u>, Vol. 5, No. 4, Dec. 1976.

51. R. L. Kashyap, "Pattern Recognition and Data Base", Proc. 1977 IEEE Computer Society Conference on Pattern Recognition and Image Processing, June 6-8, Troy, N.Y.

52. S. K. Chang, "Syntactic Description of Pictures for Efficient Storage Retrieval in a Pictorial Data Base", Proc. 1977 IEEE Computer Society Conference on Pattern Recognition and Image Processing, June 6-8, Troy, N.Y.

53. R. DeMori, P. Laface and M. Sardella, "Use of Fuzzy Algorithms for Phonetic and Phonemic Labelling of Continuous Speech", Instituto di Scienze dell' Informazione, Universita di Torino, Italy.

54. S. Y. Lu and K. S. Fu, "A Syntactic Approach to Texture Analysis", <u>Computer Graphics and Image Processing</u>, Vol. 7, No. 3, June 1978.

ERROR-CORRECTING MATCHING AND PARSING FOR SYNTACTIC PATTERN RECOGNITION

Shin-yee Lu

Dept. of Electrical and Computer Engineering
Syracuse University
Syracuse, NY 13210

ABSTRACT. Several definitions of the distance between two strings and between two trees are introduced. These distance measures can be employed as classification criteria for syntactic pattern recognition. When a class of patterns is characterized by a pattern grammar, error-correcting parsers can be used for recognizing ungrammatical patterns. Error-correcting parsing algorithms for string languages and tree languages are presented. These techniques can be applied to lower-level processing, such as texture segmentation, or at a higher level classification, where errors induced from lower-level processing can be detected, enumerated and corrected. The distance measures enables the syntactic approach to use existing cluster-seeking procedures. Applications of error-correcting matching and parsing techniques are discussed and illustrated by examples.

1. INTRODUCTION

In syntactic approach to pattern recognition, the classification can be determined by a template matching scheme or a parsing procedure [1]. The simplest form of recognition is probably template matching; the datum of an input pattern is matched against the data of prototypes or reference patterns based on some similarity criterion. The linguistic or syntactic pattern recognition has explored the use of non-numerical and structured representations that describe patterns by parts (primitives). The first part of this paper discusses distance measures on representations of string or tree structure that provide matching criteria for syntactic patterns.

Based on the production rules of a given grammar, a parser analyzes an input pattern in terms of pattern primitives and relations between them. The analysis is to generate a signal indicating grammatical correctness, as well as the syntax structure of the pattern. In formal language theory, if a sentence is syntactically correct according to a grammar then it belongs to the class described by the grammar. A pattern grammar, in which the hierarchical structures of patterns are described, level by level, by production rules of the grammar, is constructed for a class of patterns. Thus, the membership of a given pattern can be decided by using a parser. In the meantime, the description of hierarchical structure of the pattern is extracted.

In a practical application, pattern distortion often leads to misrecognition of primitives. Errors occurring at lower-level processing sometimes cause an extracted pattern representation ungrammatical. The use of error-correcting parsers has been proposed to recognize ungrammatical sentences [2,3]. We shall introduce some error-correcting parsing schemes and their applications to recognizing noisy patterns.

2. ERROR-TRANSFORMATIONS AND DISTANCE MEASURES

In this section, distance measures between two sentences are presented. A sentence could be a string, an array, a tree or a graph that describes a pattern.

The approaches used in defining the distance between two sentences include the following two:

(1) To decompose a sentence in terms of sub-sentences for which adequate metrics are available. These simpler metrics are then used to induce a distance on the sentence. Findler and Van Leeuwen have proposed a similarity measure between two strings. Given two strings x and y, where $x, y \in \Sigma^*$. The similarity between x and y, $S(x,y)$, is defined as [4],

$$S(x,y) = \frac{\sum_{\omega \in \Sigma^*} \min \{p(x:\omega), p(y:\omega)\} \cdot |\omega|}{\sum_{\omega \in \Sigma^*} \max \{p(x:\omega), p(y:\omega)\} \cdot |\omega|} \quad (1)$$

where $p(x:\omega)$ denotes the number of times substring ω occurs in x, including partial overlaps. $|\omega|$ represents the length of ω,

(2) To use the idea of language transformation. Wagner and Fisher have defined errors between two strings in terms of the three types of transformations, namely, substitution, deletion and insertion transformations. A distance

between two strings is defined to be the least cost sequence of error transformations to derive one from the other [5]. This approach has been extended to two arrays [6], and to two trees [7].

2.1 Distance on Strings

In [2], errors in a string are considered to be the three types: substitution, deletion and insertion errors, and treated as syntax errors by defining transformations from Σ^* to a subset of Σ^*.

Definition 1. For two strings, x, y $\in \Sigma^*$, define a transformation T: $\Sigma^* \to \Sigma^*$ such that y $\in$ T(x). T has the following three types:

(1) substitution error transformation,

$$\omega_1 a \omega_2 \quad \vdash^{T_\sigma} \quad \omega_1 b \omega_2, \text{ for all } a, b \in \Sigma, \ a \neq b$$

(2) deletion error transformation

$$\omega_1 a \omega_2 \quad \vdash^{T_\varepsilon} \quad \omega_1 \omega_2, \text{ for all } a \in \Sigma$$

(3) insertion error transformation

$$\omega_1 \omega_2 \quad \vdash^{T_\phi} \quad \omega_1 a \, \omega_2, \text{ for all } a \in \Sigma$$

where $\omega_1, \omega_2 \in \Sigma^*$ are substrings of x.

Definition 2. The distance between two strings x, y, d(x,y), is defined as the smallest number of transformations required to derive y from x.

Example 1. Given a sentence x = cbabdbb and a sentence y = cbbabbdb, then, x = cbabdbb

$$\vdash^{T_\sigma} \ \text{cbabbbb} \ \vdash^{T_\sigma} \ \text{cbabbdb} \ \vdash^{T_\phi} \ \text{cbbabbdb} = y$$

The minimum number of transformations required to transform x into y is three, thus, d(x,y) = 3.

We further define a weighted distance that would reflect the difference of the same type of error made on different terminals.

Let the weights associated with error transformations on terminal a in a string ω_1 a ω_2 be defined as follows [8]:

(1) $\omega_1 \, a \, \omega_2 \;\; \vdash\!\!\xrightarrow{T_\sigma, \sigma(a,b)} \;\; \omega_1 \, b \, \omega_2$, where $\sigma(a,b)$ is the cost of substituting a for b. Let $\sigma(a,a) = 0$.

(2) $\omega_1 \, a \, \omega_2 \;\; \vdash\!\!\xrightarrow{T_\varepsilon, \, \varepsilon(a)} \;\; \omega_1 \, \omega_2$, where $\varepsilon(a)$ is the cost of deleting a.

(3) $\omega_1 \, a \, \omega_2 \;\; \vdash\!\!\xrightarrow{T_\phi, \phi(a,b)} \;\; \omega_1 \, b \, a \, \omega_2$, where $\phi(a,b)$ is the cost of inserting b in front of a.

(4) $x \;\; \vdash\!\!\xrightarrow{T_\phi, \phi'(b)} \;\; x \, b$, where $\phi'(b)$ is the cost of inserting b at the end of a string x.

Let J be a sequence of transformations used to derive y from x, and $|J|$ be defined as the sum of the weight associated with transformations in J. The weighted distance between x and y, $d_w(x,y)$, is defined as

$$d_w(x,y) = \min_J \{|J|\} \tag{2}$$

An algorithm of computing the distance between two strings based on Definition 2 has been proposed by Wagner and Fisher [5]. The algorithm employs dynamic programming techniques. The time and space complexities are both at the order of O(n,m), where n and m are the length of the two strings. The algorithm can easily be modified to compute the weighted distance [8].

2.2 Distance on Trees

The idea of using language transformation in defining distance can also be extended to tree languages. In this section, a definition of tree-to-tree distance is presented [7].

Definition on trees is briefly reviewed.

<u>Definition 3</u>. Let N^+ be the set of positive integers. Let U be the free monoid generated by N^+. Let $\cdot$ be the operation and 0 be the identity of U. Then, D is a tree domain iff D is a finite subset of U satisfying:

(1) $a \cdot x \in D$, where $a \in U$ and $x \in U$, implies $a \in D$,

(2) $a \cdot j \in D$ and $i < j$, $i, j \in N^+$, implies $a \cdot i \in D$

(3) 0 is in D.

Definition 4. A labeled tree defined on D is a mapping $\alpha{:}D \rightarrow \Sigma$, where Σ is a finite set of symbols.

Example 2. Given D = {0, 1, 2, 1·1, 1·2}, Σ = {\$, s, t} and α:

$$\begin{aligned} \alpha(0) &= \$ \\ \alpha(1) &= s \\ \alpha(2) &= t \\ \alpha(1\cdot 1) &= t \\ \alpha(1\cdot 2) &= t \end{aligned}$$

then the tree α is,

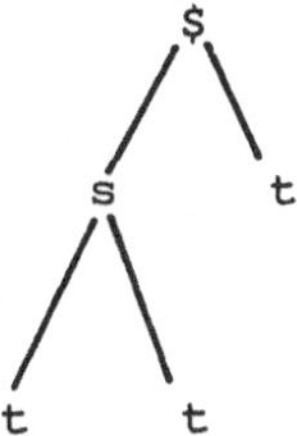

For convenience, we shall denote the tree domain of a tree, say α, D_α.

The following definition describes a transformation between trees such that for any two trees, we can always derive one from the other by repeatedly applying transformations.

Definition 5. Let α and β be two trees. Define a transformation $T : D_\beta \rightarrow D_\alpha$. T consists of the following three types:

$$T_\sigma(b) \rightarrow a \quad , \quad \sigma$$

$$T_\varepsilon(b) \rightarrow \lambda \quad , \quad \varepsilon$$

$$T_\phi(\lambda) \rightarrow a \quad , \quad \phi$$

where $b \in D_\beta$, $a \in D_\alpha$, λ is a null symbol, and σ, ε, and ϕ are costs associated with T_σ, T_ε and T_ϕ respectively.

An ε-, or a ϕ- transformation would cause one deletion, or one insertion error respectively. A σ- transformation, $T_\sigma(b)\rightarrow a$, would cause a substitution error if the node label of b in β and that of a in α are not identical, i.e., $\beta(b) \neq \alpha(a)$. We shall let $\sigma > 0$ if the σ transformation causes a substitution error, and $\sigma=0$ otherwise.

Definition 6. The distance between two trees α and β, denoted $d(\alpha,\beta)$, is the minimum cost necessary to derive α from β using a series of transformations satisfying the following criteria:

(1) the predecessor-descendant relation does not change,
(2) nodes in β do not split or merge,
(3) the sequence of postfix ordering does not change,

after applying the series of transformations.

Assume that the costs associated with ε- and φ- transformations are 1, and the cost associated with a φ-transformation is 1 if the transformation causes a substitution error and 0 otherwise. Then the distance in Definition 6 becomes the minimum number of deletion, insertion and substitution necessary to derive α from β.

<u>Example 2</u>. Given two trees α and β

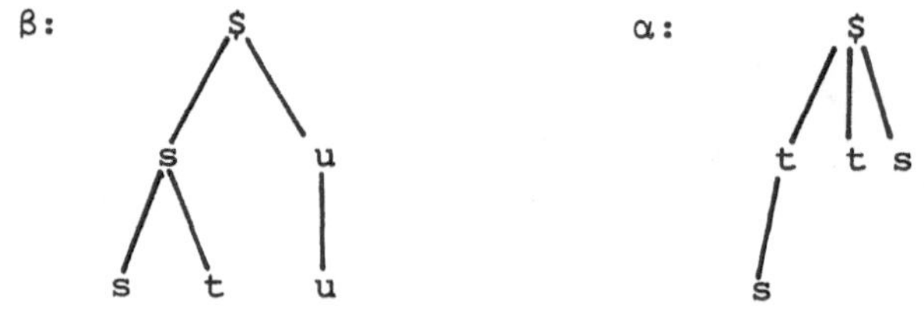

β: $; s u ; s t u

α: $; t t s ; s

Consider a set of transformations T_1, T_1 : $D_\beta \to D_\alpha$

$T_\sigma(0) \to 0$, 0

$T_\varepsilon(1) \to \lambda$, 1 (deletion)

$T_\sigma(1{\cdot}1) \to 1.1$, 0

$T_\sigma(1{\cdot}2) \to 2$, 0

$T_\phi(\lambda) \to 1$, 1 (insertion)

$T_\sigma(2) \to 3$, 1 (substitution)

$T_\varepsilon(2{\cdot}1) \to \lambda$, 1 (deletion)

Assume that the costs of a deletion, an insertion, and a substitution are 1, respectively. Then T_1 is the minimum cost sequence that satisfies Definition 6. Therefore, $d(\alpha,\beta) = 4$. Consider another set of transformations, T_2,

T_2: $D_\beta \to D_\alpha$

$T_\sigma(0) \rightarrow 0 \quad , \quad 0$

$T_\sigma(1) \rightarrow 1.1 \quad , \quad 0$

$T_\sigma(1 \cdot 1) \rightarrow 3 \quad , \quad 0$

$T_\sigma(1 \cdot 2) \rightarrow 1 \quad , \quad 0$

$T_\sigma(2) \rightarrow 2 \quad , \quad 1$ (substitution)

$T_\varepsilon(2 \cdot 1) \rightarrow \lambda \quad , \quad 1$ (deletion)

Although the cost associated with T_2 is only 2, T_2 does not satisfy Definition 6. The predecessor-descendent relation between nodes 1 and 1·2 in β is altered after T_2 is applied.

An algorithm that generates the distance described by Definition 6 for any given two trees is presented in [10]. The algorithm has time and space complexity both at the order of O(nm) where n and m are the number of labeled nodes of the two trees.

2.4 Size Normalization

In order to compute a distance between two syntactic patterns that takes the variations of pattern size into consideration, the following normalized distance is introduced [11],

$$d_N(x,y) = \frac{d(x,y)}{\max\{|x|, |y|\}} \qquad (3)$$

where $|x|$ denotes the length or the number of symbols in x.

3. A NEAREST NEIGHBOR SYNTACTIC RECOGNITION RULE

In the decision theoretic approach, one commonly used classification criteria is the similarity between patterns. For example, in the minimum distance classification, or the nearest neighbor recognition rule, similarities between patterns are measured in terms of distance between their feature vectors. Generally, the classification result depends on the selection of features and the definition of distance measure [12]. We have proposed that the same classification criteria is applicable in the syntactic approach [13]. Here, symbolic representations called sentences, are used to describe and process patterns. Like the use of feature vectors, they also reduce the dimensionality of patterns. Although no theory or quantitative measure has been proposed for the goodness of primitive and representation selection, the general criterion will be similar to feature selection: pattern representations have to be informative and discriminant.

Several available distance measures have been presented in Section 2. Let $d(x,y)$ be the distance between two sentences x and y in general. The nearest neighbor recognition rule for a two-category case is described as follows. Let prototypes in class C_1 and C_2 be represented by the two sets of sentences, $X_1 = \{x_1^1, x_2^1, \ldots, x_s^1\}$ and $X_2 = \{x_1^2, x_2^2, \ldots, x_t^2\}$, respectively. Let y be the representation of an input pattern. Then y is classified to C_1 if

$$\min_i \{d(x_i^1, y) \mid 1 \leq i \leq s\} < \min_j \{d(x_j^2, y) \mid 1 \leq j \leq t\} \quad (4)$$

and y is classified to C_2 otherwise.

In syntactic approach, one often characterizes a sample set by using a grammar such that a compact representation of the set can be formed. Let G_1 and G_2 be the grammars inferred for the sets X_1 and X_2, respectively. In this case, the class C_i is represented by a language $L(G_i)$, i=1, 2. The nearest neighbor syntactic recognition rule decides that y is in class C_1 if

$$\min_x \{d(x,y) \mid x \in L(G_1)\} < \min_z \{d(z,y) \mid z \in L(G_2)\} \quad (5)$$

and y is in class C_2 otherwise.

As for computational aspect, (4) requires the distance between y and every prototype be computed one by one. In (5), the nearest neighbor to y in $L(G_i)$ is determined by parsing y with respect to G_i. However, the algorithm designed to analyze y by given G_i is not a parser in an ordinary sense. We call it an error-correcting parser, since it is designed to process any sentence regardless whether the sentence is in the language of the given grammar. Let y be the input sentences and G_i be the given grammar, an error-correcting parser is to search for a sentence, x, in $L(G_i)$ such that the distance between x and y is the minimum among all the sentences in $L(G_i)$. The distance $d(x,y)$ and the parse of x will be generated by the algorithm. We further define the distance between a sentence, y, and a language, L(G) to be,

$$d(L(G), y) = \min_x \{d(x,y) \mid x \in L(G)\} \quad (6)$$

In the following section we shall describe error-correcting parsing algorithms for several types of languages based on the distance measures presented in Section 2.

4. ERROR-CORRECTING PARSING ALGORITHMS (ECP)

Using the notion of language transformations in defining distance between sentences has the advantage that it can easily accommodate to the syntax of a language. For example, one approach of constructing error-correcting parser is to expand the original grammar into a covering grammar in which error transformations are written as error-production rules. The covering grammar generates not only the syntactically correct sentences, but also all the erroneous sentences. The parser itself is an ordinary parser with a provision added to count the number of error production rules used in parsing a given sentence. This number will be the distance between the input and its nearest neighbor in the language generated by the original grammar [2]. We shall describe some error-correcting parsing algorithms for the following types of languages: regular languages, context-free languages and tree languages.

4.1 An ECP for Regular Languages

Wagner has proposed an error-correcting parser for regular languages that requires time linear to the input string length [3].

A regular language is characterized by the fact that its sentences are precisely the set of sentences acceptable to some finite-state automation (FSA). Suppose the FSA has scanned the first j symbols of an input string. Then, the only information retained about the symbols already scanned is contained in the "state" of the automation. The error-correcting algorithm for regular language has taken the advantage of these properties.

Let the input string be $y=a_1 a_2 \ldots a_m$ and the language under consideration be characterized by a FSA denoted by A. We shall define E(j, Q) to be the minimum number of error transformations needed to change substring $a_1 a_2 \ldots a_j$ into some substring α. The process will cause A to enter state Q after reading α. E(j,Q) can be computed by given E(j-1, R), a_j, and A:

$$E(j, Q) = \min_R\{E(j-1, R) + V(R, Q, a_j)\} \text{ for } j \geq 1 \tag{7}$$

where $V(R,Q, a_j)$ is equal to the smallest number of error-transformations which will change the single symbol a_j into a substring that can force A from state R to state Q. The initial condition is,

$$E(0,Q) = \begin{cases} 0 & \text{if S is the start state of A} \\ \infty & \text{otherwise} \end{cases} \tag{8}$$

The computation proceeds until E(m, Q) is available for all states in A, then the number

$$E(m, T) = \min_{Q \in F} E(m, Q) \tag{9}$$

where F is the set of final states of A, gives the distance from y to the nearest string acceptable to A.

The numbers V(R, Q, c) depend only on the FSA. They shall be computed and stored before using the algorithm for recognizing noisy inputs. The storage of V(R, Q, c) values is required for each symbol c in the terminal set and each ordered pair of states R and Q of the FSA.

4.2. An ECP for Context-free Languages

The approach of constructing covering grammar by incorporating error-productions into the original grammar is briefly described in this section. The work is originated from Aho and Peterson [2].

<u>Algorithm 1</u>. The construction of covering grammar. Input: A context-free grammar $G = (N, \Sigma, P, S)$. Output: A context-free grammar $G' = (N', \Sigma', P', S')$, where P' is a set of weighted productions.

Method:

<u>step 1</u>. $N' = N \cup \{S'\} \cup \{E_a | a \in \Sigma\}$, $\Sigma' \supseteq \Sigma$.

<u>step 2</u>. If $A \to \alpha_o b_1 \alpha_1 b_2 \ldots b_m \alpha_m$, $m \geq 0$ is a rule in P such that $\alpha_i \in N^*$ and $b_i \in \Sigma$, then add $A \to \alpha_o E_{b_1} \alpha_1 E_{b_2} \ldots E_{b_m} \alpha_m$, 0, to P', where each E_{b_i} is a new non-terminal, $E_{b_i} \in N'$, and 0 is the weight associated with this production.

<u>step 3</u>. Add the following rules to P',

(a) $S' \to S, 0$
(b) $S' \to Sa$, $\phi'(a)$, for all $a \in \Sigma'$
(c) $E_a \to a$, 0 , for all $a \in \Sigma$
(d) $E_a \to b$, $\sigma(a,b)$, for all $a \in \Sigma'$, $b \in \Sigma'$ and
(e) $E_a \to \lambda$, $\varepsilon(a)$, for all $a \in \Sigma$
(f) $E_a \to bE_a$, $\phi(a,b)$, for all $a \in \Sigma$, $b \in \Sigma'$, where σ, ε, ϕ and ϕ' are weights associated with error transformations as defined in Section 2.1.

In Algorithm 1 the production rules added in Step 3(b), 3(d), 3(e) and 3(f) are called error productions. Each error production corresponds to one type of error transformation on a symbol in Σ.

Therefore, the distance measured in terms of error transformations can be measured by counting the number of error productions used in a derivation. The parser used in [2] is a modified Earley's algorithm with a provision added to accumulate the weights. The algorithm always chooses the derivation that associated with the least weight.

4.3 Structure Preserved Error-correcting Tree Automata

In [14], we have proposed an error-correcting parser for tree languages called error-correcting tree automata (ECTA). Unlike the string case, where the only relation between symbols is left-right concatenation, a tree structure would be deformed under error transformations. Therefore, in addition to substitution and deletion transformations, we define insertion errors to be three types, namely split, stretch, and branch transformations, depending on where the insertion is made. In the formulation of ECTA, each type of transformation on each terminal symbol is added to the system in the form of a transition function. The ECTA counts the number of error-transition required to process a noisy input and searches for the minimum. The detail can be found in [14].

In this paper, we shall only describe the simplest type of ECTA, the structure preserved error-correcting tree automata (SPECTA). A SPECTA takes only substitution errors into consideration [15].

<u>Definition 7</u>. A grammar $G_t = (V,r,P,S)$ over $< \Sigma, r >$ is a tree grammar in expansive form where V is a set of terminal and nonterminal symbols, Σ is a set of terminal symbols, $r: \Sigma \to N$ where N is the set of non-negative integers, is the rank associated with symbols in Σ, S is the starting symbol, and P is a set of production rules in the form of

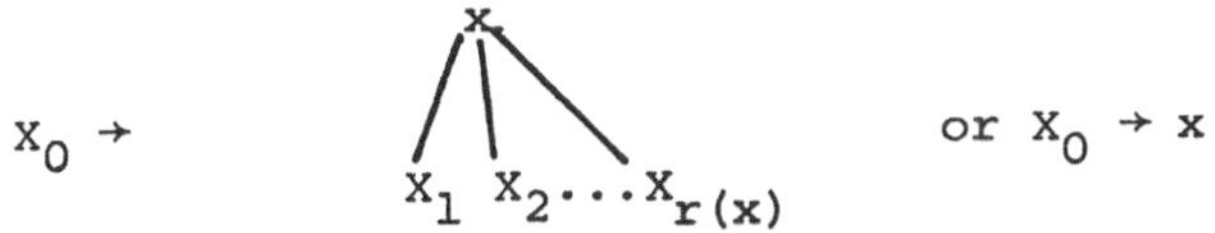

where $x \in \Sigma$, and $X_0, X_1, \ldots, X_{r(x)} \in V-\Sigma$.

The following is an example of using tree languages in modeling digitized pictures.

<u>Example 3</u>. A simple and practical tree system approach for digitized pictures is to divide the picture into windows. Gray levels of a single pixel or a small array of pixels are chosen to be pattern primitives. Each primitive in a window corresponds to a labeled node in the tree representation of the window. For implementation, a tree structure can be arbitrarily chosen and is fixed

for grammars written for this structure. An illustration of the relation between a window and a chosen tree structure is shown in Figure 1.

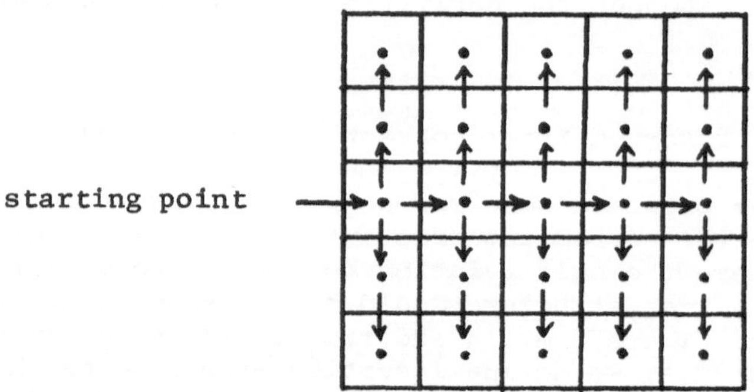

Fig. 1. The tree construction of a 5 by 5 window.

A binary window shown in Figure 2(a) has the tree representation shown in Figure 2(b) if the tree structure given in Figure 1 is used. Node label "1" represents a dark pixel and label "0" a white one.

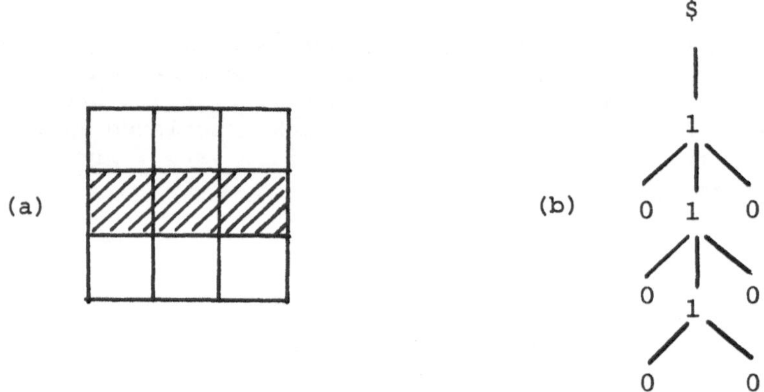

Fig. 2. A window and its tree representation

<u>Example 3</u>. Given a tree grammar G_0 = (V, r, P, S) over < Σ, r > where V = {Z, U, Y, X, A, S, $, 0, 1}

Σ = { ▨, □ }
1, 0

$r(1) = r(0) = \{0, 1, 2, 3\}$

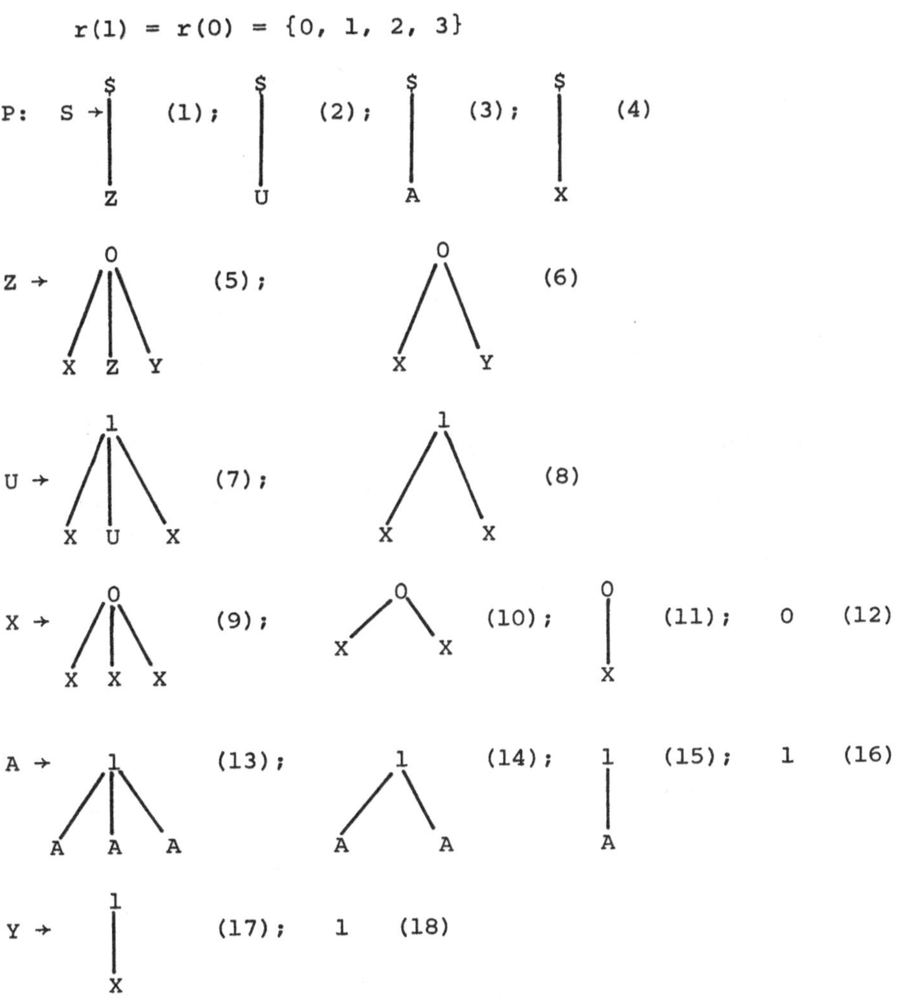

Then, G_0 generates the following four patterns:

1. only the pixels in the middle row are dark.
2. only the pixels one row above the middle row are dark.
3. all the pixels are dark.
4. all the pixels are white.

Assume that α' is an input tree, the operation of a SPECTA is a backward procedure of constructing a tree-like transition table from the frontiers to the root of α'. For each node $a \in D_{\alpha}'$, there is a set of triplets, denoted as t_a, in the transition table. Each triplet (X, σ, k) is added to t_a if X is a candidate state for node a, where σ is the minimum number of errors in subtree α'/a when node a is represented by state X, and k specifies the production rule used. The algorithm is given as follows.

<u>Algorithm 2</u>. SPECTA

input: A tree grammar $G_t = (V, r, p, S)$ over $< \Sigma, r >$ and a tree, α'.

output: Transition table of α' and $d(L(G_t), \alpha')$

Method:

<u>step 1</u>. For all nodes $a \in D_{\alpha}'$ such that $r[\alpha'(a)] = 0$, i.e. a is a frontier of α', add to t_a all the possible triplets of the form (X_0, σ, k) if $X_0 \to x$ is the kth rule in P, where $\sigma = 0$ if $\alpha'(a) = x$, or $\sigma = 1$, otherwise

<u>step 2</u>. If $r[\alpha'(a)] = n$, and the tables $t_{a\cdot 1}, t_{a\cdot 2}, \ldots, t_{a\cdot n}$ have been constructed already, then add to t_a all the possible triplets (X_0, σ, k) if

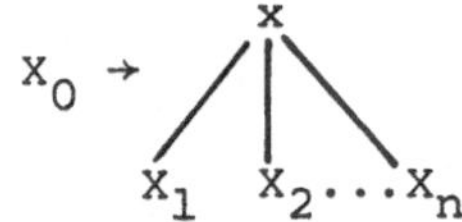

is the kth rule in P and $(X_1, \sigma_1, k_1) \in t_{a\cdot 1}, \ldots, (X_n, \sigma_n, k_n) \in t_{a\cdot n}$, where $\sigma = \sigma_1 + \sigma_2 + \ldots + \sigma_n$ if $\alpha'(a) = x$, or $\sigma = 1 + \sigma_1 + \sigma_2 + \ldots + \sigma_n$ otherwise

<u>step 3</u>. Whenever more than one item in a list t_a has the same state, delete the item with the larger number of errors.

<u>step 4</u>. If $(S, \sigma, k) \in t_0$, then $d(L(G_t), \alpha') = \sigma$. If no item in t_0 is of the form, then the input tree is rejected.

The minimum-distance correction of α' can easily be traced out from the transition table.

<u>Example 4</u>. Given a noisy pattern of size 3x3 as shown in Figure 3(a), with tree representation, called tree β, the transition table generated for β with respect to G_0 by using SPECTA is given in Figure 3(b), in which the correction of β is traced out and marked with asterisks. The corrected pattern is the pattern shown in Figure 2(a).

One advantage of applying the tree system approach and the SPECTA to picture processing is the parallel nature of the approach; all the windows can be parsed parallelly. A simulation program on parallel processing of the proposed tree system approach is given by Chang and Fu [16].

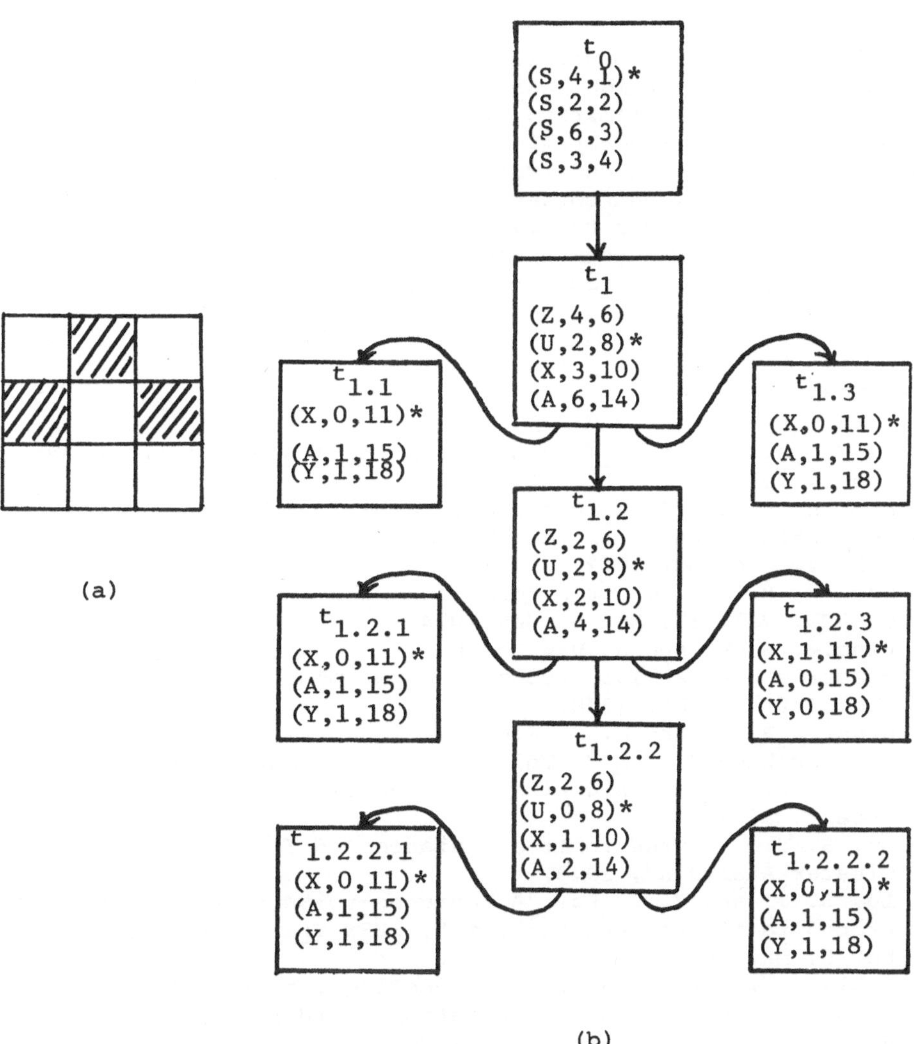

Fig. 3. The parsing of a noisy pattern by using SPECTA.

5. APPLICATIONS

The idea of error-correcting matching and parsing has been applied to noisy pattern classification [17, 18], speech recognition [19], cluster analysis for structured data [13], picture processing [15], and texture segmentation [20].

6. CONCLUDING REMARKS

Several schemes of error-correcting matching and parsing have been introduced. The technique can be used at the lower-level processing: segmentation, primitive extraction, as illustrated by the example of texture segmentation; or, at a higher-level classification, where errors induced from lower-level processing can be detected, enumerated, and corrected. It also provides non-supervised learning procedures for the syntactic approach to pattern recognition. The distance measures and classification based on the distance measures have provided clustering procedures for structured data.

REFERENCES

1. Fu, K. S., Syntactic Methods in Pattern Recognition, Academic Press, 1974.
2. Aho, A. V. and T. G. Peterson, "A Minimum Distance Error-Correcting Parser for Context-Free Languages," SIAM J. Comput. Vol. 4, December 1972.
3. Wagner, R. A., "Order-n Correction for Regular Languages", C. ACM, Vol. 17, No. 5, May 1974.
4. Findle, N. V. and J. Van Leeuwen, "A Family of Similarity Measures between Two Strings", Department of Computer Science, SUNY Buffalo, NY, 1978.
5. Wagner, R. A. and M. J. Fisher, "The String to String Correction Problem," J. ACM, Vol. 21, No. 1, January 1974.
6. Moore, R. K., "An Algirithm for the Distance between Two Finite Areas", 1977.
7. Lu, S. Y., "A Tree to Tree Distance and Its Application to Cluster Analysis", 1978.
8. Lu, S. Y. and K. S. Fu, "A Sentence-to-Sentence Clustering Procedure for Pattern Analysis", IEEE Trans. on SMC, Vol. SMC-8, No. 5, May 1978.
9. Bahl, L. R. and F. Jekinek, "Decoding for Channels with Insertions, Deletions, and Substitutions with Applications to Speech Recognition", IEEE Trans. on Information Theory, Vol. IT-21, No. 4, July 1975.
10. Fung, L. W. and K. S. Fu, "Stochastic Syntactic Decoding for Pattern Classification", IEEE Trans. on Computers, Vol. C-24, No. 6, June 1976.

11. Fu, K. S. and S. Y. Lu, "Size Normalization and Pattern Orientation Problems in Syntactic Clustering", 1978.
12. Duda, R. O. and P. E. Hart, Pattern Classification and Scene Analysis, Wiley, New York, 1972.
13. Fu, K. S. and S. Y. Lu, "A Clustering Procedure for Syntactic Patterns", IEEE Trans. on SMC, Vol. SMC-7, No. 8, October 1977.
14. Lu, S. Y. and K. S. Fu, "Error-Correcting Tree Automata for Syntactic Pattern Recognition", Proceedings 1977 IEEE Computer Society Conference on Pattern Recognition and Image Processing, June 6-8, RPI, Troy, NY.
15. Lu, S. Y. and K. S. Fu, "Structure - Preserved Error-Correcting Tree Automata for Syntactic Pattern Recognition," Proceedings of the 1976 IEEE Conference on Decision and Control, Dec. 1-3, Clearwater Beach. FL.
16. Chang, N. S. and K. S. Fu, "Parallel Parsing of Tree Languages", Proceedings of the IEEE Computer Society Conference on Pattern Recognition and Image Processing; Chicago, June 1978.
17. Lu, S. Y. and K. S. Fu, "Stochastic Error-Correcting Syntax Analysis for Recognition of Noisy Patterns", IEEE Trans. on Computers, Vol. C-26, No. 12, December 1977.
18. Thomason, M. G. and R. C. Gonzales, "Syntactic Recognition of Imperfectly Specified Patterns," IEEE Trans. on Computers, January 1975.
19. Kashyap, R. L. and M. C. Mittal, "A New Method for Error Correction in Strings with Applications to Spoken Word Recognition", Proceedings 1977 IEEE Conference on Pattern Recognition and Image Processing, June 6-8, Troy, New York.
20. Lu, S. Y. and K. S. Fu, "Syntactic Approach to Texture Analysis", Computer Graphics and Image Processing, Vol. 7, June 1978.

FEATURE SET SEARCH ALBORITHMS*

J. Kittler

Department of Electronics and Physics
E. N. S. des Télécommunications, Paris, France

ABSTRACT. Given a set Y of N measurements representing a pattern belonging to one of m classes, the feature selection problem is one of selecting a subset $X \in Y$, called feature set, of n of these measurements containing more discriminatory information than any other set of n measurements in Y. Since the number of subsets of cardinality n that can be constructed from the elements of Y is $\binom{N}{n}$ which even for moderate N and n takes a very large value, it is necessary to organize the feature set search so as to reduce the number of sets that need be evaluated. Various optimal and suboptimal feature set search algorithms are discussed, including the branch and bound method, sequential forward and backward selection algorithms and their generalized versions, the "plus ℓ - take away r" procedure, the max - min approach and algorithms based on dynamic programming. The algorithms are compared on the basis of computational complexity and reliability.

1. INTRODUCTION

In the problem of automatic classification of elements of the universe one of the major tasks is to acquire an efficient representation of objects or phenomena to be classified. Very informally, we shall understand that, to be efficient, a representation must be provided by a very small number of descriptors or measurements containing sufficient information for discriminating between elements of different classes.

---

* This work was supported by the Royal Society, U. K.

In many pattern recognition applications it is relatively easy to obtain a discriminatory information preserving representation. The more difficult task is to determine which subset of the possible descriptors contains most of the relevant information. These important descriptors will be referred to as features.

The purpose of feature selection is multifold. First of all, by minimizing the number of pattern descriptors the cost of hardware implementation of the data acquisition system (sensors, A - D convertors, data preprocessing system) can be substantially reduced. A smaller number of features will also be reflected on the reduced complexity of the decision making processor and its implementation. For a given size of the design set the reduction of dimensionality of pattern description may, in addition, have a beneficial effect on the system performance as a result of an improved set size to dimensionality ratio [1,2,3].

In mathematical terms the feature selection problem can be formulated in the following way. Let $Y = \{y_j\}$ be a set of N possible descriptors providing adequate representation of any element, Z, to be classified. Further, denote by $X_n \subset Y$ the subset of n of these descriptors which contain more discriminatory information about Z than any other set, χ_n, of n items in Y. Let J be a criterion of discriminatory information. Then feature set X_n must satisfy

$$J(X_n) = \max_{\chi_n} J(\chi_n) \tag{1}$$

We shall not elaborate here on the choice of criterion J. Ideally, of course, we should use as a feature selection criterion the function

$$J(\chi_n) = 1 - e(\chi_n) \tag{2}$$

where $e(\chi_n)$ is the Bayes error corresponding to set χ_n [4,5].

Measurements y_j, $j = 1, \ldots, N$ and features x_k, $k = 1, \ldots, n$ can be considered respectively as elements of the pattern representation vector $\underline{y}$ and feature vector $\underline{x}$, i.e.

$$\underline{y} = [y_1, \ldots, y_N]^T \tag{3}$$

$$\underline{x} = [x_1, \ldots, x_n]^T \tag{4}$$

In terms of these vectors the problem of feature selection is to find a linear mapping F satisfying

$$J(\underline{x}) = \max_{F} J(F^T\underline{y}) \tag{5}$$

where the elements of each column of the (N x n) matrix F are allowed to assume values either zero or one with only one element in each column of F having unity value. From this second formulation of the feature evaluation problem it is apparent that optimization of the criterion function J leads to a nonlinear integer programming problem.

In this paper we shall discuss various algorithms for finding optimal and suboptimal solutions of the above optimization problem. In section 2 the necessary notation will be introduced. In section 3 the optimal branch and bound feature set search algorithm will be described. Subsequently the sequential forward and sequential backward suboptimal search algorithms will be reviewed and their generalized versions given including the plus ℓ and take away r algorithm. The max - min feature selection procedure will be described in section 5. Finally in section 6 the experimental results of a comparative study of these algorithms will be discussed.

2. GENERAL REMARKS AND NOTATION

It is apparent that finding matrix F that optimizes criterion J is a combinatorial problem, for each column of F has N elements and we have to choose which one of these should be assigned a value unity. Since there are n such columns there are $\binom{N}{n}$ candidate matrices F that need be considered. Evidently, even moderate values of N can exclude an exhaustive search for the optimum solution of (5). For instance, taking N = 20 and n = 10 the corresponding number of candidate matrices is greater than 1.8×10^5. When considering that criterion J,which usually involves integration of the class conditional probability density functions of variable $F^T \underline{y}$, will have to be evaluated for every possible F,it is evident that in many practical situations this approach will not be feasible.

The first attempts to counteract this computational problem by making simplifying assumptions were reported by Lewis [5]. He assumed the measurements to be statistically independent and postulated that in such a situation the selection of features can be made by assessing each measurement individually and then setting those elements of the columns of F to unity which correspond to the n best individual measurements.

The belief that under the assumption of statistical independence the optimal feature set can be acquired in such a simple manner was not to last. The most well-known counter example demonstrating the flaw of this conjecture is due to Cover [6] who showed that even when measurements are statistically independent the two best are not the best two. But, as a matter of fact, the paper of Elashoff et al [7] published in 1967 already exposed the subtleties of the feature selection problem for the binary valued variables. The authors showed that given three independent variables y_1, y_2, y_3 satisfying the following ordering of error probabilities

$$e(y_1) < e(y_2) < e(y_3)$$

it is possible that

$$e\,(y_1, y_3) < e(y_1, y_2)$$

In 1971 Toussaint [8] extended this result to show that the set of the best two variables need not contain the best single variable.

These disturbing observations inspired further work on the integer programming problem leading to a number of interesting algorithms. The most important one of these is the branch and bound algorithm of Fukunaga and Narendra [9] which finds an optimum solution of (5). The other procedures are suboptimal methods of various degree of sophistication which must be resorted to when the optimal branch and bound algorithm is not applicable on computational grounds [10 - 16].

In all the search algorithms to be discussed in the sequel, the best feature set is constructed by adding to and/or removing from the current feature set a small number of measurements at a time until the required feature set, X_n, of cardinality n is obtained. In particular, the starting point can be either an empty set which is then gradually built up or, alternatively, we can start from the complete set of measurements, Y and successively eliminate inferior observations. The former approach is known as the "bottom up" method while the latter is referred to as the "top down" search.

In order to describe the various algorithm, we shall denote by X_k a set containing k elements $\xi_1, \xi_2, \ldots, \xi_k$, from the set of available measurements, Y, i.e. $X_k = \{\xi_i | i = 1, 2, \ldots, k, \xi_i \in Y\}$. Further, denote by $\overline{X}_k$ the set of N - k features obtained by removing k attributes $\xi_1, \xi_2, \ldots, \xi_k$, from the complete set of measurements, Y, i.e. $\overline{X}_k = \{y_i | y_i \in Y, 1 \leq i \leq N, y_i \neq \xi_j, \forall j\}$. Obviously, by choosing different descriptors ξ_i we shall construct different sets X_k and $\overline{X}_k$.

Note that

$$X_0 \equiv X_0 \equiv \emptyset$$

$$\overline{X}_0 \equiv \overline{X}_0 \equiv Y \qquad \text{and} \qquad X_n \equiv \overline{X}_{N-n}$$

All bottom up and top down search algorithms are based on the assumption of monotonicity of the feature selection criterion function. This condition requires that for nested feature sets $X_1, X_2, \ldots, X_k$, i.e.

$$X_1 \subset X_2 \subset \cdots \subset X_k \tag{6}$$

criterion J satisfies

$$J\,(X_1) \leq J\,(X_2) \leq \ldots \leq J\,(X_k) \tag{7}$$

All the commonly used criteria of feature selection are known to possess this monotonicity property. It is, however, essential that the measurements y_j are noninterfering [17].

It has been already mentioned that apart from the magnitude of numbers N and n the complexity of the optimization problem in (5) depends also on the criterion used. In particular, numerical evaluation of criterion J will be more involved if we deal with nonparametric distributions than if the underlying distributions are parametric, which usually allow analytical simplification. The evaluation of parametric criteria can be often simplified even further by recursive calculation [16, 18], that is by expressing $J(\chi_k)$ in terms of $J(\chi_{k-1})$ or $J(\chi_{k+1})$ depending on whether the search algorithm is of the bottom up or top down type. Here sets χ_k, χ_{k-1} and χ_{k+1} must, of course, satisfy

$$\chi_{k-1} \subset \chi_k \subset \chi_{k+1}$$

The majority of algorithms assume that at any stage of the search process it is feasible to evaluate the criterion function for the relevant candidate sets. This assumption, however, is not always satisfied. Often we have to base our search strategy on considerably more limited information such as second order distribution moments. A family of algorithms suitable for this type of a problem are known as max - min algorithms and these will be dealt with in section 5.

3. BRANCH AND BOUND ALGORITHM

The only optimal search method in which all the possible subsets, χ_n, of n out of N attributes are implicitely inspected without the exhaustive search is the branch and bound algorithm. It is basically a "top down" search procedure but with a backtracking facility which allows all the possible combinations of features to be examined. The computational efficiency of the method lies in an effective organization of the search process. By virtue of this process detailed enumeration of many candidate feature sets can be avoided without undermining optimality of the feature selection procedure. This is achieved by utilizing the monotonicity property (7) which for the top down search can be rewritten as

$$J(\bar{X}_1) > J(\bar{X}_2) > \dots J(\bar{X}_k) \tag{8}$$

if

$$\bar{X}_1 \supset \bar{X}_2 \supset \cdots \supset \bar{X}_k \tag{9}$$

By a straightforward application of this property, many combinations of features can be rejected from the set of candidate feature sets.

In any branch and bound algorithm the search is organized so that nested sets, $\overline{X}_k$, are constructed at each stage of the algorithm. First a number of $\overline{X}_1$ sets are constructed by removing selected attributes from the complete set of measurements, Y. At the second stage, from each set $\overline{X}_1$, a number of $\overline{X}_2$ sets are acquired in a similar manner and this process is continued until the cardinality of the sets being constructed is reduced to n. In effect this process amounts to generating a solution tree where at the k-th stage of the algorithm the features removed from each $\overline{X}_{k-1}$ set constitute nodes of the tree at the k-th level. In fig. 1 an example of this construction process is illustrated for a case where 2 features out of 6 measurements are to be selected. At most three branches (3 successive nodes) are allowed from all the nodes at each stage of the algorithm.

Suppose that some branches of a solution tree have already been explored to the final (N-n)th level, and that the criterion function for the current best feature set, $\overline{X}_{N-n}$, equals B, i.e. $J(\overline{X}_{N-n}) = B$. Now consider a node at the k-th level of an unexplored section of the tree where the value of the criterion function is less than threshold B, i.e.

$$J(\overline{X}_k) < B$$

Then, obviously, it would be pointless to evaluate the feature sets corresponding to the portion of the tree branching from this node since by virtue of (8) all these sets will be inferior to the current best feature set. Returning to the example in figure 1, if A is such a point then four candidate sets need not be evaluated. This is the main principle of reducing the computational burden of the exhaustive search.

Given a pair of numbers (N, n) a large number of different solution trees could, of course, be constructed. However, for a search to be efficient, certain rules should be observed when a solution tree is being generated. First of all the generating process should be systematic to ensure that the algorithm is easily programable. Second, since the section of a tree branching from a node associated with a low value of the criterion function is more likely to be rejected than the branches originating from a node where the criterion function takes a large value, it is advantageous to relate the number of successive nodes to each node to the magnitude of the criterion function at that node. In other words, the tree should be asymmetrical. Third, a tree must have N-n levels to ensure that the cardinality of the candidate feature sets corresponding to the terminal nodes of the tree in n.

A solution tree satisfying the above requirements is shown in figure 2 (for N = 6 and n = 2). Each node has a number of successors depending on the node level and the number of available measurements from which attributes to be discarded can be chosen. For instance, at the zero node all N measurements are available for selecting the first level nodes but at least N-n-1 (the number of

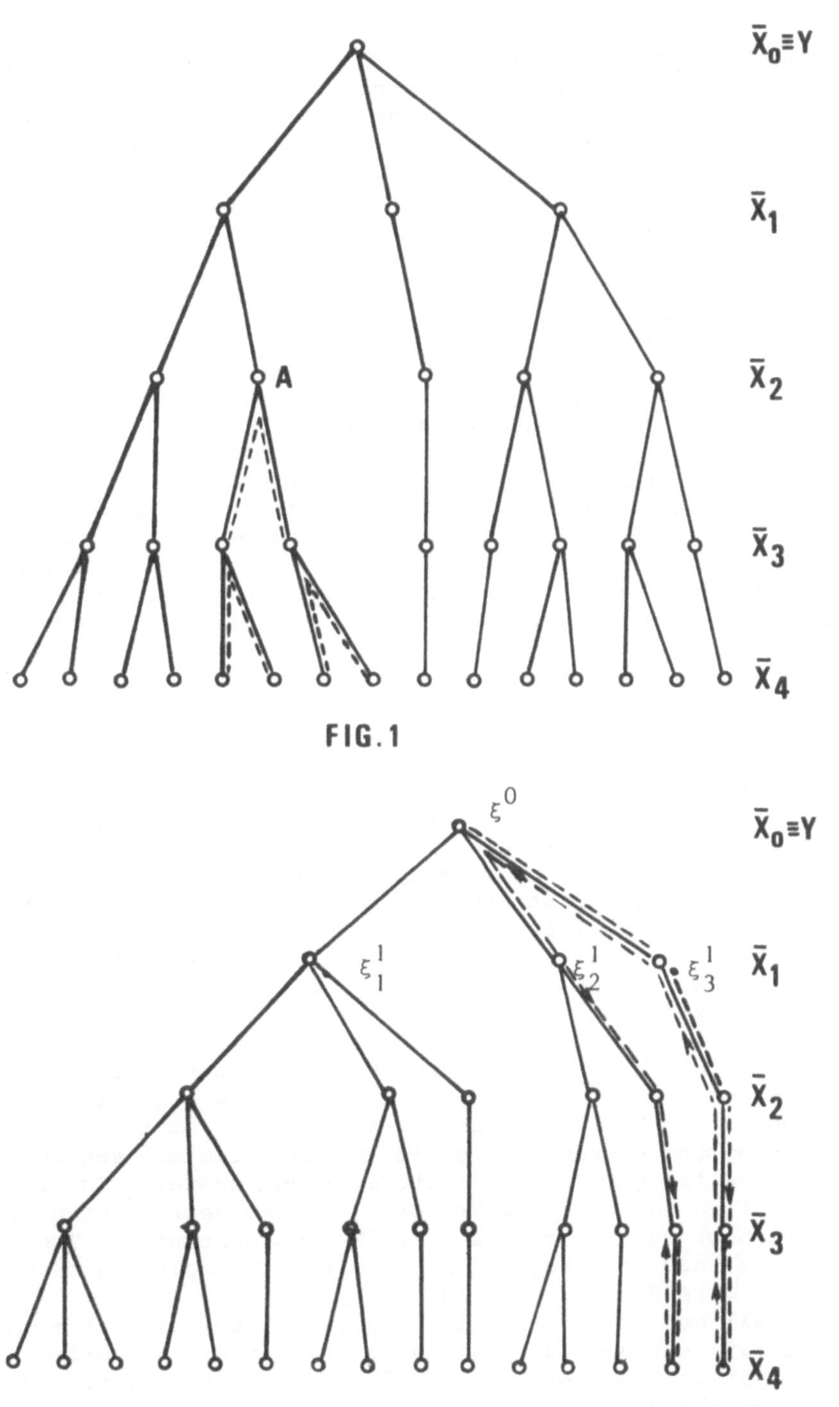

$\bar{X}_0 \equiv Y$
$\bar{X}_1$
A
$\bar{X}_2$
$\bar{X}_3$
$\bar{X}_4$

FIG.1

ξ^0
$\bar{X}_0 \equiv Y$
ξ_1^1
ξ_2^1
ξ_3^1
$\bar{X}_1$
$\bar{X}_2$
$\bar{X}_3$
$\bar{X}_4$

FIG. 2

remaining levels) attributes must be retained in the set of available descriptors to allow for completion of the tree to the final level. Thus, here the number of successive nodes to the zero node, q (0), is given as q (0) = N - (N-n-1) = n + 1 which in the example illustrated in figure 2 equals, q (0) = 3.

In accordance with our earlier discussion the actual successors, ξ_i^{k+1}, i = 1, 2, ..., q (k), to each node at the k-th level are selected so that the feature sets, $\overline{\chi}_k - \xi_i^{k+1}$, corresponding to these successor nodes satisfy

$$J\ (\overline{\chi}_k - \xi_i^{k+1}) < J\ (\overline{\chi}_k - \xi_j^{k+1}),\ i < j \tag{10}$$

since in this particular solution tree the nodes ξ_i^{k+1} with lower indices have a large number of succeeding points. As a matter of fact, it is advantageous to select ξ_i^{k+1}, i = 1,2, ..., q (k), as those q (k) attributes in the set of available measurements which yield the lowest values of the criterion function, for the lower the criterion function at a node the more probable it is that the node will be rejected. This strategy may seem paradoxical since our aim is to find the feature set with the largest possible measure of class separability. However, recalling that we are implicitly inspecting all the possible feature sets anyway, there is no cause for concern.

The solution tree is actually not constructed level by level but from the least dense part of the tree to the part with most branches (from right to left). First the single right most branch is generated and the magnitude of the separability measure at the terminal node is taken as the current threshold. We then return to the nearest branching node and generate the next right most branch of the tree. If at any node the criterion function takes on a value that is smaller than the threshold, the search of the section of the tree originating from this node is abandoned and the algorithm backtracks to the nearest branching point in a lower level. The next right most branch is then generated. This process is continued until the whole tree is constructed.

If the final level is reached while a branch is being generated this means that the set of features corresponding to the terminal node of the branch yields a greater value of the criterion function than the current threshold and this feature set must, therefore, be the best set of features found so far. The threshold is then updated and after backtracking to the nearest branching point, the next section of the tree is then generated. The path illustrating this construction process for the solution tree in Figure 2 is marked in dashed line.

We shall now give a formal statement of the algorithm which should clarify any ambiguity of the above informal description of the search procedure. Also it should facilitate software implementation of the algorithm.

The algorithm

Suppose that we are at an arbitrary node at level k. That means that for all the preceding nodes at levels $i = 0, 1, 2, \ldots, k-1$, we have sequences $Q_i = (\xi_1^{i+1}, \xi_2^{i+1}, \ldots, \xi_{q(i)}^{i+1})$ of q (i) successor points to these respective nodes. The current feature set after k measurements have been discarded is $\overline{X}_k$. The set of available measurements, Ψ_k, is of cardinality ν_k.

Step 1

Determine sequence of q (k) succeeding elements $Q_k = (\xi_1^{k+1}, \ldots, \xi_{q(k)}^{k+1})$ where q (k) is given as

$$q(k) = \nu_k - (N - n - k - 1)$$

and elements ξ_j^{k+1} satisfy

$$J(\overline{X}_k - \xi_1^{k+1}) \leq J(\overline{X}_k - \xi_2^{k+1}) \leq \ldots \leq J(\overline{X}_k - \xi_{q(k)}^{k+1}) \leq$$

$$\leq \ldots \leq J(\overline{X}_k - \xi_{\nu_k}^{k+1}), \; \xi_j^{k+1} \in \Psi_k$$

Remove Q_k from Ψ_k and update r_k, i.e.

$$\Psi_{k+1} = \Psi_k - Q_k$$

$$\nu_{k+1} = \nu_k - q(k)$$

Step 2

(Check bound for the candidate successive point $\xi_{q(k)}^{k+1}$). If q (k) = 0, then go to Step 4. If $J(\overline{X}_k - \xi_{q(k)}^{k+1}) < B$, then set $l = q(k)$ and go to Step 3. Otherwise discard $\xi_{q(k)}^{k+1}$ from $\overline{X}_k$ to form a new feature set $\overline{X}_{k+1}$, i.e.

$$\overline{X}_{k+1} = \overline{X}_k - \xi_{q(k)}^{k+1}$$

Now if $k + 1 = N - n$, jump to Step 5 else set $k = k + 1$ and return to Step 1.

Step 3

Return $\xi_{q(k)}^{k+1}$ to Ψ_{k+1}, i.e.

$$\Psi_{k+1} = \Psi_{k+1} + \xi_{q(k)}^{k+1}$$

and update r_{k+1} by one, i.e. $r_{k+1} = r_{k+1} + 1$. Set $q(k) = l$ and $l = l - 1$. If $l = 0$, then go to Step 2. Otherwise go to Step 3.

Step 4

(Backtracking). Set $k = k - 1$. If $k = -1$, terminate the algorithm. Otherwise return $\xi_{q(k)}^{k+1}$ to the current feature set, i.e.

$$\bar{X}_k = \bar{X}_{k+1} + \xi_{q(k)}^{k+1}$$

Set $r_k = r_{k+1}$, $\Psi_k = \Psi_{k+1}$, $l = 1$ and return to Step 3.

Step 5

(Bound updating). Set $B = J(\bar{X}_{N-n})$. Store $\bar{X}_{N-n}$ as X_n, i.e. the current best feature set. Set $l = q(k)$ and return to Step 3.

Note that to initialize the algorithm ν_o and Ψ_o should be set respectively to

$$\nu_o = N$$

$$\Psi_o = Y$$

The initial feature set $\bar{X}_o$ is, of course, the complete set of measurements Y.

In comparison with the exhaustive search, the branch and bound algorithm affords substantial computational saving. As a rule only a fraction of all the possible candidate feature sets need be enumerated to find the optimal set of features, with the reduction factor of computation costs being most dramatic for $n \simeq N/2$. Note, however, that in addition to the sets, $\bar{X}_{N-k}$, which correspond to the terminal nodes of the fully generated branches, their supersets, $\bar{X}_{N-n+j}$, $j = 1, 2, \ldots, N-n$, must also be evaluated. Thus, the actual number of inspected sets is somewhat higher. This should be born in mind when we are deciding whether to use the branch and bound algorithm or opt for the exhaustive search which for very small values of n or values of n approaching N may be less involved.

4. SUBOPTIMAL SEARCH

n Best Features

The simplest and, of course, the most unreliable method of constructing a feature set, X_n, is to select n individually best measurements in Y. More specifically, let us rank measurements $y_j \in Y$ so that

$$J(y_1) \geq J(y_2) \geq \ldots . \geq J(y_n) \geq \ldots . \geq J(y_N)$$

Then the feature set, X, is defined as

$$X_n = \{y_i \mid \forall\, i \leq n\}$$

The method may lead to a suboptimal feature set even if all the measurements in Y are statistically independent and, consequently, it should be used only if no other alternative is feasible.

Sequential Forward Selection (SFS)

SFS is a simple bottom up search procedure where one measurement at a time is added to the current feature set. At each stage the attribute to be included in the feature set is selected from among the remaining available measurements so that the new enlarged set of features yields a maximum value of the criterion function.

Suppose k features have been selected to form feature set X_k. Rank the elements, ξ_j, of the set of available measurements, $Y - X_k$, so that

$$J(X_k \cup \xi_1) \geq J(X_k \cup \xi_2) \geq \ldots \geq J(X_k \cup \xi_{N-k})$$

Then feature set X_{k+1} is given as $X_{k+1} = X_k \cup \xi_1$. The algorithm is initialized by setting $X_o \equiv \emptyset$.

The SFS algorithm selects successive attributes with reference to the current feature set, but by restricting the number of measurements to be added at each step of the algorithm to one, it is impossible to take into consideration statistical dependence between elements of the set of available measurements. The second main drawback of the SFS method is that once an attribute is included in the feature set there is no mechanism for deleting it from the feature set even if at a later stage, when more measurements have been added, this attribute becomes superfluous.

Sequential Backward Selection (SBS)

The SBS algorithm is the "top down" counterpart of the SFS method. Starting from the complete set of measurements, Y, we discard one (the worst) feature at a time until N-n measurements have been deleted. Algorithm: assume that k feature have been discarded from $\overline{X}_o \equiv Y$ to form feature set $\overline{X}_k$. Now to obtain reduced feature set $\overline{X}_{k+1}$ rank the elements ξ_j of set $\overline{X}_k$ so that

$$J\ (\overline{X}_k - \xi_1) \geq J\ (\overline{X}_k - \xi_2) \geq \ldots \geq J\ (\overline{X}_k - \xi_{N-k})$$

Then

$$\overline{X}_{k+1} = \overline{X}_k - \xi_1$$

At each stage of the algorithm the least informative element of the current feature set is determined by investigating statistical dependence of the features in the set. The statistical relationship of the elements already discarded is ignored. Also, as in the SFS algorithm, the feature sets $\overline{X}_k$, $k = 0, 1, 2, \ldots, N-n$, yielded by the SBS method are nested. However, the main difference between these two methods is that the SBS procedure provides as a by product a measure of maximum achievable class separability with the given set of measurements, Y, which can be used then to assess the amount of information lost in the feature selection process. Note, however, that the SBS is computationally more demanding than the SFS since in the former case criterion J (.) is evaluated in spaces of dimensionality greater or equal to n while in the latter case the dimensionality of the inspected feature spaces is at most equal to n.

"Plus ℓ - Take Away r" Algorithm

The problem of nesting in the SFS and SBS algorithms can be overcome by incorporating a low level backtracking in the feature selection process. In particular, at the k-th stage of the algorithm we first enlarge the current feature set by ℓ features using the SFS method. From the resulting set r features are then discarded applying the SBS algorithm. If $\ell > r$, the (ℓ, r) algorithm is a bottom up search method while if $\ell < r$, the (ℓ, r) algorithm is a top down procedure.

Although nesting in the (ℓ, r) algorithm is avoided, the search procedure still suffers from other drawbacks of the SFS and SBS algorithms, namely that groups of features are added and removed from the current set irrespective of their mutual relationship. This problem is resolved in the generalized SFS, SBS and (ℓ, r) algorithms introduced in the following section.

Statistical dependence of attributes added to or discarded from the current feature set, depending on the type of the search procedure employed, can be taken into account by adding to or subtracting from the feature set more than one measurement at a time. Let the number of measurements being considered at any one time be r. Then the SFS and SBS algorithms can be logically generalized as follows:

Generalized Sequential Forward Selection Algorithm GSFS (r)

Suppose k features have been selected to form feature set X_k. Generate all the possible sets, χ_{k+r}, by adding to X_k various combinations of r measurements from the set of available measurements $Y - X_k$. Then select as X_{k+r} that set χ_{k+r} that maximizes the class separability measure J (.), i.e.

$$J(X_{k+r}) = \max_{\chi_{k+r}} J(\chi_{k+r})$$

The GSFS (r) algorithm is more reliable than the SFS method but, of course, also more costly in computational terms, for at each stage $\binom{N-k}{r}$ feature sets χ_{k+r} have to be inspected to find set X_{k+r}. Note that as in the SFS procedure, nesting of successive feature sets is not prevented.

Generalized Sequential Backward Selection Algorithm GSBS (r)

Suppose k features have been discarded from Y to form feature set $\overline{X}_k$. Now form all the possible sets, $\overline{\chi}_{k+r}$, by removing various combinations of r attributes from $\overline{X}_k$. Then select as $\overline{X}_{k+r}$ the candidate feature set $\overline{\chi}_{k+r}$ that maximizes class separability measure J (.), i.e.

$$J(\overline{X}_{k+r}) = \max_{\overline{\chi}_{k+r}} J(\overline{\chi}_{k+r})$$

This greater sophistication of the feature set selection procedure is achieved once again only at the expense of extra computations, for at each stage $\binom{N-k}{r}$ sets must be evaluated in comparison with N-k candidate sets required by the SBS algorithm. Likewise, the successive feature sets are nested.

Generalized "Plus ℓ - Take Away r" Selection Algorithm

The only essential difference between the generalized "plus ℓ - take away r" selection method and the (ℓ, r) algorithm is that the former approach employs the GSFS (ℓ) and GSBS (r) algorithms instead of the basic SFS and SBS procedures. However, we shall take the generalization process a step further and allow each number ℓ and r to be composed of several integer components ℓ_i, i = 1, 2, ..., s_ℓ, and r_j, j = 1, 2, ..., s_r, satisfying

$$0 \leq \ell_i \leq \ell \qquad 0 \leq r_j \leq r$$

$$\sum_{i=1}^{s_\ell} \ell_i = \ell \qquad \sum_{j=1}^{s_r} r_j = r$$

Now, at each stage of the algorithm, instead of applying the GSFS (ℓ) and GSBS (r) methods we acquire the feature set of appropriate cardinality by applying successively first the GSFS (ℓ_i) procedure for all i and then the GSBS (r_j) procedure for all j. By splitting the forward and backward selections into a number of substeps we shall be able to curb computational complexity of the algorithm.

In order to give a formal statement of the algorithm let us designate by S_ℓ and S_r the sequences of elements ℓ_i and r_j respectively satisfying conditions (5) and (6) i.e.

$$S_\ell = (\ell_1, \ell_2, \ldots, \ell_{s_\ell})$$

$$S_r = (r_1, r_2, \ldots, r_{s_r})$$

The set of parameters (S_ℓ, S_r) defines the following generalized "plus ℓ - take away r" selection method:

Suppose k features have been selected to generate set X_k.

Step 1

Enlarge X_k by successive application of the GSFS (ℓ_i) algorithm for i = 1, 2, ..., s_ℓ, where ℓ_i are elements of the sequences S_ℓ. Set

$$k = k + \sum_{i=1}^{s_\ell} \ell_i \text{ and } \bar{X}_{N-k} = X_k.$$

Step 2

Reduce $\bar{X}_{N-k}$ by successive application of the GSBS (r_i) algorithm for j = 1, 2, ..., s_r, where r_j are elements of sequence S_r. Set

$$k = k - \sum_{j=1}^{s_r} r_j.$$

If k = n then terminate the algorithm. Otherwise set

$$X_k = \bar{X}_{N-k}$$

and return to Step 1.

Initialization of the algorithm is identical to that of the (ℓ, r) procedure discussed above.

It is interesting to note that all the search algorithms discussed in this section can be considered as special cases of the (S_ℓ, S_r) algorithm as shown in table 1.

(S_ℓ, S_r) algorithm		equivalent algorithm
S_ℓ = (1)	S_r = (0)	SFS, (1,0) algorithm
S_ℓ = (0)	S_r = (1)	SBS, (0,1) algorithm
S_ℓ = (n)	S_r = (0)	exhaustive search
S_ℓ = (ℓ)	S_r = (0)	GSFS (ℓ)
S_ℓ = (0)	S_r = (r)	GSBS (r)
S_ℓ = (1,1,1,...,1)	S_r = (1,1,...,1)	(ℓ,r) algorithm

Table 1

The (S_ℓ, S_r) algorithm can be also viewed as a search method utilizing the dynamic programming principle of optimization with the successive transition states defined by the numbers ℓ_i and r_j of sequences S_ℓ and S_r. In this sense the (S_ℓ, S_r) method can be considered as a generalization of dynamic programming search algorithms of Chang [10] and Michael and Lin [12].

5. MAX - MIN ALGORITHM

The main distinguishing characteristic of the suboptimal optimization approach to be discussed in this section is that here feature selection will be based on very limited information. In particular, we shall assume that we have available only the information on the individual and pairwise effectiveness of measurements, that is values $J(y_j)$ and $J(y_j, y_k)$, j, k $\neq$ j. Since there are only $\frac{N(N+1)}{2}$ such values, these can be easily precomputed and stored which will greatly facilitate the search process.

Suppose that we have already selected k measurements in Y to form feature set $X_k = \{x_i \mid i = 1, 2, \ldots, k\}$ and we would like now to select an additional feature, x_{k+1}, from among the elements of the set of remaining available measurements $\{\xi_j | j=1,2,\ldots N-k, \xi_j \in Y-Y_k\}$. Having available only the values of $J(\xi_j, x_i)$, j = 1,2,..., N-k, i = 1, 2, ..., k to make our judgement it is reasonable to argue that measurement ξ_j will be worthwhile including in the feature set only if it conveys additional information to that of feature x_i. In other words, the increase in the value of the criterion function due to ξ_j, $\Delta J(\xi_j, x_i)$ which is given as

$$\Delta J(\xi_j, x_i) = J(\xi_j, x_i) - J(x_i)$$

should be as large as possible. The inclusion of measurement ξ_j

will be justified however only if it is useful to the current feature set as a whole. In other words, ξ_j should not be correlated with any other feature in X_k. These requirements can be jointly satisfied by the max - min strategy [19]. We shall select as the (k + 1) s feature that measurement ξ_j which satisfies

$$J(x_{k+1}, x_i) = \max_{\forall \xi_j \in Y - X_k} \; \min_{\forall x_i \in X_k} \; \Delta J(\xi_j, x_i)$$

The first two features can be selected using the GSFS (2) algorithm which will yield, as a byproduct, all the information required for the max - min selection process.

6. EXPERIMENTAL RESULTS

The feature set search algorithms discussed in the paper were compared on a two class problem in 20 dimensional space. The classes were distributed normally with means $\underline{\mu}_i$, i = 1, 2 and covariance matrix Σ. The Mahalanobis distance between the two populations was used as a criterion of feature set effectiveness. The criterion function was evaluated by the means of the recursive formulae given in [16]. The results are summarized in Tables 2 and 3 where each row gives the values of the criterion function for feature sets of equal cardinality as yielded by the search algorithms. The blanks correspond to the optimal values of the criterion function given in the first column of Table 2.

From the results we can observe that as expected the generalized sequential forward and backward selection algorithms GSFS (r) and GSBS (r) perform better than SFS and SBS procedures but only at the expense of increased computer time. The combination of GSFS (2) and GSBS (1) or GSBS (2) gives optimal results almost everywhere with the total running time of 5-10 secs. which compares very favourably with the branch and bound algorithm. The (ℓ, r) algorithm gives good results only for $\ell \simeq r$. When the values of ℓ and r are too different the algorithm is incapable of preventing nesting.

The performance of the max - min algorithm in general is very poor reflecting the limited information used in the search process.

7. DISCUSSION

The suboptimality of the algorithms discussed in sections 4 and 5 must be understood in the sense that neither these algorithms provide a guarantee of finding the best set of features nor the more sophisticated algorithms necessarily yield better feature sets than the simple ones. This latter somewhat surprising point has been demonstrated by Cover and Van Campenhout [17]. They showed

	Optimal	GSFS (1)	GSFS (2)	GSFS (4)	GSBS (1)	GSBS (2)	GSBS (4)	Max - Min
$J(X_2)$	.471				.133	.133		
$J(X_4)$	.818	.596			.181	.260	.260	.550
$J(X_6)$	.966	.780			.576	.576		.568
$J(X_8)$	1.244	1.053	1.236		.927	.927	0.927	.629
$J(X_{10})$	2.53	1.145	1.842		2.45	2.45		.659
$J(X_{12})$	3.772	1.539	3.486	3.486				.698
$J(X_{14})$	4.11	3.493	4.026		4.099	4.099		.717
$J(X_{16})$	4.463	4.148	4.349	4.349				.990
$J(X_{18})$	4.639	4.399						4.437
Time (S)	65	2	5	32	3	8	115	1

Table 2: Sequential Forward and Backward Algorithms

	$S_\ell=(2,2)$ $S_r=(2)$	$S_\ell=(1,1)$ $S_r=(1)$	$S_\ell=(1,1,1)$ $S_r=(1)$	$S_\ell=(1,1,1)$ $S_r=(1,1)$	$S_\ell=(4)$ $S_r=(3)$	$S_\ell=(4)$ $S_r=(1)$	$S_\ell=(1)$ $S_r=(2)$	$S_\ell=(2)$ $S_r=(1)$
$J(X_2)$								
$J(X_4)$		.596	.596	.596				
$J(X_6)$			.780				.90	
$J(X_8)$		1.12	1.12	1.23			.927	
$J(X_{10})$		1.892	1.269				2.458	
$J(X_{12})$	3.743		2.555	3.743	3.743		2.833	3.743
$J(X_{14})$	4.041		3.977	4.041	4.099	4.041		4.041
$J(X_{16})$			4.351					
$J(X_{18})$								
Time (S)	16	7	5	12	185	41	13	12

Table 3: (S_ℓ, S_r) Algorithms

for a two class case that for any given ordering of feature sets satisfying the monotonicity condition it is always possible to find normal class conditional distributions which will exhibit this ordering. Thus, the reliability of the algorithms should be considered to be probabilistic rather than absolute.

It should be noted that a small sample size can affect the monotonicity properly of criterion functions. This should be remembered particularly in connection with the top down algorithms in which the search commences in the complete pattern representation space and the sample size to dimensionality ratio is small. However, the top down search has the advantage of providing information on the achievable performance in the pattern representation space which can then be used to assess the information loss due to feature selection.

REFERENCES

1. Foley, D. H., "Consideration of Sample and Feature Size," IEEE Trans. Information Theory, IT - 18, pp. 618-626, 1972.
2. Raudys, S., "On Dimensionality, Learning Sample Size and Complexity of Classification Algorithms", Proc. 3rd Int. Conf. Pattern Recognition, pp. 168-169, Coronado, Calif., Nov. 1976.
3. Kanal, L. and B. Channdrasekaran, "On Dimensionality and Sample Size in Statistical Pattern Classification", Pattern Recognition, 3, 1971.
4. Fukunaga, K., Introduction to Statistical Pattern Recognition, Academic Press, New York, 1972.
5. Lewis, P. M., "The Characteristic Selection Problem in Recognition System", IRE Trans. Inf. Theory, IT - 8, pp. 171-178, 1962.
6. Cover, T. M., "The Best Two Independent Measurements Are Not the Two Best," IEEE Trans. Systems, Man and Cybernetics, SMC - 4, pp. 116-117, 1974.
7. Elashoff, J. D., R. M. Elashoff and G. E. Goldman, "On the Choice of Variables in Classification Problems With Dichotomous Variables", Biometrika, 54, pp. 668-670, 1967.
8. Toussaint, G. T., "Note on the Optimal Selection of Independent Binary Features for Pattern Recognition", IEEE Trans. Infor. Theory, IT - 17, p. 618, 1971.
9. Narenda, P. M., and K. Fukunaga, "A Branch and Bound Algorithm for Feature Subset Selection", Proc. Cybernetics and Society International Conf., pp. 497-503, Washington, D. C., 1976.
10. Chang, C. Y., "Dynamic Programming as Applied to Feature Selection in Pattern Recognition Systems", IEEE Trans. Syst., Man and Cybern., SMC - 3, pp. 166-171, 1973.
11. Meisel, W. S., Computer Oriented Approaches to Pattern Recognition, Academic Press, New York, 1972.

12. Michael, M., and W. C. Lin, "Experimental Study of Information Measures and Inter-intra Class Distance Ratios on Feature Selection and Ordering", IEEE Trans. Syst., Man and Cybern., SMC-3, pp. 172-181, 1973.
13. Stearns, S. D., "On Selecting Features for Pattern Classifiers", Proc. 3rd Int. Conf. on Pattern Recognition, pp. 71-75, Coronado, Cal., 1976.
14. Whitney, A., "A Direct Method of Nonparametric Measurement Selection", IEEE Trans. Computers, C - 20, pp. 1100 - 1103, 1971
15. Kittler, J., "Une Généralisation de quelques Algorithmes Sous-Optimaux de Recherche d'Ensembles d'Attributs", Proc. Congrès AFCET/IRIA Reconnaissance des formes et Traitement des Images, Paris, Feb. 1978.
16. Devijver, P. A., and J. Kittler, "Selected Topics in Statistical Pattern Recognition", Technical Report CAMS/77/1, Cambridge University, 1977.
17. Cover, T. M., J. M. V. Campenhout, "On the Possible Orderings in the Measurement Selection Problem", Trans. Systems, Man and Cybernetics, SMC - 7, pp. 657-661, 1977.
18. Kittler, J., "A Fast Method of Evaluating Parametric Probabilistic Distance Measures for the Top Down Search of Feature Sets," Proc. IEEE Conf. on Pattern Recognition and Image Processing, Troy, New York, 1977.
19. Backer, E., and J. A. De Shipper, "On the max - min Approach for Feature Ordering and Selection", Proc. Seminar on Pattern Recognition, Liege, Nov. 1977.

NONPARAMETRIC ESTIMATION OF FEATURE EVALUATION CRITERIA

Pierre A. Devijver

MBLE Research Laboratory
Avenue van Becelaere, 2
B-1170 Brussels BELGIUM

ABSTRACT. This paper reviews various nonparametric estimation methods with application to the estimation of the Bayes error probability, nearest neighbor error-rates, the logarithmic equivocation, the Bhattacharyya coefficient and the divergence measure. The focus is placed on binary hypothesis problems and on estimators derived from nearest neighbor techniques. Several new asymptotic relationships are presented. The material is illustrated by results from simulation experiments.

1. INTRODUCTION

Signal and feature selection is a long standing problem of upmost importance in such fields as communication, radar systems, and pattern recognition. In recent years the current state of the art was very aptly reviewed by Chen [4], Kanal [24], and Hanakata [21]. In [25], Kittler also has a fairly exhaustive bibliography.

Although the ultimate objective is to choose amidst a number of candidate sets the signal or feature set that yields minimum error-probability of decision making, alternative criteria such as information and distance measures are often preferred because *i)* they may be easier to calculate than the error-probability, *ii)* they generally provide upper and/or lower bounds of the error-probability, and *iii)* their optimization is generally concomitant to a decrease of the error probability for some set of a priori probabilities of the hypotheses under consideration [20], [23].

The choice of a suitable, possibly suboptimal criterion is only the first part of the signal selection problem. Another particularly important one is concerned with the investigation of feature selection algorithms. This topic is discussed elsewhere

[26]. Let us note, however, that during the last two years, significant advances have been made. First, it was shown by Cover and van Campenhout that the selection problem is of unrestricted combinatorial complexity with the consequence that no sequential algorithm, in the sense of [26], can be guaranteed to produce the optimal feature set [7]. Almost simultaneously, Fukunaga and Narendra have applied a branch and bound algorithmic technique which guarantees the selection of the optimal set while avoiding the exhaustive search over the entire space of possible candidate sets [18].

One further step in the signal selection problem is concerned with the quantitative evaluation of the chosen selection criterion. In the parametric case, some measures such as the logarithmic entropy, the divergence or the Bhattacharyya coefficient reduce to closed-form expressions, and by now their use is fairly well documented in the literature [23], [29], [30]. By contrast, the study of the nonparametric case is in a far less advanced stage. It is therefore the topic which we have chosen to review.

In this paper, we shall thus be concerned with the nonparametric estimation of featurc/signal evaluation criteria. Estimates will be presented for five widely used criteria, namely i) the Bayes error-probability, (Sec. 3), ii) Nearest-Neighbor error-rates, (Sec. 4), *iii)* the logarithmic equivocation, (Sec. 5), *iv)* the Bhattacharyya coefficient, and *v)* the divergence measure, (Sec. 6). The error-bounds with these measures are recalled in Section 2 where some new bounds are given. Also some new asymptotic relationships involving a number of known estimators will be presented in due time. Some experimental results with the estimators of Sections 3-6 are reported in Section 7. These are not intended to provide for a comparative evaluation of the performance of the estimators involved. The need for such an investigation is underlined in Section 8.

In order to gain in unity and, hopefully, readability we concentrate on estimation methods based on nearest neighbor techniques. One further limitation arises from the fact that the discussion is confined to binary-hypothesis problems.

2. DISTANCE MEASURES AND ERROR-BOUNDS

2.1. Notations in this paper will be as follows: Let θ_1, θ_2 denote the hypotheses, (Pattern classes); P_i be the a priori probability of hypothesis θ_i; X be a n-dimensional vector random variable taking values in sample-space Ω; $p_i(x)$ be the probability density function of x conditioned on θ_i; $\eta_i(x)$ be the a posteriori probability of hypothesis $\theta=\theta_i$ conditioned on $X=x$; $p(x) = \Sigma_i P_i p_i(x)$ be the mixture probability density, and E be the expectation operator with respect to p(x). Also, for given x with unknown θ, let $\hat{\theta}^*$ be the optimal (Bayes) estimate of θ, $\hat{\theta}_k$ and $\hat{\theta}_{k,\ell}$ be the estimates of θ with the k-nearest neighbor rule, (k-NNR) [6], and the

(k,ℓ)-NNR [22], respectively. For any positive integer k let k' and k" be the least and greatest integers such that $k' \geq k/2 \geq k''$.

2.2. In the following we shall be concerned with the derivation of nonparametric estimates for

2.2.1. the Bayes conditional error probability $R^*(x) = \min(\eta_1(x), \eta_2(x))$ and average error-rate $R^* = E\{R^*(x)\}$;

2.2.2. the k-NN error-probabilities

$$R_k(x) = \eta_1 \sum_{j=0}^{k''} \binom{k}{j} \eta_1^j \eta_2^{k-j} + \eta_2 \sum_{j=k'}^{k} \binom{k}{j} \eta_1^j \eta_2^{k-j}, \text{ k odd,}$$

$$= \eta_1 \sum_{j=0}^{k''-1} \binom{k}{j} \eta_1^j \eta_2^{k-j} + \eta_2 \sum_{j=k'+1}^{k} \binom{k}{j} \eta_1^j \eta_2^{k-j}$$

$$+ \frac{1}{2} \binom{k}{k'} (\eta_1\eta_2)^{k'} \text{, k even,} \tag{1}$$

and $R_k = E\{R_k(x)\}$;

2.2.3. the (k,ℓ)-NN error-probabilities

$$R_{k,\ell}(x) = \eta_1 \sum_{j=0}^{k-\ell} \binom{k}{j} \eta_1^j \eta_2^{k-j} + \eta_2 \sum_{j=\ell}^{k} \binom{k}{j} \eta_1^j \eta_2^{k-j}$$

all k and ℓ with $\ell > k'$ (2)

and $R_{k,\ell} = E\{R_{k,\ell}(x)\}$;

2.2.4. the (logarithmic) equivocation $H = E\{H(x)\}$ with $H(x) = -\Sigma_i \eta_i(x) \ln \eta_i(x)$;

2.2.5. the Bhattacharyya coefficient $B = E\{B(x)\}$ with $B(x) = [\eta_1(x)\eta_2(x)]^{1/2}$;

2.2.6. the divergence measure $J = E\{J(x)\}$ with

$$J(x) = \Sigma_i \eta_i(x) \ln \frac{\eta_i(x)}{1-\eta_i(x)} \tag{3}$$

It is to be noted that this definition of the divergence is consistent with that given by Chen [4]. In the above form, it has the advantage that it can be readily extended to multihypothesis problems [12].

Remark. Occasionally, ambiguity will be avoided by using the notations $H(\Theta/X)$ and $H(\theta/x)$ instead of H and $H(x)$. The same remark applies to B and J.

2.3. Error-Bounds.

The derivation of tight bounds of R^* has been the subject of substantial research work over the last decade, especially in pattern recognition circles. For the measures given above, the best known bounds, some of which have never appeared before, are listed hereafter.

It has been known for some time that for any $k \geq 1$, $R^* \leq R_k$ and the bound becomes tighter as k increases [6]. Lower-bounds of R^* with the same property have been found by the author [13], and a general result is as follows: Let $R_k' = R_{k,k'+1}$ for k even and $R_k' = (R_k + R_{k,k'+1})/2$ for k odd, then

$$R_k' \leq R^* \leq R_k \qquad \text{for all } k. \tag{4}$$

Besides, the bounds obtained with k=2q and k=2q-1 coincide [10], [13]. The upper and lower bounds in (4) converge to R^* as $k \to \infty$. Thus arbitrarily tight bounds can be obtained by choosing k large enough.

The best known bounds in terms of H are

$$(2\ln 2)R^* \leq H(\Theta/X) \leq H(R^*, 1-R^*) \tag{5}$$

where $H(R^*, 1-R^*) = -[R^* \ln R^* + (1-R^*) \ln (1-R^*)]$. For the Bhattacharrya coefficient we have:

$$R^* \leq B(\Theta/X) \leq B(R^*, 1-R^*) \tag{6}$$

where $B(R^*, 1-R^*) = [R^*(1-R^*)]^{1/2}$, and the upper bound in (6) can be seen to be equivalent to the tightest known bound with B [31]. The author has recently obtained the following lower error-bound in terms of the multihypothesis generalization of the divergence [12]

$$J(\Theta/X) \geq J(R^*, 1-R^*) + R^* \log \frac{1-R^*}{m-1-R^*}, \quad (R^* > 0) \tag{7}$$

where $J(R^*, 1-R^*) = R^* \ln [R^*/(1-R^*)] + (1-R^*) \ln [(1-R^*)/R^*]$, and m is the number of hypotheses. Furthermore, the bound in (7) can be shown to be as tight as possible. For the binary hypothesis problem, (7) becomes $J(\Theta/X) \geq J(R^*, 1-R^*)$. This latter bound can be seen to coincide with the tightest bound given by Chen [4].

2.4. The estimation methods described hereafter are all based, to some extent, on nearest neighbor techniques. Thus it will be assumed that there is available a set X consisting of N iid samples $\{x^i, \theta^i\}_1^N$, where θ^i is the true class of x^i and the samples are drawn according to the (unknown) underlying distributions. Furthermore, it is assumed that X is partitioned into a design set X_d with $|X_d| = N_d$ and a test set X_t with $|X_t| = N_t, N_t + N_d = N$. Let also N_d^1 and N_d^2 denote the numbers of samples in X_d with $\theta = \theta_1$ and $\theta = \theta_2$ respectively. For any $x \in X_t$ let k_i be the number of NN's with $\theta = \theta_i$ among the first k NN's to x from X_d. We shall take r,h,b and j to denote finite-sample estimates of R,H,B, and J respectively. To allow for easy comparison of the estimates they will be indexed according to their order in the presentation. Eventually, when taking limits for $N \to \infty$ the following will be implicitly assumed: *i)* both N_d and N_t tend to infinity, *ii)* $N_d^i \to \infty$ while $N_d^i / N_d \to P_i$, i=1,2, and *iii)* the same property holds in test sample X_t.

3. ESTIMATION OF THE BAYES ERROR-RATE

Many estimates which are claimed in the literature to estimate the Bayes error-rate are, in fact, estimates of some NN error-rates. Consequently they are deferred to the following section. One notable exception, somewhat akin to the Lofstgaarden and Quesenberry method of probability-density estimation was studied by Fukunaga and Hostetler [17]. Let k_1 and k_2 be fixed and let v_i be the volume out to the k_i th NN of x. Then, assuming absolute continuity of the conditional probability densities it can be shown that the finite-sample maximum likelihood (ML) estimate of $p_i(x)$ is $k_i/(N_d^i v_i)$, and since N_d^i/N_d is the M.L. estimate of P_i, the ML estimate $r_1^*(x)$ of the conditional Bayes error probability $R^*(x)$ is given as

$$r_1^*(x) = \frac{1}{2}\left(1 - \frac{|k_1 v_2 - k_2 v\ |}{k_1 v_2 + k_2 v_1}\right), \tag{8}$$

while the estimate r_1^* of R^* is

$$r_1^* = N_t^{-1} \sum_{x \in X_t} r_1^*(x) . \tag{9}$$

The asymptotic (large-sample) expected value and mean-squared error of $r_1^*(x)$ can be found in [17]. It can be seen that $r_1^*(x)$ is a biased estimate of $R^*(x)$ and that the bias is a function of $R^*(x)$. As a consequence nothing is known about the large-sample behaviour of r_1^* with respect to R^*. If, however, departing somewhat from the Fukunaga-Hostetler approach we let k_i depend on N_d^i in such a way that $k_i \to \infty$ as $N_d^i \to \infty$ while $k_i/N_d^i \to 0$ we can invoke general results by Chen and Fu [5] who demonstrate that the expectation of the estimate in (8) converges to the Bayes risk in probability and in the kth mean. Chen and Fu note, however, that the choice of k_i is quite critical in that an improper choice may lead to poor results [5]. It should also be pointed out that the estimates in (8) and (9) do not use the class information of test samples. Hence, the method remains applicable when X_t consists of samples x for which θ is unknown.

4. ESTIMATION OF NEAREST NEIGHBOR ERROR-RATES

4.1. Error-counting techniques.

Probably the simplest way to estimate R_k or $R_{k,\ell}$ is by classifying the samples in X_t and determining the frequency of decision error. Thus, given (x,θ) let

$$r_2^k(x) = \begin{cases} 1 & \text{if } \hat{\theta}_k \neq \theta \\ 0 & \text{otherwise} \end{cases} \tag{10}$$

$$r_3^k(x) = \begin{cases} 1 & \text{if } [(\hat{\theta}_{k,k'+1}=\theta_1 \text{ and } \theta=\theta_2) \text{ or } (\hat{\theta}_{k,k'+1}=\theta_2 \text{ and } \theta=\theta_1)] \\ 0 & \text{otherwise} \end{cases} \tag{11}$$

$$r_j^k = N_t^{-1} \sum_{X_t} r_j^k(x)\ , \qquad j=2,3. \tag{12}$$

These definitions lead to asymptotically unbiased estimates r_2^k and r_3^k of R_k and $R_{k,k'+1}$ respectively. Moreover, let

$$r_3^{'k} = \begin{cases} r_3^k & k \text{ even} \\ (r_2^k + r_3^k)/2 & k \text{ odd} \end{cases} \tag{13}$$

then, in view of (4), and when N grows arbitrarily large we have

$$\lim_{N\to\infty} r_3^{'k} = R'_k \leq R^* \leq R_k = \lim_{N\to\infty} r_2^k \tag{14}$$

and the bounds can be made arbitrarily tight by choosing k large enough. Contrasting with all the other estimators to be discussed in this paper, r_2 and r_3 require the knowledge of θ for the test samples. One simple way to remedy this situation was devised by the author who showed that one can exchange the need to know θ for the (k+1)st NN, [9]. Hence if $\theta^{'j'}$ denotes the class of the jth NN to x, we also have

$$r_4^k(x) = \begin{cases} 1 \text{ if } \hat{\theta}_k \neq \theta^{'k+1'} \\ 0 \text{ otherwise} \end{cases} \tag{15}$$

$$r_5^k(x) = \begin{cases} 1 \text{ if } [\,(\hat{\theta}_{k,k'+1}=\theta_1 \text{ and } \theta^{'k+1'}=\theta_2) \text{ or} \\ \qquad (\hat{\theta}_{k,k'+1}=\theta_2 \text{ and } \theta^{'k+1'}=\theta_1)\,] \\ 0 \text{ otherwise} \end{cases} \tag{16}$$

with r_4^k and r_5^k being defined as in (12). The estimates r_4 and r_5 enjoy the same asymptotic properties as r_2 and r_3 and they give rise to the same error bounds as in (14).

All error-counting estimation schemes are known to suffer from a large variance which is mainly due to the coarse discretization in Eqs. (10-11) and (15-16). The following methods attempt to overcome this shortcoming.

4.2. Direct k-NN estimates

4.2.1. Direct k-NN estimates have been investigated in great detail by Fukunaga and Kessel, and Fukunaga and Hostletler, [16], [17]. Subsequently, their analysis was significantly simplified by the author [13]. For fixed k and under the assumption of absolute continuity of the conditional probability densities it can be shown that the finite sample ML estimate of $\eta_i(x)$ is k_i/k, [17]. This prompts us to use

$$r_6^k(x) = \min_i (k_i/k) \tag{16}$$

as an ML estimate of $R^*(x)$. However this estimate is biased for it is, in fact, an estimate of a lower bound to R^* [16]. The interpretation of this bound has been laid bare in [13] where it is shown that the asymptotic expected value of $r_6^k(x)$ is nothing but $R'_{k-1}(x) \leq R^*(x)$. There follows that when N grows arbitrarily large, we have

$$\lim_{N\to\infty} [r_6^k = N_t^{-1} \sum_{X_t} r_6^k(x)] = R'_{k-1} \leq R^* \tag{17}$$

The variance of r_6^k can be found in ([16], Eq. 32) while the mean-squared error of using $r_6(x)$ as an estimate of $R^*(x)$ is given in ([17], Eq. 49).

4.2.2. A closely related estimate was devised by Lissack and Fu [27], [28]. They use the Parzen density estimation technique with a hypercubic kernel function, (see [15], p. 88). Specifically, they specify a neighborhood V(x) centered at x and they count the number k_v of samples falling in V(x). Let k_{iv} of them have $\theta=\theta_i$. Then their estimate can be written

$$r_7^v(x) = \min_i (k_{iv}/k_v) \tag{18}$$

The major difference with r_6 is that here k_v is also a random variable, one fact which dramatically complicates the analysis. Nevertheless r_7 is seen to be an estimate of $E\{R'_{kv}\}$ where the expectation is over the distribution of k_v. It turns out that r_7 is also an estimate of a lower bound of R* but the bound is not quite as tight as that with r_6. Lissack and Fu note that the proper choice of V is not very clear [28]. Also they do not specify the value of $r_7(x)$ when $k_v=0$.

4.2.3. The simplest upper-bound of R* appears to be R_1 and several schemes have been derived for estimating it. Using the approach of Section 4.2.1. it is tempting to take $2k_1k_2/k^2$ as an (M.L.) estimate of $R_1(x) = 2\eta_1(x)\eta_2(x)$. Devijver and Dekesel have shown, however, that this estimate is asymptotically biased in the sense that

$$E\{ \frac{2k_1k_2}{k^2} / X = x \} = \frac{k-1}{k} R_1(x). \tag{19}$$

Hence a simple asymptotically unbiased estimate of R_1 is given by

$$r_8^k = N_t^{-1} \sum_{X_t} r_8^k(x), \text{ with [14],}$$

$$r_8^k(x) = 2k_1k_2/k(k-1). \tag{20}$$

The variance of $r_8^k(x)$ is [14],

$$\text{var}\,\{r_8^k(x)\} = 2k^{-1}R_1(x)[1-(k-1)^{-1}(2k-3)R_1(x)] \tag{21}$$

Under the large-sample assumption, we have

$$\frac{1}{2} \lim_{N\to\infty} r_8^k = R_1/2 \le R^* \le R_1 = \lim_{N\to\infty} r_8^k. \tag{22}$$

4.2.4. The approach of Lissack and Fu has also been applied to the estimation of R_1 yielding [28]

$$r_9^v(x) = 2k_{1v}k_{2v}/k_v(k_v-1). \tag{23}$$

This estimate is again asymptotically unbiased [28]. Obviously, our comments relative to r_7^v also apply here.

4.2.5. Direct k-NN methods have also been used to estimate the k-NN error-rate R_k, $k > 1$ [16]. Let $\alpha^i_{k,j}(x)=1$ if $k_i=j$ and 0 otherwise, i=1,2, and take

$$r_{10}^k(x) = \sum_{i=1}^{2} \sum_{j=k'}^{k} (1-j/k)\, \alpha^i_{k,j}, \quad k \text{ even.} \tag{24}$$

For k odd, let

$$\alpha_{k,j} = \Sigma_i\, \alpha^i_{k,j}, \text{ then}$$

$$r_{10}^k(x) = \sum_{j=k'+1}^{k} (1-j/k)\, \alpha_{k,j} + \frac{1}{2}E(1+\frac{k''}{k})\, \alpha_{k,k''}. \tag{25}$$

By using simple results from [16] or [9] it is easy to show that

$$\begin{aligned} \lim_{N\to\infty} [r_{10}^k = N_t^{-1} \sum_{X_t} r_{10}^k(x)] &= R_{k-1} \quad k \text{ even,} \\ &= R_{k-2} \quad k \text{ odd,} \\ &\geq R^* . \end{aligned} \tag{26}$$

It is interesting to note that, in the limit, the difference between the upper and lower bounds in (26) and (17) reduces to $\frac{1}{2}E\{\binom{k}{k'}\eta_1^{k'}(x)\eta_2^{k'}(x)\}$, one quantity which tends to zero as $k\to\infty$. Hence, theoretically speaking, the interval between the upper and lower bounds of R* can be made arbitrarily small by choosing k large enough.

4.3. ONNSS estimate.

A very different approach first used by Garnett and Yau,[19] and formalized by the author [11] makes use of Ordered Nearest Neighbor Sample Sets (hence the ONNSS method) and is based on a series expansion of the error-probability. It has been shown that [11]

$$R^* = 2^{-1} \sum_{j=1}^{\infty} (2j-1)^{-1} A_{2j} \tag{27}$$

where $A_{2j} = E\{A_{2j}(x)\}$ and $A_{2j}(x)$ is the conditional probability of observing j NN's from either class among the 2j NN's to x. From $A_{2j} \geq 0$ (all j) and $A_{2j'} \leq A_{2j}$ if $j' > j$, it is easily seen that (27) yields the following error-bounds [11],

$$2^{-1} \sum_{j=1}^{k} (2j-1)^{-1} A_{2j} \leq R^* \leq 2^{-1} \sum_{j=1}^{k} (2j-1)^{-1} A_{2j} + 2^{-1} A_{2k}. \quad (28)$$

Consequently, from here on it suffices to estimate the probabilities A_{2j}, $j=1,\ldots,k$ to obtain estimates of the error-bounds in (28). Thus, let $a_k(x)=1$ if $k_1=k_2=k/2$ (k even) and $a_k(x)=0$ otherwise, and let

$$r_{11}^{2k}(x) = 2^{-1} \Sigma_{j=1}^{k} (2j-1)^{-1} a_{2j}(x) \quad (29)$$

$$r_{12}^{2k}(x) = 2^{-1} \Sigma_{j=1}^{k} (2j-1)^{-1} a_{2j}(x) + 2^{-1} a_{2k}(x) \quad (30)$$

and

$$r_j^{2k} = N_t^{-1} \sum_{X_t} r_j^{2k}(x),\ j=11,\ 12\ . \quad (31)$$

One interesting observation that went unnoticed in [19] and [11] is that

$$\lim_{N\to\infty} r_{11}^{2k} = R'_{2k-1} \leq R^* \quad (32)$$

and

$$\lim_{N\to\infty} r_{12}^{2k} = R_{2k-1} \geq R^* \quad (33)$$

Here again, bounds which have been derived by a purely mathematical process are seen to correspond to the k-NN error-rates in (4) for one suitable value of k.

4.4. Comments. In this section eleven formally different error-estimates have been reviewed. It is clear that each of them uses the information conveyed by nearest neighbors in a different manner. Consequently, they can be expected to exhibit different finite-sample behaviour, and extensive experimental investigations are very much needed. On the other hand, they all converge to some k-NN error-rate as $N \to \infty$. These convergence can be summarized as follows, (k odd),

$$\frac{1}{2} r_8^{\text{all } k} \leq r'^{k}_{3} = r'^{k+1}_{3} = r'^{k}_{5} = r'^{k+1}_{5} = r_6^{k+2} = r_{11}^{k+1}$$

$$= R'_k \leq R^* \leq R_k$$

$$= r_{12}^{k+1} = r_{10}^{k+2} = r_4^k = r_4^{k+1} = r_2^k = r_2^{k+1} \leq r_8^{\text{all } k} . \qquad (34)$$

Quite evidently, these relationships only hold under the large-sample assumption.

One final remark is in order. Long as it is the above enumeration of error-estimates is far from being exhaustive: We have only selected those schemes which we believe are most representative of each individual approach. In particular, it is shown in [9] that direct k-NN estimates are plentiful. Experimental investigation with these should also deserve consideration.

5. ESTIMATION OF THE LOGARITHMIC EQUIVOCATION

Many of the approaches of the pervious sections can be used for the purpose of estimating the logarathmic equivocation $H = E\{H(x)\}$ with $H(x) = -\Sigma_i \, \eta_i(x) \ln \eta_i(x)$. For example, if as in Section 3 we use the estimate $k_i/(N_d v_i)$ of $p_i(x)$ we obtain the following M.L. estimate h_1 of H:

$$h_1 = N_t^{-1} \sum_{X_t} h_1(x)$$

with

$$h_1(x) = \sum_{i \neq j} \frac{k_i v_j}{k_1 v_2 + k_2 v_1} \ln \frac{k_i v_j}{k_1 v_2 + k_2 v_1} . \qquad (35)$$

The asymptotic properties of this estimate do not seem to be known. There are however some reasons to believe that $h_1(x)$ is an asymptotically biased estimate of H(x), with a bias that depends on the true value of H(x).

In a like manner, it is tempting to use the direct k-NN estimate k_i/k of $\eta_i(x)$ to obtain

$$h(x) = - \Sigma_i \, (k_i/k) \ln (k_i/k) . \qquad (36)$$

This estimate has been shown to be biased [1], (and consistent, and asymptotically normal) and to give grossly optimistic values of H(x) especially when k is small [2], [3]. Various solutions have been proposed to overcome these shortcomings. Basharin suggests correcting the bias by using an additive constant which only depends on k and the number of hypotheses, [1]. On the other hand, Butler and Ritea use the probability density of $\eta_i(x)$ conditioned on k_1 and k_2 to show that the expected value of η_i is $(k_1+1)/(k+2)$ thus yielding, [3],

$$h_2(x) = -\Sigma_i \frac{k_i+1}{k+2} \ln \frac{k_i+1}{k+2} \tag{37}$$

This in turn appears to be a rather pessimistic estimate of H(x), especially when k is small.

Still another approach was developed by Devijver and Dekesel who showed that the bias of h(x) in (36) is nearly proportional to $(k-1)^{-1}$ H(x), [14]. Hence the following estimate:

$$h_3(x) = -\Sigma_i \frac{k_i}{k-1} \ln \frac{k_i}{k} \tag{38}$$

The Parzen-type approach of Lissack and Fu [28] can also be combined with the Devijver-Dekesel method of bias correction to yield

$$h_4(x) = -\Sigma_i \frac{k_{iv}}{k_v-1} \ln \frac{k_{iv}}{k_v} . \tag{39}$$

One further method belonging to the ONNSS family (Cfr. Sec. 4.3. above) was developed recently by the author who noted that H can be expanded in an infinite series according to [11],

$$H = 1 - \sum_{j=2}^{\infty} \frac{1}{j(j-1)} Q_j \tag{40}$$

where $Q_j = E\{Q_j(x)\}$ and $Q_j(x)$ is the probability that all of the first j NNs to x are from the same class. From (40) and $Q_j \geq 0$ and $Q_j \leq Q_{j'}$ if $j \geq j'$ one obtains

$$1 - \sum_{j=2}^{k} \frac{1}{j(j-1)} Q_j - k^{-1} Q_k \leq H \leq 1 - \sum_{j=2}^{k} \frac{1}{j(j-1)} Q_j . \tag{41}$$

Consequently, for any $x \in X_t$ let $q_j(x)=1$ if all of the first j NN's to x from X_d belong to the same class, and $q_j(x)=0$ otherwise. Let further

$$q_j = N_t^{-1} \sum_{X_t} q_j(x) .$$

By substitution in (40) we obtain

$$h_5 = 1 - \sum_{j=2}^{k} \frac{1}{j(j-1)} q_j \tag{42}$$

$$h_6 = 1 - \sum_{j=2}^{k} \frac{1}{j(j-1)} q_j - k^{-1} q_k,$$

with the properties that

$$\lim_{N\to\infty} h_6 \leq H \leq \lim_{N\to\infty} h_5 \tag{43}$$

and the bounds converge to H as k grows arbitrarily large. Multi-hypothesis extensions of this technique can be found in [11].

6. ESTIMATION OF THE BHATTACHARYYA COEFFICIENT AND THE DIVERGENCE

The reader who wants to try his hand at writing down estimates of B and J based on k-NN estimates of $\eta_1(x)$ and $\eta_2(x)$ will soon notice that many problems arise and that there is scant hope that asymptotic properties of such estimators will ever be derived. (The specific difficulties of estimating the divergence are discussed at some length in [11]). One alternative is offered by the ONNSS method which exchanges the need to estimate local probabilities for that of estimating average probabilities of NN sequences properties. Here, a brief outline of the derivation of the ONNSS estimate of the Bhattacharyya coefficient is presented since it does not seem to have ever appeared before. Next we shall conclude this section with the ONNSS estimate of the divergence.

By using a simple transformation, the Bhattacharyya coefficient can be written as

$$B = \frac{1}{2} E\{[1-(\eta_1(x)-\eta_2(x))^2]^{1/2}\}, \tag{44}$$

and the expression between the braces can be expanded in a classical, uniformly convergent series expansion, namely

$$B = \frac{1}{2} E\{1 - \sum_{j=1}^{\infty} c_{2j} (\eta_1(x)-\eta_2(x))^{2j}\}, \tag{45}$$

where $c_{2j}=(2j-3)!!/(2j)!!$. It is readily verified that all conditions are satisfied to write

$$B = \frac{1}{2} [1 - \sum_{j=1}^{\infty} c_{2j} E\{(\eta_1(x)-\eta_2(x))^{2j}\}]. \tag{46}$$

Upon expansion, the term $E\{(\eta_1(x)-\eta_2(x))^{2j}\}$ can be interpreted as the difference between the average probability that the first 2j NN's cast an even number of votes for either class, and the average probability that the first 2j NN's cast an odd number of votes for either class. Putting this formally, let

$$E\{(\eta_1(x)-\eta_2(x))^{2j}\} = S_{2j}$$

with

$$S_{2j} = 2\, E_{2j} - 1 \tag{47}$$

and E_{2j} is the average probability of even numbers of votes from the 2j NN's. Then

$$B = \frac{1}{2}\{1 - \sum_{j=1}^{\infty} c_{2j}S_{2j}\}. \tag{48}$$

And from here on it is a simple matter to show that

$$\frac{1}{2}\{1- \sum_{j=1}^{k-1} c_{2j}(S_{2j}-S_{2k})-S_{2k}\} \leq B \leq \frac{1}{2}\{1- \sum_{j=1}^{k} c_{2j}S_{2j}\} \tag{49}$$

and the bounds converge to B as $k \to \infty$. It can further be seen that the upper bound is tight when B is large whereas the lower bound is tight when B is small.

In the line of the previous section, the finite-sample implementation of the ONNSS approach is now obvious: Let $e_{2j}(x)=1$ if $\{k_1$ and k_2 are even $/k = 2j\}$ and $e_{2j}=0$ otherwise, $e_{2j} = N_t^{-1} \sum_{X_t} e_{2j}(x)$ and $s_{2j}= 2\, e_{2j}-1$. Then, we have

$$b_1 = \frac{1}{2}\{1 - \sum_{j=1}^{k} c_{2j}\, s_{2j}\}, \tag{50}$$

and

$$b_2 = \frac{1}{2}\{1 - \sum_{j=1}^{k-1} c_{2j}(s_{2j}-s_{2k})-s_{2k}\} \tag{51}$$

with the property that

$$\lim_{N\to\infty} b_2 \leq B \leq \lim_{N\to\infty} b_1 \tag{52}$$

As before, the bounds can be made arbitrarily tight by choosing k large enough.

Finally, we turn to the consideration of the divergence measure. It is shown in [11] that under the assumption that the underlying class conditional probability densities do not vanish on Ω, the divergence

$$J = E\{-\Sigma_i \, \eta_i(x) \, \ln \frac{\eta_i(x)}{1-\eta_1(x)} \}$$

admits of the following series expansion:

$$J = -1 + \sum_{j=2}^{\infty} \frac{j+1}{j(j-1)} Q_j \tag{53}$$

where Q_j is as defined in Section 5. In the finite sample case, we have to content ourselves with a lower bound to J, namely

$$j_1 = -1 + \sum_{j=2}^{k} \frac{j+1}{j(j-1)} q_j \, , \tag{54}$$

(Cfr. Sec. 5 for definition of q_j) with

$$\lim_{N\to\infty} j_1 \leq J, \tag{55}$$

and

$$\lim_{k\to\infty}(\lim_{N\to\infty} j_1) = J \tag{56}$$

where we have assumed that $\lim(k/N)=0$.

We conclude by noting that the estimates in (50), (51) and (54) are asymptotically unbiased, but their variance is unknown.

7. SIMULATION EXPERIMENTS

In order to demonstrate the performance of the estimators discussed in this paper, some results from simulation experiments are now given. The simulated data were drawn from two-dimensional, normal distributions with equal covariance matrix and with a Mahalanobis distance of 2.45. Each class was chosen both in the design and test sets with probability 0.5. Thus, in this example the Bayes error probabilities is 11 percent, the 1-NN error-rate is of the order of 16 percent, the Bhattacharyya coefficient is 0.236, the divergence is 3.01 and the equivocation is unknown but is comprised between R_1 and $H(R^*, 1-R^*)=.347$. For each trial $N_d=N_t=500$. This was repeated for ten trials with the average results shown in Figures 1-6.

Experimental results with r_1^* are shown in Figure 1 for $k_1=k_2$, and they show a significant pessimistic bias. It has been our

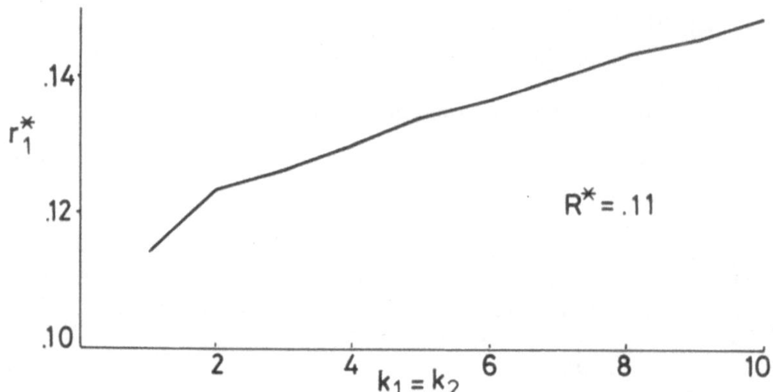

Figure 1. The Fukunaga-Hostetler estimate r_1^*.

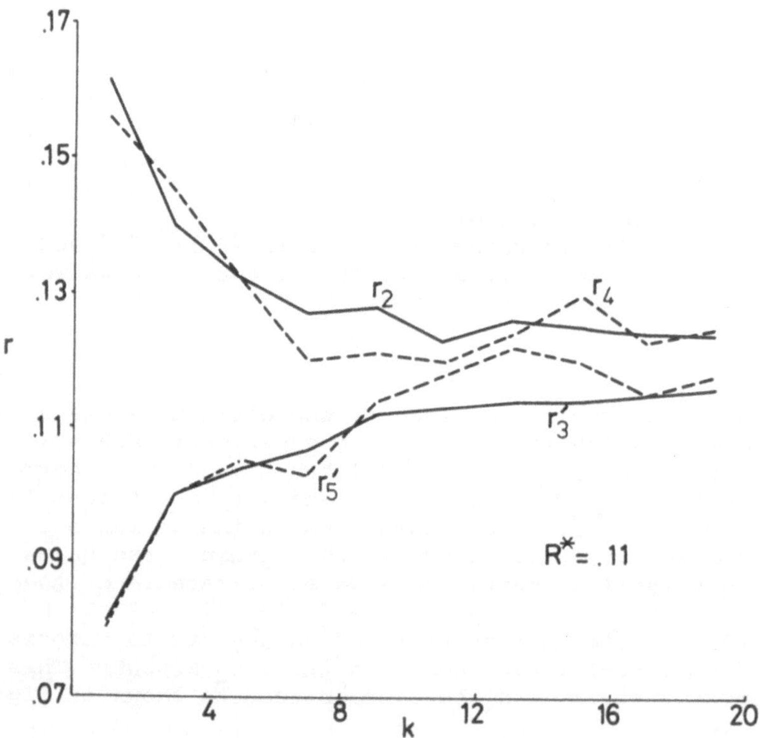

Figure 2. The error-counting estimates r_2-r_5.

experience that r_1^* exhibits a definite tendency to increase with k and this phenomenon is clearly seen here. Besides, no general rule for choosing k_i is known. These observations make the use of this estimator quite critical. The sample-variance of r_1^* was of the order of 10^{-4}.

The results with the error-counting estimates r_2-r_5' are displayed in Figure 2. They show the tendency of these estimators to converge to 0.12. Thus they all tend to slightly overestimate R*. Estimators r_2 and r_3' which make use of the information about θ are more stable than the estimates r_4 and r_5' for which this information is ignored. This is reflected in the sample-variances with var $\{r_4,r_5'\} \approx 1.6$ var$\{r_2,r_3'\}$, and var $\{r_2\} \approx 2.3\ 10^{-4}$.

In Figure 3, the direct k-NN estimates r_6 and r_{10} also display a slight tendency to overestimate R*, while r_8 is seen to be fairly independent of k. The results with r_7 and r_9 have to be understood in the following sense: For each choice of volume V(x) the average of the number k_V was determined. The curves of Figure 3 represent r_7 and r_9 as function of k = ave $\{k_V/V=v\}$. The low values of r_7 and r_9 for small k are due to the fact that we used $r_7(x)=r_9(x)=0$ whenever $k_V=0,1$. We note that, as expected, $r_7 < r_6$ and $r_9 < r_{10}$. The sample-variance of r_6, r_7 and r_{10} was also of the order of 10^{-4}. That of r_8 and r_9 was about twice as large.

The experiments with the ONNSS method were very successful as can be seen from Figure 4. While these estimates appear fairly well unbiased, their variance is of order of that of the error-counting estimates with the variance of the upper bound r_{12} constantly decreasing as k increases.

Results of experimental estimation of the logarithmic entropy can be seen in Figure 5. The results with h_1 and h_2 are not shown because in these experiments they produced not very realistic results which exceeded the upper bound of 0.347. Since the true value of H is unknown the estimators h_3-h_6 cannot be judged on the basis of bias. The variance of h_3 and h_4 was of the order of 4.10^{-4}. That of h_5 and h_6 was about twice as large. The interested reader may refer to [11] and [14] for additional experimental results with h_3, h_5 and h_6.

Application of the ONNSS method to the estimation of the Bhattacharyya coefficient produced the results displayed in Figure 6. Again, they appear fairly well unbiased. The sample-variance of b_1 and b_2 was of the order of 4.10^{-4}.

In these experiments, the results with estimate j_1 of divergence J (not shown) have indicated a slow convergence of the series expansion (51), (or (52)). To produce satisfactory results would have required using a number of NN's larger than 20. This is not a rule however as can be seen from the experimental results in [11].

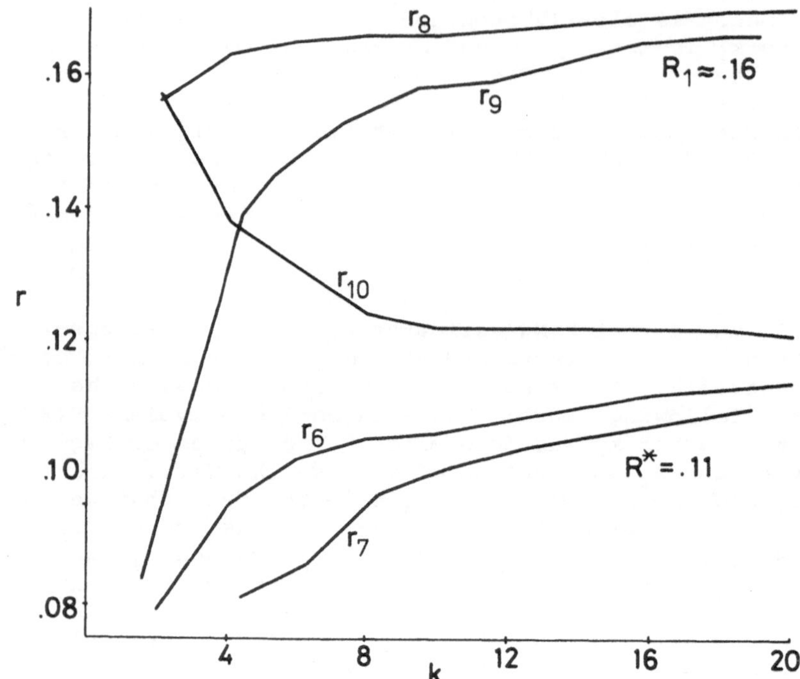

Figure 3. The direct k-NN estimates r_6-r_{10}.

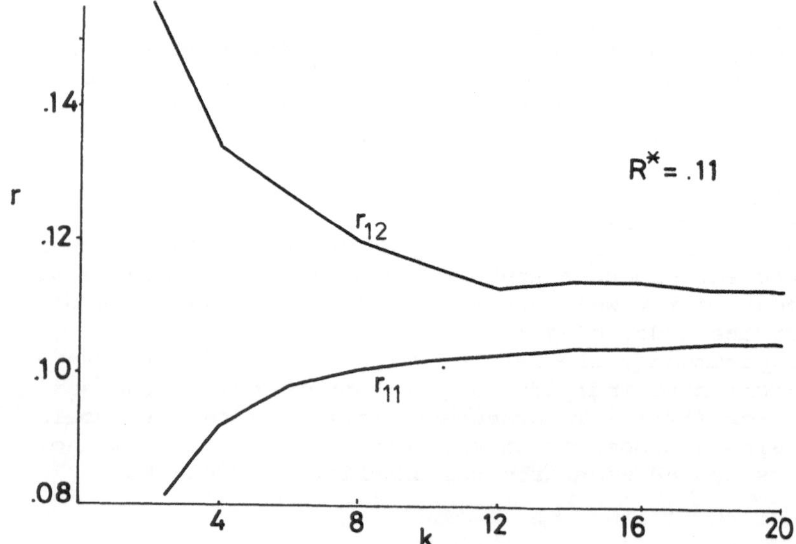

Figure 4. The ONNSS error-bounds r_{11}-r_{12}.

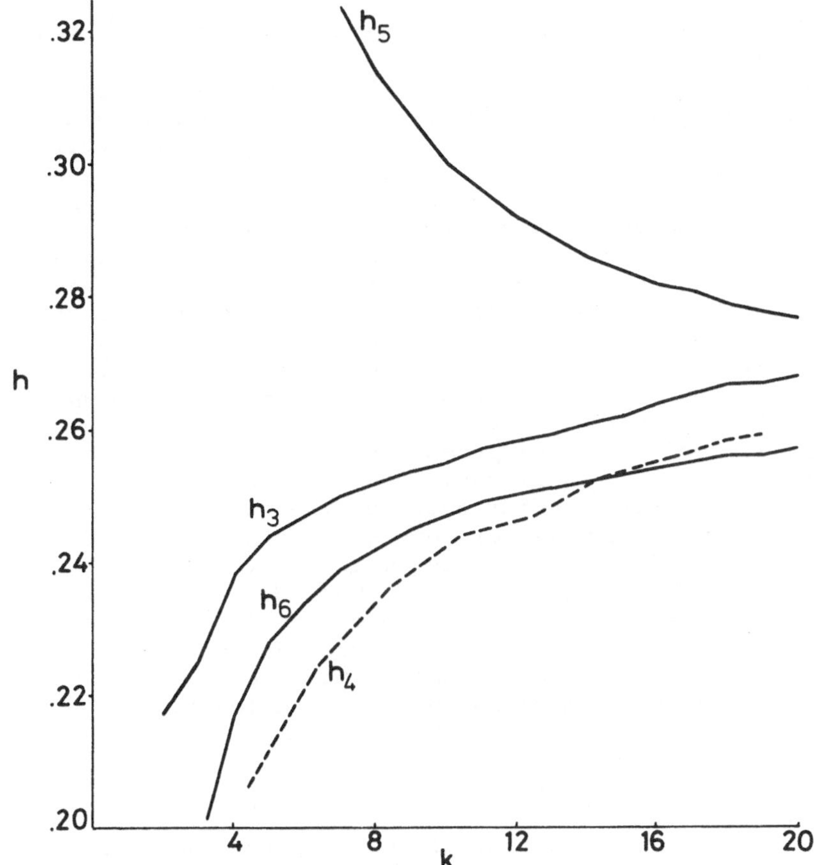

Figure 5. Logarithmic entropy estimates h_3-h_6.

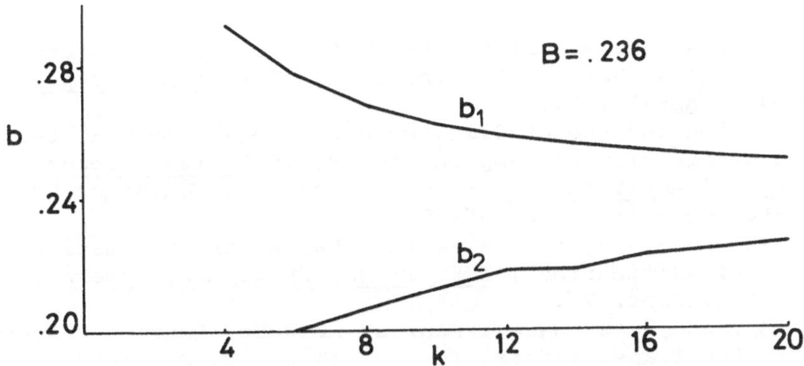

Figure 6. ONNSS bounds of the Bhattacharyya coefficient.

8. CONCLUSIONS AND FURTHER COMMENTS

In this paper several nonparametric estimation schemes have been reviewed and more than twenty different estimators were discussed. A number of new asymptotic relationships have been pointed out. Our enumeration of estimators could have grown much longer had we not (purposely) decided to concentrate on the most popular criteria, the binary hypothesis problem and the nearest-neighbor approach.

At the present state of development of the theory it seems fair to say that although we have acquired a fairly good understanding of the large-sample properties of nonparametric estimators, much remains to be done to clarify the finite-sample case. Probably the greatest deficiency in the current state of the art is the lack of comparative finite-sample experimental results. Only when such results are available and well understood does it become possible to uncover practical guidance for making an apt choice amidst the bewildering array of alternatives.

As far as the estimators discussed in this paper are concerned, pertinent questions are: How is their behaviour affected by sample size and dimensionality? What could be the best choice of k? How do these estimators perform for very small and very large values of the criterion function to be estimated? It is our belief that an even more prominent question is: How does the behaviour of finite-sample estimators relate to the probability distribution of the conditional Bayes risk? It is our contention that laying bare the existence of such a relationship would be of great help when attacking problems of sample size and dimensionality.

REFERENCES

1. Basharin, G. P., "On a Statistical Estimate for the Entropy of a Sequence of Independent Random Variable", _Theory of Probability and Applications_, _4_, 333-336, 1959.
2. Butler, G. A., "Evaluating Feature Spaces for the Two Class Problem", in _Annu. Symp. Rec. IEEE Syst. Man, Cybern._, 119-125, Oct. 1971.
3. Butler, G. A., and H. B. Ritea, "Estimation of Mutual Information in Two-Class Pattern Recognition", _IEEE Trans. Comput._, C-23, 410-420, April 1974.
4. Chen, C. H., "On the Use of Distance and Information Measures in Pattern Recognition and Applications", in _Pattern Recognition Theory and Applications_, K. S. Fu and A. B. Whinston Eds., Noordhoff-Leyden - 1977, pp. 45-60.
5. Chen, Z., and K. S. Fu, "Nonparametric Bayes Risk Estimation for Pattern Classification", _IEEE Trans. Syst. Man, Cybern._, SMC-7, 651-656, Sept. 77.
6. Cover, T. M., and P. E. Hart, "Nearest Neighbor Pattern Classification", _IEEE Trans. Inform. Theory_, IT-13, 21-27, 1967.

7. Cover, T. M. and J. M. Van Campenhout, "On the Possible Orderings in the Measurement Selection Problem", IEEE Trans. Syst. Man, Cybern., SMC-7, 657-661, 1977.
8. Devijver, P. A., "Decision Theoretic and Related Approaches to Pattern Classification", in Pattern Recognition Theory and Applications, K. S. Fu and A. B. Whinston Eds., Noordhof-Leyden - 1977, pp. 1-34.
9. Devijver, P. A., "Reconnaissance des Formes par la Méthode des Plus Proches Voisins", Doctoral Dissertation, Univ. Paris VI, Paris, June 1977.
10. Devijver, P. A., "A Note on Ties in Voting with the k-NN Rule", Pattern Recognition, 10, (4), 1978.
11. Devijver, P. A., "Nonparametric Estimation by the Method of Ordered Nearest Neighbor Sample Sets", submitted to IEEE Trans. Comput.
12. Devijver, P. A., "A Generalized Divergence Measure for Signal Selection", Proc. 1st AFCET-SMF Conf. Applied Mathematics, Paris, Sept. 1978.
13. Devijver, P. A., "New Error Bounds with the Nearest Neighbor Rule", submitted to IEEE Trans. Inform. Theory.
14. Devijver, P. A., and M. M. Dekesel, "k-NN Estimates of Information Measures", Proc. Sem. Pattern Recognition, Liege, 1977, pp. 941-946.
15. Duda, R. O. and P. E. Hart, Pattern Classification and Scene Analysis, New-York: Wiley, 1973.
16. Fukunaga, K. and D. L. Kessel, "Nonparametric Bayes Error Estimation Using Unclassified Samples", IEEE Trans. Inform. Theory, IT-19, 434-440, July 1973.
17. Fukunaga, K. and L. D. Hostetler, "k-Nearest-Neighbor Bayes Risk Estimation", IEEE Trans. Inform. Theory, IT-21, 285-293, May 1975.
18. Fukunaga, K. and P. M. Narendra, "A Branch and Bound Algorithm for Feature Subset Selection", IEEE Trans. Comput., C-26, 917-922, Sept. 1977.
19. Garnett, J. M. III, and S. S. Yau, "Nonparametric Estimation of the Bayes Error of Feature Extractors Using Ordered Nearest Neighbor Sets", IEEE Trans. Computers, C-26, 46-54, Jan. 1977.
20. Grettenberg, T. L., "Signal Selection in Communication and Radar Systems", IEEE Trans. Inform. Theory, IT-9, 265-275, 1963.
21. Hanakata, K., "Feature Selection and Extraction for Decision Theoretic Approach and Structural Approach", in Pattern Recognition Theory and Application, K. S. Fu and A. B. Whinston Eds., Noordhof-Leyden, 1977, pp. 133-168.
22. Hellman, M. E., "The Nearest-Neighbor Classification Rule with a Reject Option", IEEE Trans. Syst. Sci. Cybern. SSC-6, pp. 179-185, 1970.
23. Kailath, T., "The Divergence and Bhattacharyya Distance Measures in Signal Selection", IEEE Trans. Communication Technology, Vol. COM-15, pp. 52-60, Feb. 1967.

24. Kanal, L., "Patterns in Pattern Recognition", IEEE Trans. Inform. Theory, IT-20, (6), 697-722, Nov. 1974.
25. Kittler, J., "Statistical Feature Selection and Extraction", in Data Structure Analysis and Applications, J. Kittler Ed., ENST Tech Rept C-78002, 1978, pp. 16-28.
26. Kittler, J., "Feature Set Search Algorithm", in this Proceedings.
27. Lissack T., and K. S. Fu, "A Separability Measure for Feature Selection and Error Estimation in Pattern Recognition", Purdue Univ., Lafayette Ind. Tech. Rep. TR-EE 72-15, 1972.
28. Lissack, T., and K. S. Fu., "Error Estimation in Pattern Recognition via L^{α}-Distance Between Posterior Density Function", IEEE Trans. Inform. Theory, IT-22, 34-45, 1974.
29. Schweppe, F., "On the Distance Between Gaussian Processes: the State Space Approach", Information and Control, 11, (4), 373-395, 1967.
30. Tou, J. T. and R. P. Heydorn, "Some Approaches to Optimum Feature Extraction", in Computer and Information Sciences - II, J. T. Tou Ed., New-York: Academic Press, 1967, pp. 57-89.
31. Toussaint, G. T., "Comments on the Divergence and Bhattacharyya Distance Measures in Signal Selection", IEEE Trans. Communication Technology, Vol. COM-20, p. 485, June 1972.

FINITE LEARNING SAMPLE SIZE PROBLEMS IN PATTERN RECOGNITION

L. F. Pau

Department of Electronics and Physics
E. N. S. Télécommunications
46, rue Barrault
75634 PARIS Cédex 13 FRANCE

SUMMARY. This paper surveys some major problems, and solutions, relating finite learning sample constraints to the design of a statistical pattern classifier. These problems are also put in relation to some application areas where such constraints are critical.

NOTATION. (see also Table 1)

x : $(x_1, \ldots, x_p)$, pattern vector of dimension p.

p : dimension of the pattern vectors

K : number of classes

$\omega(x)$: class of pattern vector x

ω_j : class number j, $j = 1, \ldots, K$

$\|\cdot\|$: norm in E^p

f, g, h : probability density functions defined in Table 1

F, G, H : cumulated probability functions associated to f, g, h

c_j : number of patterns classified into ω_j

M : number of quantization levels

N : number of unsupervised pattern vectors

n : test set cardinality, or cardinality of any other then specified sample set

λ : Lebesgue probability measure, or other parametric probability measure

D : Mahalanobis distance

$m, \hat{m}$: mean vector, and estimated mean

σ, s : standard deviation, and estimate

Σ, S : covariance matrix, and estimate

1. INTRODUCTION

In practice the design of almost all pattern recognition systems is sample-based, i.e. all classes, features, clusters, decision rules, etc... have to be inferred from finite sets of patterns, although the system itself is supposed to operate on infinite numbers of unclassified patterns. Within the framework of statistical pattern recognition and decision theory, sample-based rules are substitutes for unknown optimal rules.

As soon as the number K of classes, the dimension p of the space, the number of learning samples and the number M of quantization levels, jointly become so large, that partition of E^p into K classes is no longer possible, the finiteness of the sample sets makes the problem untractable. It should be stressed, however, that it is interesting to investigate what happens when the number of samples decreases from infinity to large finite values; this is called asymptotic behavior of sample-based pattern recognition systems. It should, however, be realized that there is a large void space, as far as results are concerned, between the small (K, p, N, M) case, and the large (N, M) case corresponding to asymptotic properties.

The motivation for such a study lies in a number of practical problems, to be reviewed later, for which it is essential jointly:

a) to cope with the unavailability of large numbers of design samples
b) to minimize the time needed to classify a pattern x
c) to minimize the storage capacity needed for a pattern recognition system
d) to minimize the measurement and classification costs, especially those due to classification errors, while putting constraints on some of the latter
e) to minimize the hardware complexity, also in terms of coding and quantization levels.

Moreover, while in the classical pattern recognition approach, each pattern being observed is classified individually, there are many cases where it is necessary (for the reasons stated above) to classify a set of N patterns (a lot) at a time on the basis of the measurements on a small subset of n patterns (test patterns), $n \leq N$.

2. MAJOR FINITE SAMPLE CONSTRAINTS ON A CLASSIFICATION SYSTEM

In the following we will review some effects of the finite sample constraint, so that better pattern recognition systems can be designed.

The following learning pattern sets will be considered:

a) $N = (N_1 + \ldots + N_K)$ unsupervised learning vectors x

b) N_j supervised patterns, building the design (training) set of class ω_j

c) N_o unsupervised patterns, building a lot

d) n unsupervised patterns, called sample set, which is a subset of the lot.

Some effects of the finite sample constraint will only be mentioned shortly, either because of insufficient research in this area, or because of the availability of good surveys.

2.1 Imperfect teacher or clustering procedure (Section 3)

One of the effects of a finite sample assumption, is in practice to replace the teacher or clustering operator, generally assumed to be perfect and learning set-independent, by an imperfect teacher. The imperfect teacher, working on N learning patterns only, does not have enough information available to label these correctly, which means independently of the learning set considered.

2.2 Coding and quantization effects (Section 4)

If a M-region quantizer is considered in E^p, the measure of performance of such a quantizer depends not only on p, M, on the range T of possible measurements, but also on the class conditional p.d.f. f.

Because T and f can only be inferred from design samples, the selection of a proper quantizer will depend, for each class ω_j, upon the number N_j of design samples.

2.3 Tolerance regions of each class (Section 5)

To know whether the feature space E^p can be partitioned or not into K classes, it is necessary to know the size of the extension of each class.

For the finite sample case, one must determine the β-expectation tolerance region T of each class in E^p.

2.4 Differences among all class-conditional sample based probability density functions, and cumulated distribution functions

Table 1 summarizes the various sample sets used at the different stages of a classification procedure. It is essential to remark that:

a) all p.d.f. and c.d.f. of supervised learning patterns must be used separately, in order to obtain K class conditional p.d.f. and c.d.f.
b) all class conditional p.d.f. and c.d.f. are sample-based, and based on different sets of patterns, thus resulting in differences even for the same class.

Set	Cardinality	Probability density function
Population ω_j $j = 1, \dots K$	$N_j^{\infty} \to +\infty$	$f_j\ (x)$
Design (training) set from ω_j	N_j	$\hat{f}_j\ (x)$
Unsupervised learning vectors	$N_1 + N_2 + \dots + N_K$	
Lot	N_o	$\hat{g}_{N_o}\ (x)$
Test (sample)set taken from the lot	$n << N$	$\hat{h}_n\ (x)$
Pattern x in E^p	1	$\delta\ (x)$
Space E^p	Dimension p	

Table 1 : Sample sizes and p.d.f.'s of the various sample sets.

More precisely, the three main effects of a finite sample assumption are the following:

1. Because of sampling problems, the class conditional p.d.f. $\hat{f}_j$ (x) is not necessarily identical in nature with that f_j (x) of the population ω_j
2. the p.d.f. $\hat{g}_{N_o}$ (x) of the patterns in the lot is not necessarily identical with that of the unsupervised learning set of size $N = N_1 + \ldots + N_K$, nor with any mixture of the f_j (x); the usual assumption $N_o = n = 1$, with only one pattern classified at a time, eliminates this difficulty.
3. the p.d.f. $\hat{h}_n$ (x) of the sample set taken from the lot will, in general, differ from that of the lot, due to, once again, sampling effects.

Simplifying assumptions which may be considered are the following:

A. the p.d.f. $\hat{g}_{N_o}$ (x) of the lot differs from that of the design data in a priori class probabilities only: (see KITTLER - PAU [18]):

$$\hat{g}_{N_o}(x) = \sum_{j=1}^{K} \hat{P}_A(\omega_j)\, \hat{f}_j(x)$$

B. the p.d.f. $\hat{h}_n$ (x) of the test set differs from that of the design data in a priori probabilities only

$$\hat{h}_n(x) = \sum_{j=1}^{K} \hat{P}_B(\omega_j)\, \hat{f}_j(x)$$

C. $\hat{g}_{N_o}$ (x) has some general form which is completely different from that of the design sets, and eventually of the population p.d.f.

As a consequence, all of the following are random variables, thus making the distribution of each quantity very complex:

i a priori class probability $P(\omega_j)$

ii class-conditional p.d.f. $\hat{f}_j(x)$

iii lot p.d.f. $\hat{g}_{N_o}(x)$

iv test set p.d.f. $\hat{h}_n(x)$

v all classification errors.

One method in which to cope with this complexity, is to use, for design purposes, distribution-free bounds on p.d.f. and c.d.f. estimators (see Section 6). They have been found to be most useful in practice, although these bounds are not very tight.

2.5 Optimization of a p.d.f. estimator vs sample size (Section 7)

Finally, any estimation of these p.d.f. or c.d.f. will require specified parameters, although the estimation itself may be non-parametric. The problem is then to select these parameters as to minimize the estimation error, as function of the dimension p, the sample size, and the errors.

2.6 Optimization of estimators of statistical moments vs sample size (Section 8)

When finite sample sizes are considered, it is essential to use improved estimators of the means and variances appearing in distance measures and parametric classification rules; these estimators may be Bayesian, or take into account the non-independence of the samples.

2.7 Finite sample perturbations of a linear feature extraction transformation (Section 9)

The distribution of the eigenvalues of the covariance matrix, or of the inverse of the covariance matrix, as well as the perturbations on the directions of the eigenvectors (CHANDLER [54]), are essential to estimate. Results have been obtained assuming p-dimensional normal distributions.

See also ANDERSON [66], LI et al [67], GURLAND [40].

2.8 Finite sample distance measures and information measures

Distance and information measures are useful for feature selection, classification, and for error bounds computations. To determine the finite sample behaviour, a maximum likelihood estimation must be used. Quadratic forms, the divergence, the Bhattacharyya distance and equivocation are studied by GURLAND [40], C. H. CHEN [30], HOEL [12].

2.9 Nearest neighbor classification rule (Section 10)

The asymptotic distribution of the L^2 distance to the closest neighbor is essential to know in order to set decision thresholds, and introduce a possible reject threshold in the small sample case.

2.10 Finite sample discriminant analysis (Section 11)

A number of procedures have been proposed for assigning an individual to one of two or more groups on the basis of a multivariate observation. The usual procedures are:

(1) The linear discriminant function. If the parameters are known, this is the optimum rule for assigning to one of two normal populations with the same covariance matrix. In practice, an observation is assigned to the first population if

$$D(x) = (x - \frac{1}{2}(\hat{m}_1 + \hat{m}_2))^t S^{-1} (\hat{m}_1 - \hat{m}_2) > C$$

where $\hat{m}_i$ is the sample mean vector in ω_i, S is the sample covariance matrix, and C is a threshold.

(2) The quadratic discriminant function. If the parameters are known, this is the optimum rule for assigning to one of two normal populations with different covariance matrices. In practice, an observation is assigned to the first population if

$$Q(x) = (x - \hat{m}_2)^t S_2^{-1} (x - \hat{m}_2) - (x - \hat{m}_1)^t S_1^{-1} (x - \hat{m}_1) > C$$

(3) The K - class discriminant function. This rule is theoretically optimal when there are K multivariate normal populations with the same covariance matrix. In practice, an observation is assigned to the population ω_i if

$$M_j(x) = (x - \frac{1}{2}\hat{m}_j)^t S^{-1} \hat{m}_j + C_j$$

is maximized for $j = i$, and C_j is a constant depending on the a priori probability of the jth population.

(4) <u>The eigenvector, or canonical vector</u>, approach to the K - class problem. This is an extension of Fisher's original approach to discriminant analysis. The variables are transformed to eigenvectors, and assignment may be made using any number of the transformed variables as in the multiple discriminant method.

The assumptions underlying these techniques are not always evident to the user, nor are the consequences of their violation. The assumptions include multivariate normality, common covariance matrices, and the correct assignment of the initial learning patterns. Small sample size problems arise via their effects on:

a) unequal covariance matrices (see GILBERT [21], MARKS-DUNN [22])
b) continuous non-normal distributions (see LACHENBRUCH [19])
c) contamination of distributions (PÖPPL [39])
d) initial misclassification of samples (see LACHENBRUCH [20], MC LACHLAN [23])

It is however unfortunate that the small sample results reported in the literature are few and inconsistent, thus making computer simulation necessary (especially for different covariance matrices)

2.11 Sample-based estimation of the probability of error

Estimating the probability of error due to a particular classification rule, is discussed in many papers, e.g. in the surveys by COVER [61], TOUSSAINT [59], RAUDYS [35], DEVIJVER [45].

The emphasis is on the existence of distribution-free bounds for these estimates of the probability of error.

For the finite sample problem, MUISE - BOORSTYN [58] establish that the time-varying two hypothesis decision rule essentially stores a quantized version of the likelihood ratio, although the quantization is not of any simple form; using the proposed detectors will result in the fastest decay of error probability with increasing sample size. COVER-FREEDMANN et al [60] also demonstrate that knowledge of the sample size can be of use in lowering the error probability, and that the optimized rule is deterministic.

DUIN [32] has established an upper bound on the error probability in the two-class case, which is sample size dependent. Let $f_1(x)$, $f_2(x)$ be the two class conditional p.d.f.; the Bayes error ($n = \infty$ case), with equal class probabilities is:

$$\varepsilon^* = \frac{1}{2} \int_{E^p} \mathrm{Min}\ (f_1(x),\ f_2(x))\ dx$$

The actual expected error probability ε, can be bounded by a function of the sample size $n = N_1 = N_2$, of the (unknown) Bayes error ε^*, and of the measurement complexity M, defined as the number of possible quantized values of x:

$$E(\varepsilon) \leq \varepsilon^* + (2\pi n)^{-1/2}([(2\varepsilon^*(1 - 2\varepsilon^*/M) / M]^{1/2} + [2(1-\varepsilon^*)(1 - 2(1-\varepsilon^*)/M) / M]^{1/2})$$

In some other cases, including exponential densities, good theoretical approximations of error probability given finite sample size is possible.

2.12 Sequential classification

The small sample behaviour of two-class sequential classification schemes is summarized in Section 12. Most sequential classification procedures do not still allow for the selection of optimum sample size dependent decision thresholds, moreover, they are restricted to one-dimensional measurements. The computation of the expected sample size needed for classification gives however an estimate of the memory size needed for implementations.

2.13 Classification of lots using finite test sets from the lots (Section 13)

Approximate results are indicated, mostly in the case of classification by attributes. The general problem, and especially classification by variables are very difficult and require simplifying assumptions due to the problems mentioned above under 2.4.

2.14 Small sample properties of time-dependent process models (Section 14)

The asymptotic distribution of the parameters in an ARMA model is very useful, when a classification system uses these parameters as features. This asymptotic distribution is sample size, and rank dependent.

3. IMPERFECT TEACHER OR CLUSTERING PROCEDURE

Let $\{x_i\}$ $i = 1, \ldots, N$ be the N unsupervised sample vectors; in order to obtain the N supervised design samples one may either apply an unsupervised clustering procedure based on some very simple assumptions, or use a teacher which assigns each sample x_i into one, $\omega(x_i)$, of the K possible classes. This allocation is implicitly assumed to be perfect, i.e. independent of the learning set considered.

An important variation corresponds to an imperfect teacher [49][56], due:

- either to measurement errors on the samples x_i
- or insufficient information in terms of sample size N for a perfect teacher to assign correctly the samples to a class, had the true class relationship been available; this assignment may be incorrect, because of the dependence of the learning set considered.

As a consequence, the probabilities of error for this teacher will generally not be constant, being not only a function of the true classes derived from a full information case, but also of the values of x_i, $i=1,\dots,N$.

Two simplifying approaches to this problem have been investigated, none of them however stressing the small sample issue:

a) learning with a probabilistic teacher, by which the teacher derived class membership $\omega(x_i)$ is assigned by drawing an independent random variable with probabilities equal to the class posterior bayesian probabilities of the x_i's. This scheme has been shown to converge (AGRAWALA [48]).
b) assume the probabilities of error for the imperfect teacher to be constant, and known; the class conditional probabilities $f(x_i)$ are then averaged over all K possible assignments ω_j, $j = 1, \dots, K$ when weighted by the teacher's confusion matrix [51].

4. CODING AND QUANTIZATION EFFECTS

4.1 Let $X = (x_1, x_2, \dots, x_p)$ be a p-dimensional random variable with probability measure μ defined on λ-Lebesgue measurable subsets of the p-dimensional space E^p.

Let R_i, $1 \leq i \leq M$ be a set of M λ-measurable disjoint subsets of E^p, with:

$$\sum_{i=1}^{M} \mu(R_i) = 1$$

We define the M-region quantizer Q with quantization regions R_i, as a mapping of the portion of E^p covered by the union of the R_i into the integers 1 to M, given by:

$$x \in R_i \iff Q(x) = i \in \{1, \dots, M\}$$

Such a quantizer Q may be used as a model for the grouping of p variate data (statistics), the quantization of signals or pictures, and analog-digital conversion. It maps each x into the integer index i, $1 \leq i \leq M$, which labels the region R_i in which x falls, and saves only the value of i for further processing.

The quantizer Q is thus the optimum classifier in the case of non-overlapping pattern classes R_i. The main problem, is the minimization of a measure of the error introduced by a quantizer Q, having a fixed number M of quantizing regions, by varying the shapes of the regions.

4.2 Measures of performance of a quantizer Q

Given a M region/class quantizer Q with regions R_i, one measure of performance is the width $\Delta_i\ (x_k)$ of R_i in the k-th coordinate:

$$\Delta_i\ (x_k) \stackrel{\Delta}{=} \sup_{x \in R_i} \{x_k\} - \inf_{x \in R_i} \{x_k\} \qquad k = 1, \ldots, p$$

The performance of Q with respect to the probability measure μ can then be measured by $\underline{M}_r\ (Q)$, the r-th mean of the quantization errors $\Delta_i\ (x_k)$ averaged over the p coordinates of each R_i with equal weights, and over the different R_i with weights $\mu\ (R_i)$ (ELIAS [63] [64]):

$$\underline{M}_r\ (Q) \stackrel{\Delta}{=} \left(\sum_{i=1}^{M} \mu\ (R_i)\ p^{-1} \sum_{k=1}^{p} \Delta_i^r (x_k) \right)^{1/r} , \quad o < r < \infty$$

A second measure of performance is given by:

$$\underline{M}_r^*\ (Q) = \left(\sum_{i=1}^{M} \mu\ (R_i)\ \lambda(R_i)^{r/p} \right)^{1/r} , \quad o < r < \infty$$

4.3 Bounds on the performances of a quantizer Q

a) If μ is absolutely continuous, and $f \stackrel{\Delta}{=} d\mu/dx$ is bounded, ZADOR [65] prooves that:

$$E\ ((x - \hat{x}\ (Q(x))^r) \sim (C_{pr}/M^{r/p})\ x\ (\int_{E^n} f^{p/(p+r)} d\lambda)^{(p+r)/p}$$

where C_{pr} are constants not known for $p > 1$.

b) If μ has a compact support T, vanishing outside some cube in E^p, lower bounds can be obtained which are also asymptotic estimates in M, for all $o \leq r \leq \infty$:

$$M^{1/p}\ \underline{M}_r \geq M^{1/p}\ \underline{M}^*_r \geq (\int_T f^{p/(p+r)}\ d\lambda)^{(p+r)/pr}$$

$$M^{1/p}\ \underline{M}_o \geq M^{1/p}\ \underline{M}^*_o \geq 0 \qquad\qquad \mu\ (T) > 0$$

$$M^{1/p}\ \underline{M}_o \geq M^{1/p}\ \underline{M}^*_o \geq \exp\ (-(1/p)\ \int_T \mathrm{Log}\ f\ .\ d\lambda) \qquad \mu(T) = 0$$

$$M^{1/p}\ \underline{M}_\infty \geq M^{1/p}\ \underline{M}^*_\infty \geq \lambda\ (T)^{1/p}$$

These four lower bounds can be approached for large M in all cases. Given μ, r, it is possible to construct a sequence (Q_m) of quantizers (M_m), M_m being an increasing sequence, such that $M_m^{1/p}\ \underline{M}_r\ (Q_m)$ converges to the expression on the right as $m \to +\infty$.

5. TOLERANCE REGIONS OF EACH CLASS

Let T (x_1, ..., x_N) be a statistical tolerance region in E^p, that is a statistic defined on the N-th power of the sample set, with values in the space of subsets of E^p; the size of this tolerance region will be determined by the probability measure λ of this set, also called coverage (see WILKS [12], TUKEY [14]).

A tolerance region is distribution free iff the induced probability distribution of λ ($T(x_1, ..., x_N)$) is independent of the parameters $\theta \in \Omega$ of the measure λ.

T is a β-expectance tolerance region iff $E(\lambda(T(x_1,...,x_N))) \leq \beta$ for all $\theta \in \Omega$; for such a region, the average probability content is at most β.

Consider sampling from a multivariate normal distribution N (m, Σ) for which the density function is:

$$f\ (x) = a\ \exp\ (-\frac{1}{2}\ (x - m)^t\ \Sigma^{-1}(x - m))$$

Let the parameter space Ω be given by $m \in R^p$, and Σ belonging to the space of (p x p) symmetric positive definite matrices.

The tolerance factors c_β for p-variate normal distributions with unknown mean, unknown covariance matrix, and sample size N is:

$$c_\beta = (1 + 1/N)\ (N - 1)\ (P/N - p)\ F_{(1 - \beta)}$$

where:

F_α is the point exceeded with probability α using Fisher's F-distribution with (p, N - p) degrees of freedom

c_β defines the tolerance region T by the ellipsoidal subset:

$$\{x \mid (x - \hat{m})^t A^{-1} (x - \hat{m}) \leq c_\beta\}$$

$$\hat{\mu} \overset{\Delta}{=} N^{-1} \sum_{\ell=1}^{N} x_\ell$$

$$A \overset{\Delta}{=} (N-1)^{-1} \sum_{\ell=1}^{N} (x_\ell - \hat{m}) (x_\ell - \hat{m})^t$$

Numerical results are given in FRASER-GUTTMAN [70].

When the nature of the true p.d.f.f is not known, it may be difficult to determine tolerance regions, although it may be reasonable to assume a functional form of f (including series expansion into e.g. Gaussian or Poisson kernels).

6. DISTRIBUTION FREE ESTIMATION OF A SAMPLE-BASED CUMULATED DISTRIBUTION FUNCTION F

Let $\{x_i\}$ be a sequence of n independent identically distributed random patterns with a common c.d.f. F.. Suppose F unknown, while F_n is the estimated c.d.f. from $x_1, \ldots, x_n$, e.g. $F_n(x)$ is the frequency of the n first samples less or equal to x in E^p.

If ε is given , and c_o, c(p) are universal constants which do not depend on F, DVORETSKY et al [57] have established the following distribution free bound:

$$\text{Prob } ((\sup_x \left| F_n (x) - F (x) \right|) \geq \varepsilon) \leq c_o \exp (-2 \varepsilon^2 c (p) n)$$

where c (p) decreases with the dimension p, according to COVER et al [60].

An analog result can be derived in terms of distribution free bounds on a probability density function estimator.

The use of distribution free bounds for local discrimination rules is illustrated by ROGERS-WAGNER [43], CHEN-FU [8].

7. OPTIMIZATION OF A P.D.F. ESTIMATOR VS SAMPLE SIZE

Let $\{x_1, \ldots, x_n\}$ be a set of n independent and identically distributed p-dimensional random vectors, all belonging to the same class conditional p.d.f. f (x).

7.1 k - nearest neighbor estimate ([15], [44])

The k-NN estimate $\hat{f}_n$ (x) of the density f (x) is defined as:

$$\hat{f}_n (x) = \frac{k - 1}{n \, v (x)}$$

where v (x) is the volume of the region about x containing its k nearest neighbours.

FUKUNAGA-HOSTETLER [38] have indicated a functional form of the smallest k_o, which depends on n, p, and f, and optimizes the mean-square error criterion on $\hat{f}_n(x)$:

$$k_o = n^{4(p+4)} . [\ p\ (p+2)^2\ \pi^2 \int f^2\ (x)\ dx\ ]^{p/(p+4)} .$$

$$. \left[\Gamma^{4/p} \left(\frac{p+2}{2}\right) \int f^{-4/p} (x)\ Tr^2 \left[\left[\frac{A}{|A|^{1/n}}\right]^{-1} \left[\frac{\partial^2 f(x)}{\partial x^2}\right] dx \right] \right]^{-p/(p+4)}$$

where A is the positive definite quadratic form associated to the metric d:

$$d^2\ (x,\ y) = (y - x)^t\ A\ (y - x) = ||\ x - y\ ||^2$$

This result on k_o is shown to be independent on linear transformations in the feature space.

If f is a p-dimensional gaussian distribution, the optimum k_o is given by:

p	4	8	16	32
k_o	$0,75\ n^{1/2}$	$0,94\ n^{1/3}$	$0,62\ n^{1/5}$	$0,42\ n^{1/9}$

7.2 Parzen estimators ([44])

The Parzen estimate $\hat{f}_n$ (x) of the density f (x) is defined as:

$$\hat{f}_n\ (x) = \frac{1}{n} \sum_{i=1}^{n} K\ (x - x_i)$$

where the kernel function K is centered about each sample point x_i, and :

$$K\ (y) = \frac{1}{v(y)}$$

where v (y) is the volume of the set of all x's such that d^2 (x, y) $\leq h^2$ for given h.

Several authors (PARZEN, CACOULLOS, FUKUNAGA et al [38]) have indicated which was the optimum step size h_o, which depends on n, p and f, x and optimizes the mean-square error criterion on $\hat{f}_n(x)$:

$$h_o = n^{-1/(p+4)} \cdot [p(p+2)^2 |A|^{1/2} \Gamma((p+2)/2) f(x)]^{2/(p+4)} \cdot$$

$$\cdot \left[\pi^{p/2} \, Tr^2 \left[A^{-1} \left[\frac{\partial^2 f(x)}{\partial x^2} \right] \right] \right]^{-2/(p+4)}$$

Moreover, in this case, following RETJÖ et al [62]:

$$\text{Prob } (\sup_x | \hat{f}_n(x) - f(x) | < \varepsilon) \leq \exp(-c_o \, n \, h_o)$$

$$\text{Prob } (\lim_{n \to +\infty} \frac{\text{Min } (\sqrt{n \, h_o^p} \, ; \, 1/h_o^2)}{\text{Log } n} \sup_x | \hat{f}_n(x) - f(x) | = 0) = 1$$

8. OPTIMIZATION OF ESTIMATORS OF STATISTICAL MOMENTS VS SAMPLE SIZE (SHENTON ET AL [31])

8.1 Acceleration of the Bayes estimation for the mean $\hat{\mu}$ of a N-dimensional normal distribution

8.11 Let the n independent samples x_i, i = 1, ..., n be distributed according to N (m, σ^2 I), with σ unknown and scalar.

Define:

$$s_n^2 \triangleq \sum_{i=1}^{n} \sum_{j=1}^{p} (x_{ij} - x_{i.})^2 \qquad s^2 = (\widehat{\sigma^2})$$

$$x_i = (x_{ij}, \, j = 1, \ldots, p) \qquad i = 1, \ldots, n$$

$$x_{i.} = (\sum_{j=1}^{p} x_{ij}) \, / \, (p-1)$$

$$\bar{x} = (\sum_{i=1}^{n} x_i) \, /n$$

Then ALAM [2], BARANCHIK [3] have exhibited an extended Bayes estimator:

$$\hat{m} = \bar{x}\ \psi\ (a\ ||\bar{x}||\ /\ s_n^2$$

of the population mean m, where each of the classical estimators $\bar{x}$ is modified by a correction factor depending on all n sample vectors x_i, i = 1, ..., n. This estimator is admissible, minimax, and dominates the estimator $\bar{x}$, when ψ and a fulfill some conditions.

8.1.2 Correction function ψ

The correction function ψ is given by confluent hypergeometric functions, parametrized by ν:

$$\psi\ (z) = \left(\frac{2\nu}{p}\right) M\ \left(\nu + 1, \frac{p}{2} + 1, \frac{z}{2}\right) / M\ \left(\nu, \frac{p}{2}, \frac{z}{2}\right)$$

$$\psi\ (o) = 2\ \nu\ /\ p$$

$$\nu \text{ free, such that } 1 > \nu \geq [2p + 5 - (4p^2 + 8p - 7)^{1/2}\ /\ 4]$$

$$M\ (\nu, b, z) = \sum_{k=o}^{\infty} \frac{(\nu)_k}{(b)_k} \frac{z^k}{k!} \qquad M\ (\nu, b, o) = 1$$

$$(\nu)_k = (\nu + 1)\ \ldots\ (\nu + k - 1)$$

$$z\ (b - \nu)\ M\ (\nu, b+1, z) = b\ [\ (z+b - 1)\ M\ (\nu, b, z) - (b - 1)\ M\ (\nu, b-1, z)\ ]$$

$$\nu\ M\ (\nu+1, b+1, z) = (\nu-b)\ M\ (\nu, b+1, z) + bM\ (\nu, b, z)$$

$$z\ M\ (\nu, b+1, z) = b\ [M\ (\nu, b, z) - M\ (\nu-1, b, z)]$$

ABRAMOWITZ-STEGUN [1] [6], GRADSHTEYN [5], have a survey of the functions and tables for M, although the recursive relations above are more efficient.

8.1.3. Correction factor a

The estimation $\hat{m}$ is minimax for a satisfying the condition, parametrized by ν :

$$a \geq [1 + \frac{1 - \nu}{\nu} \frac{p}{p - 2}]\ \left(\frac{n + 2}{2}\right)$$

8.1.4. Performances of the Bayes estimator $\hat{m}$

The risk of $\hat{m}$ is found to vary considerably smaller than p for $||\hat{m}||$ small, and the advantage over $\bar{x}$ is increasing with p. As

$||\hat{m}||$ increases up to 3, risk and bias converge to those of $\bar{x}$. For $||\hat{m}||$ small, this $\hat{m}$ estimate is heavily biased towards 0.

8.2 Finite sample variance estimation for correlated samples

If x_i, i = 1, ..., n is a sample of a 1-dimensional population with a power law spectral density $F(f) = h\,|f|^{\alpha}$ unit (f) = unit (1/x), then there is generally a bias to the estimated variance s_n^2 of x, due to the correlation between these n samples:

$$s_n^2 = \frac{1}{n-1} \sum_{k=1}^{n} (x_k - \bar{x})^2$$

The expected value of s_n^2 is then proportional to τ^{γ}, where τ is the spacing between samples in E, and γ is a constant related to α, the exponent in the spectral density:

$$\gamma = \begin{cases} -2, & \text{if } \alpha \geq 1 \\ -(\alpha+1), & \text{if } -3 < \alpha \leq 1 \\ \text{not defined otherwise} \end{cases}$$

BARNES [9] provides numerical computations of the correction factor B_1, by which the n-sample case is compared to the 2-sample case:

$$B_1(n, \gamma, \tau) = E(s_n^2) \,/\, E(s_2^2)$$

The assumption about a power law spectral density can be replaced by an equivalent assumption in terms of the corresponding autocorrelation function of the process $\{x_i\}$.

9. FINITE SAMPLE PERTURBATIONS OF A LINEAR FEATURE EXTRACTION TRANSFORMATION

9.1 Eigenvalues of a sample covariance matrix S

The distribution of the eigenvalues of a sample covariance matrix S depends on a definite integral over the group of orthogonal matrices. This integral involves the eigenvalues of both the population and sample matrices; this integral is expressed as a hypergeometric series involving zonal polynomials, and converges slowly unless the eigenvalues of the argument matrices lie in very limited ranges.

ANDERSON [66] has obtained an asymptotic expansion for the distribution of the eigenvalues of the sample covariance matrix S of patterns $x \in E^p$ if $x \in N(0, \Sigma)$. In this case of a p-dimensional gaussian population, S is itself distributed as Wishart (n, p, Σ) where n is the sample size.

9.2 Eigenvalues of the two-sample matrix $S_1 S_2^{-1}$ from the Karhunen-Loeve transform and the Mahalanobis distance

Consider now two samples from gaussian populations $N(0, \Sigma_1)$, $N(0, \Sigma_2)$, of sample sizes n_1, n_2 respectively, in E^p. The sample covariance matrices S_1, S_2 appear in the combined $S_1 S_2^{-1}$ form in the following cases:

<u>i</u> $\Sigma_1 = I \quad n_1 = \infty$: Karhunen-Loeve transform
The principal components and corresponding loadings are the eigenvectors and eigenvalues of S_2^{-1}, and the K-L transform writes (after rearranging these eigenvectors by decreasing eigenvalues):

$$y = S_2^{-1} x$$

<u>ii</u> $\Sigma_1 = I \quad n_1 = \infty$: Mahalanobis distance
The Mahalanobis distance D (x) of a pattern x to the class with population mean 0, and population covariance Σ_2, is:

$$D(x) = x^t S_2^{-1} x$$

Here again, the eigenvalues of $S_1 S_2^{-1}$ are complex to compute for similar reasons to the one sample case.

Define:

a_k, $k = 1, \ldots, p$ eigenvalues of $(\Sigma_1 \Sigma_2^{-1})^{-1}$,
$o < a_1 < a_2 < \ldots < a_p$

b_k, $k = 1, \ldots, p$ eigenvalues of $(S_1 S_2^{-1})$,
$o < b_p < \ldots < b_2 < b_1$

$A \triangleq \text{diag}(a_1, \ldots, a_p)$

$B \triangleq \text{diag}(b_1, \ldots, b_p)$

$n = n_1 + n_2 \qquad r_j \triangleq a_j / (1 + a_j b_j)$

The asymptotic distribution of the eigenvalues $b_k > o$ of $S_1 S_2^{-1}$ for large degrees of freedom $n = n_1 + n_2$, is then given by F :

$$F \triangleq C \cdot 2^p \left[\prod_{j=1}^{p} a_j^{(1/2)n_1} b_j^{(1/2)(n_1-p-1)} (1 + a_j b_j)^{(-1/2)n} \cdot \prod_{j<k}^{k=p} (b_j - b_k) \right] \cdot \left[\prod_{j=1}^{p} a_j b_j \times \prod_{j<k}^{k=p} (2\pi / nc_{jk})^{1/2} \cdot [1 + (2n)^{-1} (\sum_{j<k} c_{jk}^{-1} + \alpha (p) + \ldots)] \right]$$

where C, c_{jk}, α (p) are given by:

$$C \triangleq \Gamma_p (\tfrac{1}{2} n) [2^p \Gamma_p (\tfrac{1}{2} n_1) \Gamma_p (\tfrac{1}{2} n_2)]^{-1}$$

$$\Gamma_\ell (x) \triangleq \pi^{(1/4)\ell(\ell-1)} \prod_{j=1}^{\ell} \Gamma (x - \ell + \tfrac{1}{2})$$

$$c_{jk} \triangleq c_{kj} = (r_{kj} - r_j r_k b_{jk}) b_{jk}$$

$$r_{jk} \triangleq r_j - r_k \qquad\qquad b_{jk} \triangleq b_j - b_k$$

$$\alpha (N) \triangleq p (p - 1) (4p + 1) / 12$$

These results assume the eigenvalues a_k of $(\Sigma_1 \Sigma_2^{-1})^{-1}$ to be all distinct. If the samples x are complex variables, then LI [67] gives the generalized distribution of the eigenvalues b_k.

10. SMALL SAMPLE NEAREST NEIGHBOR CLASSIFICATION RULE

10.1 Small sample nearest neighbor decision rule (COVER-HART [69])

The k-nearest neighbor decision rule (k-NNDR) is attractive in the sense that the error probability ε is asymptotically upper bounded by a function of the Bayes risk ε^* ([44]):

$$\varepsilon \leq 2\varepsilon^* - [K / (K-1)] \varepsilon^{*2}$$

The small sample behavior is, however, little understood from a theoretical point of view. The finite sample NNDR behavior is important because the data storage and computational requirements can only be met in this case: all $(N_1 + \ldots + N_K)$ design samples have to be stored and retrieved for the classification; it has been investigated in special cases by FIX-HODGES, KANAL et al [36],

LEVINE et al [17], GLICK [41]. Again under very special assumptions about the a priori class probabilities, it is shown that for a given sample size, the 1-NNDR is uniformly better than the k-NNDR.

In many cases, there is a small subset of good design samples that dominate the performance, and in other words the performance would be insensitive to sample sizes for good design samples.

10.2 Distribution of the L^2-distance to the closest neighbor

Let f(x) denote the probability density function of $x \in E^p$, and let x_i, i = 1, ..., n be independent samples.

Define:

$$C_{n,\ p} = \underset{1 \leq i \leq n}{\mathrm{Min}} \quad (\| x - x_i \|)$$

Then:

$$\Gamma_{n,\ p}(c) = \mathrm{Prob}\ (C_{n,p} \leq c) = 1 - \int (1 - R(x,c))^n f(x)\ dx$$

$$R\ (x,\ c) \triangleq \int_{|u| < c} f\ (x - u)\ du$$

TRYBUS [4] provides the asymptotic distributions of $\Gamma_{n,\ p}$ (c) for $n \to \infty$, for f uniform or exponential, as well as the moments of $\Gamma_{n,\ N}(c)$.

11. SMALL SAMPLE DISCRIMINANT ANALYSIS

Regardless of the true performance ε^* of a two-class classifier, COVER [34][49] shows that, if the total number of samples $(N_1 + N_2)$ is less than twice the dimension p, there exists a linear hyperplane such that the probability of error on the design set is always zero. The major difficulties thus arise if $(N_1 + N_2) > 2\ p$ in the two-class problem, and much earlier in the K - class case.

FOLEY [33] then demonstrates that the design set error rate is an extremely biased estimate of either the Bayes or the test set error rates, if $(N_j\ /\ p) < 3$, j = 1, 2. The variance of the design set error rate is approximated by a function which is bounded by E (N_j) / 8.

For the case of non-linear discriminant analysis, it is wise to use design sets for which the ratios $(N_j\ /\ p)$ are much larger than 3, e.g. 10.

An expansion of the classification errors for linear discriminant analysis is given by DEEV [37] for the case $N_j\ /\ p$ small.

12. SEQUENTIAL CLASSIFICATION

The idea behind sequential classification (FU [52]) is to compute a sample-additive test statistic, and sequentially to decide among the three hypothesis:

H_1 : classify into class ω_1

H_2 : classify into class ω_2

H_o : do not classify yet, but request a new sample

Sequential classification is essentially restricted to one-dimensional measurements x_i, $i = 1, \ldots, n$.

12.1 Wald's likelihood based sequential test (WALD [53])

This 2-class test uses the following statistic:

$$t_n = \sum_{i=1}^{n} x_i$$

If risks of 1st and 2nd kind are specified, and normal or exponential distributions of the measurements are assumed, approximate expected sample sizes E (n) can be computed, needed for a classification into ω_1 or ω_2. The small-sample behaviour of the decision thresholds is very little understood (WALD [53], GHOSH [24], DE GROOT [25]).

12.2 Non-parametric sequential test

Generalizing Wald's test, the test statistic is here made non-parametric for each given n, (FRASER [26], CARLYLE [27]), thus allowing for a formula for the expected sample size E (n) needed for a classification.

Let R (x_k) be the rank of measurement x_k, relatively to $x_1, \ldots, x_n$ ranked by increasing values. A threshold a on x is introduced to define the signed rank z_k of x_k:

$$z_k = R\,(|x_k - a|)\ \mathrm{Sign}\,(x_k - a)$$

If the c.d.f. F_1, F_2 of the measurements in classes ω_1, ω_2 each fulfill the assumption:

$$\begin{cases} F\,(a - x) = F\,(a)\,[1 - F\,(x - a) + F\,(a - x)] \\ x \geq a \end{cases}$$

and the test statistic is:

$$t_n = \sum_{i=1}^{n} z_i/i \qquad \begin{cases} t_n < A & : H_1 \\ t_n > B & : H_2 \\ A \leq t_n \leq B & : H_o \end{cases}$$

then (PAU [28]) :

$$E(n) = \mathrm{Max}\,(E_1(n), E_2(n))$$

$$E_j(n) = [2B + 2(A-B)\,\mathrm{Prob}_j(t_n \geq A)] / (1 - 2F_j(a)), \; j = 1,2$$

$$\mathrm{Prob}_j(t_n \geq A) = (1 - \exp(Bh_j)) / (\exp(Ah_j) - \exp(Bh_j)]$$

$$F_j(a)(1 - \exp(-h_j)) / h_j + (1-F_j(a))(\exp(h_j)-1)/h_j = 1$$

13. CLASSIFICATION OF LOTS USING FINITE MULTIDIMENSIONAL TEST SETS EXTRACTED FROM THE LOTS

13.1 Framework

Considering N_i design samples of class ω_i, $i = 1, \ldots, K$, we study the small sample properties of classification rules applying to lots of N p-dimensional patterns, given only $n \leq N$ p-dimensional measurements. Under the assumption of normality within the K-classes ω_i, $i = 1, \ldots, K$, the classification rule takes into account sample-size dependent bounds on the probabilities of misclassification.

The problem is jointly to select the n-sample decision rule/threshold and the relative sample size n/N, in order to fulfill constraints on the probabilities of misclassification.

In a signal classification framework, for K = 2 classes, the problem is to select the sampling rate as a function of n/N, which should be minimum, given constraints on the false alarm and misclassification rates applying to slices of N signal samples.

When the classification rule applying to the lots of size N is derived from the estimated class memberships of the individual patterns in the n-sample set, then the classification is said to be by attributes.

When the classification rule applying to the lots of size N is derived from statistics of the n p-dimensional measurements in the n-sample set, then the classification is said to be by variables.

13.2 Classification by attributes using the k-nearest neighbor rule (PAU [29])

Considering the two-class problem, the classification statistic is the frequency P_n of test samples assigned to the class ω_2:

$$P_n \triangleq \text{Card}\,\{i \mid \hat{f}_1(x_i) < \hat{f}_2(x_i);\ i = 1, \ldots, n\} / n$$

$$P_n \triangleq c_2 / n$$

If $P_c(N)$ is a threshold value of P_n, this test spells out as:

$$N - \text{lot} \in \omega_1 \qquad \text{iff } P_n \leq P_c\ (N)$$

$$N - \text{lot} \in \omega_2 \qquad \text{iff } P_n > P_c\ (N)$$

where $\hat{f}_j\ (x_i)$ is the k-nearest neighbor estimator of the probability density of class ω_j at location x_i, $j = 1, 2$ $i = 1, \ldots, n$. The integer parameter $\hat{k}$ is determined by formula (Section 7.1), assuming f_j to be a p-dimensional normal distribution.

According to FUKUNAGA et al [42], the conditional classification error is:

$$r_k\ (x) = r^*\ (x) \sum_{\ell=o}^{(k-1)/2} \binom{k}{\ell} (r^*\ (x))^{\ell}\ (1-r^*(x))^{k-\ell} +$$

$$+\ (1-r^*(x)) \sum_{\ell=(k+1)/2}^{k} \binom{k}{\ell}\ (r^*(x))^{\ell} (1-r^*(x))^{k-\ell}$$

where $r^*(x) = \text{Min}\ (f_1(x),\ f_2(x))$ is the conditional Bayes risk. Using the estimation of the expected unconditional Bayes risk $E\ (\widehat{r^*\ (x)})$ proposed by MAC LACHLAN [16], one gets ([77]):

$$\text{Prob}\ (\omega_2|\omega_1) < r_1 = \sum_{\ell=o}^{(k-1/2)} \binom{k}{\ell} (G(-D/2))^{\ell+1} (1-G(-D/2))^{k-\ell}$$

$$\text{Prob}\ (\omega_1|\omega_2) \leq r_2 = \sum_{\ell=(k+1)/2}^{k} \binom{k}{\ell} (G(-D/2))^{\ell} (1-G(-D/2))^{k-\ell+1}$$

where

$$G\ (y) = (1/\sqrt{2\pi}) \int_{-\infty}^{y} \exp\ (-t^2/2)\ dt$$

and, assuming normal distributions $N\ (m_j,\ \Sigma_j)$ for the classes ω_j, with pooled covariance estimate $\hat{\Sigma}$:

$$\hat{D}^2 = (\hat{m}_1 - \hat{m}_2)^t\ \hat{\Sigma}^{-1}\ (\hat{m}_1 - \hat{m}_2)$$

$\hat{D}^2$ is the design sample estimated Mahalanobis distance D^2 between ω_1 and ω_2; LACHENBRUCH [19] mentions that this sample based estimate may be misleading for small sample sizes n, and suggests alternative estimators of Prob $(\omega_2|\omega_1)$, Prob $(\omega_1|\omega_2)$, which are compared by MAC LACHLAN [16], and summarized by KITTLER-PAU [18], for given dimension p of the space and interclass distance $\hat{D}^2$.

The constraints on the risks of 1st and 2nd kind are specified as:

$$\hat{r}_1 \leq \alpha \qquad \hat{r}_2 \leq \beta$$

The classification statistic P_n is such that the samples assigned to the class ω_2 may either be incorrectly classified items from ω_1, or items from ω_2 correctly classified into ω_2. The true test statistic to be used <u>instead</u> of P_n is:

$$Q_n = \frac{P_n - \hat{r}_1}{1-\hat{r}_1-\hat{r}_2}$$

This result implies that for the model (A) of Section 2.4, we can calculate the true rate of patterns in ω_2 in a lot from the rates of samples assigned by the classification system into ω_2, and the unweighted errors r_1, r_2 which can be determined once for all during the design stage of the classification system.

The threshold P_c on Q_n is derived in [28] using classical results from quality control by attributes, by assuming the true rate P (ω_2) to be known. In general however, for a finite sample size P (ω_2) is unknown, and only estimates $\hat{P}$ (ω_2) can be obtained, which are random variables, the distribution of which is dependent on n, N, P (ω_2). The only random variable whose realization will be known is the number $\hat{c}_2$ of items classified into class ω_2

$$c_2 = P(\omega_2)(1 - r_2) + r_1(1 - P(\omega_2))$$

The probability distribution of $\hat{c}_2$ is computed in KITTLER-PAU [18], taking into account all joint occurences of the random variables P (ω_2), r_1, r_2.

This result demonstrates that, in order to achieve the α and β consstraints, the sample size n will have to be substantially increased; even classification errors of 1% have enormous effects.

In the arguments above, $\hat{r}_1$ and $\hat{r}_2$ have been assimilated with sample-based estimates of the risks of 1st and 2nd kind. It is obvious that the constraints $\hat{r}_1 \leq \alpha$ and $\hat{r}_2 \leq \beta$ could be replaced by others, in which better parametric risk estimates would be used, if all pattern vectors in ω_j were assumed N (m_j, Σ_j). The reason for not using said better parametric estimators, was only to propose a framework which would remain formally unchanged if the pattern measurements were not normally distributed.

13.3 Classification by attributes using a discriminant function ([28])

The classification rule is made parametric, the discriminant function ψ being sample-based:

$$P_n = \text{Card}\ \{i \mid \hat{\psi}\ (x_i) < 0;\ i = 1, \ldots, n\}\ /\ n$$

$$P_n = c_2\ /\ n$$

If f_j are $N\ (m_j, \Sigma_j)$:

$$\hat{\psi}\ (x) = \frac{p}{2} \log\ (\ |\ \hat{\Sigma}_2| / |\ \hat{\Sigma}_1|) + \frac{1}{2}\ (x - \hat{m}_2)^t (\hat{\Sigma}_2)^{-1}\ (x - \hat{m}_2) - \frac{1}{2}\ (x - \hat{m}_1)^t (\hat{\Sigma}_1)^{-1} (x - \hat{m}_1)$$

Analytical results similar to those of the previous subsection are established in [28].

13.4 Classification by variables using Parzen kernels

Let K (y) be a Parzen - type kernel in E^p; the two-class classification statistic t_n, will here depend explicitly on:

- the design patterns x^i_ℓ $\ell = 1, \ldots, N_j$ $\qquad j = 1, 2$
- the sample patterns x_i $i = 1, \ldots, n$

$$t^j_n = \frac{1}{n \cdot N_j} \sum_{i=1}^{n} \sum_{\ell=1}^{N_j} k(x_i - x^j_\ell) \qquad j = 1, 2$$

The classification rule is then:

$$N - \text{lot} \in \omega_1 \text{ iff } (t^1_n\ /\ t^2_n) \leq \gamma\ (n, N)$$

$$N - \text{lot} \in \omega_2 \text{ iff } (t^1_n\ /\ t^2_n) > \gamma\ (n, N)\ ,$$

subject to constraints on the risks of 1st and 2nd kind.

Approximate results can be obtained assuming f_1, f_2, to be normal.

14. SMALL SAMPLE PROPERTIES OF TIME-DEPENDENT PROCESS MODELS

Consider a linear autoregressive process $\{x_k\}$ of order m, defined by the scalar difference equation:

$$x_k = a_1 x_{k-1} + \dots + a_m x_{k-m} + z_k$$

where $\{z_k\}$ is a stationary white noise process, uncorrelated with x_ℓ for $k > \ell$, with variance σ_z^2 I.

If $\{x_k\}$ is a gaussian process, the asymptotic distribution of the (m+1) - vector:

$$\delta = \begin{bmatrix} (1 / \sqrt{n}) \sum_{i=1}^{n} x_i \\ \\ \sqrt{n} (\hat{a}_i - a_i) \quad i = 1, \dots, m \end{bmatrix}$$

as n tends to infinity, is (m+1) - dimensional gaussian with zero mean and covariance matrix (AKAIKE [11]):

$$\sigma_z^2 \begin{bmatrix} A^{-2} & 0 \\ \\ 0 & R_m^{-1} \end{bmatrix}$$

where:

$$A = 1 - \sum_{i=1}^{m} a_i$$

$$R_m = \begin{bmatrix} 1 & \rho(1) & \dots & \rho(m) \\ \rho(1) & 1 & \rho(1) & \rho(m-1) \\ \vdots & \vdots & \vdots & \vdots \\ \rho(m) & \rho(m-1) & \dots & 1 \end{bmatrix}$$

$$\rho(\ell) = E(x_{k+\ell} x_k) / E(x_k^2)$$

15. APPLICATIONS

This section will review some examples of applications, where small sample considerations are crucial.

1. Radar target cross section analysis ([7], [55])

Unavailability of enough learning samples.

2. Signal classification

Unavailability of enough learning samples, or selection of the number n of test patterns for the classification of windows of N consecutive samples ([10]).

3. Mechanical signatures

Unavailability of enough learning samples.

4. Failure diagnosis ([71])

Unavailability of enough learning samples for the classes of failure / degradation modes.

5. Medical diagnosis

As well tolerance considerations, as the case of rare deseases, deserve attention.

6. Classification of relatively well-discriminated targets

The concern is about minimizing the number of returns, thus in some cases justifying a sequential classification scheme.

7. Multiframe classification ([47])

In a number of passive observation problems, false alarm/ classification events may be due to background events or to spurious phenomena. The sensor and classification system must then make a series of classifications on successive scans to confirm a target classification, thereby discriminating e.g. a genuine moving target from a comparatively static one.

8. Subpicture / Zonal classification ([46])

In some applications, it is essential to find the number n of zones used for feature measurements within a picture of N zones to be classified globally. Small sample results are then useful, along with histograms of gray level jumps. However, the missing account for contextual information, makes often the purely statistical selection of n/N somehow hazardous.

9. Quality control: acceptance sampling with multidimensional measurements ([28])

The account for constraints on the misclassification probabilities, while minimizing the relative sample size n/N is here essential. Another difficulty originates in missing learning measurements from classes of bad items.

10. Microprocessor based classification systems

Due to generally slow processing speeds, and high access times to memory in most microprocessor systems, it is essential to minimize the number of design samples or design statistics called in the processing unit during the feature extraction and classification stages.

CONCLUSION. The issue of finite sample based classification leads to many interesting, but very difficult, problems. The results available in the literature are often inconsistent, and often widely scattered. In order to obtain progress in this area, it is essential to realize and influence progress in mathematical statistics, signal theory and data analysis.

Those sub-problems deserving most attention are those related to the design of "robust" pattern recognition systems, which means those operating in various modes (users, locations, environment), and for which the training stage should be as little crucial as possible in terms of optimal classification performances ([56]). This robustness depends crucially on the way in which the finite sample effects are corrected.

REFERENCES

1. Abramowitz, M., I. A. Stegun, Handbook of Mathematical Functions, Dover, 1965.
2. Alam, K., "A Family of Admissible Minimax Estimators of the Mean of a Multivariate Normal Distributions", Ann. Statist., Vol. 1, 1973, 517-525.
3. Baranchik, A. J., "A Family of Minimax Estimators of the Mean of a Multivariate Normal Distribution", Ann. Math. Statist. Vol. 41, 1970, 642-646.
4. Trybus, G., "On the Distance Random Variable", Zastosowania Matematyki (Applicationes Mathematicae), Vol. 14, n° 2, 1974, 237-243.
5. Gradshteyn, I. S., Tables of Integrals, Series and Products, Academic Press, N. Y., 1965.
6. Abramowitz, Stegun, Handbook of Mathematical Functions, National Bureau of Standards, 1965.

7. Hansen, V. G., B. A. Olsen, "Nonparametric Radar Extraction Using a Generalized Sign Test", IEEE Trans. AES, Vol. AES-7, Sept. 1971, 942-950.
8. Chen, Z., K. S. Fu., "Nonparametric Bayes Risk Estimation for Pattern Classification", IEEE Trans. SMC, Vol. SMC-7, n° 9, Sept. 1977.
9. Barnes, J. A., "Tables of Bias Functions B_1 and B_2 for Variances Based on Finite Samples of Processes with Power Law Spectral Densities", NBS Technical note n° 375, National Bureau of Standards, January 1969.
10. Gerdin, K., "On Histogram Uncertainty for Normal, Stationary Signals with an Application to Paper Machines Disturbances", Acta Polytechnica Scandinavica, Vol. MA - 23, Royal Swedish Academy of Engineering Sciences, Stockholm, 1971.
11. Akaike, H., "Statistical Predictor Identification", Ann. Inst. Statist. Math., Vol. 22, 203 - 217, 1970.
12. Hoel, P. G., "Small Sample X^2 - Distance", Ann. Math. Statist., Vol. 14, 1943, 155-162.
13. Wilks, S. S., "Determination of Sample Sizes for Setting Tolerance Limits", Ann. Math. Statist., Vol. 12, 1941, 91-96.
14. Tukey, W., "Non Parametric Estimation II", Ann. Math. Stat., Vol. 18, 1947, 529 - 538.
15. Patrick, E. A., F. P. Fischer, "A Generalized k - nearest Neighbor Rule", Information and Control, Vol. 16, n° 2, April 1970, 128-152.
16. Mac Lachlan, G. J., "Estimates of Errors of Misclassification on the Criterion of Asymptotic Mean Square Error", Technometrics, Vol. 16, n° 2, 255-260, May 1974.
17. Levine, A., L. Lustick, B. Saltzberg, "The Nearest-Neighbor Rule for Small Samples Drawn from Uniform Distributions", IEEE Trans. Information Theory, Vol. IT - 19, n° 5, 697 - 699, Sept. 1973.
18. Kittler, J., L. F. Pau, "Small Sample Properties of a Pattern Recognition System in Lot Acceptance Sampling," Proc. 4th Int. J. Conf. on Pattern Recognition, Kyoto, November 1978.
19. Lachenbruch, P. A., Discriminant Analysis, Hafner Press, N.Y. 1975.
20. Lachenbruch, P. A., "Discriminant Analysis When the Initial Samples are Misclassified", Technometrics, Vol. 8, p. 657, 1966; and Technometrics, Vol. 16, 1974.
21. Gilbert, E., "The Effect of Unequal Variance - Covariance Matrices on Fisher's Linear Discriminant Function", Biometrics, Vol. 25, 505-516, 1969.
22. Marks, S., O. J. Dunn, "Discriminant Functions When Covariance Matrices are Unequal", JASA, Vol 69, 555 - 559, 1974.
23. Mc Lachan, "Asymptotic Results for Discriminant Analysis When the Initial Samples are Misclassified", Technometrics, Vol 14, 415 - 422, 1972.
24. Ghosh, B. H., Sequential Tests of Statistical Hypothesis, Addison-Wesley Publ., 1970.

25. De Groot, M. H., *Optimal Statistical Decisions*, McGraw Hill, N. Y., 1970.

26. Fraser, D. A., *Non-parametric Methods in Statistics*, Wiley, N. Y., 1967.

27. Carlyle, J. W., *Non-parametric Methods in Detection Theory*, Academic Press, N. Y. 1972.

28. Pau, L. F., C. H. Chen, G. Weber, "Contrôle Statistique de Qualité", *Monographie B.N.M.*, Editions Chiron, Paris, 1978.

29. Pau, L. F., "Properties of Small Sample Classification Rules", Proc. J. Int. Analyse des données et Informatique, IRIA, Rocquencourt, Sept. 1977.

30. Chen, C. H., "Finite Sample Considerations in Statistical Pattern Recognition", Proc. 1978 PRIP Conference, May 1978.

31. Shenton, L. R., K. O. Bowman, "Maximum Likehood Estimation and Small Samples", *Statistical Monograph* n° 38, Charles Griffin Co., London, 1977.

32. Duin, R. W., "A Sample Size Dependent Error Bound", 3rd Int. J. Conf. Pattern Recognition, *IEEE 76* - CH 1140 - 3C, 156-160, 1976.

33. Foley, D. H., "Considerations of Sample and Feature Size", *IEEE Trans. Inform. Theory*, Vol. IT - 18, 618-626, Sept. 1972.

34. Cover, T. M., "Geometrical and Statistical Properties of Systems of Linear Inequalities with Applications in Pattern Recognition", *IEEE Trans. Electronic Computers*, Vol. EC - 14, 326-334, June 1965.

35. Raudys, S., "On Dimensionality, Learning Sample Size and Complexity of Classification Algorithms", *3rd IJCPR*, 166-169, 1976.

36. Kanal, L., B. Chandrasekaran, "On Dimensionality and Sample Size in Statistical Pattern Classification", *Proc. NEC, 1969; and Pattern Recognition*, Vol. 3, 1971, 225-234.

37. Deev, A. D., "Representation of Statistics of Discriminant Analysis and Asymptotic Expansions When Space Dimensions are Comparable with Sample Size", *Dokl. Akad, Nauk SSSR*, Vol. 195, 1970, n° 4; or *Soviet Math. Dokl.*, Vol 11, 1970, n° 6, 1547-1550.

38. Fukunaga, K., L. Hostetler, "Optimization of k - nearest Neighbor Density Estimates", *IEEE Trans. Information Theory*, Vol IT., May 1973, 320-326.

39. Poppl, S. J., "Optimal Training Sets", in F. T. De Dombal & F. Gremy (Ed.), *Decision Making and Medical Care*, North Holland, Amsterdam, 1976.

40. Gurland, J., "Distribution of Quadratic Forms and Ratios of Quadratic Forms", *Ann. Math. Stat.*, Vol. 24, 416-427, 1953.

41. Glick, N., "Sample-based Classification Procedures Derived From Density Estimators", *JASA*, Vol. 67, 116-122, March 1972.

42. Fukunaga, K., L. D. Hostetler, "K - nearest Neighbor Bayes-risk Estimation", *IEEE Trans. Inform. Theory*, Vol IT - 21, 285-293, May 1975.

43. Rogers, W. H., T. J. Wagner, "A Finite Sample Distribution-free Performance Bound for Local Discrimination Rules," Ann. Statist., 1975.
44. Young, T. Y., T. W. Calvert, Classification, Estimation and Pattern Recognition, Elsevier, N. Y., 1974.
45. Devijver, P. A., "Decision-Theoretic and Related Approaches to Pattern Classification", in K. S. Fu & A. B. Whinston (Ed.), Pattern Recognition Theory and Applications, Noordhoff, Leyden, 1977, 1-34.
46. Lagneau, M., "Utilisation des Propriètes Statistiques du Signal d'image pour la Detérmination de la Dimension du Quadrillage Optimal de Cette Image en vue de la Classification des Zones Obtenues", Institut de Programmation, Univ. Paris 6, 28 June 1976.
47. Miller, B., "US Moves to Upgrade Missile Warning", Aviation Week and Space Technology, December 2, 1974, 16 - 18.
48. Agrawala, A. K., "Learning with a Probabilistic Teacher", IEEE Trans. Information Theory, Vol IT - 16, n° 4, 373-379, July 1970.
49. Cover, T. M., "Learning in Pattern Recognition", in M.S. Watanabe (Ed.) , Methedologies of Pattern Recognition, Academic Press, 1969, 111 - 132.
50. Hellman, M. E., T. M. Cover, "Optimal Learning with Finite Memory", Ann. Math. Statist., Vol. 39, n°, October 1968, 1793-1794.
51. Shanmugam, K., "A Parametric Procedure for Learning with an Imperfect Teacher", IEEE Trans. Inform. Theory, Vol IT - 18, March 1972, 300 - 302..
52. Fu, K. S., Sequential Methods in Pattern Recognition, Academic Press, N. Y., 1968.
53. Wald, A., Sequential Analysis, Wiley, N. Y., 1947.
54. Chandler, D., W. M. Kahan, "The Rotation of Eigenvectors by a Perturbation", SIAM J. Numer Analysis, Vol 7, n° 1, March 1970.
55. Therrien, C. W., "The Application of Linear Prediction to Sequential Classification of Radar Target Signatures", T. R. MIT/Lincoln Laboratory, #517, 25 March 1976, ESD - TR - 76 - 73.
56. Pau, L. F., "Optimization d'une Métrique en Reconnaissance des Formes", in: Proc. GALF - 1973, Presses Univ. Libre de Bruzelles, Brussels, 1973.
57. Dvoretsky, A., J. Kiefer, J. Wolfowitz, Ann. Math. Statist, Vol. 27, p 642, 1956.
58. Muise, R. W., P. R. Boorstyn, "Detection with Time-varying Finite-memory Receivers", Abstract of papers, 1972 Int. Symp. Information Theory, 1972.
59. Toussaint, G. T., "Bibliography on Estimation of Misclassification", IEEE Trans. Information Theory, Vol. IT-20, p. 472, July 1974.

60. Cover, T., M. Freedman, M. Hellman, Information and Control, 1977.

61. Cover, T. M., T. J. Wagner, "Topics in Statistical Pattern Recognition", in K. S. Fu (Ed.), Digital Pattern Recognition, Springer Verlag, 1976.

62. Rejtö, L. P. Révész, "Density Estimation and Pattern Classification", Problems of Control and Information Theory, Vol. 2, n° 1, 67 - 80, 1973.

63. Elias, P., "Bounds and Asymptotes for the Performance of Multivariate Quantizers", Annals of Mathematical Statistics, Vol 41, n° 4, 1970, 1249 - 1259.

64. Elias, P.,"Bounds on the Performance Quantizers", IEEE Trans. Information Theory, Vol IT - 16, 1970, 172-184.

65. Zador, P., "Development and Evaluation of Procedures for Quantizing Multivariate Distributions", PhD Thesis, Stanford Univ., 1964.

66. Anderson, G. A., "An Asymptotic Expansion for the Distribution of the Latent Roots of the Estimated Covariance Matrix", Ann. Math. Statist., Vol 36, 1965, 1153 - 1173.

67. Li, H. C., K. C. S. Pillai, T. C. Chang, "Asymptotic Expansions for Distributions of the Roots of Two Matrices from Classical and Complex Gaussian Populations", Annals of Mathematical Statistics, Vol. 41, n° 5, 1970, 1541 - 1556.

68. Drake, A. W., Fundamentals of Applied Probability Theory, McGraw Hill, N. Y., 1967.

69. Cover, T. M., P. E. Hart, "Nearest Neighbor Pattern Classification", IEEE Trans. IT, Vol IT - 13, n° 1, 1967, 21 - 27.

70. Fraser, D. A. S., I. Guttman, "Tolerance Regions", Ann. Math. Statist., Vol 27, 1956, 162 - 179.

71. Pau, L. F., Failure Diagnosis and Performance Monitoring, Marcel Dekker Inc., N. Y., 1979.

A REVIEW OF STATISTICAL PATTERN RECOGNITION*

C. H. Chen

Department of Electrical Engineering
Southeastern Massachusetts University
North Dartmouth, MA 02747, U.S.A.

1. INTRODUCTION

After more than twenty years of progress, the theory and applications of statistical pattern recognition are now well developed. A number of textbooks (e.g. [1-11]) have been available. The limitations of the statistical pattern recognition are also evident: the patterns are not characterized by the statistical information alone and many useful statistical properties cannot be fully exploited with available mathematical statistics. Like in many other fields there is a wide gap between theory and practice. The limitation of the finite sample size is mainly responsible for such a gap. The finite sample size effect is the one among ten problem areas [12] in statistical pattern recognition for which the solutions are much needed.

In this paper the current status of the statistical pattern recognition is reviewed by topics including classification rules, feature extraction, contextual analysis, supervised and unsupervised learning and clustering, finite sample size effects, and computational recognition complexity. Other important but unsolved problem areas are examined. The relationship between the statistical pattern recognition and signal processing is also considered. Some experimental results on preprocessing for recognition are provided in the Appendix.

---

* This work was supported in part by Grant AFOSR 76-2951.

2. THE CLASSIFICATION RULES

Statistical pattern recognition makes use of the decision theoretic approach to pattern recognition. The fundamental assumption is that the patterns are random in nature and thus can be described statistically in parametric or nonparametric forms. The recognition problem essentially consists of preprocessing, feature extraction and selection, and classification (decision making) along with training or learning process. A good classification is almost always the main objective of a recognition system. Two most well known statistical classification rules are the Bayes decision rule and the nearest-neighbor decision rule.

Let x be a vector measurement of a pattern sample, and m be the number of classes. The Bayes decision rule minimizes the average risk with respect to the given a priori probabilities P_i, $i = 1,2,\ldots,m$. For equal loss functions, the Bayes decision rule reduces to the maximum likelihood decision rule (MLDR) which chooses the class that maximizes the function

$$P_i \; p(x/\omega_i); \; i = 1,2,\ldots,m$$

where the conditional probability densities $p(x/\omega_i)$ must be known or estimated. The optimal property of the Bayes decision rule is not always realized in practice because the required a priori knowledge is either unavailable or inaccurate. For two multivariate Gaussian densities with mean μ_i and covariance matrix Σ_i, $i = 1,2$, the MLDR is to assign x to the class for which

$$(x - \mu_i)' \, \Sigma_i^{-1} \, (x - \mu_i) - \ln(P_i^2/|\Sigma_i|) \tag{1}$$

is the minimum. It is not unusual to find in practice [13] that a modified MLDR which chooses the minimum of the form,

$$(x - \mu_i)' \, \Sigma_i^{-1} \, (x - \mu_i) \tag{2}$$

can perform better than the MLDR. This is an example of the gap between theory and practice. The performance of the Bayes decision rule or the Bayes error probability in general cannot be expressed with a closed form. The error estimate which critically depends on the sample size is by itself a fundamental problem in statistics (see e.g. [14]).

The nearest neighbor decision rule (NNDR) identifies the vector sample x with the class of its nearest neighbor; nearness being measured by the Euclidean distance. For k-NNDR, the decision is based on the majority vote of k nearest neighbors. The advantage of the NNDR is that its asymptotic error rate is upper bounded by twice of the Bayes error. The NNDR is nonparametric because the information on probability densities is not needed. Obvious drawbacks of the NNDR are the large amount of distance computation and the storage requirement of learning samples. Procedures to reduce the computation and storage requirements include the condensed NNDR, edited NNDR, selection of training samples, and the use of branch and bound algorithms. Other modifications of the NNDR include the distance weighted NNDR which can provide better recognition result in practice than the unweighted NNDR discussed above. Replacement of the Euclidean distance by a quadratic form similar to that given by Eq. (2) also demonstrated a superior recognition performance in practice [15]. The performance of the NNDR at small sample size is not clear as the limited available theoretical results are inconclusive. For moderate to large sample size, the NNDR performance is comparable to the MLDR.

The reject option has been considered for both Bayes decision rule and the NNDR. The errors can be reduced at the expense of some rejects. The error-reject trade-off is an additional consideration in the reject option (see [16] for some recent result).

Linear, piecewise linear, and quadratic discriminant functions have been extensively investigated especially in the statistical literatures. However, the closed form error probability expressions are generally unavailable except in the simple case of multivariate Gaussian densities with unequal mean and equal covariance matrices. The use of the MLDR is implied for the parametric discriminant analysis and the optimization criterion is the minimum error probability. The Fisher's linear discriminant is a nonparametric technique that maximizes the ratio of between-class scatter to within-class scatter in the one-dimensional space on which the vector measurements are projected. This projection is a many-to-one mapping and in theory cannot possibly reduce the minimum attainable error probability. However the Fisher's linear discriminant has been a very effective procedure in practice for classifying high dimensional vector measurements.

For complex patterns such as images, a multi-stage decision-tree classifier has been shown experimentally to have a better overall performance than the conventional single-stage classifier (see e.g. [17][18]). However, the classification time increases due to the complexity of computation. A linear binary tree classifier can be used [19] to take advantage of the accuracy of a decision-tree classifier and to use the linear discriminant function at decision stages to reduce the classification time. With predesigned tree structure, the overall computation time can then be less than ten percent of that of a single-stage classifier. Although different feature subset may be used at each decision stage, the search for an optimum feature subset requires additional computation. The problem of optimizing the decision tree structure has been considered (see e.g. [20]). The methods of reducing the computational complexity considered include clustering the decision rules, and the use of branch and bound procedure to find efficient decision rule and for feature assignment, etc. The decision tree classifier is the most promising classification mechanism for increasingly complex recognition problems in the future. Features can be mathematical, structural or various combinations.

Although the sequential decision procedure is, theoretically speaking, suitable mainly for independent identically distributed feature measurements, the flexibility allowed by feature ordering or even on-line feature ordering is the most attractive capability of the sequential decision procedure. The advantage of the sequential decision procedure in requiring a minimum average number of feature measurements is particularly important in some applications where the feature measurements are costly.

The table look-up decision rule stores the decision rule itself rather than the densities. The vector measurement x is used as an address to a table which can look-up the class assignment for x. The table which is stored in the memory assigns a class to each (quantized) vector in the measurement space. Procedures to reduce the memory requirements and to speed-up the decision assignment time have been considered ([21][22]).

Other generalization of the conventional decision theory framework is the simultaneous membership of a measurement in several classes which has the origin of "degree of membership" from fuzzy set theory. The compound decision rules and the finite sample size effects in sample-based classification rules will be discussed in later sections.

3. FEATURE EXTRACTION

The mathematical features as well as the structural features are best suited for automatic recognition although they may not necessarily have physical meaning or may be quite different from features derived by human recognition process. A fundamental approach to extract features in statistical pattern recognition is by evaluating a number of available features to select a small subset of good features. Such evaluation can be based on the direct estimate of the error probability or an indirect measure of feature effectiveness as provided by the class of distance and information measures (see e.g. [23][24]). These measures are very effective even though they do not always choose the feature set that has the smallest error. The relative effectiveness of various measures has been considered [25]. These measures are also very useful as error bounds (see e.g. [26-28]).

Another useful approach is the linear transformation methods. If a pattern can be completely described by the second order statistics, the Karhunen-Loeve transform is optimal in the mean square error sense. In addition to the fact that the second order statistics is not adequate for most patterns, the transform also requires excessive computation. It is a misconception that feature extraction is nothing more than dimensionality reduction and that the Karhunen-Loeve transform solves all mathematical feature extraction problems.

A realistic solution to feature extraction must take into consideration the nature of patterns, the a priori knowledge available, and the specific requirements and constraints of the given recognition task. Although exhaustic search is about the only way to find the best feature set, efficient feature set search procedures have been considered [29] to provide a computationally feasible solution.

Feature extraction and selection is important not only for pattern recognition but also useful to signal processing and communications. Properly selected feature subset, for example, in high altitude or on-board pattern recognition can represent a compression of discriminating information for the original signals so that the transmission requirement such as the bandwidth can be greatly reduced. However, feature selection differs from signal selection in communications in one important aspect; the additive white noise usually assumed in communications does not apply to real pattern recognition problems.

As a final remark of this section, we would like to emphasize that to extract the right features that truly characterize a pattern is a real challenge to human intelligence. Although much has been studied, feature extraction will remain to be a key problem in pattern recognition. Experimental methods should be relied on whenever the theoretical mechanism is inadequate.

4. CONTEXTUAL ANALYSIS

A major weakness of statistical pattern recognition is the difficulty to take the contextual relations into account in the recognition process. The compound decision theory appears to be the closest statistical theory that can take the contextual information into account. When a statistical decision problem is repeated n times, with no relationships among the individual problems, the compound decision rule makes use of the information from all measurements from the n repetitions to make decisions on individual problems. In character recognition of a text, for example, decisions have to be made on individual characters. The contextual information in terms of transition probabilities among characters can be utilized to improve the recognition for individual characters. Similarly in image recognition, individual picture elements or subimages may have to be classified. The information on the correlation among picture elements or subimages should be used for better classification. Although very little theoretical result is available to measure the amount of performance improvement due to the use of contextual information, experimental results have all demonstrated the available improvement. To implement the compound decision rule, Markov chain, model of stationary stochastic process for the pattern, and coding of spatial correlation parameters [30] are among the useful tools.

Consider the recognition of each subimage of an image. By assuming dependence only on four adjacent subimages, the compound decision rule is to choose the class which maximizes

$$p(x_o/\omega_k)P(\omega_k) \prod_{j=1}^{4} p(x_j/\omega_k) \tag{3}$$

where $\omega_k = 1,2,\ldots,m$ and x_o is the vector measurement of the subimage under consideration. If we assume the dependence on all eight neighboring subimages, then the expression inside the product sign should have the conditional probability densities of all eight neighbors. Experimental result has demonstrated [31] that there is very little performance difference between four and eight neighbors.

While there is very much to be done in image recognition using the contextual information to classify a whole image or individual subimages (or picture elements), there has been very significant progress in the character recognition area (see e.g. [32][33]).

5. SUPERVISED AND UNSUPERVISED LEARNING AND CLUSTERING

Learning is needed in pattern recognition to establish the required statistical knowledge, from samples, such as the statistical parameters, probability densities, or even the decision boundaries. When the samples are of known classification, learning is supervised; otherwise it is unsupervised. In terms of the statistical framework, the supervised learning follows exactly the classical Bayesian and maximum likelihood estimation theories. The mixture estimation and decomposition in statistics is one approach to unsupervised learning. To overcome some difficulties with the Bayes procedures, a quasi-Bayes approximate learning procedure has been considered [34]. Much details on the learning algorithms as well as the decision-directed learning are available in pattern recognition texts [1-11]. It is important to note that the criterion of minimizing the mean-square error between the estimated and true parameters is used almost exclusively in learning and estimation. While the objective of classification is the minimum error probability, there is no guarantee that the learning algorithms will result in minimum classification error. Some effort has been made to design learning algorithms using window functions to minimize directly the classification error [35]. However, the convergence rate may be slow. In addition to properly selecting the window parameter, other procedures should be examined to speed up the convergence. A good understanding of the relationship between estimation and decision [36] is helpful. More flexible structures for the learning process should be considered. For example, the initial learning phase may be the conventional minimum mean-square error criterion. The subsequent learning phase can be based on the minimum error probability criterion. Another example is that a supervised learning process can be switched to unsupervised learning or vice versa. Of course the optimum usage of each learning phase would be a new problem to be examined [37].

Clustering is an important subject by itself in statistical data analysis, although it may be considered as unsupervised learning in pattern recognition. Clustering can be defined as a partition of the set of vector measurements such that each measurement will be assigned to one and only one set among a collection of disjoint sets. A recent discussion on the subject is in [38], [39], in addition to the texts [1-11]. The problem of clustering individuals can be considered within the context of a mixture of distributions [40]. Discussion of the cluster validity problem is in [41].

6. FINITE SAMPLE SIZE EFFECTS

In practical recognition problems the sample size is limited. The actual recognition performance may be quite different from that theoretically predicted based on infinite sample size. Indeed the finite sample size and its associated dimensionality problem is fundamental to all pattern recognition problems. For example, the decision rules in practice are sample-based. Expected errors of the sample-based classification rules generally do not have closed form solution at small sample size. Distance and information measures evaluated under finite sample size may be highly inaccurate. Detailed discussion of the finite learning sample size problem is in [42-45] among others.

The best way to reduce the finite sample size effect is to increase the sample size with respect to the dimensionality. However, certain decision rules perform better than the others at the small sample size (see e.g. [46]). There are distributions, such as the uniform, which are less sensitive to the sample size. For images the dimensionality includes the numbers of picture elements and quantization levels as well as the number of image samples. The relationships among the performance, sample size, and dimensionality are highly nonlinear, even for the Gaussian case [47][48]. The fundamental issue of the "peaking" behavior between the performance and dimensionality (or sample size) is not quite resolved [49]. A thorough study of the subject of finite sample size in pattern recognition is much needed as it will certainly be helpful to design a more reliable recognition systems for a given set of features.

7. COMPUTATIONAL RECOGNITION COMPLEXITY

The term "computation complexity" has a different meaning at different situations and is not well defined for pattern recognition researchers. The Kolmogorov information-theoretic computational complexity is defined as the minimum length of the program to obtain an object from data. While in linear discrimination the complexity of the classifier is usually identified with the dimensionality of the vector measurement, the discriminating capability of Boolean classifiers is determined not only by dimensionality of the feature vectors but also by the type of combinations these features are permitted to undergo. In this case we talk about the combinational complexity of the decision rule. Intuitively, the complexity concept can give us a feeling of what is complex and what is less complex. So the complexity should be a relative not an absolute measure. A more familiar complexity definition to engineers is the amount of computational effort including time and cost to accomplish a recognition task. To be machine independent, the complexity will include mainly the number of manipulations such as the multiplication and comparison operations. The recognition complexity based on this definition can be reduced by proper implementation techniques such as the use of sequential-parallel operations, etc. It is important to note that fast algorithms have been developed to compute the quadratic forms resulting from Gaussian patterns [50][51].

To determine the overall computational complexity of a recognition system, the trade-off between feature extraction and classification must be considered. A complicated feature extraction process results in a few but good features. The resulting classifier can be a very simple one. If no feature extraction effort is made so that a large number of features are used, the required classification and learning processes will be very complicated. The problem of determining an overall recognition time should be considered. The solution to this problem will be particularly useful for real-time pattern recognition.

8. OTHER PROBLEM AREAS

In addition to the topics considered above, there are a number of other problem areas where the solutions are partially available or completely unavailable.

(1) Learning and classification of nonstationary patterns. Only limited special cases were examined.
(2) A truly optimal recognition system that simultaneously optimizes the preprocessing, feature extraction, and classification and learning. The solution is not available. Right now each stage is performed independently of the others.
(3) Statistical and syntactic mixed model (see e.g. [52]). Much has been said but the real progress is slow.
(4) Automatic generation of recognition rules, i.e. to have the machine determine the classification and learning procedures without human interaction. The solution is not available.
(5) Interactive pattern recognition. A very significant progress has been made to provide man-machine interaction in pattern recognition.

There are certainly other areas not listed above. The statistical pattern recognition, however, has found applications in numerous fields with varying degrees of success. It has provided an important "bag of tricks" to solving many increasingly complex recognition problems.

9. RELATIONSHIPS WITH SIGNAL PROCESSING

Digital signal processing techniques have been much needed for preprocessing and feature extraction in statistical pattern recognition. For example, in classification problems with time series, the partial correlation and other parameters of the signal model [53] [54] are very useful features. In fact, signal modelling and representation is a common problem area to both statistical pattern recognition and signal processing. Proper modelling will make a good use of contextual information for classification. The statistical pattern recognition is also useful to signal processing. For example, in image segmentation and boundary extraction of image processing, statistical classifications may be involved. Also, in on-board processing as mentioned earlier, preliminary classification may be necessary for data reduction. The potential uses of statistical pattern recognition in signal processing remain to be further explored.

APPENDIX
EXPERIMENTAL RESULTS ON PREPROCESSING

In our pattern recognition research, extensive effort has been made in preprocessing for the recognition of seismic waveforms and the reconnaissance imagery. For the seismic data, the classification involved is the discrimination of nuclear explosions from natural earthquakes. By using a set of short-term spectral features, obtained by nonrecursive digital filtering and FFT operations, a 94.17% correct recognition was reported [15]. Features based on the coefficients of an autoregressive (AR) and moving average (MA) model of the waveform also showed a good recognition performance. Experimental results have indicated that the use of the AR (i.e., all-pole) model alone is not adequate as far as the classification and spectral matching are concerned. Figure 1 provides a typical comparison of spectral matching based on an AR model (Figure 1a) and an ARMA model (Figure 1b) for a 15th order model with an explosion event. The ARMA model clearly does better. To determine accurately the coefficients of the ARMA model is an important problem to both signal processing and pattern recognition.

For the reconnaissance imagery, the classification objective is to detect and locate the objects (targets) of interest. Useful features include the texture measures, average gradient values, and statistical parameters, etc. [55]. Decision tree schemes provided nearly perfect recognition for all imagery studied [56]. This is the result of many effective preprocessing operations including the modified gradient [57], a rotationally invariant local operator and an adaptive Kalman filtering scheme for image enhancement [58][59]. The adaptive filtering scheme uses a Kalman filter with no state "jumps" plus a second system that uses a generalized likelihood ratio test to detect the significant change in gray level due to the edge or boundary of an object and adjust the filter accordingly. The scheme is very effective in image enhancement whether the original image is perturbed by additive noise or not. Figure 2a shows the filtered image of a tank scene (real imagery) without additive noise. If the original image is added by Gaussian noise with signal-to-noise ratio of 2.2, the filtered image as shown in Figure 2b also has significant enhancement. Even though only a minimum amount of preprocessing is needed if target detection is the only objective, the example illustrates the importance of preprocessing for various image recognition tasks.

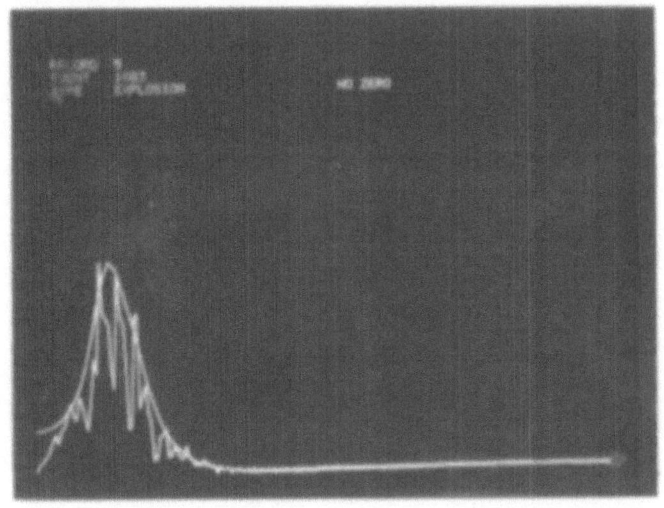

Figure 1a

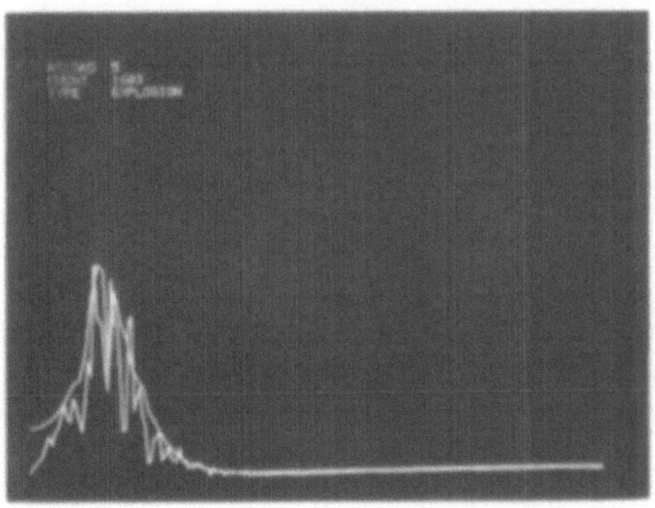

Figure 1b

Figure 2a

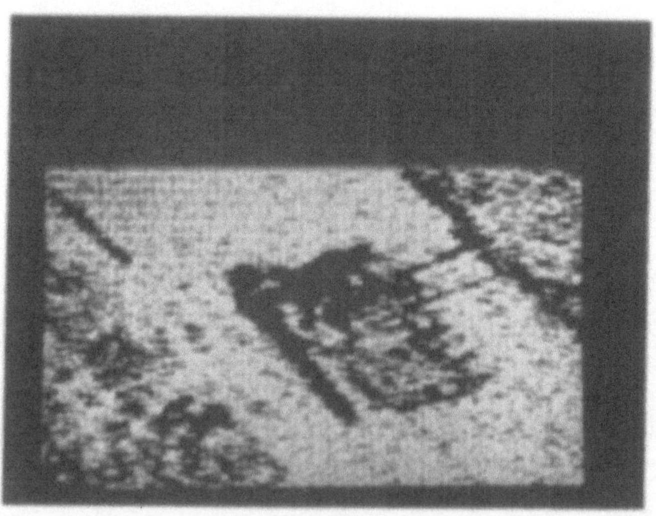

Figure 2b

REFERENCES

(Author's note: No attempt is made to provide a complete list of all contributions in statistical pattern recognition. It is hoped that the references may give the readers an idea of many current activities in the field. The ASI refers to the 1978 NATO Advanced Study Institute on Pattern Recognition and Signal Processing.)

1. Sebestyen, G. S., _Decision Making Processes in Pattern Recognition_, Macmillan, 1962.
2. Fu, K. S., _Sequential Methods in Pattern Recognition and Machine Learning_, Academic Press, 1968.
3. Andrews, H. C., _Introduction to Mathematical Techniques in Pattern Recognition_, Wiley, 1972.
4. Fukunaga, K., _Introduction to Statistical Pattern Recognition_, _Academic Press_, 1972.
5. Meisel, W., _Computer-Oriented Approaches to Pattern Recognition_, Academic Press, 1972.
6. Patrick, E. A., _Fundamentals of Pattern Recognition_, Prentice-Hall, 1972.
7. Duda, R. O., and P. E. Hart, _Pattern Classification and Science Analysis_, Wiley, 1973.
8. Chen, C. H., _Statistical Pattern Recognition_, Hayden Book Company, 1973.
9. Young, T. Y., and T. W. Calvert, _Classification Estimation, and Pattern Recognition_, American Elsevier, 1973.
10. Ullmann, J. R., _Pattern Recognition Techniques_, Crane, Russak & Co., 1973.
11. Tou, J. T., and R. C. Gonzales, _Pattern Recognition Principles_, Addison-Wesley Publishing Co., 1974.
12. Chen, C. H., "Statistical Pattern Recognition: Review and Outlook", _TR EE-75-4_, June 1975.
13. Chang, J. K., "Modified Maximum Likelihood Decision Rule and Minimax Bayes Decision Rule", Proc. of the Third International Joint Conference on Pattern Recognition, November 1976.
14. Toussaint, G. T., "Bibliography on Estimation of Misclassification", _IEEE Trans. on Information Theory_, Vol. IT-20, pp. 472-479, July 1974.
15. Chen, C. H., "Seismic Pattern Recognition", _Geoexploration Journal_, April 1978.
16. Devijver, P. A., "Error-reject Relationships in Nearest Neighbor Decision Rules", Proc. of the Third International Joint Conference on Pattern Recognition, November 1976.
17. Wu, C. L., "The Decision Tree Approach to Classification", Ph.D. thesis, School of Electrical Engineering, Purdue University, May 1975.
18. Hauska, H., and P. H. Swain, "The Decision Tree Classifier: Design and Potential", Proc. of Symposium on Machine Processing of Remotely Sensed Data, June 1975.

19. You, K. C., and K. S. Fu, "An Approach to the Design of a Linear Binary Tree Classifier", Proc. of Symposium on Machine Processing of Remotely Sensed Data, June 1976.
20. Kulkarni, A. V., and L. N. Kanal, "An Optimization Approach to Hierarchical Classifier Design", Proc. of the Third International Joint Conference on Pattern Recognition, November 1976.
21. Eppler, W. G., "An Improved Version of the Table Look-up Algorithm for Pattern Recognition", Proc. of the Ninth International Symposium on Remote Sensing of Environment, April 1974.
22. Haralick, R. M., "The Table Look-up Rule", Proc. of the Third International Joint Conference on Pattern Recognition, November 1976.
23. Kanal, L. N., "Patterns in Pattern Recognition", IEEE Trans. on Information Theory, Vol. IT-20, pp. 697-722, November 1974.
24. Chen, C. H., "On Information and Distance Measures, Error Bounds and Feature Selection", Information Sciences Journal, Vol. 10, pp. 159-173, 1976.
25. Chen, C. H., "Theoretical Comparison of a Class of Feature Selection Criteria in Pattern Recognition", IEEE Trans. on Computers, Vol. C-20, pp. 1054-1056, September 1971.
26. Devijver, P. A., "On a New Class of Bounds on Bayes Risk in Multihypothesis Pattern Recognition", IEEE Trans. on Computers, Vol. C-23, pp. 70-80, January 1974.
27. Devijver, P. A., "Nonparametric Estimation of Feature Evaluation Criteria", in this Proceedings.
28. Toussaint, G. T., "Probability of Error, Expected Divergence, and the Affinity of Several Distributions", IEEE Trans. on Systems, Man, and Cybernetics, vol. SMC-8, no. 6, pp. 482-485, June 1978.
29. Kittler, J., "Feature Set Search Algorithms" in this Proceedings.
30. Yu, T. S., and K. S. Fu, "Statistical Pattern Recognition Using Contextual Information", TR-EE-78-17, Purdue University, 1978.
31. Welch, J. R. and K. G. Salter, "A Context Algorithm for Pattern Recognition and Image Interpretation", IEEE Trans. on Systems, Man and Cybernetics, vol. SMC-1, pp. 24-30, January 1971.
32. Toussaint, G. T., "The Use of Context in Pattern Recognition", Proc. of Pattern Recognition and Image Processing Conference, June 1977.
33. Suen, C. Y., "Handprint Recognition and Standardization", presented at the ASI.
34. Makov, U. E., "Approximations to Sequential Bayes Procedures", presented at the ASI.
35. Dotu, H., "Learning Algorithms for Discriminant Function Solution of the Minimum Error Classification Problem", Proc. of the Third International Joint Conference on Pattern Recognition, November 1976.

36. Hodgkiss, W., and L. W. Nolte, "On Relationships Between Detection and Estimation Theory", in this proceedings.
37. Cooper, D. B., "When Should a Learning Machine Ask for Help", IEEE Trans. on Information Theory, Vol. IT-20, pp. 455-471, July 1974.
38. Diday, E., and J. C. Simon, "Clustering Analysis" in Digital Pattern Recognition edited by K. S. Fu, Springer-Verlag, 1976.
39. Backer, E., Cluster Analysis by Optimal Decomposition of Induced Fuzzy Sets, Delft University Press, 1978.
40. Sclove, S. L., "Population Mixture Models and Clustering Algorithms", Communications in Statistics, Vol. A6, pp. 417-434, 1977.
41. Jain, A. K., "Cluster Validity", presented at the ASI.
42. Chen, C. H., "Finite Sample Considerations in Statistical Pattern Recognition", Proc. of Pattern Recognition and Image Processing Conference, May 1978.
43. Pau, L. F., "On Finite Learning Sample Size Problems in Pattern Recognition", in thie proceedings.
44. Raudys, S., and V. Pikelis, "On Dimensionality, Sample Size and Classification Error in Discriminant Analysis", submitted for publication, 1978.
45. Jain, A. K., "On an Estimate of the Bhattacharyya Distance", IEEE Trans. on Systems, Man, and Cybernetics, vol. SMC-6, no. 11, pp. 763-766, Nov. 1976.
46. Bailey, T., and A. K. Jain, "A Note on Distance-Weighted k-nearest Neighbor Rules", IEEE Trans. on Systems, Man, and Cybernetics, vol. SMC-8, no. 4, pp. 311-313, April 1978.
47. Jain, A. K., and W. G. Waller, "On the Optimal Number of Features in the Classification of Multivariate Gaussian Data", to appear in Pattern Recognition Journal.
48. Trunk, G. V., "A Problem of Dimensionality: a Simple Example", submitted for publication, 1978.
49. Waller, W. G., and A. K. Jain, "On the Monotonicity of the Performance of Bayesian Classifiers", IEEE Trans. on Information Theory, vol. IT-24, no. 3, May 1978.
50. Böhme, J. F., "Computation of Quadratic Forms Used in Statistical Pattern Recognition", Proc. SITEL-ULG Seminar on Pattern Recognition, Liege, Belgium, Nov. 1977.
51. Böhme, J. F., "Fast Computation of Quadratic forms with Applications to Signal Detection, Classification, and Estimation", in this Proceedings.
52. Chen, C. H., "On Statistical and Structural Feature Extraction", in Pattern Recognition and Artificial Intelligence, ed. by C. H. Chen; Academic Press, Inc. 1976.
53. Chen, C. H., "An Introduction to Seismic Signal Processing", presented at the ASI.
54. Kashyap, R. L., "Optimal Feature Selection and Decision Rules in Classification Problems with Time Series", IEEE Trans. on Information Theory, vol. IT-24, no. 3, pp. 281-288, May 1978.

55. Chen, C. H., "Theory and Applications of Imagery Pattern Recognition", Proc. of the Fourth International Congress for Stereology, National Bureau of Standards Special Publication 431, January, 1976.

56. Chen, C. H., and C. K. Lau, "Feature Set Formulation for Reconnaissance Image Analysis", Seventh Annual Automatic Imagery Pattern Recognition Symposium, College Park, MD, May, 1977.

57. Chen, C. H., "Note on a Modified Gradient Method for Image Analysis", Pattern Recognition Journal, June, 1978.

58. Willsky, A. S., and H. L. Jones, "A Generalized Likelihood Ratio Approach to the Detection and Estimation of Jumps in Linear Systems", IEEE Trans. on Automatic Control, pp. 108-112, February, 1976.

59. Chen, C. H., "Some New Results on Image Processing and Recognition", presented at the IRIA Pattern Recognition and Image Processing Seminar, Le Chesnay, France, June 22, 1978.

ON RELATIONSHIPS BETWEEN DETECTION AND ESTIMATION THEORY

W. S. Hodgkiss

Marine Physical Laboratory
Scripps Institution of Oceanography
San Diego, CA 92132, U.S.A.

L. W. Nolte

Departments of Electrical and Biomedical Engineering
Duke University
Durham, NC 27706, U.S.A.

1. INTRODUCTION AND OVERVIEW

Within the communications context, the two broad subjects of signal detection and estimation theory are concerned with decision making based upon operations performed on some received data. In the first, only a decision about the presence or absence of a certain signal, or subset of signals, in the data is required. In the second, the decision involves estimating one or several parameters which are contained in the data. Processors for these tasks are designed based on some criterion of goodness or optimality.

This paper takes a global approach to the processing of information. The viewpoint will be Bayesian where any uncertain parameters are modeled as random variables and knowledge about them is summarized by _a priori_ probability density functions. Essentially, the unprocessed data are considered as the observables. Processor structure then is allowed to evolve freely out of the mathematics of the particular problem under consideration with the sole restriction being the criterion of optimality.

Specifically, the processors discussed here must decide if the random processes observed consist of a signal obscured by noise or noise alone. The resulting detectors are optimum in the sense of making a least risk decision. The general form of the likelihood ratio is derived in Section 2.

Once the mathematics of the likelihood ratio has been written, the optimal processor can be implemented in various structures. Four such canonical implementations are discussed: 1) one shot, 2) pseudo estimator, 3) sequential, and 4) two step. The pseudo estimator structure is shown to be the optimal counterpart of an appealing ad hoc approach to processor design where any uncertain parameters are first estimated, then plugged into the parameters known likelihood ratio as if they were known exactly. Although an estimate and plug structure is appealing due to its explicit adaptive characteristics, it is shown that the optimal processor exhibits learning or adaptive features naturally when implemented sequentially.

Research results from the area of array processing are presented in Section 3. Of particular interest are two specific problems involving either signal or noise source location uncertainty: 1) signal known except for direction in Gaussian noise which is independent from sensor to sensor (SKED) and 2) signal known exactly in noise with an additive directional noise component of uncertain direction (SKE in NUD). In the first, a performance comparison is made between the Bayes optimal detector and an appealing estimate and plug structure utilizing the MAP estimate of signal source location. In the second, computer simulation runs of the SKE in NUD processor are used to illustrate the Bayesian updating which occurs as an integral part of the sequential structure.

2. BAYESIAN SIGNAL DETECTION THEORY

The following sections will provide the mathematical framework of signal detection theory from which we will work. Important concepts to be emphasized are: (1) the likelihood ratio, (2) processor structure, and (3) performance. Parts of the presentation will closely follow an excellent paper by Birdsall and Gobien (Birdsall and Gobien, 1973).

2.1 The Likelihood Ratio

Consider the binary decision problem where there are two mutually exclusive and exhaustive hypotheses, H_0 and H_1. Assume a vector of observations $\underline{R}$ is made from a space $\mathcal{R}$. Under H_0, the distribution on $\mathcal{R}$ is characterized by a probability density conditioned on a vector of parameters $\underline{\theta}_0$ which belongs to the family $\{p(\underline{R}|\underline{\theta}_0, H_0)\ ;\ \underline{\theta}_0 \in \Theta_0\}$. Under H_1, the density belongs to the family $\{p(\underline{R}|\underline{\theta}_1,H_1)\ ;\ \underline{\theta}_1 \in \Theta_1\}$. There may be components of $\underline{\theta}_0$ and $\underline{\theta}_1$ which represent the same parameters. In summary

$$H_0 : p(\underline{R}|\underline{\theta}_0, H_0) \quad , \quad \underline{\theta}_0 \in \Theta_0$$
$$H_1 : p(\underline{R}|\underline{\theta}_1, H_1) \quad , \quad \underline{\theta}_1 \in \Theta_1 \, . \tag{2.1}$$

Based upon the observation vector, the processor must make a decision (D_0 or D_1) as to which hypothesis it believes is true. Classical detection theory has shown that decisions based upon the likelihood ratio are optimum for a wide range of goodness criteria (Peterson, Birdsall, and Fox, 1954; Middleton and Van Meter, 1954)

$$\Lambda(\underline{R}) \triangleq \frac{p(\underline{R}|H_1)}{p(\underline{R}|H_0)} \underset{D_0}{\overset{D_1}{\gtrless}} \eta \tag{2.2}$$

Birdsall has shown more generally that any optimality criterion based on "detection probability" $P(D_1|H_1)$ and "false alarm probability" $P(D_1|H_0)$ where good decisions are preferred over bad leads to the calculation of $\Lambda(\underline{R})$ as the decision statistic (Birdsall, 1973). Thus, a separation is achieved between the processing of $\underline{R}$ and the actual optimality criterion chosen which arises in the selection of a threshold value η.

The situation may arise where one or several of the conditioning parameters in either or both $\underline{\theta}_0$ and $\underline{\theta}_1$ are uncertain. These are then modeled as random variables and any prior knowledge about them is summarized by *a priori* probability density functions $p(\underline{\theta}_0)$ and $p(\underline{\theta}_1)$. The desired decision statistic now becomes the likelihood ratio of marginal probability density functions on $\mathcal{R}$

$$\Lambda(\underline{R}) = \frac{\int_{\Theta_1} p(\underline{R}|\underline{\theta}_1, H_1) p(\underline{\theta}_1) d\underline{\theta}_1}{\int_{\Theta_0} p(\underline{R}|\underline{\theta}_0, H_0) p(\underline{\theta}_0) d\underline{\theta}_0} \tag{2.3}$$

2.2 Processor Structure

Once the mathematics of the likelihood ratio has been written, any realization of $\Lambda(\underline{R})$ will achieve identical results. None the less, structuring the processor in various ways often can be advantageous from the standpoint of any potential insight gained, comparison with non-Bayesian approaches to a similar problem, or the desire of greater feasibility and flexibility of implementation. Already considered, the one shot processor simply calculates $\Lambda(\underline{R})$ as shown in (2.3) and illustrated in Figure 2.1. In this structure, all of the data is processed at the same time and the likelihood ratio is obtained directly. Three other specific structures to be discussed are: (1) pseudo estimator, (2) sequential, and (3) two step.

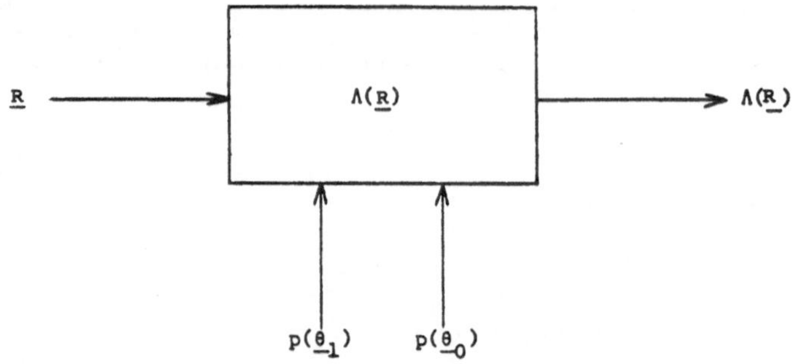

Figure 2.1 One Shot Processor.

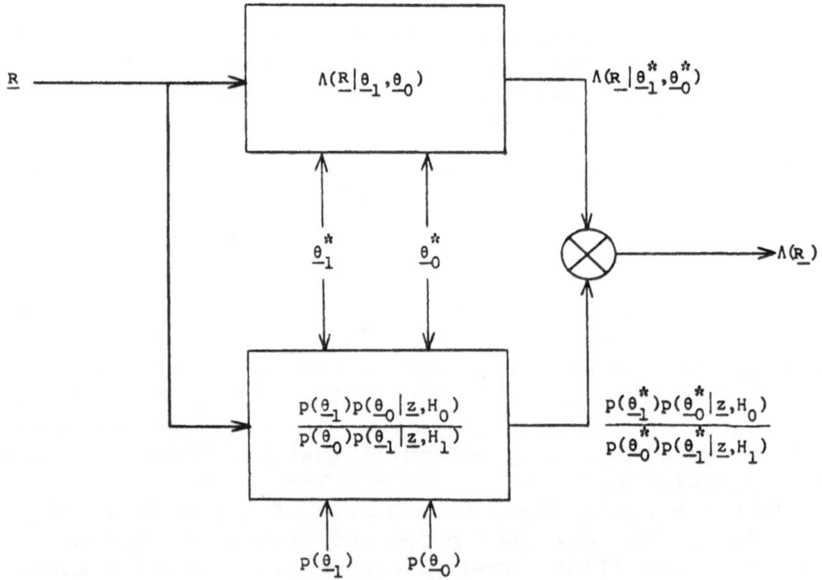

Figure 2.2 Pseudo Estimator Processor.

The pseudo estimator structure is actually a particular case of a more general class of structures resulting from the application of Bayes' rule

$$p(\underline{\theta}|\underline{R}) = \frac{p(\underline{R}|\underline{\theta})p(\underline{\theta})}{p(\underline{R})} . \tag{2.4}$$

Here the *a posteriori* probability density function of θ is calculated based on the observation vector $\underline{R}$ and the prior knowledge $p(\underline{\theta})$. The marginal density appearing in the denominator of (2.4) is simply a normalizing constant

$$p(\underline{R}) = \int_{\Theta} p(\underline{R}|\underline{\theta}')p(\underline{\theta}')d\underline{\theta}'. \tag{2.5}$$

Utilizing Bayes' rule in (2.3), the likelihood ratio may be written as

$$\Lambda(\underline{R}) = \frac{p(\underline{\theta}_1)p(\underline{\theta}_0|\underline{R},H_0)}{p(\underline{\theta}_0)p(\underline{\theta}_1|\underline{R},H_1)} \Lambda(\underline{R}|\underline{\theta}_1,\theta_0) \tag{2.6}$$

and its computation is illustrated in Figure 2.2 where

$$\Lambda(\underline{R}|\underline{\theta}_1,\underline{\theta}_0) = \frac{p(\underline{R}|\underline{\theta}_1,H_1)}{p(\underline{R}|\underline{\theta}_0,H_0)}$$

i.e., the parameters known likelihood ratio. Note that *any* convenient, admissible value of the parameters may be used to evaluate (2.6).

A pseudo estimator structure results if values for $\underline{\theta}_0$ and $\underline{\theta}_1$ may be found such that

$$\frac{p(\underline{\theta}_1)p(\underline{\theta}_0|\underline{R},H_0)}{p(\underline{\theta}_0)p(\underline{\theta}_1|\underline{R},H_1)} = 1. \tag{2.7}$$

Solution values thus chosen for $\underline{\theta}_0$ and $\underline{\theta}_1$ are called pseudo estimates (Jaarsma, 1969). One value of such a structure is for the purpose of comparison with other *ad hoc* processors where some estimation scheme is joined with the parameters known likelihood ratio to yield a suboptimal, but perhaps easily implementable, receiver design (Hatsell and Nolte, 1974).

An appealing approach to a detection problem where uncertain parameters exist is to estimate these parameters and then plug them into the conditional likelihood ratio as if they were known exactly. When "good" estimators are used, such a structure as illustrated in Figure 2.3 appears to be operating in an optimal fashion. It is not clear, however, that piecing together locally optimal techniques (i.e., "good" estimators and a solution optimal when all parameters are known) will yield global optimality when the overall goal is good detection performance. Note that the structures in Figures 2.2 and 2.3 are equivalent only when $\hat{\underline{\theta}}_1 = \underline{\theta}_1^*$ and $\hat{\underline{\theta}}_0 = \underline{\theta}_0^*$ are the pseudo estimates of $\underline{\theta}_1$ and $\underline{\theta}_0$. Recalling from (2.7), utilizing the pseudo estimates results in the lower branch of Figure 2.2 equaling unity. Thus, the optimal processor can be realized in an estimate and plug structure. It should be noted that the pseudo estimate is generally not equal to a well-known estimate.

A second approach to processing the observation vector is to do so sequentially, i.e., the processor operates on one or a small block of observations at a time in a serial fashion until all the data has been exhausted. Consider a vector of dimension K

$$p(\underline{R}_K) = \prod_{i=1}^{K} p(R_i|\underline{R}_{i-1}). \tag{2.8}$$

An expression for $p(R_i|\underline{R}_{i-1})$ is desired. Assuming parameter conditional independence of the R_i

$$\begin{aligned} p(R_i|\underline{R}_{i-1}) &= \int_\theta p(R_i|\underline{R}_{i-1},\underline{\theta})p(\underline{\theta}|\underline{R}_{i-1})d\underline{\theta} \\ &= \int_\theta p(R_i|\underline{\theta})p(\underline{\theta}|\underline{R}_{i-1})d\underline{\theta} \end{aligned} \tag{2.9}$$

where $p(\underline{\theta}|\underline{R}_{i-1})$ is an updated version of the <u>a priori</u> probability density function of $\underline{\theta}$

$$p(\underline{\theta}|\underline{R}_{i-1}) = \frac{p(R_{i-1}|\underline{\theta})p(\underline{\theta}|\underline{R}_{i-2})}{p(R_{i-1}|\underline{R}_{i-2})} . \tag{2.10}$$

The expressions (2.8), (2.9), and (2.10) when conditioned on H_1 and H_0 are the design equations used to obtain the marginal distributions in (2.2) for K iterations. Figure 2.4 illustrates the corresponding processor structure. At the end of each incremental observation period, the prior knowledge for that period is updated under H_1 and H_0 to reflect the processing of an additional increment of data. These updated densities then are used as prior knowledge for the next incremental observation period. In general, the numerator and denominator expressions in (2.2) must remain separated in the updating sequence.

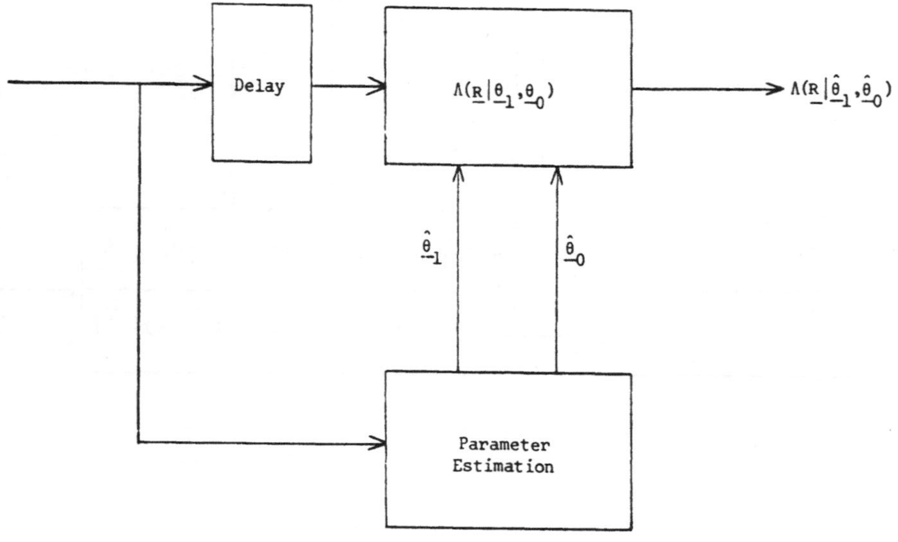

Figure 2.3 Estimate and Plug Processor.

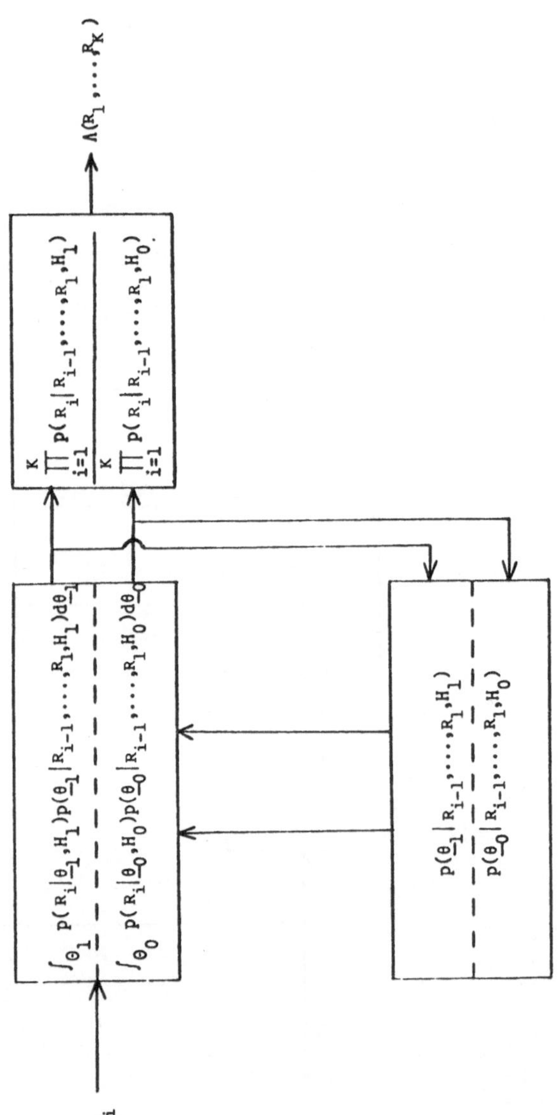

Figure 2.4 Time Sequential Processor.

Principal advantages gained in structuring a processor sequentially include no need to specify the actual total observation length and the inherent learning or adaptive nature of the processor through the iterative updating of its a priori knowledge of the uncertain parameters. Furthermore, Nolte has shown that it may be necessary to implement some processors sequentially in order to avoid feasibility problems such as a growing memory requirement (Nolte, 1965; Nolte, 1966).

The final processing structure to be discussed concerns a two step approach proposed by Birdsall (Birdsall, 1968; Birdsall and Gobien, 1973). In the primary processor, the observation vector is processed in conjunction with any convenient densities (subject to minor restrictions) substituted for the actual a priori knowledge. The output of the primary processor is utilized by the secondary processor along with the true a priori knowledge to calculate the likelihood ratio. A more detailed description will require introducing the concepts of sufficient statistics and reproducing densities.

From a Bayesian point of view, a fixed finite dimensional sufficient statistic of the observation for an unknown parameter vector would be defined as

$$p(\underline{\theta}|\underline{R}_K) = p(\underline{\theta}|\underline{\delta}(\underline{R}_K)) \quad , \text{ for all } \underline{R}_K. \tag{2.11}$$

Thus, the statistic $\underline{\delta}(\underline{R}_K)$ contains as much information about $\underline{\theta}$ as do the observations themselves. Furthermore, the dimension of $\underline{\delta}(\underline{R}_K)$ remains constant even as the dimensionality of the observation vector increases.

The classical factorization theorem provides conditions for identifying a sufficient statistic.

Theorem 2.1 Let $p(\underline{R}_K|\underline{\theta})$ be the conditional density and $\underline{\delta}(\underline{R}_K)$ the fixed finite dimensional statistic as defined previously. Then $\underline{\delta}(\underline{R}_K)$ is sufficient for $\underline{\theta}$ if there exist:

(1) a function $g[\underline{\delta}(\underline{R}_K),\underline{\theta}]$ which depends on the observation only through $\underline{\delta}(\cdot)$, and
(2) a function $G(\underline{R}_K)$ which does not depend on $\underline{\theta}$, such that

$$p(\underline{R}_K|\underline{\theta}) = g[\underline{\delta}(\underline{R}_K),\underline{\theta}]\; G(\underline{R}_K). \tag{2.12}$$

Consider the following example to illustrate this concept. The observations R_i consist of an unknown scalar θ added to independent samples n_i drawn from a distribution $N(0,\sigma^2)$.

$$R_i = \theta + n_i \qquad i = 1,2,\ldots$$

Since the observations are parameter conditionally independent, their joint distribution may be written and factored as follows

$$p(\underline{R}_K|\theta) = \prod_{i=1}^{K} p(R_i|\theta)$$

$$= (2\pi\sigma^2)^{-K/2}\exp[-\frac{1}{2\sigma^2}\sum_{i=1}^{K} R_i^2] \cdot \exp[-\frac{1}{2\sigma^2}(K\theta^2-2\theta\sum_{i=1}^{K} R_i)]$$

$$= G(\underline{R}_K)\ g[\underline{\delta}(\underline{R}_K),\theta] \tag{2.13}$$

where

$$\underline{\delta}(\underline{R}_K) = [K, \sum_{i=1}^{K} R_i]^T \quad .$$

By simply applying Bayes' rule, we can show that if $\underline{\delta}(\underline{R}_K)$ is sufficient for θ, then the a posteriori density of $\underline{\theta}$ given $\underline{\delta}(\underline{R}_K)$ is independent of $\underline{R}_K$. Substituting (2.12) into (2.4) we see that $G(\underline{R}_K)$ cancels between the numerator and denominator leaving

$$p(\underline{\theta}|\underline{R}_K) = \frac{g[\underline{\delta}(\underline{R}_K),\underline{\theta}]p(\underline{\theta})}{\int_\Theta \text{(numerator)}\ d\underline{\theta}} \ . \tag{2.14}$$

Next, we wish to consider a definition and theorem on reproducing density functions (Birdsall and Gobien, 1973).

Definition 2.1 Let $\mathcal{H}_\Gamma(\Theta) = \{h(\underline{\theta};\gamma)\ ;\ \underline{\gamma}\in\Gamma\subset R^m\ ,\ \underline{\theta}\in\Theta\}$ be a family of pdf's on Θ which is indexed by the m-dimensional parameter γ. $\mathcal{H}_\Gamma(\Theta)$ is said to be a reproducing class of probability densities under $p(\underline{R}_K|\underline{\theta})$ if, for any K, whenever the a priori pdf on Θ is

$$p'(\underline{\theta}) = h(\underline{\theta};\underline{\gamma}^0) \quad , \quad \underline{\gamma}^0\in\Gamma$$

there exists a

$$\underline{\gamma}^K = \underline{\gamma}^K(\underline{\gamma}^0,\underline{R}_K)\in\Gamma$$

such that the a posteriori pdf is

$$p'(\underline{\theta}|\underline{R}_K) = h(\underline{\theta},\underline{\gamma}^K) \quad , \quad \underline{\gamma}^K\in\Gamma \ .$$

Thus, the a posteriori pdf remains in the same family of functions as the a priori pdf, differing only in the values of the parameters characterizing members in the family. Primes will be used to signify that we are within the class of natural reproducing densities.

Theorem 2.2 Suppose $p(\underline{R}_K|\underline{\theta})$ admits a sufficient statistic of fixed dimension $\underline{\delta}(\underline{R}_K)$ for $\underline{\theta}$ and hence can be factored as in (2.12); let the function $g(\cdot,\cdot)$ be as defined there, and, provided the integral exists, put

$$p'(\underline{\theta};\underline{\gamma}) = \frac{g[\underline{\gamma},\underline{\theta}]}{\int_{\Theta} g[\underline{\gamma},\underline{\theta}']d\underline{\theta}'} \quad , \underline{\gamma} \in \Gamma \tag{2.15}$$

where Γ is the image of the space of observations under $\underline{\delta}(\cdot)$. Then $\{p'(\underline{\theta};\gamma) \ ; \ \underline{\gamma} \in \Gamma\}$ is a reproducing class of densities under $p(\underline{R}_K|\underline{\theta})$. The class thus defined is called the natural conjugate class of pdf's under $p(\underline{R}_K|\underline{\theta})$; existence of a sufficient statistic implies existence of such a class.

It is possible that the natural conjugate class may not contain a member suitable for describing the true a priori knowledge. However, suppose the true a priori pdf on Θ can, for some $\underline{\gamma}^0 \in \Gamma$, be written

$$p(\underline{\theta}) = r(\underline{\theta})p'(\underline{\theta};\underline{\gamma}^0) \tag{2.16}$$

where $r(\underline{\theta})$ is a nonnegative function defined on Θ and $p(\underline{\theta})$ is absolutely continuous with respect to $p'(\underline{\theta};\gamma^0)$. Since $p'(\underline{\theta};\gamma^0)$ is reproducing, a simple application of Bayes' rule (2.4) reveals that $p(\underline{\theta})$ also reproduces with parameter $\underline{\gamma}$

$$p(\underline{\theta}|\underline{R}_K) = \frac{r(\underline{\theta})p'(\underline{\theta};\underline{\gamma}^K)}{\int_{\Theta} (\text{numerator})\ d\underline{\theta}} \tag{2.17}$$

Utilizing (2.17) and a Bayes' rule substitution for $p'(\underline{\theta};\gamma^K)$, the marginal distribution of $\underline{R}_K$ given the true a priori knowledge can be written in terms of the natural conjugate class of densities and $r(\underline{\theta})$

$$p(\underline{R}_K) = p'(\underline{R}_K) \int_{\Theta} r(\underline{\theta})p'(\underline{\theta};\underline{\gamma}^K)\ d\underline{\theta}\ . \tag{2.18}$$

The mathematical description of the two step approach to processor structure now can be completed. The primary processor uses a convenient description of a priori knowledge out of the class of natural conjugate densities (if such a class exists). The secondary processor utilizes the likelihood ratio calculated on the basis of these densities together with the resulting sufficient statistics characterizing the a posteriori pdf's and the true a priori knowledge to calculate the true likelihood ratio

$$\Lambda(\underline{R}_K) = \Lambda'(\underline{R}_K) \frac{\int_{\Theta_1} r(\underline{\theta}_1)p'(\underline{\theta}_1;\underline{\gamma}_1^K)\, d\underline{\theta}_1}{\int_{\Theta_0} r(\underline{\theta}_0)p'(\underline{\theta}_0;\underline{\gamma}_0^K)\, d\underline{\theta}_0} \tag{2.19}$$

where

$$\Lambda'(\underline{R}_K) = \frac{\int_{\Theta_1} p(\underline{R}_K|\underline{\theta}_1)p'(\underline{\theta}_1)\, d\theta_1}{\int_{\Theta_0} p(\underline{R}_K|\underline{\theta}_0)p'(\underline{\theta}_0)\, d\underline{\theta}_0}$$

Figure 2.5 illustrates the overall configuration. The benefit of such an approach to receiver design is that potentially a major portion of the processor can be designed without knowing the exact _a priori_ knowledge. Furthermore, the mathematical tractability of the natural conjugate priors may simplify the design of the primary processor.

2.3 Block Diagrams and Sufficient Statistics

From the preceding discussion, it has been shown that once the likelihood ratio was determined, $\Lambda(\underline{R})$ could be implemented in various structures which might look quite different. Perhaps the epitome of this is in Birdsall's two step approach where the primary processor might take on any one of an infinite number of structures depending on the tractable prior chosen. Nevertheless, it is quite common for those of us engaged in optimum receiver design to look for structural pieces in our particular realization of $\Lambda(\underline{R})$ that might be either the same as, or in contrast to, structure arrived at by other _ad hoc_ approaches to the same problem. The contention here is that perhaps too much emphasis has been placed in the past on the overall structure of $\Lambda(\underline{R})$ instead of some more fundamental component pieces. A solution for the likelihood ratio not in closed form always has appeared only half completed.

The expression in (2.3) shows that the fundamental components of $\Lambda(\underline{R})$ before the introduction of a particular _a priori_ knowledge are the conditional densities of $\underline{R}$ under H_1 and H_0. Assuming $p(\underline{R}|\underline{\theta})$ admits a sufficient statistic for $\underline{\theta}$, (2.12) shows that the two basic building blocks are the sufficient statistic $\underline{\delta}(\underline{R})$ and $G(\underline{R})$. Substituting (2.12) into (2.3)

$$\Lambda(\underline{R}) = \frac{G_1(\underline{R})\int_{\Theta_1} g_1[\underline{\delta}_1(\underline{R}),\underline{\theta}_1]\ p(\underline{\theta}_1)\ d\underline{\theta}_1}{G_0(\underline{R})\int_{\Theta_0} g_0[\underline{\delta}_0(\underline{R}),\underline{\theta}_0]\ p(\underline{\theta}_0)\ d\underline{\theta}_0} \underset{D_0}{\overset{D_1}{\gtrless}} \eta \tag{2.20}$$

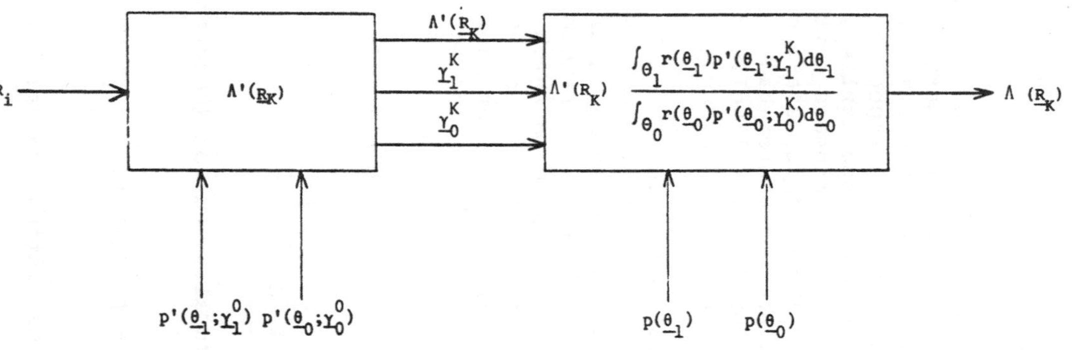

Figure 2.5 Two Step Processor.

In general, no matter what a priori knowledge is chosen, $\underline{\delta}(R)$ and G(R) will have to be calculated under H_1 and H_0. Any final structure of $\Lambda(\underline{R})$ will always be a function of these basic parts and it is their structure which will provide the most fundamental basis from which comparison with other detectors can be made. As an example, a recent paper by Adams and Nolte derives and discusses the interpretation of these basic components for optimum array processors in fluctuating ambient noise fields (Adams and Nolte, 1975).

2.4 Performance

The complete description of a detection device includes both the processor itself (i.e., the mathematical transformation from observation space to decision statistic) and the performance of the processor evaluated with respect to the goodness criterion initially chosen. As mentioned earlier, the likelihood ratio has been shown optimum for any goodness criterion based on "detection probability" $P(D_1|H_1)$ and "false alarm probability" $P(D_1|H_0)$ where good decisions are preferred over bad. Thus, the appropriate description of performance for a likelihood ratio computing device is its detection and false alarm probabilities as a function of decision threshold.

The likelihood ratio is simply a transformation of random variables (the observation vector $\underline{R}$, typically of large dimension) to a one dimensional decision statistic $(\Lambda(\underline{R}) \in [0,\infty)]$. Thus, the likelihood ratio itself is a random variable whose probability density function will depend on which hypothesis (H_0 or H_1) is actually active on $\mathcal{R}$. It is the conditional distribution of Λ under H_1 and H_0 from which processor performance is calculated. Recalling from (2.2) that the threshold η divides the decision space, define

$$P_D \triangleq p(D_1|H_1) \triangleq \int_\eta^\infty p(\Lambda|H_1)\, d\Lambda \tag{2.21}$$

$$P_F \triangleq p(D_1|H_0) \triangleq \int_\eta^\infty p(\Lambda|H_0)\, d\Lambda \,. \tag{2.22}$$

Peterson, Birdsall, and Fox (Peterson, Birdsall, and Fox, 1954) have shown that

$$p(\Lambda|H_1) = \Lambda\, p(\Lambda|H_0). \tag{2.23}$$

Thus, P_D may be written equivalently as

$$P_D = \int_\eta^\infty \Lambda\, p\,(\Lambda|H_0)\, d\Lambda. \tag{2.24}$$

The expression (2.24) is particularly valuable when the densities of Λ under the two hypotheses cannot be determined analytically. Typically, one then carries out a Monte Carlo simulation of the optimum processor and from his results forms estimates $\hat{p}(\Lambda|H_0)$ and $\hat{p}(\Lambda|H_1)$ of the desired densities. We see that (2.24) implies that only the density under H_0 actually need be obtained. One clear benefit of this approach is eliminating the need to simulate the signal (which may contain uncertain parameters) along with the noise to provide receiver input under H_1. Here we make the distinction between the optimum processor which contains both signal and noise parameters and the simulated input to the processor which may consist of noise alone or signal plus noise depending on whether the true hypothesis is H_0 or H_1. Several important concepts in the calculation of performance for likelihood ratio processors via computer simulation are discussed by Hodgkiss and Nolte (Hodgkiss and Nolte, 1975).

Peterson, Birdsall, and Fox introduced a graphical representation of P_D versus P_F as a function of η known as the ROC (receiver operating characteristic) curve (Peterson, Birdsall, and Fox, 1954). The ROC curve will be the means by which performance of the detection receivers discussed in this paper will be evaluated and compared.

In general, the entire ROC curve is necessary to completely specify performance. However, in the classic SKE in WGN problem (H_1: signal known exactly + white Gaussian noise vs. H_0: white Gaussian noise alone), performance is summarized by a single number known as the detectability index d^2. In this case, the distribution of $\ell(\underline{R}) = \ln\Lambda(\underline{R})$ is Gaussian under H_1 and H_0 with equal variances of $2E/N_0$ and means separated by $2E/N_0$ (E = received signal energy; $N_0/2$ = noise power spectrum height). By definition (Van Trees, 1968)

$$d^2 \triangleq \frac{[E(\ell|H_1) - E(\ell|H_0)]^2}{\mathrm{var}(\ell|H_0)} \tag{2.25}$$

$$= \frac{2E}{N_0}$$

Detection and false alarm probability expressions corresponding to (2.21) and (2.22) are

$$P_D = \mathrm{erfc}_*\left(\frac{\ln \eta}{d} - \frac{d}{2}\right) \tag{2.26}$$

$$P_F = \mathrm{erfc}_*\left(\frac{\ln \eta}{d} + \frac{d}{2}\right) \tag{2.27}$$

where

$$\text{erfc}_* = \int_x^\infty \frac{1}{\sqrt{2\pi}} \exp(-\frac{x^2}{2})\, dx \,. \tag{2.28}$$

The SKE performance curves are illustrated in Figure 2.6 on normal-normal paper. Note that performance increases linearly on the negative diagonal as a function of d.

2.5 Performance and Sufficient Statistics

As mentioned previously, $\Lambda(\underline{R})$ can be viewed as a transformation of random variables. To compute performance for a particular pair of priors $p(\underline{\theta}_1)$ and $p(\underline{\theta}_0)$, the transformation through Λ is completed to yield $p(\Lambda|H_1)$ and $p(\Lambda|H_0)$ from which the ROC can be determined. The expression in (2.20) suggests that a canonical intermediate step in the calculation of performance would be the space formed by the random variables $G_1(\underline{R})/G_0(\underline{R}), \underline{\delta}_1(\underline{R})$, and $\underline{\delta}_0(\underline{R})$. This step occurs just prior to the averaging over $p(\underline{\theta}_1)$ and $p(\underline{\theta}_0)$ and is illustrated in Figure 2.7. Recall that only the transformation conditional to H_0 need be carried through due to the formula in (2.24). Two observations can be made:

(1) Uncertain parameters under H_1 only.
In this special case of the general problem, the observation statistics conditional to H_0 stay the same when new prior knowledge is assumed. Thus, the proposed intermediate step is a valid point from which calculations of performance for any prior knowledge can be started.

(2) Uncertain parameters under H_1 and H_0.
Since the observation statistics under H_0 essentially have $p(\underline{\theta}_0)$ embedded in them, the proposed intermediate step will be valuable only when $p(\underline{\theta}_0)$ remains fixed and $p(\underline{\theta}_1)$ alone is allowed to change.

A major benefit of a natural intermediate step between observation space and the likelihood ratio is found in performance calculation via computer simulation. When only the prior knowledge under H_1 is arbitrary, the intermediate step shown in Figure 2.7 is a valid point from which performance calculations can begin. This step can represent a significant reduction in dimensionality from that of the observation space. The procedure would be to generate observation vectors $\underline{R}$ conditional to H_0. For each $\underline{R}$, the values of $G_1/G_0, \underline{\delta}_1$, and $\underline{\delta}_0$ would be calculated and retained (in effect, generating a discrete version of $p(G_1/G_0, \underline{\delta}_1, \underline{\delta}_0|H_0)$). Then, for every $p(\underline{\theta}_1)$ of interest, the new observation vectors

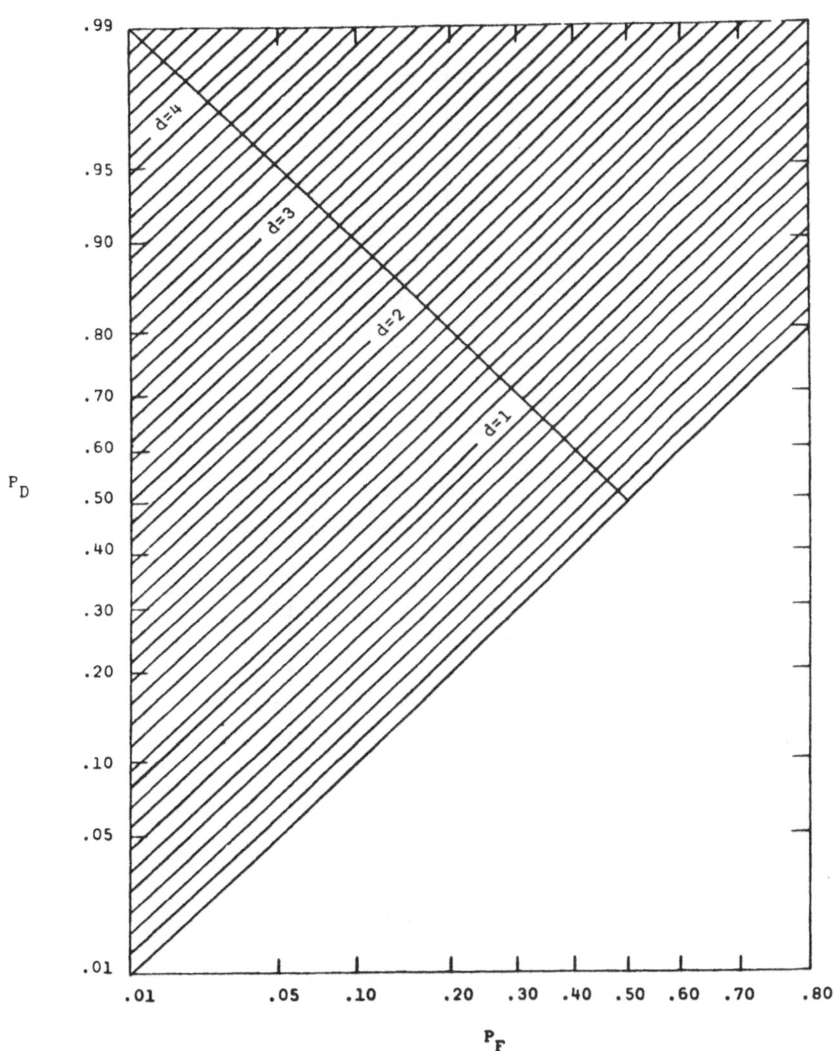

Figure 2.6 Performance of the SKE Processor.

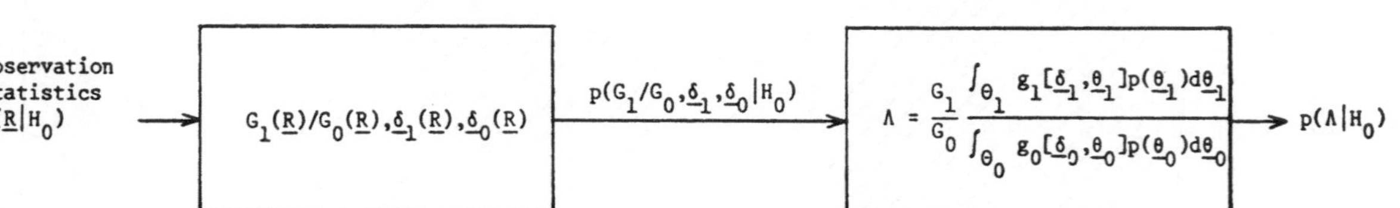

Figure 2.7 Two Step Sequence for the Calculation of Performance.

consisting of $(G_1/G_0, \underline{\delta}_1, \underline{\delta}_0)$ would be used as input to the secondary processor shown in Figure 2.7. The resulting collection of Λ's would be used to form an approximation of $p(\Lambda|H_0)$ and performance calculated.

2.6 An Example: SKEP

As a specific illustration of the concepts discussed to this point, consider the problem of detecting a signal known except for phase (SKEP) in additive white Gaussian noise bandlimited to W Hz. (Roberts, 1965). Since there is a single uncertain parameter and it exists under the H_1 hypothesis only, the likelihood ratio can be written as

$$\Lambda(\underline{R}) = \exp[-d^2/2]\int_\phi \exp[d(x(\underline{R})\cos\phi + y(\underline{R})\sin\phi)]p(\phi)d\phi \underset{D_0}{\overset{D_1}{\gtrless}} \eta \quad (2.29)$$

where

$$d^2 = \frac{2E}{N_0} \quad (2.30)$$

$$E = \frac{1}{2W}\sum_{i=1}^{2WT} s^2(t_i) = \frac{1}{2W}\sum_{i=1}^{2WT} [m(t_i)\cos(\omega t_i - \phi)]^2 \quad (2.31)$$

$$N_0/2 = \text{height of the noise power spectrum} \quad (2.32)$$

$$x(\underline{R}) = \frac{1}{W\sqrt{2EN_0}}\sum_{i=1}^{2WT} R(t_i)m(t_i)\cos\omega t_i \quad (2.33)$$

and

$$y(\underline{R}) = \frac{1}{W\sqrt{2EN_0}}\sum_{i=1}^{2WT} R(t_i)m(t_i)\sin\omega t_i\ . \quad (2.34)$$

When a priori knowledge is chosen from the class of densities

$$p(\phi) = \frac{\exp[A_0 \cos(B_0 - \phi)]}{2\pi\, I_0(A_0)}, \quad 0 \leq \phi \leq 2\pi \quad (2.35)$$

$$= 0 \quad , \text{otherwise}$$

Roberts (Roberts, 1965) has shown that the likelihood ratio becomes

$$\Lambda = \frac{\exp[-d^2/2]\ I_0(A_1)}{I_0(A_0)} \quad (2.36)$$

where

$$A_1^2 = [d \cdot x(\underline{R}) + A_0 \cos B_0]^2 + [d \cdot y(\underline{R}) + A_0 \sin B_0]^2. \qquad (2.37)$$

In addition, a natural by-product of the detector calculations is all the information required to calculate the a posteriori distribution of the uncertain parameter, Ø

$$p(\phi|\underline{R},H_1) = \frac{\Lambda(\underline{R}|\phi)}{\Lambda(\underline{R})} p(\phi)$$

$$= \frac{\exp[A_1 \cos(B_1 - \phi)]}{2\pi I_o(A_1)}, \quad 0 \leq \phi \leq \pi \qquad (2.38)$$

$$= 0 \quad , \text{otherwise}$$

where

$$B_1 = \tan^{-1}\left[\frac{d \cdot y(\underline{R}) + A_o \sin B_o}{d \cdot x(\underline{R}) + A_o \cos B_o}\right]. \qquad (2.39)$$

The processor block diagram is shown in Figure 2.8. Note that the statistic $[x(\underline{R}), y(\underline{R})]^T$ is calculated just prior to the incorporation of the a priori knowledge.

A sequential realization of the SKEP processor is developed as follows (Roberts, 1965). Utilizing (2.8-2.10) for this singly composite hypothesis problem

$$\Lambda(\underline{R}) = \prod_{i=1}^{K} \Lambda(R_i|\underline{R}_{i=1}) \qquad (2.40)$$

where

$$\Lambda(R_i|\underline{R}_{i-1}) = \frac{\exp[-d'^2/2]\, I_o(A_i)}{I_o(A_{i-1})} \qquad (2.41)$$

$$A_i^2 = [d' \cdot x(R_i) + A_{i-1} \cos B_{i-1}]^2 + [d' \cdot y(R_i) + A_{i-1} \sin B_{i-1}]^2 \qquad (2.42)$$

$$B_{i-1} = \tan^{-1}\left[\frac{d' \cdot y(R_{i-1}) + A_{i-2} \sin B_{i-2}}{d' \cdot x(R_{i-1}) + A_{i-2} \cos B_{i-2}}\right]. \qquad (2.43)$$

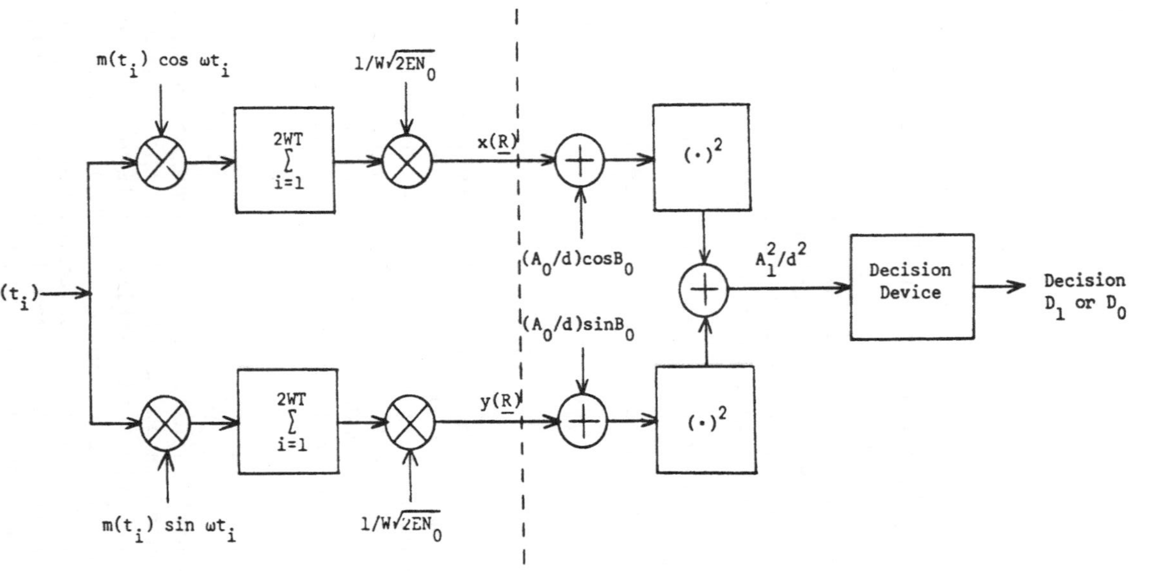

Figure 2.8 SKEP Processor. (Roberts, 1965)

Note the natural learning or adaptive feature which arises out of the sequential Bayesian updating of the a priori knowledge of the uncertain parameter, ϕ.

In the notation of sufficient statistics, (2.29) becomes

$$\Lambda(\underline{R}) = \int_{\Phi} g[\underline{\delta}(\underline{R}),\phi]p(\phi)d\phi \underset{D_0}{\overset{D_1}{\gtrless}} \eta \tag{2.44}$$

where

$$g[\underline{\delta}(\underline{R}),\phi] = \exp[-d^2/2 + d(\delta_{01}(\underline{R})\cos\phi + \delta_{02}(\underline{R})\sin\phi)] \tag{2.45}$$

and

$$\underline{\delta}(\underline{R}) = [\delta_{01}(\underline{R}),\delta_{02}(\underline{R})]^T = [x(\underline{R}),y(\underline{R})]^T. \tag{2.46}$$

Note that the sufficient statistic $\underline{\delta}(\underline{R}) = [x(\underline{R}),y(\underline{R})]^T$ always is calculated regardless of the actual a priori knowledge. In terms of the performance calculations discussed previously, the joint density $p(x(\underline{R}),y(\underline{R})|H_0)$ is a convenient low dimensional starting point from which processor performance can be computed for any $p(\phi)$. Conditional to H_0, $x(R)$ and $y(R)$ are independent Gaussian random variables with zero mean and unit variance. Thus, the desired intermediate density as indicated in Figure 2.9 is

$$\begin{aligned} p(\underline{\delta}|H_0) &= p(x(\underline{R}),y(\underline{R})|H_0) \\ &= N(0,1) \cdot N(0,1) \end{aligned} \tag{2.47}$$

where $N(0,1)$ denotes a Gaussian density with zero mean and unit variance. We may now view $p(\underline{\delta}|H_0)$ as new observation statistics and performance for any prior p can be obtained by completing the indicated transformation of random variables. Note that a significant dimensionality reduction has taken place without losing the ability to compute performance for an arbitrary prior.

3. RESEARCH RESULTS

Specific research results from the area of array processing now will be presented as illustrations of the concepts discussed in the previous section. Of particular interest are two problems involving either signal or noise source location uncertainty: 1) signal known except for direction in Gaussion noise which is independent from sensor to sensor (SKED) and 2) signal known exactly in noise with an additive directional noise component of uncertain direction (SKE in NUD). The first will be used to discuss performance of an estimate and plug structure; the second will be

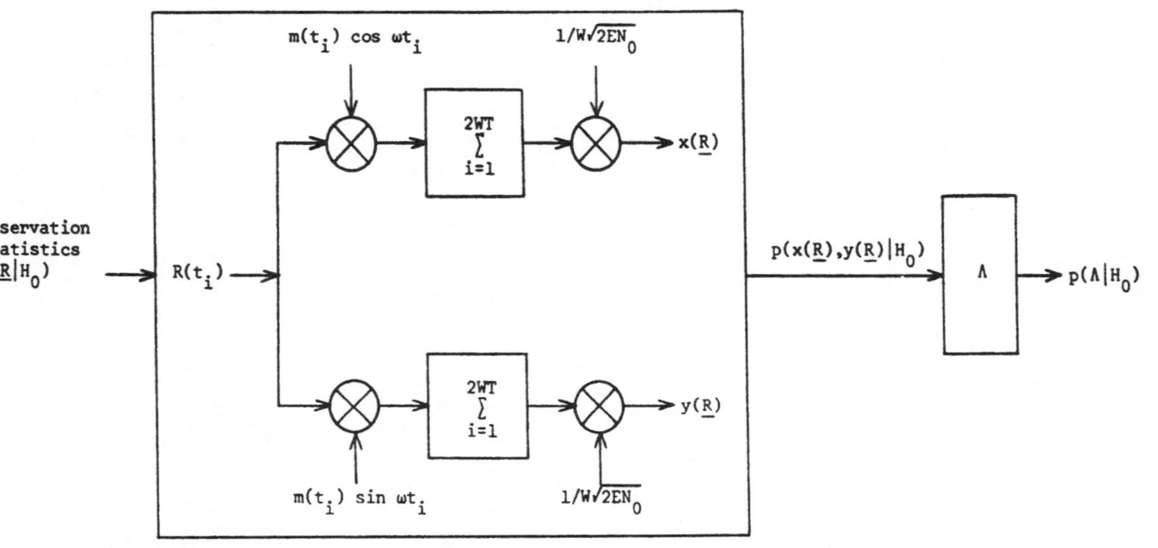

$$\Lambda = \exp[-d^2/2] \int_{\phi} \exp[d\cdot(x(\underline{R})\cos\phi + y(\underline{R})\sin\phi)]p(\phi)d\phi$$

Figure 2.9 Two Step Sequence for SKEP.

used to discuss the time evolving characteristics of an optimal processor implemented sequentially. The presentation will be brief. Details can be found in two papers by Hodgkiss and Nolte (Hodgkiss and Nolte, 1977; Hodgkiss and Nolte, 1978).

An appealing approach to an array detection problem where uncertain parameters exist is first to estimate these parameters, then use the estimated values as if they were known exactly in the processor which is optimal when all parameters are known. When an estimation scheme is used which is generally considered "good," the estimate and plug structure in Figure 2.3 appears to be operating in an optimal fashion. The underlying assumption is that the parameters known solution (i.e. the conditional likelihood ratio, $\Lambda(\underline{R}|\underline{\theta}_1, \underline{\theta}_0)$), provides a canonical "optimum" processor structure for any environment of uncertainty. Should uncertain parameters be present under either or both hypotheses as denoted by the parameter vectors $\underline{\theta}_1$ and $\underline{\theta}_0$, the erroneous assumption is made that with the addition of an estimator appendage, the overall structure remains optimal.

For simplicity of conceptualization and implementation, often there is interest in suboptimal processors. Although the array processor structure in Figure 2.3 is intuitively appealing, it is not clear that piecing together techniques which may be optimal locally (i.e. a "good" estimation scheme joined with a solution optimal when all parameters are known) will yield global optimality when the overall goal is good detection performance. An important question is how closely the performance of such an ad hoc system approaches that of the truly optimal processor.

Here, performance for the SKED problem will be compared between the Bayes optimal detector and the estimate and plug structure where the maximum a posterior (MAP) estimate of signal source location is used. The MAP estimate is determined by utilizing knowledge of the environmental uncertainty with Bayes' rule to form the a posteriori probability density function.

Each array processor sees location uncertainty reflected in terms of an uncertain time delay of the directional signal source between adjacent elements. And, in turn, this corresponds to an uncertain phase delay in the frequency domain where the processing is actually carried out. Thus, location uncertainty will be summarized by the probability density function $p(\omega_0\tau)$. As Figure 3.1 indicates, varying the parameter A_0 from zero to infinity models a wide range of uncertainty from diffuse to highly localized. For all cases where performance is reported here, $L = 8$ and $B_o = 0$. The array elements are assumed one half wavelength apart at frequency $L\omega_0/2\pi$ Hz. Thus, Figure 3.1 corresponds to a location uncertainty over $\pm 90^o$ in physical angle from broadside to the array or $\pm\pi$ in phase at frequency $L\omega_0/2\pi$ Hz.

The MAP estimate corresponds to the global maximum of the density $p(\omega_0\tau|\underline{R}, H_1)$. Since $p(\underline{R}|H_0)$ is independent of the uncertain parameter, an equivalent computation is to determine the global maximum of the conditional likelihood ratio multiplied by the *a priori* knowledge

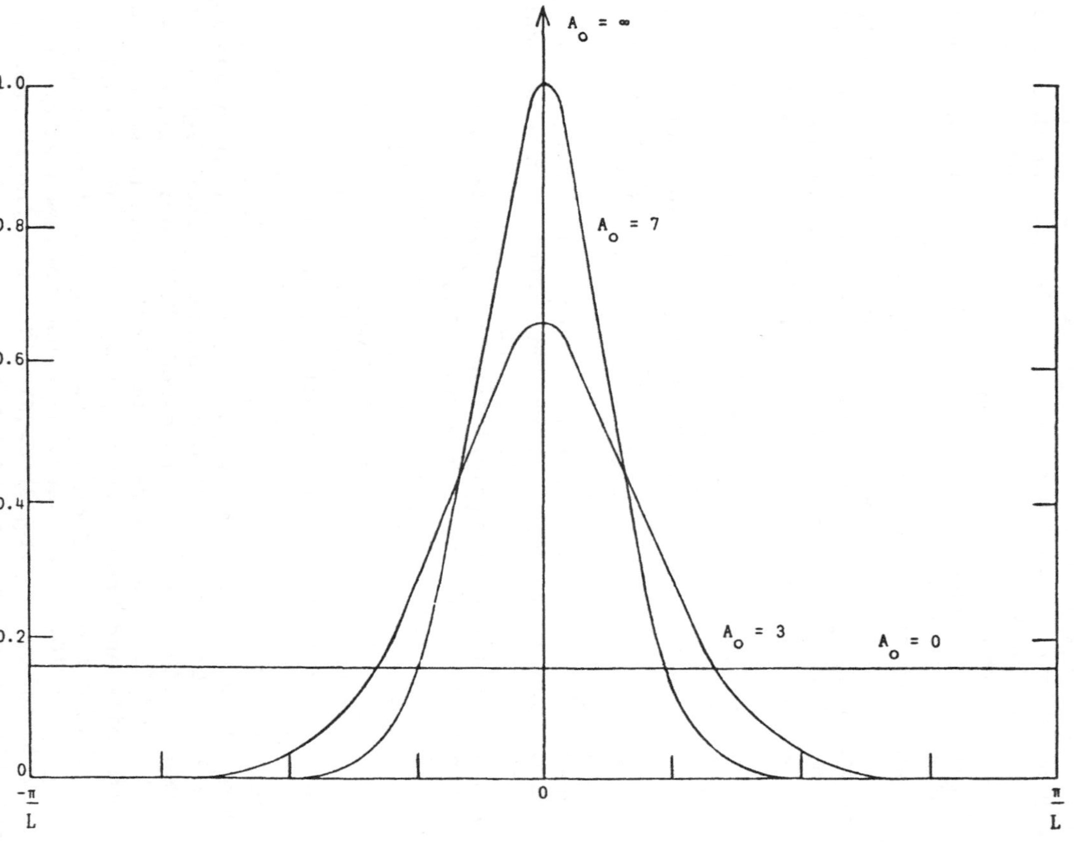

Figure 3.1 $p(\omega_0\tau)$ vs. $\omega_0\tau$; $B_0 = 0$.

$$\max_{\omega_0\tau} \Lambda(\underline{R}|\omega_0\tau)\ p(\omega_0\tau) \rightarrow \omega_0\tau_{MAP}\ .$$

Detection performance in terms of the ROC (receiver operating characteristic) curve is presented in Figure 3.2-3.3 for the two array processor structures. Each ROC is labeled by the A_0 value corresponding to a particular level of environmental uncertainty. The array size investigated was for nine elements as denoted by the parameter K. A single frequency signal is assumed at $L\omega_0/2\pi$ Hz. Its energy over the observation interval is given by $E = 2\ b_0(L)^*b_0(L)$. The Gaussian noise which is independent from sensor to sensor has a power spectrum which is white and bandlimited to $L\omega_0/2\pi$ Hz. Its spectral height is denoted by N.

These performance results point out the necessity of properly incorporating *a priori* knowledge into the array processor design. Even though a "good" estimation scheme has been coupled with the parameters known detector, performance in the global sense has suffered.

In the next problem to be considered, computer simulations of the SKE in NUD processor will be used as illustrations of the natural adaptive feature of an optimal array processor when implemented sequentially. The uncertain parameter in this case is the Gaussian noise source's location reflected in the phase term $\underline{\theta} = \omega_0\tau_n$. The signal, when present, has a known location.

Figures 3.4-3.5 illustrate single computer simulation runs of the processor consisting of 13 iterations each. Since the uncertain parameter exists under both hypotheses, two columns of the sequentially computed *a posteriori* density corresponding to (2.10) under H_0 and H_1 are recorded. Each probability density function displays P (WOTN) = $p(\omega_0\tau_n|R_i,\ldots,R_1)$ versus WOTN = $\omega_0\tau_n$ where ITER = i. The first density in each column (ITER = 0) is the *a priori* knowledge. A uniform prior of $p(\omega_0\tau_n) = 4/\pi$ for $-\pi/8 \leq \omega_0\tau_n < \pi/8$ was assumed in all cases. The "MAX" value indicates the maximum value of the particular density recorded. Between the two columns of densities is a third column which indicates the iteration number ("ITER") and the value of the likelihood ratio ("L") at the completion of that iteration as given by (2.2) utilizing (2.8) - (2.10). The lower left-hand corner of each figure records the incremental signal energy processed ("E"), noise power spectra heights ("N" and "D"), and array size ("K"). The noise spectra are assumed white and bandlimited to $8\omega_0/2\pi$ Hz. The signal consists of a single frequency at $8\omega_0/2\pi$ Hz and location in terms of phase of $\omega_0\tau_s = \pi/16$. Its energy over one incremental observation interval is given by $E = 2b_0(8)^*b_0(8)$. The spacing between the array elements is assumed one half wavelength at the frequency of the signal. The lower right-hand corner of each figure records the *a priori* probability density function parameters ("AO" and "BO",

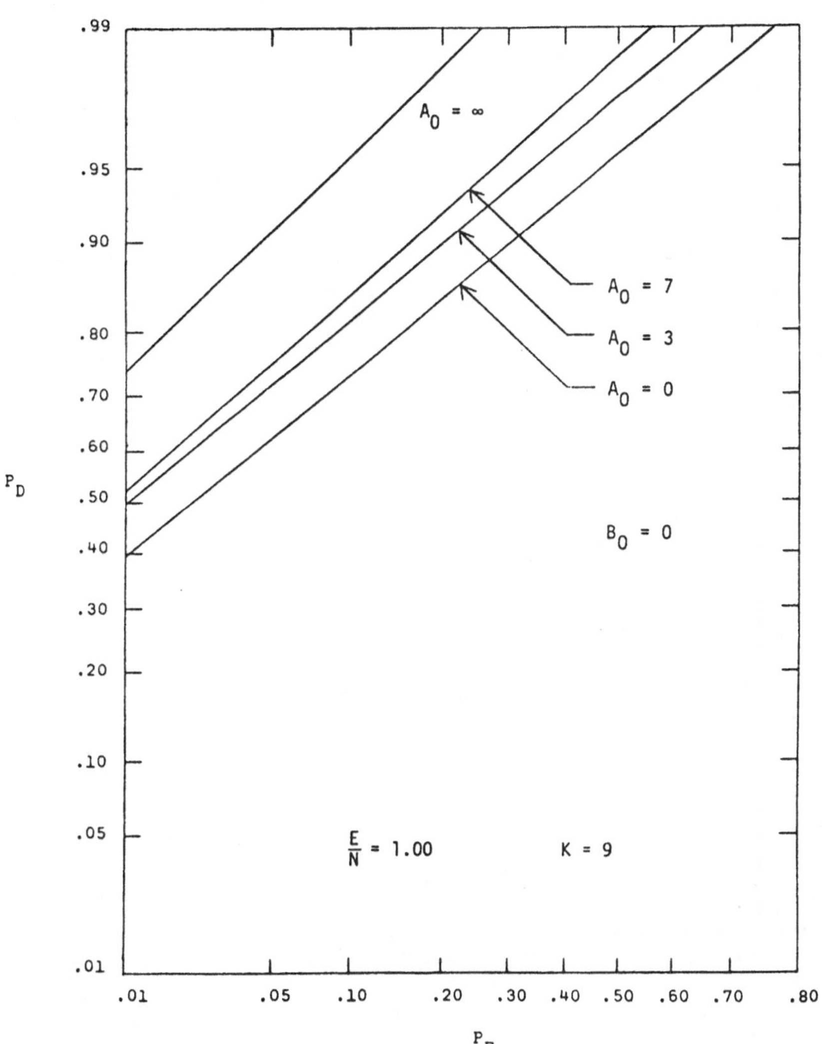

Figure 3.2 Performance of the Optimal SKED Array Processor.

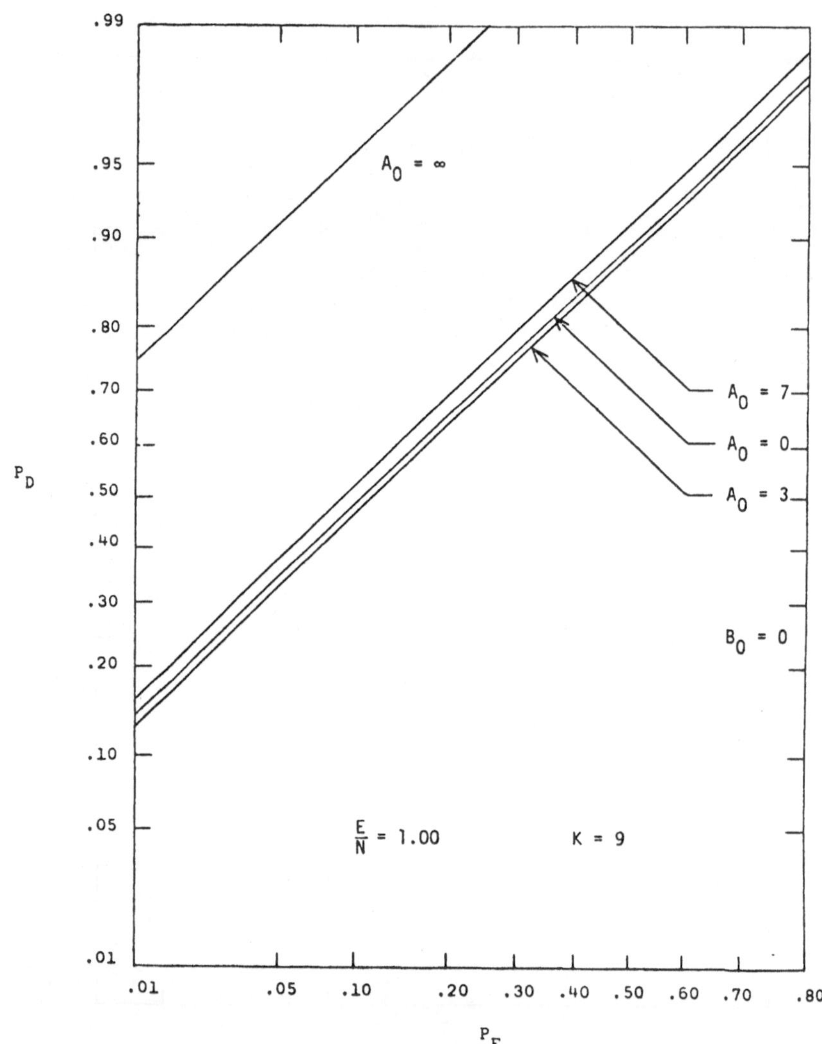

Figure 3.3 Performance of the Suboptimal SKED Array Processor. MAP Estimate

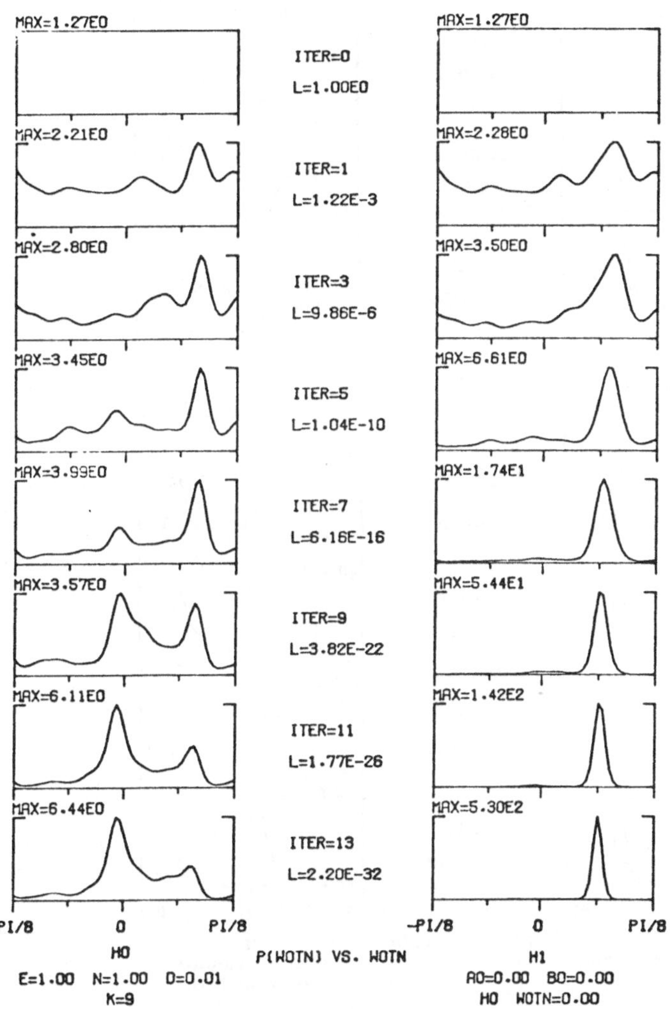

Figure 3.4 Sequential SKE in NUD Simulation.

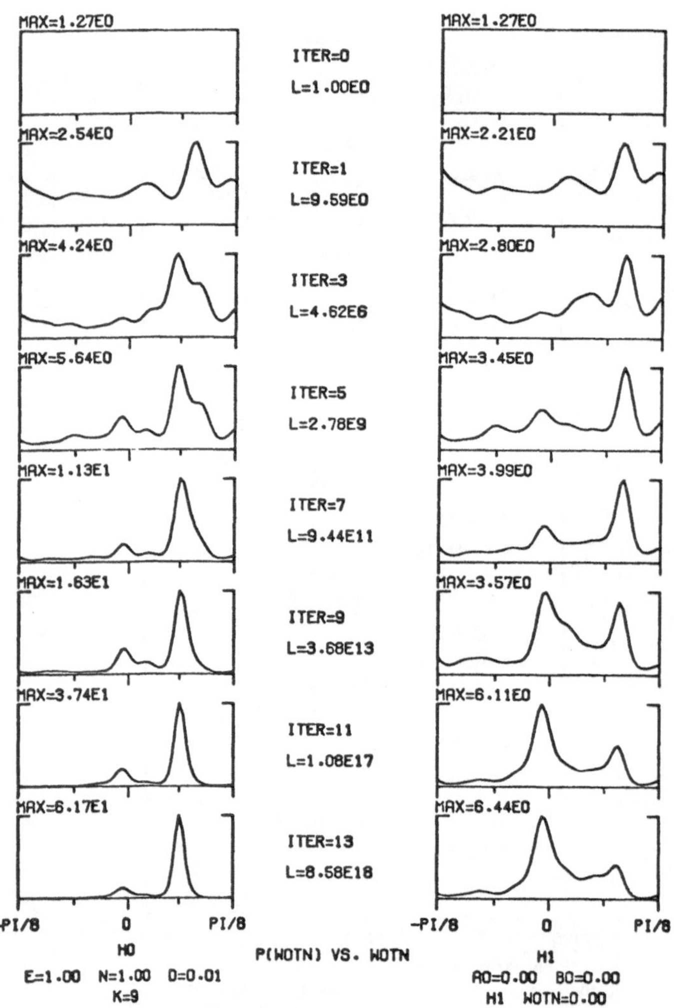

Figure 3.5 Sequential SKE in NUD Simulation.

see Figure 3.1 and let τ represent τ_n), the true hypothesis in force during the simulation ("H1" or "HO"), and the true location of the noise source reflected in terms of phase ("WOTN"). In all cases WOTN = $\omega_0\tau_n = 0$.

The two figures report single processor runs for a nine element array. In the first, the true hypothesis in force is H_0; in the second, H_1. All figures have a single-to-noise ratio of E/N = 1 and the noise-to-noise ratio (D/N) has the value .01. Note that even though the noise-to-noise ratio investigated was relatively low, the optimal processor was able to learn the noise source's location (under the correct hypothesis). As these figures indicate, the optimal array processor exhibits natural learning or adaptive features when implemented sequentially.

4. SUMMARY

This paper has taken a global approach to the processing of information. Essentially, the unprocessed data have been considered as the observables. The processor structure was allowed to evolve freely with the sole restriction being the criterion of optimality. Such an approach has been emphasized since it is not clear that the imposition of a structure on a processor which appears optimal locally (such as the use of a "good" estimation technique in an estimate and plug configuration) will facilitate the overall goal of good signal processing.

Specifically, the processors discussed were to decide if the random processes observed consisted of a signal obscured by noise or noise alone. Any uncertain parameters in the problems considered were treated as random variables and knowledge about them was summarized by _a priori_ probability density functions. The resulting detectors were optimum in the sense of making a least-risk decision.

Once the mathematics of the likelihood ratio has been written, the optimal processor can be implemented in various structures. Four such canonical implementations were discussed: 1) one shot, 2) pseudo estimator, 3) sequential, and 4) two step. The pseudo estimator structure was shown to be the optimal counterpart of an appealing ad hoc approach to processor design where any uncertain parameters are first estimated, then plugged into the parameters known likelihood ratio as if they were known exactly. Although an estimate and plug structure is appealing due to its explicit adaptive characteristics, it was shown that the optimal processor exhibits learning or adaptive features naturally when implemented sequentially.

Specific research results from the area of array processing were presented which involved either signal or noise source location undertainty. It was demonstrated that array processor performance can suffer significantly in an estimate and plug configuration even though an estimator generally considered as "good" (the MAP estimate) is utilized. In addition, computer simulation runs were used to illustrate the Bayesian updating which occurs as an integral part of the sequential structure.

REFERENCES

1. Adams, S. L. and Nolte, L. W., 1975. Bayes Optimum Array Detection of Targets of Known Location. J. Acoust. Soc. Am., Vol. 58, No. 3, pp. 656-669.
2. Adams, S. L. and Nolte, L. W., 1975. Optimum Array Detection in Fluctuating Ambient Noise Fields. J. Acoust. Soc. Am., Vol. 58, No. 3, pp. 670-677.
3. Birdsall, T. G., 1968. "Adaptive Detection Receivers and Reproducing Densities." Technical Report No. 194, Cooley Electronics Laboratory, University of Michigan, Ann Arbor, Michigan.
4. Birdsall, T. G., 1973. "The Theory of Signal Detectability: ROC Curves and Their Character." Technical Report No. 177, Cooley Electronics Laboratory, University of Michigan, Ann Arbor, Michigan.
5. Birdsall, T. G. and Gobien, J. O., 1973. "Sufficient Statistics and Reproducing Densities in Simultaneous Sequential Detection and Estimation". IEEE Transactions on Information Theory, Vol. IT-19, No. 6, pp. 760-768.
6. Hatsell, C. P. and Nolte, L. W., 1974. "A New Generalized Likelihood Ratio Formula." IEEE Trans. on Systems, Man, and Cybernetics, Vol. SMC-4, No. 4, pp. 389-392.
7. Hodgkiss, W. S. and Nolte, L. W., 1975. "A Note on the Calculation of Performance for Likelihood Ratio Processors via Computer Simulation." IEEE Trans. on Information Theory, IT-21, No. 6, pp. 695-698.
8. Hodgkiss, W. S. and Nolte, L. W., 1977. Adaptive Optimum Array Processing. J. Acoust. Soc. Am., Vol. 61, No. 3, pp. 763-775.
9. Hodgkiss, W. S. and Nolte, L. W., 1978. Bayes Optimal Versus Estimate and Plug Array Processor Performance When There is Directional Uncertainty. Accepted for publication in the IEEE Trans. on Aerospace and Electronic Systems.
10. Jaarsma, D., 1969. "The Theory of Signal Detectability: Bayesian Philosophy, Classical Statistics, and the Composite Hypothesis." Technical Report No. 200, Cooley Electronics Laboratory, University of Michigan, Ann Arbor, Michigan.
11. Middleton, D. and Van Meter, D., 1954. "Modern Statistical Approaches to Reception in Communication Theory." IRE Transactions, PGIT-4, pp. 119-145.
12. Nolte, L. W., 1965. "Adaptive Realizations of Optimum Detectors for Synchronous and Sporadic Recurrent Signals in Noise." Technical Report No. 163, Cooley Electronics Laboratory, University of Michigan, Ann Arbor, Michigan.

13. Nolte, L. W., 1966. "An Adaptive Realization of the Optimum Receiver for a Synchronous Recurrent Waveform." IEEE Trans. on Information Theory, Vol. IT-12, No. 1, pp. 78-80.
14. Peterson, W. W., Birdsall, T. G., and Fox, W. C., 1954. "The Theory of Signal Detectability." IRE Transactions, PGIT-4, pp. 161-211.
15. Roberts, R. A., 1965. "On the Detection of a Signal Known Except for Phase." IEEE Trans. on Information Theory, Vol. IT-11, No. 1, pp. 76-82.
16. Van Trees, H. L., 1968. Detection, Estimation, and Modulation Theory: Part I (book). John Wiley and Sons, Inc., New York, N. Y.

FAST COMPUTATION OF QUADRATIC FORMS WITH APPLICATIONS TO SIGNAL DETECTION, CLASSIFICATION, AND ESTIMATION

J. F. Böhme

Forschungsinstitut für Hochfrequenzphysik
5307 Wachtberg-Werthhoven, West Germany °)

ABSTRACT. Noise signals have to be detected, classified, or estimated in many applications. Usual decision and estimation rules require the computation of quadratic forms in sampled data. We investigate methods for efficiently computing the forms considering restrictions to the data and to the matrices defining the forms which are typical in applications, e.g. narrowband data, Toeplitz matrices, and matrices constructed by recursive filters. We discuss the complexity of the computational problems and exact algorithms for different situations. An approximate technique for Toeplitz forms and the behavior of the computational error are studied. Finally, we touch numerical problems and integer computations in residue rings.

1. INTRODUCTION

Common problems in passive sonar, radar, radio astronomy and so on are detection of noise signals disturbed by additive noise, parameter or spectral estimation, and classification of noise signals. The statistical inference is typically done by means of some energy estimations from finite pieces of sampled data. These estimations require the calculation of quadratic forms. The problem we are interested in is not the discussion of inference techniques, this has been done since more than thirty years as documented in [2], [3], [6], [27] etc., but are methods for efficiently computing the quadratic forms, since we assume that the inference problem arises again and again. Especially, we are searching for fast

°) Formerly Institut für Informatik der Universität Bonn

algorithms considering restrictions to the matrices which define the quadratic forms and to the data to be expected in applications. In the next paragraphs, we discuss four examples of inference techniques and the corresponding signal processing tasks and indicate the restrictions. We presume zero mean random data for the sake of simplicity.

a) Detection of weak Gaussian signals additively disturbed by Gaussian noise: The covariance matrix of noise is K_o and $\mu\bar{K}_1$ that of signal. All we know are the matrices K_o and $\bar{K}_1$ and that the signal-to-noise ratio $\mu \geq 0$ is small. A favorable decision rule resulting in a false-alarm probability β is given by a locally best test, cf. [11], which decides "signal" if $q \leq \lambda$ and "no signal" otherwise. Herein, $q = \underline{x}K_o^{-1}\bar{K}_1K_o^{-1}\underline{x}'$ is a quadratic form in measured data $\underline{x}=(x_o,\ldots,x_{n-1})$ written in usual matrix notation and assuming K_o to be regular, and λ is a threshold depending on β, K_o, and $\bar{K}_1$. Since we may assume to know the matrix $A=K_o^{-1}\bar{K}_1K_o^{-1}$ of q as reference data, the problem is the computation of a nonnegative form $\underline{x}A\underline{x}'$ in a vector $\underline{x}$ of measured data.

b) Bayes classification of a Gaussian signal in one of m+1 classes: The class i is described by the positive covariance matrix K_i and by the a priori probability p_i $(i=0,\ldots,m)$. The well-known classification rule, cf. [2], decides "class 0" if $\max_{i=1,\ldots,m}(q_i+e_i) \leq 0$ and otherwise "not class ℓ" if $q_\ell + e_\ell < \max_{i=1,\ldots,m}(q_i+e_i)$, where we have quadratic forms $q_i=\underline{x}(K_o^{-1}-K_i^{-1})\underline{x}'$ in measured data $\underline{x}=(x_o,\ldots,x_{n-1})$ and constants $e_i=\log((p_i/p_o)^2\det(K_oK_i^{-1}))$ $(i=1,\ldots,m)$. Supposing the matrices $A_i=K_o^{-1}-K_i^{-1}$ and the constants e_i to be known, we have to compute m indefinite forms $q_i=\underline{x}A_i\underline{x}'$

$(i=1,\ldots,m)$ in measured data $\underline{x}$ or m+1 positive forms $\underline{x}K_i^{-1}\underline{x}$ followed by m differences if the K_i are known.

c) Estimation of the spectral density of a stationary signal by use of smoothed spectrograms: For a frequency $\lambda, \hat{f}(\lambda) = \Sigma_{t,s=0}^{n-1} v_{t-s} \cos\lambda(t-s)x_t x_s$ is a reasonable point estimate for the spectral density of the stationary process from measured data $x_o,\ldots,x_{n-1}$, where v_t describes a suitable window, cf. [3]. If we need estimates at the points $\lambda_1,\ldots,\lambda_m$, the quadratic forms $q_i=\hat{f}(\lambda_i)$ $(i=1,\ldots,m)$ with known Toeplitz matrices have to be computed for measured data $\underline{x}$.

d) Detection of an unknown signal possibly received at the same time in m-1 receivers which are additively disturbed by mutually-independent and identically-distributed Gaussian noise processes: The only knowledge about the inverse of the covariance matrix of noise is summarized in a nonnegative matrix D. A reasonable detector, cf. [6], compares $\underline{x}.D\underline{x}.'/\Sigma_{i-1}^{m-1}(\underline{x}_i-\underline{x}.)D(\underline{x}_i-\underline{x}.)'$ with a threshold, where $\underline{x}_i=(x_{io},\ldots,x_{i,n-1})$ are the measured data from receiver i and $\underline{x}.=\Sigma_{i=1}^{m-1}\underline{x}_i/(m-1)$ estimates the signal. Apart from the calculations of the sum and the residues one quadratic form with a known, usually simply structurated, nonnegative matrix has to be computed for m vectors of measured data.

To obtain sufficient confidence for the inferences in the examples, the number n of measurements is large. Therefore, the computational problems are serious. In special applications, we may suppose the covariance matrices to be circulant, the data to be narrowband, or the matrices to be constructable from the parameters of a recursive filter. Such a knowledge can drastically reduce the computational complexity of the problems. The topic of this paper is the design of efficient algorithms for simultaneously computing either m quadratic forms for one vector of measured data or one quadratic form for m vectors of measurements taking into consideration restrictions as indicated before. A general investigation of this kind for detection problems is given in [6] which is a source of many ideas for this paper. A short report of similar solutions for classification problems is [5]. Circulant as well as Toeplitz forms can be calculated using the fast Fourier transform as known, cf. [17]. Efficient algorithms for computing one circulant form are proved in [7] assuming narrowband conditions and profitting by interpolation formulas related to those of Kohlenberg [15]. It is shown in [21] that the latter method may be used for approximately computing a Toeplitz form. Schweppes technique [19] can be generalized to use a Kalman filter for calculating quadratic forms for one vector, cf. [4],[5]. The ideas of Strassen and Winograd, cf. [24], [26], [29], on bounds of the computational complexity of problems under discussion have helped to find useful algorithms.

2. SEQUENTIAL COMPUTATION

In the applications mentioned in the last section, the data in a vector of measurements are sequentially put into a signal processor of the receiving system. Let us postulate that the signal processor is realized by a small computer suitable for vector processing in a pipelined manner. We presume a floating point arithmetic to avoid some numerical problems. Because of the architectural limitations, the arithmetic unit works sequentially. We however suppose that operations necessary for transferring the data into the arithmetic etc. can be done independently of the floating point operations.

Now, we are interested in algorithms for fast simultaneously computing quadratic forms by use of a vector processor as described. We then search for sequential algorithms requiring the minimum number of arithmetic operations. Since in real computers an addition or a subtraction can be executed faster than a multiplication, we evaluate the latter by γ operations and all other by one operation, where $1 \leq \gamma \leq 2$. An evaluation of divisions is unnecessary since they can be avoided for the computational problems in question which is discussed in section 3. The number of operations required by a sequential algorithm without branches is an intuitive complexity measure for this algorithm. Similar measures for more general problems have been discussed frequently. The results of an algebraic complexity theory of Strassen [23] may be applied to the problems in this paper and could show ways how to prove bounds for the smallest number of operations needed for a given computational problem. This idea is not followed in this paper. We only cite some results of complexity theorists on bounds for the number of multiplications and on algorithms requiring the minimum number of multiplications. Frequently, these algorithms are also favorable with respect to the number of operations.

Some remarks about parallel computation should be added. Let us assume the vector processor to have p identical arithmetic units which calculate mutually independently. If we have a sequential algorithm requiring ℓ operations, where $\gamma=1$, then the Brent algorithm, cf. [9], for the parallel machine requires at most $4\log_2\ell+2(\ell-1)/p$ steps of operations for the computation. Of course, this bound is not tight if the number p of parallel units is small. Only a little is known how sequential algorithms could be compiled to efficient algorithms for machines with a small number of parallel processors. Vector processors like MAP 200 and AP 120B have a multiplier and an adder working in parallel. Favorable sequential algorithms for quadratic forms as discussed in the following section frequently reduce the number of multiplications and limit the number of additions in front of usual calculation schemes. These algorithms can easily be translated to advantageous programs for those processors. In the following sections, only sequential algorithms are of interest.

3. ONE QUADRATIC FORM FOR ONE VECTOR

We first investigate the problem

$$q = \Sigma_{t,s=0}^{n-1} a_{ts}x_tx_s \tag{1}$$

for a real matrix of reference data a_{ts} $(t,s=0,\ldots,n-1)$ and a real vector of measurements x_t $(t=0,\ldots,n-1)$. Without loss of generality, we may assume $a_{ts}=a_{st}$, since quadratic forms for odd matrices are identically equal to zero. Defining $b_{tt}=a_{tt}$ and $b_{ts}=a_{ts}+a_{st}$ for $t<s$, we have

$$q = \Sigma_{t=0}^{n-1}(\Sigma_{s=t}^{n-1} b_{ts}x_s)x_t. \tag{2}$$

The usual algorithm represented by formula (2) requires $\gamma n(n+3)/2 + n(n+1)/2-1$ operations for a calculation of q not counting the calculation of the b_{ts}. If the matrix (a_{ts}) has the rank r, there exists an upper triangular matrix (d_{ts}) having nonzero elements at most in row 0 to r-1 such that

$$q = \Sigma_{t=0}^{r-1}(\Sigma_{s=t}^{n-1} d_{ts}x_s)^2. \tag{3}$$

A computation by this formula requires $\gamma r(n-r/2+3/2)+r(n-r/2+1/2)-1$ operations. The matrix (d_{ts}) can be calculated a priori using e.g. a variant of the so-called forward solution for systems of linear equations as known.

Formulas (2) and (3) are examples of preconditioning of reference data. Preconditioning means beforehand manipulations with reference data or measured data and then the calculation of q by a suitable formula. Contrary to the reference data, the calculations for preconditioning the measured data must be taken into account for the number of operations required by the algorithm. The Winograd algorithm for inner products, cf. [28], can be applied for calculating the inner products in (2) and (3) which results in an algorithm profitted by preconditioning of reference data as well as measured data. We do not explain this technique in detail but only note the Winograd algorithm for even vector length ℓ

$$\Sigma_{t=1}^{\ell} c_t x_t = \Sigma_{t=1}^{\ell/2} (c_{2t-1}+x_{2t})(c_{2t}+x_{2t-1}) - \Sigma_{t=1}^{\ell/2} c_{2t-1}c_{2t} - \Sigma_{t=1}^{\ell/2} x_{2t-1}x_{2t}.$$

A similar expression yields for odd ℓ. If the numbers c_t describe reference data, the calculation of the second sum can be executed a priori. A calculation of (2) then requires

$$V(n) = \begin{cases} \gamma(n^2+8n)/4+(3n^2+6n+4)/4 & \text{for even } n \\ \gamma(n^2+6n+1)/4+(3n^2+12n-11)/4 & \text{otherwise} \end{cases}$$

operations and that of (3) $V(n)-V(n-r)-1$ operations. Roughly speaking, the number of multiplications is reduced by fifty per cent and the number of additions is increased by the same rate.

If γ is greater than one, the number of operations is decreased.

The rest of this section is a discussion of some results of the complexity theory. Strassen proved the following statements, cf. [24], [26]:

(i) Let k be an infinite field, x_t indeterminates over k, $K=k(x_o,\ldots,x_{n-1})$ the k-field over $k\cup\{x_o,\ldots,x_{n-1}\}$ and similarly $\bar{K}=K(b_{ts}:0\leqq t\leqq s\leqq n-1)$. We consider binary operations +,-,*,/ over $\bar{K}$, scaler multiplications with elements of k, and the elements of $K\cup\{b_{ts}:0\leqq t\leqq s\leqq n-1\}$ as constants. If we count only point operations * and /, a computation of (2) over $\bar{K}$ requires at least $n(n+1)/2$ point operations. We remark that a multiplication $x_t * x_s$ is a constant in the model. This result motivates the conjecture that the usual algorithm is nearly optimal if we do not use preconditioning.

(ii) In case the characteristic of k is greater than two and the b_{ts} are elements of k and we compute (2) over K, then n-p point operations are required, where p is the dimension of a maximum nullspace of the quadratic form. The latter means for full-rank forms that the form is equivalent to $x_1x_2+\ldots+x_{2p-1}x_{2p}+c_{2p+1}x^2_{2p+1}+\ldots+c_nx_n^2$, where the c_t are elements of k. Hence, the computation of q can be executed by n-p point operations between linear forms in the x_t over k except for the (+,-) -operations and the multiplications by scalars from k.

(iii) If we compute q over the corresponding k-rings, i.e. divisions are not allowed, then the number of multiplications required are the same as the corresponding numbers of point operations. Therefore, divisions do not help for fast computations of a quadratic form.

(iv) Finally, Wonograd showed, cf. [29], that preconditioning can reduce the number of multiplications in a computation of (2) at most by fifty per cent. Winograd's inner product algorithm is approximately optimal in this sense.

4. ONE QUADRATIC FORM FOR m VECTORS

The computational problem reads

$$q_i = \Sigma_{t,s=0}^{n=1} a_{ts} x_{it} x_{is} \qquad (i=1,\ldots,m) \tag{4}$$

for a real matrix of reference data a_{ts} $(t,s=0,\ldots,n-1)$ and real vectors $(x_{io},\ldots,x_{i,n-1})$ $(i=1,\ldots,m)$ of measured data. Interpreting (4), we have, similar to (2), a matrix multiplication $y_{it} = \Sigma_{s=t}^{n-1} b_{ts} x_{is}$ $(i=1,\ldots,m;\ t=0,\ldots,n-1)$ between a triangular (nxn)-matrix of reference data and an (mxn)-matrix of measured data followed by m inner products $\Sigma_{t=0}^{n-1} y_{it} x_{it}$ $(i=1,\ldots,m)$. Generally, the computational effort for (4) is m-times that for (2) which can be reduced by preconditioning in the same way as in section 3. If the number m is approximately equal to n, the number of operations is in the order of n^3 for the matrix multiplication and in the order of n^2 for the rest. We can obtain asymptotically a reduction of the effort for the matrix multiplication to the order $n^{2.7}$ when Strassen's matrix multiplication algorithm, cf. [25], is applied. It is not known if this number can be reduced further.

When we have special forms of matrices (a_{ts}) as discussed in the next section, the number of operations for a simultaneous computation of $q_1,\ldots,q_m$ is m-times that number required by the corresponding algorithm for one form for one vector. Therefore, we stop the discussion of this problem. If the matrix (a_{ts}) is diagonal and positive and we are searching for the smallest (or greatest) value q_i, then we can interpret that the nearest neighbor of zero is to be determined. Hence, suitable modifications of the fast algorithms of Friedman et al. [12] and Yunck [30] for finding nearest neighbors may be used with profit.

5.m QUADRATIC FORMS FOR ONE VECTOR

This case is typical for classification problems:

$$q_i = \Sigma_{t,s=0}^{n-1} a_{ts}^i x_t x_s \qquad (i=1,\ldots,m) \tag{5}$$

has to be calculated for matrices (a_{ts}^i) $(i=1,\ldots,m)$ of reference data and a vector of measurements x_t.

Let us first discuss some results of the complexity theory. We assume, similarly to (ii) in section 3, that the a_{ts}^i are elements of k and the computation is executed over $k(x_o,\ldots,x_{n-1})$. We ask for representations

$$q_i = \Sigma_{\ell=1}^{r} c_\ell^i (\Sigma_{t=0}^{n-1} e_{\ell t} x_t)(\Sigma_{s=0}^{n-1} q_{\ell s} x_s) \qquad (i=1,\ldots,m) \tag{6}$$

over k with minimum r. Such an r is then the smallest number of point operations required for a simultaneous computation of q_1, $\ldots,q_m$. Strassen [26] proved that r is the minimum of ranks $rk(a_{ts}^i+d_{ts}^i)$, where the $d_{ts}^i=-d_{st}^i$ are arbitrary elements of k and $rk(h_{ts}^i)$ is the smallest numper p such that

$$\Sigma_{i=1}^{m}\Sigma_{t,s=0}^{n-1} h_{ts}^i x_t y_s z_i = \Sigma_{\ell=1}^{p}(\Sigma_{i=1}^{m} c_\ell^i z_i)(\Sigma_{t=0}^{n-1} e_{\ell t} x_t)(\Sigma_{s=0}^{n-1} g_{\ell s} y_s)$$

is a representation over k with indeterminates x_t, y_s, z_i. If $a_{ts}^i=a_{st}^i$, we have $rk\ (a_{st}^i)/2 \leq r \leq rk(a_{ts}^i) \leq \min(mn,n^2)$. It is however not known how representations (6) with minimum r could be constructed from given elements a_{ts}^i of k.

If we assume to have a representation (6) over the field of real numbers and we count arithmetic operations as above, then a computation of (6) requires $\gamma r(2n+m-1)+r(2n+m-2)$ operations with a possible reduction by use of preconditioning. A different way for finding fast algorithms for (5) is the following. We search for representations (6) with not necessarily minimum r but allowing fast computations of the linear forms $\Sigma_{t=0}^{n-1} e_{\ell t}x_t$, $\Sigma_{s=0}^{n-1} g_{\ell s}x_s$ $(\ell=1,\ldots r)$. The latter may be possible either if many of the numbers $e_{\ell t}$ and $g_{\ell s}$ have only values -1,0, or 1 or if the coefficients $e_{\ell t}$ and $g_{\ell s}$ are complex numbers and the linear forms can be computed by a fast transform algorithm like FFT, cf. [10]. In the following paragraphs, special matrices are discussed which allow fast computations of the quadratic forms via fast linear transform algorithms.

We first presume circulant matrices, i.e. $a^i_{ts}=a_{|t-s|}$ and $a^i_t=a^i_{n-t}$ $(i=1,\ldots,m;\ s,t=0,\ldots,n-1)$. Defining $w_n=\exp(-j2\pi/n)$ and using the discrete Fourier transform, we can write, as known,

$$a^i_t = n^{-1}\Sigma_{f=0}^{n-1} A^i_f w_n^{-ft} \qquad (t=0,\ldots,n-1)$$

and then

$$q_i = \Sigma_{\ell=1}^{r} c^i_\ell \left|\Sigma_{t=0}^{n-1} x_t w_n^{f_\ell t}\right|^2 \qquad (i=1,\ldots,m). \tag{7}$$

Herein, the f_ℓ are those integers f of $\{0,\ldots,[n/2]\}$, where [a] denotes the integer part of a, satisfying $A^i_f \neq 0$ for at least one i, and c^i_ℓ equals $2A^i_{f_\ell}/n$, if $f_\ell \neq 0$ or n/2, and $A^i_{f_\ell}/n$ otherwise. Clearly, $r \leq [n/2+1]$. If the discrete Fourier transform of n real data points requires μ operations, the computation of (7) needs approximately

$\mu+\gamma r(m+3)+r(m+1)$ operations. Using an FFT algorithm, we have $\mu\sim(1.4\gamma+2.1)n\log_2 n$ and if the Winograd Fourier transform algorithm, cf. [20], is applied, then $\mu\sim 2\gamma n+3n\log_2 n$.

Toeplitz matrices which satisfy $a^i_{ts}=a^i_{|t-s|}$ may be handled similarly to circulant matrices. We continue the vector of n data points by n zeros, define $a^i_{2n-t}=a^i_t$ $(t=0,\ldots n-1)$, and can thus use (7), where 2n is substituted for n. This technique is also known to signal processing workers, cf. [17].

Narrowband conditions of the data can help to accelerate the computation of a circulant form as shown in [7]. The assumptions are the following. The n-point Fourier transform of the measured data x_t $(t=0,\ldots,n-1)$ is narrowband with bandwidth α/n, where α is a submultiple of n and less than n. Let $\beta=n/\alpha$ and k be a suitable integer with $0<k<\beta$. Then, interpolations in the sense of Kohlenberg, cf. [15],[13], of the data x_t are possible, and the following formula is correct

$$q_i = \Sigma_{f=0}^{\alpha-1}\{(|z_f|^2+|z_{\alpha-f}|^2)C^i_f + (\mathrm{Im} z_f z_{\alpha-f} - j(|z_f|^2-|z_{\alpha-f}|^2)/2)D^i_f\} \quad (i=1,\ldots,m), \tag{8}$$

where Im denotes the imaginary part,

$$z_f = \Sigma_{s=0}^{\alpha-1}(x_{\beta s}+jx_{\beta s+k})w_\alpha^{sf} \qquad (f=0,\ldots,\alpha-1)$$

and C^i_f and D^i_f can be computed beforehand from the numbers A^i_f as explained in [7]. The computation of (8) requires roughly $3/2\gamma\alpha(m+2)+3/2\alpha(m+1)+\mu'$ operations, where μ' is the number of operations for the α-point Fourier transform of complex data, i.e.

$\mu' \sim (2\gamma+3)\alpha\log_2\alpha$ by use of FFT and $\mu' \sim 4\gamma\alpha+6\alpha\log_2\alpha$ by the Winograd Fourier transform.

If the data $x_o,\ldots,x_{n-1}$ are interpreted as a piece of stationary stochastic narrowband process having a bandwidth not greater than α/n, the data generally do not have a bandlimited n-point Fourier transform. When we use formula (8) for all that, we compute estimates $\hat{q}_i$ of the values q_i $(i=1,\ldots,m)$. Steimel [21] proved that for Gaussian processes and $n\to\infty$ and fixed $\beta=n/\alpha$, the bias $E(\hat{q}_i-q_i)$ and the variance $Var(\hat{q}_i-q_i)$ asymptotically behave as $\max_\ell|c^i_\ell|\log_2 n$ and $\max_\ell(c^i_\ell)^2(\log_2 n)^3$, respectively. If in addition $\max_\ell|c^i_\ell|(\log_2 n)^{3/2}\to 0$ as $n\to\infty$, $\hat{q}_i$ is a consistent estimate of q_i, i.e. $E(\hat{q}_i-q_i)^2\to 0$ as $n\to\infty$. An interesting application of this result is the estimation of the spectral density of a stationary narrowband process. As in example c) of section 1, we use smoothed spectrograms and have to compute a set of Toeplitz forms with matrix elements $a^{-i}_{ts}=\bar{a}^{i}_{|t-s|}$. First, the Toeplitz forms are approximated by circulant forms with matrix elements $a^{i}_{ts}=a^{i}_{|t-s|}$, where $a^{i}_{t}=a^{-i}_{t}+(a^{-i}_{n-t}+a^{-i}_{t})t/n$ $(t=0,\ldots,n-1)$. The circulant forms are estimated by the narrowband algorithm. Thus, $\hat{q}_i$ estimates the corresponding Toeplitz form. Let us assume that the usual conditions for the consistency of the smoothed spectrogram estimator are satisfied, as they are stated e.g. in [3]. If the window length asymptotically grows by a factor $(\log_2 n)^{-2/3}$ slower than allowed in [3], Steimel [21],[22] could prove the consistency of the estimator $\hat{q}_i$ for the spectral density at the corresponding frequency. Moreover, the asymptotic order of the expected quadratic error induced by the narrowband algorithm is approximately the same as that of the smoothed spectrogram estimator. Recent simulations with approximately band limited linear processes and corresponding estimations show the applicability of the method.

Whereas the algorithms of the beforegoing paragraphs allow asymptotically for large n a speedup of the computation in the order of $n/\log_2 n$, the last method discussed in this section results in a speedup in the order of n. The algorithm is investigated in [4] and [5] and generalizes an idea of Schweppe [19]. We first assume the matrices (a^i_{ts}) to be positive. Hence, it is shown in [5] that

$$q_i = \Sigma_{t=0}^{n-1} (x_t^i - \hat{x}_t^i)^2 / e_t^i \qquad (i=1,\ldots,m), \tag{9}$$

where $\hat{x}^i_t$ denotes the conditional expectation of the component x^i_t of a zero mean Gaussian random vector $(x^i_o,\ldots,x^i_{n-1})$ with a covariance matrix which is the inverse of the matrix (a^i_{ts}) when observations $x^i_{t-1}=x_{t-1},\ldots,x^i_o=x_o$ are given. The number e^i_t is the expectation of $(x^i_t-\hat{x}^i_t)^2$. Formula (9) can efficiently be calculated if there exists a fast algorithm for computing the predictions $\hat{x}^i_t$. The method presumes the existence of a recursive filter with a white noise vector process as input and m output processes having the property that a piece $(x^i_o,\ldots,x^i_{n-1})$ of the i-th output is a Gaussian random vector as before. By such a model, we can design a Kalman filter [14] which computes the predictions $\hat{x}^i_t$. If the recursive filter has r states, the Kalman filter computes $\hat{z}_o=0$,

$$\hat{\underline{x}}_t = \hat{z}_t V_t \qquad (t=0,\ldots,n-1),$$

$$\hat{z}_t = \hat{z}_{t-1}\hat{F}_t + x_{t-1}\hat{d}_{t-1} \qquad (t=1,\ldots,n-1). \tag{10}$$

Herein, $\hat{\underline{x}}_t = (\hat{x}_t^1,\ldots,\hat{x}_t^m)$, $\hat{z}_t$ is an r-vector, V_t an (rxm)-matrix, $\hat{F}_t$ an (rxr)-matrix, and $\hat{d}_t$ an r-vector. The parameters in (10) can be determined a priori from the parameters of the recursive filter and the covariance matrix of z_o, cf. [5]. The algorithm (10) and (9) requires less than $(n-1)(\gamma(r^2+r(m+1)+2)+r^2+m(r+1))$ operations. If the number r of states is small in comparison with n, the algorithm is efficient. The applicability of the method depends on the possibility for constructing a filter with a small number of states from the matrices defining the forms. For instance in example b) of section 1, the matrices of the quadratic forms are differences $K_o^{-1}-K_i^{-1}$ of positive matrices. If we know a filter model for the matrices K_i^{-1} (i=1,...,m), the forms $\bar{q}_i=\underline{x}K_i^{-1}\underline{x}'$ are calculated by (9), and then $q_i=\bar{q}_o-\bar{q}_i$ (i=1,...,m). This technique generalizes that in [19]. Even if a model for K_o and $\bar{K}_1$ is known in example a) of section 1, we do not know how to construct a filter for $K_o^{-1}\bar{K}_1K_o^{-1}$ from this knowledge. Generally, we first try to represent the matrices as linear combinations of a small set of positive matrices. If we have no filter model for the positive matrices from knowledge about the physical problem, the model must be constructed. Space does not permit a discussion of this problem. Some ideas for a solution may be found in [4] and [8].

7. NUMERICAL PROBLEMS AND INTEGER CALCULATIONS

Algorithms for calculating quadratic forms in floating point requiring less multiplications than the usual algorithms may have a poor numerical stability. Miller [16] investigated the problem of finding algorithms which satisfy some strong stability conditions and minimize the number of multiplications. He found for calculations of sets of inner products that the usual algorithm is stable and that faster algorithms using preconditioning are not. Normalizing can help to stabilize the algorithms, however this requires additional multiplications. A corresponding study for quadratic forms should still be done. We can however state that avoiding pre-conditioning for inner products will stabilize the algorithms for quadratic forms. Floating point computations of the fast Fourier transform may be considered to be stable in a slightly different sense, cf. [18], [17]. Therefore, calculations of circulant and Toeplitz forms using a suitable fast Fourier transform algorithm as described in section 6 are not problematic.

In some applications, only few bits of measurements and of the matrix elements are known. Suitable scaling and interpreting the numbers as integers motivates integer calculations of the quadratic forms. Numerical problems can then be neglected. The question is if there are efficient algorithms. General results are not known. The following paragraph introduces a solution for circulant forms.

Formula (5) for circulant forms can be written

$$q_i = \Sigma_{\ell=0}^{n-1} a_\ell^i \, \Sigma_{t=o}^{n-1} x_t \bar{x}_{\ell-t} \qquad (i=1,\ldots,m), \qquad (11)$$

where $\bar{x}_t = x_{n-1-t}$ and the second sum is a periodic convolution. We assume the numbers a_ℓ^i and x_t from the set $\{-[(N-1)/2],\ldots,0,\ldots,[N/2]\}$, where N is a suitable positive integer. The calculation of (11) is done mod(N), i.e. in the residue ring R_N with elements $0,1,\ldots,N$. For the calculation of the convolution, Agrarwal and Burrus [1] showed the following theorem. If and only if z is an element of R_N and n is the smallest positive integer with $z^n = 1 \bmod(N)$ and n^{-1} is an element of R_N, then the integer transform

$$x_s = \Sigma_{t=0}^{n-1} x_t z^{ts} \bmod(N) \qquad (s=0,\ldots,n-1) \qquad (12)$$

has an inverse

$$x_t = n^{-1}\Sigma_{s=0}^{n-1} X_s z^{-st} \mod(N) \qquad (t=0,\ldots,n-1)$$

and the convolution property

$$\Sigma_{t=0}^{n-1} x_t y_{\ell-t} = n^{-1}\Sigma_{s=0}^{n-1} X_s Y_s Z^{-s\ell} \mod(N) \quad (\ell=0,\ldots,n-1),$$

where x_t, y_t are from R_N and Y_S is the transform of y_t. Clearly, if we have the convolution property of an integer transform, (11) is transformed to

$$q_i = n^{-1}\Sigma_{s=0}^{n-1} A^i_{n-s} z^{n-s} X_s X_{n-s} \mod(N) \quad (i=1,\ldots,m), \qquad (13)$$

where A^i_s is the transform of a^i_t. If we had an efficient algorithm for the integer transform, formula (13) could be used for designing an algorithm which is very similar to (7). It is known that there are FFT-type algorithms if n is decomposable in many factors. If z=2, the multiplications in (12) are only shifts and may be counted as additions. An interesting question arises, namely that for a choice of N,n, and z satisfying the conditions of the theorem. If N is the Fermat number $2^{2^h}+1$ for h=5 or h=6 and z=2, we have $n=2^{h+1}$ and a word length of 2^h+1 bits, for example, h=5, n=64 and 33 bits. Of course, a vector length of 64 is poor. A few tricks are described in the literature, cf. [1], how longer vectors could be handled using the same word length, but we shall not discuss this point. The applicability of the method for convolutions is extensively investigated in [1]. The results are also correct for the problem under question. We only mention that a calculation mod(N) requires a special hardware which depends on N. Clearly, this is inconsistent with any demands for flexibility of a system.

REFERENCES

1. Agrarwal, R. C. and Burrus, C. S.: "Number theoretic transforms to implement fast digital convolutions." Proc. IEEE 63, 1975, 550-560.
2. Anderson, T. W.: An Introduction to Multivariate Statistical Analysis. J. Wiley, New York, 1958.
3. --: The Statistical Analysis of Time Series. J. Wiley, New York, 1971.
4. Bohme, J. F.: "Computation of quadratic forms by means of prediction." Proc. IEEE Conf. on Pattern Recognition and Image Processing, Chicago, Ill., 1978.
5. --: "Computation of quadratic forms used in statistical pattern recognition." Proc. SITEL-ULG Seminar on Pattern Recognition, Liège, Belgium, 1977, 9.7.1-4.
6. --: (in German) Detektion mit Rechnern. Habilitationsschrift. Informatik Berichte 15, Institut für Informatik der Universität Bonn, 1977.
7. --: "Sampling of narrowband sequences with application to digital detection of stochastic signals." Proc. IRE Conf. on Digital Processing of Signals in Communications, Loughborough, Leic., 1977, 367-374.
8. Box, G.E.P. and Jenkins, G.M.: Time Series Analysis: Forecasting and Control. Holden-Day, San Francisco, 1976.
9. Brent, P.R.: "The parallel evaluation of arithmetic expressions in logarithmic time." In Complexity of Sequential and Parallel Numeric Algorithms, ed. J.F. Traub, Academic Press, New York, 1973, 83-102.
10. Cooley, J.M. and Tukey, J.W.: "An algorithm for the machine calculation of complex Fourier series." Math. Comp. 19, 1965, 297-301.
11. Ferguson, T.S.: Mathematical Statistics: A Decision Theoretic Approach. Academic Press, New York, 1967.
12. Friedman, J.H. et al.: "An algorithm for finding nearest neighbors."IEEE Trans. C 24, 1975, 1000-1006.
13. Grace, O.D. and Pit, S.P.: "Quadrature sampling of high-frequency waveforms." J. Acoust. Soc. Am. 44, 1968, 1453-1454.
14. Kalman, R. E.: "A new approach to linear filtering and prediction problems." Trans. ASME, J. Basic Eng. 82, 1960, 35-45.
15. Kohlenberg, A.: "Exact interpolation of band limited functions." J. Appl. Physics 24, 1953, 1432-1436.
16. Miller, W.: "Computational complexity and numerical stability. SIAM J. Comput. 4, 1975, 97-107.
17. Oppenheim, A. V. and Schafer, R. W.: Digital Signal Processing. Prentice Hall, Englewood Cliffs, N. J., 1975.
18. Ramos, G.: "Roundoff error analysis of the fast Fourier transform." Math. Comp. 25, 1971, 757-768.
19. Schweppe, F. C.: "Evaluation of likelihood functions for Gaussian signals." IEEE Trans. IT 11, 1965, 61-70.

20. Silverman, H.F.: "An introduction to programming the Winograd Fourier transform." IEEE Trans. ASSP 25, 1977, 152-165.
21. Steimel, U.: "Fast computation of Toeplitz forms under narrowband conditions." Proc. SITEL-ULG Seminar on Pattern Recognition, Liège, Belgium, 1977, 9.8.1-4.
22. --: "Fast computation of Toeplitz forms under narrowband conditions with applications to statistical signal processing. Submitted to Signal Processing, 1978.
23. Strassen, V.: (in German) Berechnung und Programm. Teil I und II. Acta Informatica 1, 1972, 320-335, und 2, 1973, 64-79.
24. --: "Evaluation of rational function." In Complexity of Computer Computations, eds. R.E. Miller et al., Plenum Press, New York, 1972, 1-10.
25. --: "Gaussian elimination is not optimal." Numerische Mathematik 13, 1969, 354-356.
26. --: (in German) Vermeidung von Divisionen. Crelle J. für die Reine und Angewandte Mathematik 264, 1973, 184-202.
27. van Trees, H. L.: Detection, Estimation, and Modulation Theory, Part III. J. Wiley, New York, 1971.
28. Winograd, S.: "A new algorithm for inner products." IEEE Trans. C 17, 1968, 693-694.
29. --: "On the number of multiplications necessary to compute certain functions." Comm. Pure and Appl. Math. 23, 1970, 165-179.
30. Yunck, T. P.: "A technique to identify nearest neighbors." IEEE Trans. SMC 6, 1976, 678-683.

PATTERN RECOGNITION WITH A HYBRID PROCESSOR INCLUDING AN INCOHERENT/COHERENT LIGHT CONVERTER

L. Brock-Nannestad

Danish Defence Research Establishment
Østerbrogades Kaserne, DK-2100 Copenhagen Ø, Denmark

ABSTRACT. The hybrid processor considered in this paper consists of a digital minicomputer HP-2100 and a coherent optical processor. An incoherent/coherent light converter of the type Phototitus is optically coupled to a TV-monitor, which acts as the input device for the optical system. The TV-monitor receives its signals either from a TV-camera - for the processing of real images - or from the computer for the processing of computer generated patterns. The output of the optical processor is scanned by another TV-camera, digitized and used as input for the computer. The paper gives a description of the various elements of the complete system and presents examples of the processing of patterns for recognition purposes.

1. INTRODUCTION

The literature of image processing and pattern recognition shows the existence of two schools or lines of thought. One of these the optical school, started at a time when the laser had become available, whereas the minicomputer was not yet born and processing time on the large central computers was hard to obtain and very expensive. The most significant event in the life of the optical school is the publication of VANDER LUGT's famous paper on complex spatial filtering [1] in 1964. The years following this publication were very productive and many ingenuous optical methods and processors have been described.

The inventiveness of the optical school is still flourishing and it is still making very important contributions in the field of image processing and pattern recognition.

However, since the optical systems "only" can treat 2D brightness distributions, a large amount of pattern recognition problems in the statistical and multidimensional field must be solved by digital means. The availability of minicomputers and a more easy access to large computers created the conditions for a tremendous growth in digital methods for treating problems in pattern recognition with any degree of complexity. The digital school claims, correctly, that, given the right size of computer and enough computer time, every problem including those treated by the optical school can be solved with any desired accuracy.

It has been recognized for some time, however, that it might be advantageous to combine the two lines of thought, and this has led to what is known as a hybrid processor. In section II an example of a general hybrid processor is given. Section III describes some of the components used in this processor and in section IV its use is illustrated by giving examples of pattern recognition problems, which have been solved.

This paper will not be complete with respect to mentioning all the possibilities which exist in this field. Therefore, the list of references should not be taken as an indication of what is available, but rather that the literature is so numerous that only those papers, which have a direct relation to the text of this paper have been mentioned. The reader interested in a recent account of optical computing is referred to [2] and [13] with special reference to G.W. Strokes Keynote Address on "A New Assessment op Optical and Digital Image Processing for Real-World-Applications" [2]. This paper and others in [2] and [13] contain lists of useful literature on the subject.

2. A GENERAL HYBRID PROCESSOR

The term "hybrid computer" or "hybrid processor" was originally coined for describing the combination of a digital and an analog computer, where the last mentioned consisted of operational amplifiers, resistors and capacitors. The motivation for combining these two principles was, that at that time the analogue computer was better suited for solving differential equations than the digital computer was. Today, this kind of combination is rather rare and the term "hybrid processor" is more often used to describe combinations of digital and optical systems.

As before, this combination tries to benefit from the special characteristics of the two systems. The digital system has all the advantages in forms of actual calculations, comparisons, sortings, etc., but since the instructions are carried out one at the time, the system has basically a one channel character, which, in combination with a limited core memory, can introduce considerable processing time. The optical system is basically a multichannel system in which the processing is performed at the speed of light. The advantage of the system is, therefore, that it has a large time-bandwidth product, which can be utilized in various ways. Optical systems whould be considered as operating in analogue form, although sometimes digital information is processed. A number of disadvantages can be listed for such systems since optical components are imperfect and have limitations. However, the main problems with optical systems can be found in the input and output devices.

A very general block diagram of a hybrid processor is shown in fig. 1. Some of the general characteristics mentioned above are indicated by full line single arrows for the digital signals and multiple arrows for the optical signals. Some of the peripheral equipment needed, such as external memories and display are shown, but the figure is not complete in this respect. The figure can, however, serve as a functional diagram of practically all types of hybrid processors. Even an extremely simple processor of the type described by Häusler and Lohmann [3] can be modelled as shown, as an example, in fig. 2. If, in fig. 1, the digital system is replaced by a single switch, linking the optical-to-digital converter directly to the digital-to-optical converter and if the optical system is replaced by a beamsplitter and a mirror the equivalence between the two figures becomes evident. Fig. 2 can be simplified further, as done by Häusler and Lohmann, by moving the camera to the place of the mirror and rotating it 90°.

In addition to the full line arrows indicating the flow of information, fig. 1 shows a few dashed lines with arrows. These lines indicate the flow of control signals, which can change the characteristics of the converters and the optical system. This feature is very convenient for the interactive operation of the system.

As mentioned, many of the problems encountered in a hybrid processor can be found in the input- and output devices of the optical systems. Some general remarks can be made about these devices, which in fig. 1 have been indicated by seperate boxes. Normally the problems are different for the digital/optical-to-optical converter and the optical-to-digital converter, but in both cases there are limitations in terms of spatial resolution, gray-level resolution (dynamic range) and signal-to-noise ratio. In special situations there are trade off possibilities resulting in converters tailor made for the special application, but in case of a more general purpose hybrid processor the devices are limited. However, in this case some of the limitations introduced by the converters and the optical system can be corrected for by the digital system. Some of these points become evident from the following description of the complete hybrid processor constructed at the DDRE.

The general layout of the hybrid processor of the DDRE is shown in fig. 3. The equivalence with fig. 1 is evident from the various functional blocks outlined in fig. 3. The optical system is mounted on small optical benches which are placed on a stable

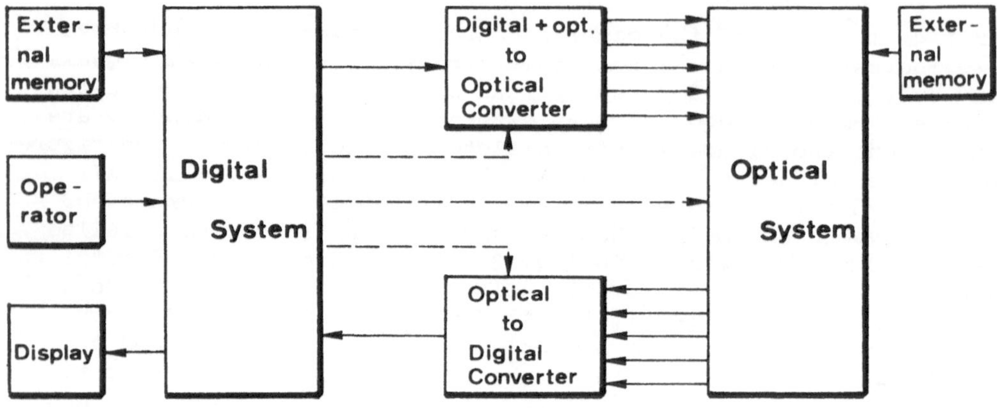

Fig. 1 Block diagram of a hybrid processor

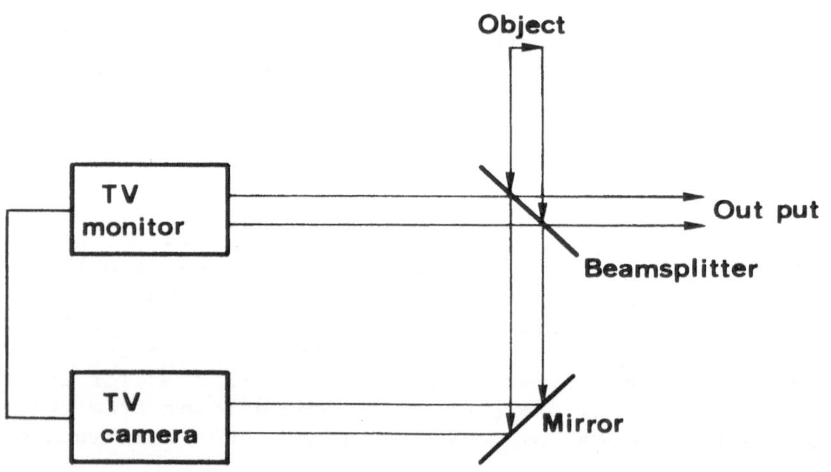

Fig. 2 Hybrid processor with electro-optical feedback [3].

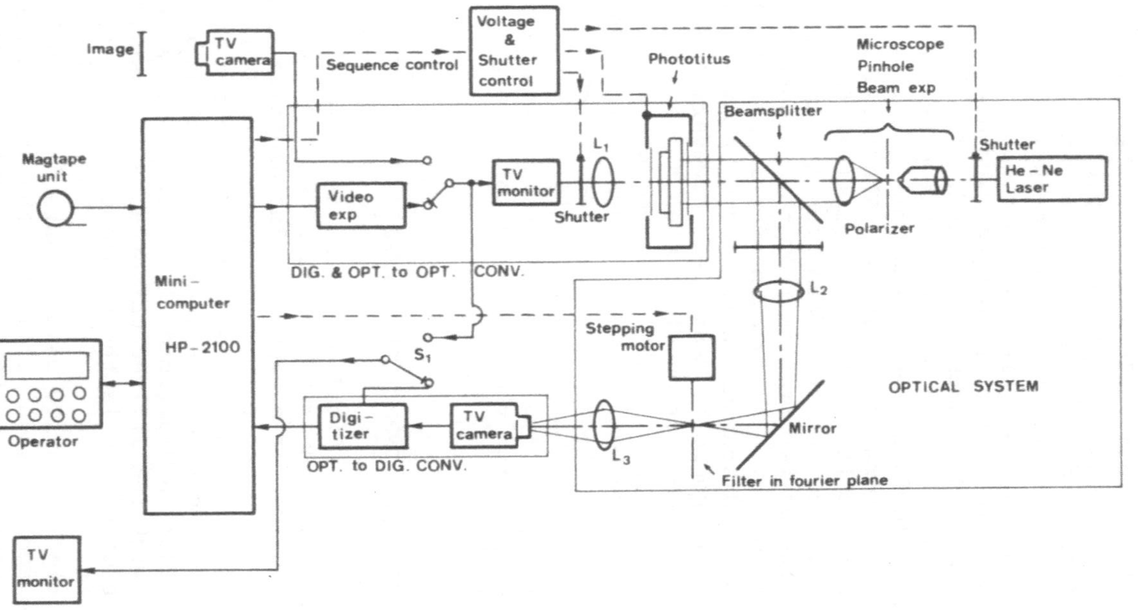

Fig. 3 General layout of DDRE's hybrid processor

optical table together with the input- and output devices. The photographs in fig. 4 give an impression of the set up.

The optical system outlined in fig. 3 is a modified classical VANDER LUGT filtering system. This system is just one example of a number of optical systems, which can be put in use. The optical benches on top of the table make it rather easy to change to other systems, but the one shown here has proved to be useful in solving a number of problems, which will be shown later.

Apart from the way in which the optical information is inserted into the optical system, the set up is classical with He-Ne-laser, pinhole filter and a beam expander which generates a parallel polarized coherent lightbeam. The lightbeam is modulated by the information to be processed and passes a polarizer. A lens L_2 with suitable focal length and position is used to perform the 2D Fourier transform of the input. The focal plane of L_2 is the Fourier plane and it is at this place that the filtering operation is performed. Another transform lens L_3 can then be used to transform the filtered information from the frequency domain back into the spatial domain. The details of this part of the optical system will not be discussed in this paper because it is rather conventional. It should be mentioned, however, that the same set up is used for the generation of complex filters by replacing the filter in the Fourier plane with a holographic film and by introducing a reference beam from the laser (not shown in fig. 3). Another simple modification will be mentioned at a later stage. The examples given above will be sufficient to indicate that the optical part of the hybrid processor is quite flexible although it has to comply with the input and output requirements of the two converters.

3. THE OUTPUT AND INPUT CONVERTERS

The output of the optical system is fed into the optical-to-digital converter. This converter consists of a TV-camera and a digitizer, both standard type of equipment. The digitizer, a CVI model 270, will accept video information from most standard video cameras [4]. The image is resolved with 512 elements on the X-axis and 580 (= 2 x 290) elements on the Y-axis for a standard European TV-image.The video is digitized with 8 bits. This resolution is in accordance with the limited performance of most standard TV cameras, which of course is the bottleneck of the optical-to-digital converter.

Some significant TV camera characteristics, which should be taken into account when estimating the performance of the output converter, are given in [4,5]. Obviously the camera introduces limitations in the "quality" of the image being stored and in some instances digital corrections can be introduced. However, the low pass filtering action, resulting from the cameras lower spatial resolution cannot be removed digitally. In cases where the higher

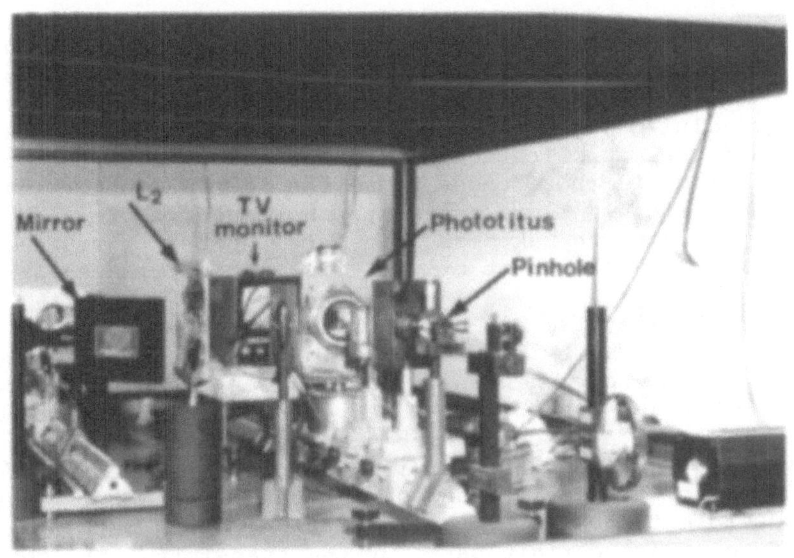

a. General view

b. Details

Fig. 4 Photograph of the optical system

resolution of the optical system has to be maintained through the output converter special tubes must be employed. DDRE experience seems to indicate that very many problems can be solved also with limited resolution by proper choice of image size, which is a question of the magnification in the optical part of the processor. This means that in those cases where the information of interest is concentrated in a limited area of the image, a proper choice of the magnification can make use of a large number of picture elements (pixels). This point of view will be illustrated with the following example.

The optical system can be modified by removing L_3 and the filter in fig. 3 and by placing the light sensitive target of the TV camera in the Fourier plane. In this way the power spectrum of the 2 D Fourier transform can be recorded in digital form. Fig. 5 shows such a spectrum within a frame corresponding to standard TV size. In order to describe the complete spectrum only one half of the display is needed, and thus by enlarging the image it is possible to obtain better resolution. This method can be utilized to select certain parts of the transform. The actual magnification obtained is determined by the focal length and the position of L_2 in fig. 3.

One special characteristic of the camera requires mentioning, and that is the blooming effect. Since the dynamic range of the camera is limited, the lightlevel can be so large that it overloads the camera, or its electronics. For cameras with automatic gain control (AGC) blooming will normally take place through overload of the electronics and could be overcome by switching off the AGC; however, in an optical system like fig. 3 blooming will normally take place at the DC-peak. Since its level normally is of no interest it is advisable to introduce a DC-stop at a proper place in front of the camera.

The process of the optical-to-digital conversion can be monitored on a special output of the digitizer and with the switch S_1 in the position shown. This monitoring process is quite useful because it permits the operator to make sure that the image being digitized has the required quality. The control of the digitization and the transfer of data is performed by the computer through proper interfaces and fairly simple software. In the present case the minicomputer is of type HP-2100. A complete data aquisition, management and processing system (FTSS) has been developed for this computer. Details are given in [6,7], but with the purpose of giving an impression of the capabilities of the system, a block diagram from [7] is reproduced here as fig. 6.

The system consists of a number of modules arranged in various groups. The communication between the modules is controlled by the MAIN-module, which is not shown in fig. 6. It gives also access to the rest of the FTSS software. Each box represents a module and has been given a name indicating its use. In addition for the purpose of clarity the actual application is also mentioned. Only a few additional remarks are required in order to show the performance of the system.

Fig. 5 Power Spectrum of a 2D brightness distribution

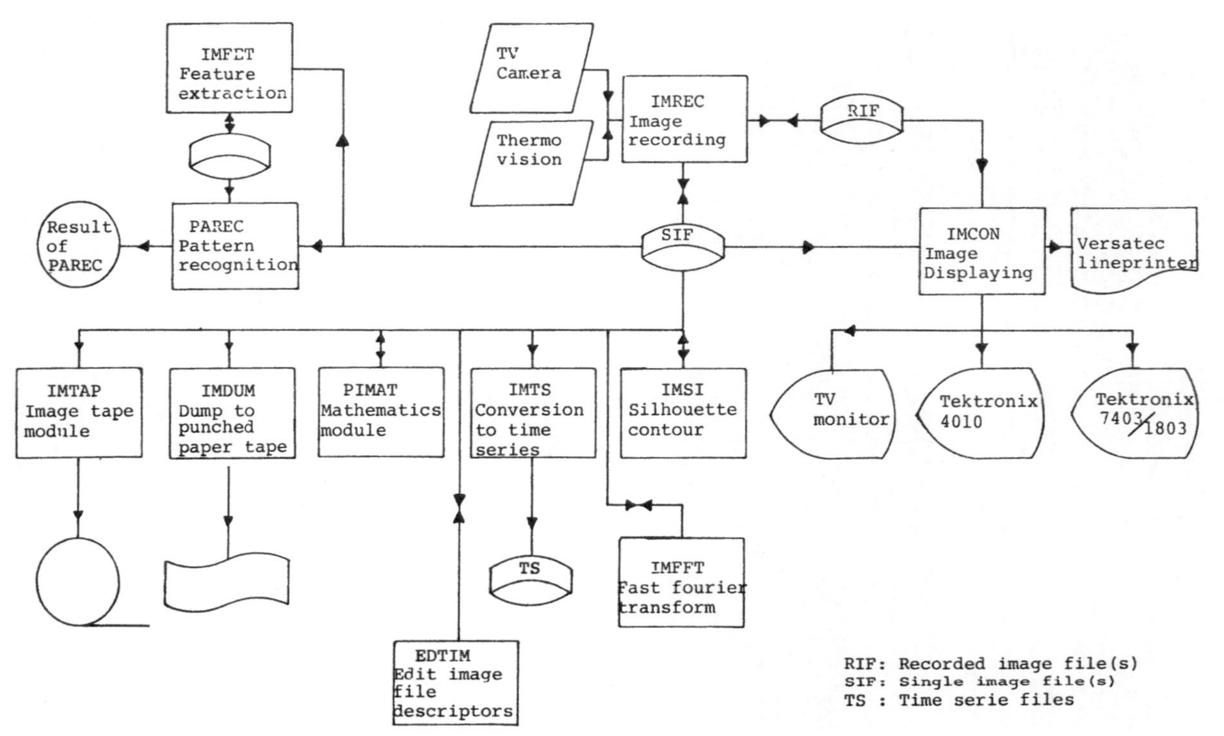

Fig. 6 FTSS-Image pattern recognition software for HP-2100 minicomputer

The recording of TV images is managed by IMREC, which controls the video digitizer and allows the selection of image size, sampling density and image position. The time required for the recording of a 256 x 256 TV-picture is approximately 6 seconds.

The IMTAP module controls the transfer of image files from disc to tape and vice versa. It is able to verify any transfer of data. The files on magnetic tape created by IMTAP comply with the NATO standard [9].

Although 2D filtering and Fourier transforms in principle are done in the optical part of the present hybrid processor it has proved useful to have some digital capabilities in this field. The modules IMFFT and PIMAT can be used for these purposes. The latter permits the interactive generative of images from mathematical expressions. The mathematical formula must obey a syntax and are interpreted by a formula interpreter in the program. The PIMAT module has application in e.g. correction of images, inversion and mirroring.

The IMSI module is able to detect boundaries and intensity gradients in an image. It can be used to generate a binary image from a general gray scale image in the form of either a contour or a silhouette. It can further be used to remove single spots in an image, such as may be caused by defects in the original or optically processed image. It has been shown by Hu, [12], that both contours and silhouettes are useful for pattern recognition purposes.

Two other modules PAREC and IMFET are used in the actual process of pattern recognition. IMFET has the important task of extracting features according to various methods of which the one described by Hu [12], the method of moment invariants, has been found to be extremely useful. The classification matrix file generated by IMFET can be analysed by a number of clustering methods, which will not be discussed in this paper.

PAREC performs classification in the classical way by comparing the features extracted with the classification matrix generated by IMFET. The module has access to a statistics module IMSTA (not shown in fig. 6) and is able to evaluate the classification.

The input converter has a more complicated structure because it should be a digital + optical-to-optical converter with special emphasis on incoherent-to-coherent light conversion. The switch S_2 can change the input of a TV monitor to the output of a TV camera and in this way any kind of image can be introduced. With S_1 in the upper position this image can be monitored before processing. It can also be switched to the input of the digitizer instead of the output camera (this connection is not shown in fig. 3) with the purpose of making a digitized record of the input.

The other input channel is the one with S_2 in the position shown in fig. 3. The TV monitor now receives its signal from a video expander, which is controlled by the computer. This equipment, CVI Video Expander type 275 A - C, is basically a RAM, which can store a digital picture of 256 x 256 x 6 bit. The way this picture is stored is completely controlled and is described in detail in [8] which explains why it is called expander. Parts or all of a digitally stored image can be read-out as a complete TV picture in standard form for presentation on a monitor.

The video expander is used in two applications. One is for read-out of digitally stored images or patterns the other is for transferring image information from a magnetic tape to the hybrid processor. In this latter application the images are stored on the tape according to the NATO tape standard, which is described in [9] and which has made the exchange of information easier. The tape is of course under the control of the FTSS as shown previously.

With the possibility of using digitally stored images as input to the optical system, the whole system is able to perform an optical feedback in a way similar to the one mentioned in [3]. The present system is, however, very flexible, because it permits the selection of which part of an image should be subject to feedback. These possibilities opens some new aspects in image processing, which are being investigated at the moment.

As mentioned above, the input information, whether it originally was in digital or analogue form, is displayed on a TV monitor, which thus forms the input plane of the converter, which makes it possible to modulate a coherent light beam with information from an incoherent light distribution. There exist a number of incoherent-to-coherent light converters working on different principles. The one chosen by the DDRE is called Phototitus and is manufactured by LEP [10]. The reasons for selecting this device for the present application will become clear from the following and only a very brief description of its operating principle will be given here. The most complete description of the theory of Phototitus and its applications are found in the collection of papers published in [11].

The multilayered structure of Phototitus is shown schematically in fig. 7. It consists of a single crystal of deuterated potassium dihydrogen phosphate (DKDP), a dielectric mirror M, a layer of amorphous selenium Se and two conducting transparent electrodes A_1 and A_2. The selenium is photoconductive and if an optical image is projected on this layer electron-hole pairs are created near A_1. Assuming that A_1 is positive with respect to A_2 some of the holes drift towards the interface with the dielectric mirror M, where they are stored. The accumulated charge is then at any point of the interface proportional to the intensity of the incident light and the optical image is thus converted into a latent charge image. If the polarity of A_1 and A_2 is reversed during the projection of an image, electrons will drift towards the interface and the resulting charge pattern will be the difference between the two corresponding patterns. This feature of subtracting two images is very useful and has a number of important applications.

The latent image at the interface is read-out with the DKDP crystal. This operation depends on the Pockels effect of DKDP,

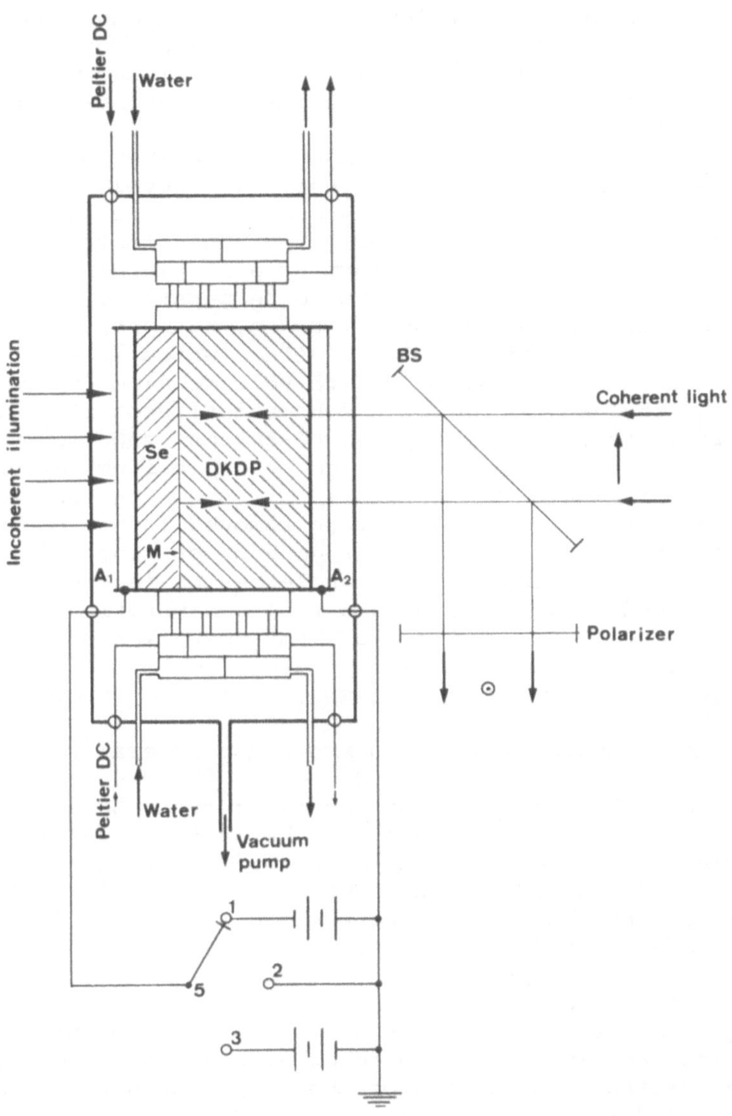

Fig. 7 Schematic view of Phototitus and its accessories

which changes linearly polarized light into elliptically polarized light if there is an electric field across the crystal. The deviation from linear polarization depends on the magnitude of the field. In this way the light beam, after having passed the polarizer in fig. 7, is modulated by the latent image with bright patterns corresponding to areas of high charge density in the latent image.

Although this principle is common to a number of other devices and although Phototitus originally was developed for large screen TV operation, this device is one of the few which does not destroy the coherence of the laser beam. This feature is present because the DKDP is used in the form of a single non-scattering crystal operating just above the Curie temperature.

The operating point of the device is, as mentioned, close to the Curie temperature of the DKDP crystal, which is at about -50°C. The cooling required is achieved with Peltier elements in thermal contact with the crystal as shown schematically on fig. 7. The cooling effect is determined by the DC-current supplied to the elements, which in turn is controlled by a platinum resistance thermometer built into the structure (not shown on fig. 7). The heat removed from the crystal must be removed from the hot side of the Peltier elements and this is done by pumping cold water through the heat exchangers. The whole device is enclosed in a container evacuated with a liquid nitrogen adsorption pump. Although this sounds very complicated, the operation of the device is quite simple and the close view of the Phototitus in place on fig. 8 shows that the device fits very well into an ordinary optical bench.

Although Phototitus has been known for several years it must be considered as an experimental device in its present configuration. The DDRE has had the opportunity to test the device at various stages of its development and has noted several improvements. Many measurements and characteristics are given in the literature [11] and will not be repeated here. A few characteristics, which have influenced the DDRE in its choice of the incoherent-to-coherent converter should be mentioned.

The size of the optically active area of Phototitus corresponds to a standard film of 24 by 36 mm, which permits the use of photographic lenses on the input side and reasonable size optical devices and lenses on the coherent side of the device. The light and spectral sensitivity of the Se photoconductor is well suited for read-in from the screen of a TV monitor. The "white" light of the screen is more blue than red, which is an advantage in this case.

The Se photoconductor has a low sensitivity in the red end of the spectrum so that the red read-out light has less influence on the storage of the latent image. Depending on stray light conditions, an image can be stored for ten minutes or more without read-out. During the read-out operation the image will fade, but for periods up to about ten seconds no serious degradation will take

Fig. 8 Close view of Phototitus on optical bench

place depending on the intensity of the read-out beam. With a TV camera as the output sensor rather low levels of laser light can be used, so that the fading is reduced to unimportant levels.

The capability to store and to subtract images in the Phototitus is quite useful in a hybrid processor. In the present set up of fig. 3, real images can be subtracted from digitally stored ones, a feature which is applicable to what is called change detection. In this kind of operation the read-out does not necessarily consist of red coherent light. On the contrary, visual inspection can be performed much better under normal light conditions. In the present processor the difference can, however, be viewed on the output monitor with the switch S_1 on fig. 3 in the position shown.

The resolution of Phototitus depends on the application. With incoherent read-out 70 lp/mm have been quoted in the literature. In the present application with coherent light read-out about 20 lp/mm can be expected. This corresponds very well with the resolution of the TV cameras incorporated in the hybrid processor and has proved to be sufficient for the type of operation performed on the system.

The operation of the Phototitus requires a certain sequence to be followed. The read-in of an image requires the laser beam to be shut off and the simultaneous application of voltages on the electrodes and illumination of the photoconductor. During read-out the input shutter is closed and the electrodes are shortcircuited. Erasure of an image is done by showing a white screen to the Phototitus with electrodes shortcircuited. All these operations are controlled from a control box in fig. 3. The sequence of events can be controlled from the computer if wanted.

This completes the description of the hybrid processor and its more special components. The application of the processor is to process various images for pattern recognition purposes and examples of this are shown in the following section.

4. EXAMPLES OF APPLICATIONS

It is evident from the previous sections that the hybrid processor presented in this paper is extremely flexible and will interactively process any kind of information. In practice it is being used for image processing with subsequent feature extraction and pattern recognition. Two examples which are typical for the application of the processor will be given in this section.

The first example illustrates the processing of Forward Looking Infra Red (FLIR) images of military targets in natural background. The images are obtained through cooperation in a NATO Research Study Group and are received on magnetic tape in the NATO tape standard format [9]. The purpose of the processing is to detect the targets, determine the number of targets and, if possible, to classify them. The signal-to-noise ratio (SNR) is usually low and before any attempt to detect and classify can be made the SNR has to be improved. This can be done by filtering processes, which are easily performed on the hybrid processor. The system is used as shown in fig. 3. The images are read from the magnetic tape and presented to the optical system in the way described in the previous sections. The filtering is done in the Fourier plane and the effect of the filter is displayed on the TV monitor with the switch S_1 in the position shown. One type of filter has been proved to be very useful for this application and that is the simple band-pass filter. The lower and upper cut off frequencies of the filter depend on the content of the image and the glass plate holding a selection of filters is moved until the proper filter has been found. Fig. 9 shows two examples of the filtering process. In fig. 9a a rather broad band-pass filter with a minimum DC-stop has been applied. The contour of a target is visible, but the background is still dominating the image which still makes the digital detection and classification difficult. In fig. 9b a filter with a large DC-stop has been used. Most of the large variations in the background have now been removed. Also the target has suffered some degradation. It turns out, however, that in spite of this degradation, which looks worse on the photograph than it is in digital form, binarization and contour extraction is possible. (Both photographs in fig. 9 show a curve between two vertical bars. This curve indicates the brightness variation in the column being digitized and is for information to the operator only).

In certain cases it is advantageous to perform the contouring operation in the analog part of the hybrid processor. As explained in section 3, this process can be carried out in the Phototitus through a subtraction of two images. The result of such an operation is shown in fig. 10, which was produced in the following way. A picture of an aircraft (Harrier) is presented to the upper TV camera in fig. 3. With the switch S_2 in the upper position the image is projected onto the Phototitus. Then the image is defocused slightly and through reversal of the polarity of the electrodes this image is subtracted on the Phototitus. The read out takes place in the way described and a simple operation can generate the contour in digital form. As seen from fig. 10, the contour appears as a very distinct outline, which permits the use of gradient or threshold operators for the extraction of contours. The silhouettes and contours can then be used in the way described by Hu [12].

It is obvious from these examples that various combinations of techniques can be used for the extraction of contours. Through the feedback capability it is possible to perform hybrid operations on the images, which have been processed before. Nonlinearities and gray level changes can be introduced by the digital part of the system, thus permitting an operator the freedom of choice for the application at hand.

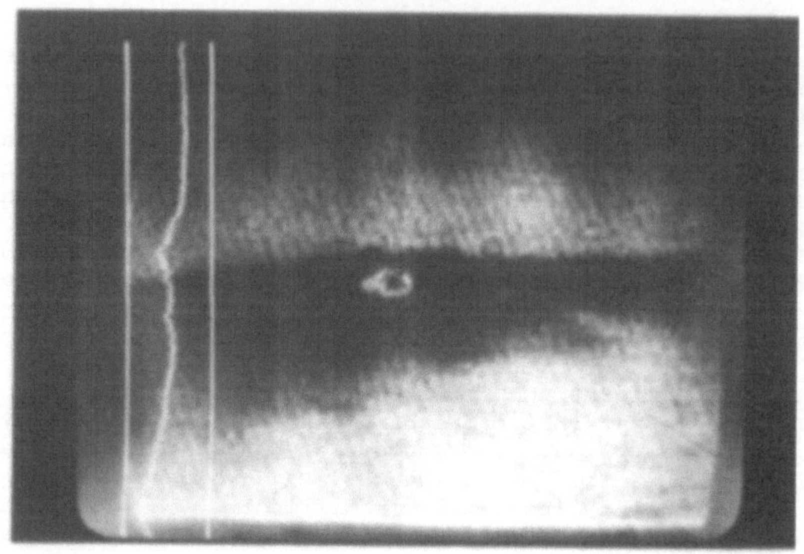

a. Wide band-pass with minimum DC stop

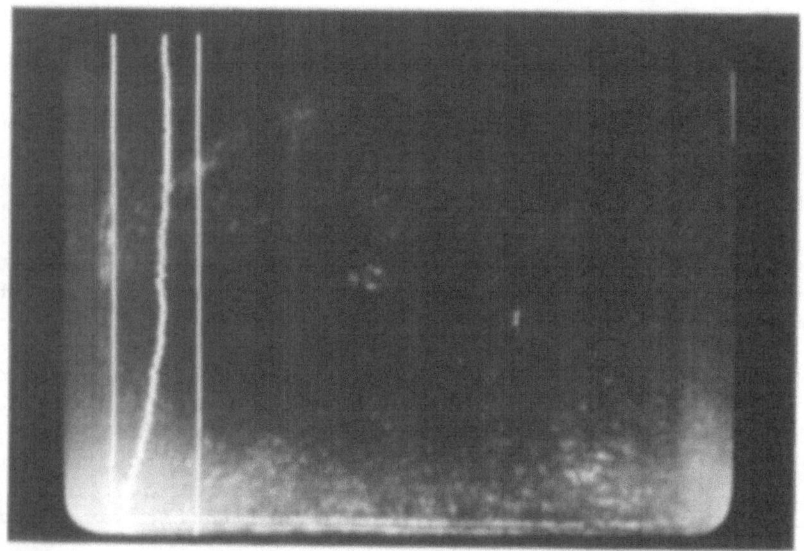

b. Narrow band-pass removes background
and permits further processing

Fig. 9 Application of band-pass filtering in the processing of FLIR scanner images

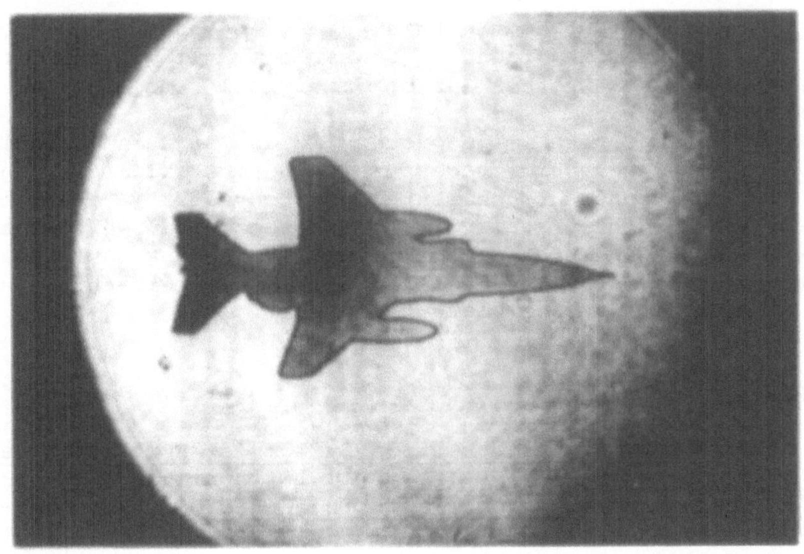

Fig. 10 Contour extraction with Phototitus

5. CONCLUSIONS

Before the system described in this paper was designed, a study of various image processing techniques was made. It was found that no single techniques whether analog or digital, was superior in every respect.

For general purpose image processing and pattern recognition, it turns out that a hybrid processor is the most flexible, especially if a good incoherent/coherent light converter is available. The usefulness of the system for pattern recognition hinges critically on the software of the minicomputer. The one outlined in this paper has been shown to be a powerful tool permitting interactive operation of the complete processor.

REFERENCES

1. A. Vander Lugt, Signal Detection By Complex Spatial Filtering IEEE Trans. Inform. Theory, Vol. IT-10, April 1964, pp 139-145.
2. 1976 International Optical Computing Conference, 31 Aug - 2 Sep 1976, Italy, Sponsored by IEEE and SPIE, IEEE Catalog No. 76CH 1100 - 7C
3. G. Häusler and A. Lohmann, Hybrid Image Processing with Feedback Optics Communications, Vol. 21, No. 3, June 1977, pp 365-368.
4. Instruction Manual, CVI Model 270, Video Digitizer, Colorado Video, Inc., Boulder, Col., 1975.
5. J. C. Richmond, Image Quality of Monochrome Television Cameras, NBS - SP - 480 - 25, Oct. 77, PB - 273 117.
6. L. Brock-Nannestad and P. B. Ring, FTSS - A System for Data Aquisition and Analysis, DDRE 1976/38, June 1976.
7. G. Hvedstrup Jensen, O. C. Mortensen, P. B. Ring, A Digital System for Image Processing and Pattern Recognition, DDRE 1976/61, Oct. 1976.
8. Instruction Manual, CVI Model 275, Video Expander Colorado Video, Inc., Boulder, Apr. 1977.
9. NATO Tape Standard for Image Information, To be published in "Computer Graphics and Image Processing".
10. Phototitus (Titus = Tube Image á Transparance Variable Spatio-temporelle), LEP - Laboratoire D'Electronique et de Physique Appliquée, Limeil-Brevannes, France.
11. Relais Optiques d'Image (Image Light-Valves), Special Issue devoted to these devices, Acta electronica, Vol. 18, no. 2, Apr. 1975, p. 81-154, and no. 3, Jun. 1975, p. 155-252.
12. M. K. Hu, Visual Pattern Recognition by Moment Invariants, IRE Trans. Inform. Theory, Vol. 8, Feb. 1962, p. 179-187.
13. Data Extraction and classification from Film Proc. SPIE, Vol. 117, 23 - 24 Aug. 1977

IMAGE PROCESSING TECHNIQUES FOR REAL-TIME IMAGERY

John S. Dehne

U. S. Army
Night Vision and Electro-Optics Laboratory
Fort Belvoir, Va. 22060

ABSTRACT. Automatic image processing techniques are currently being applied to real-time electro-optical sensors. A wide range of uses from image optimization to automatic target detection are being pursued. This paper discusses the current limitations of such real-time applications. Included are discussions of the constraints imposed by the sensors, the image processing hardware technology, and the algorithms themselves. Areas requiring further effort are also indicated.

1. INTRODUCTION

Automatic image processing techniques, long associated with the worlds of big computers and optical benches, now seem on the verge of practical application in real-time EO sensor systems. Designers are discovering that a proper choice of algorithm and hardware can result in a system which accomplishes substantial processing at real-time rates and still retains a respectably trim configuration. This is particularly surprising in lieu of the data rates (1-25 megasamples/second) and process complexities involved. It seems appropriate, at this time to assess this new technology in order to better understand its future potential and current limitations.

2. REAL-TIME ELECTRO-OPTICAL IMAGERS

Many of the opportunities and limitations of the new processing technology stem from properties of the sensors themselves. Generally, EO imaging systems are of two varieties - direct view and remote view. Direct view systems are typified by being primarily optical in nature from end to end. The observer may be thought of as looking "through" such devices and the display and processing of information is usually continuous rather than discrete in time. Such devices include binoculars, telescopes, and metascopes and image intensifiers (which apply electron optics for a portion of the path through the device).

In general, this class of devices is best suited for "hand-held" applications where simplicity of design, small size and cost, and ease of use are paramount. For this reason the practical application of image processing techniques to direct view imaging sensors appears unlikely. The direct view nature of the device itself makes it very difficult to connect or insert image processing elements. Further, most of the applications areas for such devices cannot support the additional cost, size, and complexity which would be required.

Remote view devices are typified by a non-optical (typically electronic) connection (and often a considerable distance) between the sensing elements and the observer. The observer is generally thought to be looking at a "display" in such systems and the information is generally presented in discontinuous frames. The illusion of continuous imagery is maintained by use of a frame rate above the critical flicker frequency of the eye (10-30 frames per second for typical display intensities). Such devices include television, intensified television, pyroelectric vidicons, and flir (forward looking IR). The remote view and often expensive and high performance nature of such systems makes them ideal candidates for the application of image processing. Connection of the new processing subsystem into the video data stream is relatively straightforward and the added cost and size can be tolerated if sufficient performance benefits accrue. For these reasons, practical real-time processing applications seem limited to remote view devices and must interface to a generally serial video data stream. Remote view devices are well represented by the television and the flir or thermal imagery.

While the principles of television are generally familiar, those of thermal imagers often require brief explanation. Thermal imagers depend on the thermal radiation for which all matter emits at temperatures above $0^{o}K$. This radiation is generally governed by the blackbody radiation law well known from quantum physics. The general concept of a thermal imager is shown in Fig. 1. The thermal radiation is imaged on an array of one or more thermal detectors which generate an electronic signal proportional to thermal radiation intensity as they scan the thermal image. The resultant signal is ac coupled to improve contrast (inherent thermal contrasts of a few percent per ^{o}C temperature difference are typical of blackbody radiation), amplified, and eventually displayed as a visible image.

A comparison of several relevant characteristics of television and thermal imaging is shown in Table 1. The first point which should be noted is that thermal images are much more dependent on

THERMAL IMAGERS
(CONCEPT)

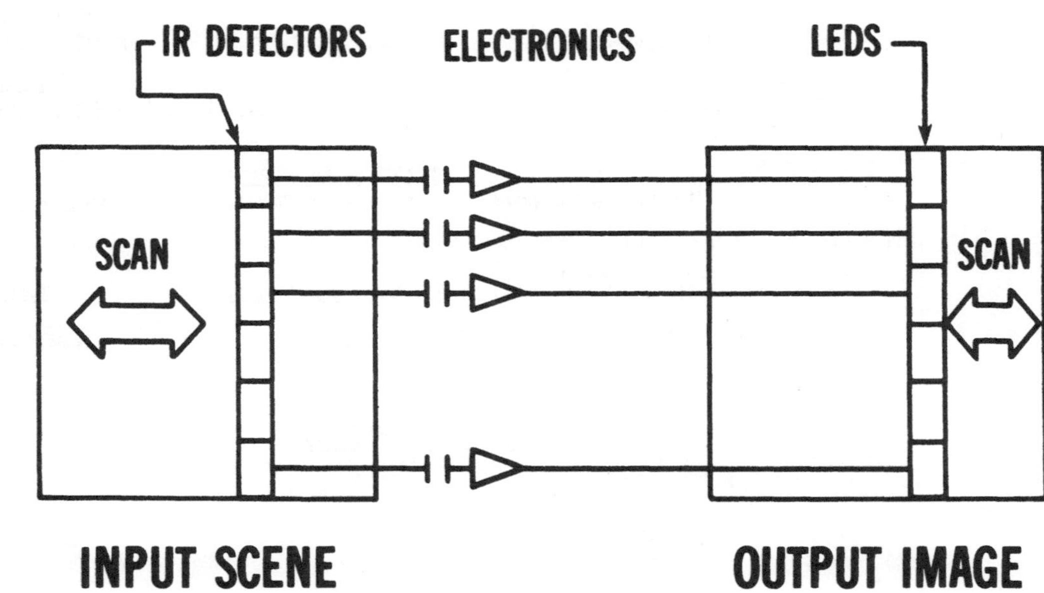

Fig. 1. Thermal Imagers (Concept)

TV/THERMAL COMPARISON

PROPERTY	TV	THERMAL
SURFACE	REFLECTANCE $[R(m,n,\lambda)]$	EMISSIVITY $[\epsilon(m,n,\lambda)]$ TEMPERATURE $[T(m,n,)]$ EMITTANCE $[W(m,n,\lambda) = k\ \lambda^{-5}\{EXP[k_2/\lambda T(m,n,)]-1\}^{-1}]$
ENVIRONMENTAL	Illumination $[I(m,n,\lambda)]$ ATTENUATION $[A(m,n,\lambda)]$	ATTENUATION $[A(m,n,\lambda)]$
SIGNAL AT SENSOR INPUT $[\chi(m,n)]$	$\int_{\lambda_1}^{\lambda^2} A(m,n,\lambda)\ R(m,n,\lambda)\ I\ (m,n,\lambda) d\lambda$	$\int_{\lambda_1}^{\lambda^2} A(m,n,\lambda)\ \epsilon\ (m,n,\lambda)\ W\ (m,n,\lambda) d\lambda$
SIGNAL AT SENSOR OUTPUT $[y(m,n)]$	$h\ (m) \circledast h_y(n) \circledast \chi(m,n)$	$h_n(m) \circledast h_y(n) \circledast \chi(m,n)$
NOISE	OFTEN POISSON-SIGNAL DEPENDENT	ALMOST ALWAYS GAUSSIAN
IMAGE ON DISPLAY $[Z(m,n)]$	$F[y(m,n)]$	$F[y(m,n)]$

Table 1. TV/Thermal Comparison

the gross properties of surfaces (i.e., emissivity and temperature) for their content than are television images which are also heavily dependent on the illumination. This factor portends a greater ease in detecting surfaces (hence objects) in the thermal regime and has implications with regard to the opportunities for image enhancement and camouflage in the two domains as well. The next point to note is that the scanned nature of both flir and tv produces a non-isotropic and separated point spread function ($h_x(m) \neq h_y(n)$). This can result in imaging artifacts particularly in the flir where the ac coupling in the horizontal dimension can introduce so called "DC restoration" artifacts. Finally, it is important to note that both TV and flir are often limited by the non-linear input/output function of the display. This is particularly important due to the mismatch in dynamic range between the sensor (60db typical) and the display (10:1 - 20:1 luminance range typical) which usually occurs in these systems.

Characteristics and limitation of EO sensor technologies such as those noted above imply some of the image processing effects to be desired. These naturally fall into two categories; aids to the human observer, and replacements for the observer. The first category is dominated by attempts to overcome some of the limitations noted above. These include: contrast optimization and enhancement to overcome the limitations of the display input/output function; restoration of system induced artifacts (e.g., DC restoration problems); and improvement of general image quality and signal to noise. Such functions require multi-dimensional signal processing. Desired operator replacement functions (at least in military systems) involve object detection, recognition, and tracking. These usually require the use of some form of pattern recognition techniques to provide the required analysis and decision making capabilities. It is also noteworthy that the relative independence of image properties from environmental factors in the thermal domain probably implies an easier application of detection and tracking processing to flir than to TV.

3. IMAGE PROCESSING HARDWARE TECHNOLOGY

As the sensors constrain the processing functions desired and the format of the image data, the available image processing hardware technology constrains the types of operations to be considered. Two processing technologies generally vie for the application - optical processing and electronic processing.

Optical processing depends on impressing the imagery as spatial modulation on a beam of light. Lenses, apertures, and phase plates are then used to process the modulated beam. These techniques are well suited to implementation of linear processing functions (e.g., correlation, convolution) and accomplish such functions in a completely parallel (hence high data rate) manner.

Unfortunately, there are several factors which limit the utility of such optical processing techniques in real-time systems. The primary problem is that the conversion from the input (generally serial) video data to the modulated optical beam requires a display. This restricts the dynamic range of the processor while limiting its throughput to the frame rate (eliminating much of the processing speed advantage of the optical approaches). Secondary problems include a relatively large size (driven by reasonable lens apertures and focal lengths), criticality of alignment tolerances (particularly in coherent systems), and relative inflexibility as to processing function repertoire.

Electronic processing is extremely flexible and highly compatible with the remote view EO systems. Further, the current explosive pace toward ever higher circuit complexities and speeds and ever lower circuit size and power seems likely to continue for some years to come. At the present, however, the primary limitation of this processing technology is the speed, size, and power limitation. Neither the 1-10 megasample per second rate typical of analog CCD technology nor the 10-100 nanosecond access times typical of high speed digital memories are sufficient to accomplish the 3-30 megasample per second processing rates desired. The result is that parallelism (e.g., simultaneous repetition of operations in duplicate hardware) and pipelining (e.g., simultaneous execution of sequential operations on sequential data elements) will be required. Minimization of the processing hardware implies some limitations on the types of processing algorithms which can easily be implemented in such structures.

Minimization of parallel processing structures is possible by constraining algorithms to operate on a "local neighborhood" at any given instant. This constraint, which limits the degree of parallelism and, hence, the duplication of hardware, is actually not very restrictive. The data is generally available only one or a few lines at a time in any case, and regions of interest are often small compared to the entire image.

Minimization of pipeline processing structures can be accomplished by constraining algorithms to minimize iterative operations (e.g., operations which repetitively apply operations to a local neighborhood). Such procedures require either enough speed to process several iterations in one sample interval or else replication of the hardware for each iteration level. Neither approach is very viable. Reduced use of iterative approaches tends to eliminate or minimize the use of region growing and relaxational type algorithms.

Finally, total hardware could be minimized by constraining the algorithm to process only the "important" part of the information in the image. This can be attempted either by "compressing" the imagery to a few high level attributes at all points and then processing these at a reduced rate, or by applying some form (typically low level) of "interest" operator to the entire image and using the results to select a limited number of sub-images for detailed analysis. The first approach demands a high speed (typically hardwired) pre-processor to measure the high level attributes in real-time. It also limits the types of information available for pattern recognition. The second approach places critical demands on the "interest" operator which must (by simple operations) detect all regions of interest while restricting false detections to a very low number (so that the high level processing which follows will not be overwhelmed). Neither approach is entirely satisfactory. The underlying dilemma is that higher level and more varied types of information can be expected to improve processing accuracy and robustness, but only at the expense of rapidly expanding hardware. Thus, these sort of algorithm/hardware trades must be made with great caution.

4. THE ALGORITHMS

Although the data format and desired processing functions are determined by the sensor and its use, and the applicable algorithm forms are constrained by available processing hardware, it is the algorithm which finally determines how and how well real-time image processing can be accomplished. At the present state of the art, design of image processing algorithms is accomplished on an almost entirely ad hoc basis with little or no theoretical foundation. The prime reason for this is the lack of a sound theoretical framework for describing images and image components and the perceptual effects on human observers.

As previously noted, the desired functions divide into those which attempt to aid the human operator by enhancing or optimizing the displayed imagery and, those which attempt to automatically detect and classify objects.

5. IMAGE ENHANCEMENT/OPTIMIZATION ALGORITHMS

Desired algorithm functions in this category include: optimization of contrast in the display; restoration of sensor induced artifacts, particularly DC restoration artifacts and; enhancement of the signal-to-noise ratio of the displayed imagery. One or more algorithms will be considered for each function.

5.1 Contrast enhancement/optimization algorithms

As previously noted, the major purpose of this class of algorithms is to "match" a high dynamic range remote view sensor to a display of limited luminance range. Clearly, all the information in the sensor signal cannot be displayed by such a process. The problem then is one of deciding what to display to the observer and what to hide from him. As the perceptual processes of the observer are only poorly and grossly understood, there is presently no firm basis for designing optimal techniques. As a result, only simple ad hoc approaches currently exist. These include histogram modification, homomorphic filtering, and local area control of display gamma factors.

Histogram modification draws its impetus from the connection between information content and entropy. The information content of a signal is maximized, it is argued, when the intensity histogram is uniform. A variant, histogram hyperbolization, attempts to present a uniform intensity histogram to the visual cortex of the observer. In either case, the operation is accomplished by first histogramming the high dynamic range sensor imagery. The output image is then computed by assigning display luminance levels to increments of the input cummulative histogram via a lookup table. This ignores the spatial content of imagery and its importance to visual perception. The largest image regions tend to diminate the input histogram and are, therefore, emphasized in the output image at the expense of spatially smaller regions. This effect only worsens the problem of detecting small targets for which the eye has a lower sensitivity already. In addition, the process requires storage of a full image frame and computation of the histogram of a wide dynamic range signal, both of which require considerable hardware. Application of these techniques to smaller local areas reduces the de-emphasis of small image detail and the storage requirement. However, this demands the computation and application of a different translation function for each point in the image if artifacts are to be avoided at the boundaries of local windows. The resultant hardware is far from simple and enhancement of noise is often the only result.

Homomorphic filtering depends on the fact that visible images result from the point-wise product of object reflectance and external illumination (see Table 1). Use of logarithmic and fourier transforms reduces the image to the sum of illumination and a reflectance based signals. If the relative frequency spectra of the two signals is known or can be guessed, a filter can be designed to attenuate the illumination signal (which is only impeding object detection) and pass the reflectance signal. Inverse transformation produces the enhanced image. In practice, the power spectra are unknown and too complex to guess and high pass filters are generally applied. This has the effect of emphasizing spatial small details in the image at the expense of large regions (e.g., details in shadowed areas are enhanced) and with proper filter design, noise is not unduely emphasized. However, the process requires a full frame of storage (at least) plus logarithmic, exponential, and two fourier transforms. The hardware is thus rather prohibitive.

Local area gamma control is a much simpler technique which accomplishes the same objective. The simplest of this algorithm family, unsharp masking, was originally implemented photographically to improve transmission of newswire service (e.g., UPI, API) imagery. By adding the blurred negative of a picture to itself, small features (which have been essentially removed in the blurred copy) are retained while large features are compressed in dynamic range. The algorithm can be viewed as locally adaptive brightness control or a high emphasis filter. An extension, the local area gain brightness control (LAGBC) algorithm, also measures the local variance and adapts the gain appropriately. The result is a further (non-linear) enhancement of detail in areas which are nearly uniform. Enhancement of noise is prevented by reducing the gain when the variance equals that of sensor noise. Both algorithms are simple to implement in real-time with CCD's and recursive low pass filters to compute the required local averages as shown in Figures 2 and 3. Results are quite similar to homomorphic filtering employing a high pass approach (as might be expected). Figures 4a and 4b illustrate the effects of LAGBC on improving visibility through smoke.

Clearly, proper manual adjustment of the display gain and brightness can always produce the same object contrast for any given object as that produced by the above techniques. The effect on observer performance is primarily due to reduction of task loading (by eliminating the need to manipulate the display controls) with secondary effects when the image contains large area variations in environmental effects. Perhaps further improvements could be made if more selection could be applied as to what to enhance. Clearly, the current approach (i.e., enhancing small detail as far as possible as long as noise does not dominate) does not account for the relative perceptual importance of various image features to differing tasks. Such improvements await a better understanding of the perceptual process and a reasonable framework for modeling non-linear processes.

5.2 Pseudo-DC restoration

DC restoration occurs in thermal imagery whenever scene temperatures change greatly from one portion of a given scan line to the adjacent scan line. This often occurs at the boundary of any object considerably hotter or colder than its background and at the horizon. The effect is caused by the required AC coupling which makes the average output of all lines to decay to zero independently. Thus, the background intensities are darkened on a line containing a bright object relative to values on an adjacent line which does not. A dark band is created which annoys the observer and may obscure dim objects.

The effect is often correctable by an image processing technique (US patent pending), which relies on a simple a priori model of imagery. Imagery is expected to consist of local areas which are typified isotropically by some property (e.g., intensity, texture, etc.). If DC restoration artifacts exist, they will tend to destroy the isotropy of these areas. The local anisotropy (on a line by line basis) can be taken as a measure of the prominence of the

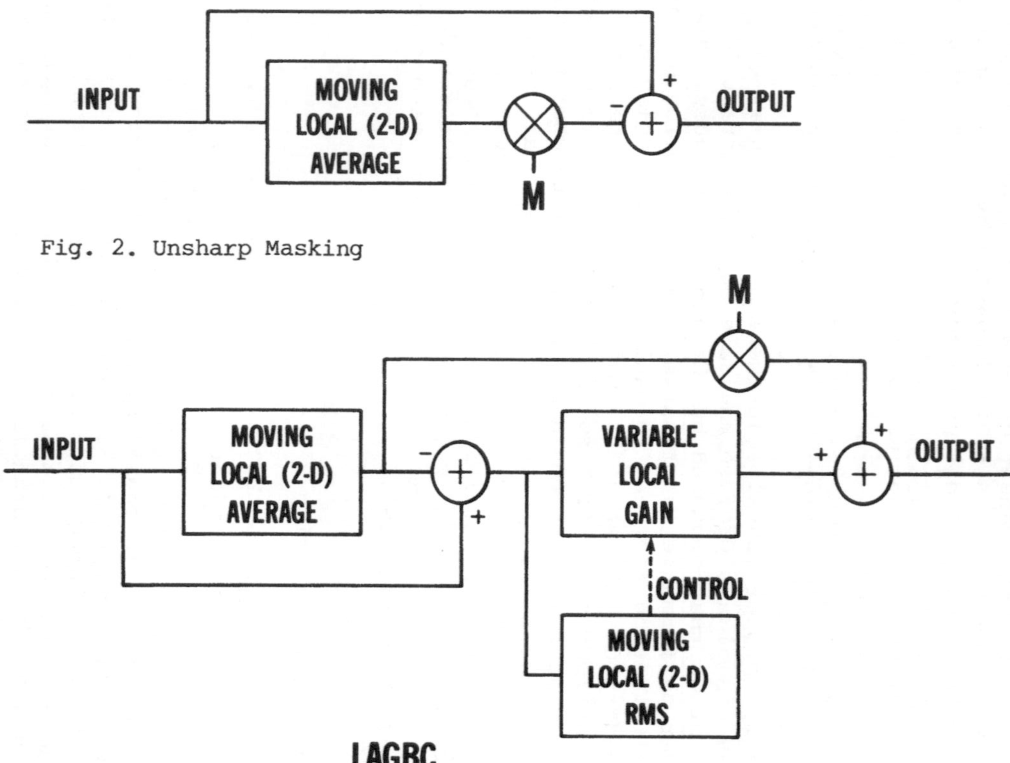

Fig. 2. Unsharp Masking

Fig. 3. Local Area Gain Brightness Control

a) Original FLIR frame from video tape

b) Enhanced by real-time hardware

a) Original FLIR from video tape

b) Enhanced by real-time hardware

FIG 4. EFFECT OF LAGBC

artifact and line by line brightness shifts which restore the local isotropy will remove the artifact.

The approach is easily implemented (using intensity as the local property) by computing the differences between adjacent samples on adjacent lines. If local intensity isotropy holds, most such pairs will have the same intensity and the mode of the histogram of the differences will be zero. When a DC restoration artifact exists, the mode of the histogram will be equal to the brightness shift between the two lines (assuming the offending object which caused the artifact subtends less than half the line). This value may then be used to correct the second line and the process is re-applied using the correct newly corrected line and the next one until the image is finished. Only a few lines of storage and one histogram development are needed (further simplifications are possible [1]) so that hardware requirements are minimal.

The effect of the approach is illustrated in Fig. 5. The apparent success of this simple technique indicates that development of improved models of images can indeed be expected to provide insight into improved algorithm development and optimization.

5.3 Signal to noise enhancements

Under sufficiently degraded visibility conditions, the performance of all EO sensors becomes limited by sensor noise. Under these conditions, it is preferable to sacrifice long range detection performance (useless anyway) to achieve improved performance at mid and short ranges. The obvious solution is some form of spatial filtering. Unfortunately, it is currently impossible to optimize the design of such a filter. The problem is twofold. First, while the noise can be adequately described as a random process with known properties, the signal (the image) is neither nicely random (e.g., its statistics are generally non-stationary in space) nor deterministic (e.g., no two images are the same and the sequence of occurrence cannot be accurately predicted a priori). Thus, one is left trying to design an optimum filter for an unknown signal. The second difficulty involves the metric for optimization which should ideally reflect the psychovisual effects of noise on the observer. The understanding of the perceptual process required to develop such a metric is not currently available and all currently available metrics (e.g., least mean square error) fail to correlate well with observer performance. As a result, the current state of the art is driven by attempts to smooth spatially small intensity changes.

Probably the simplest approach to this problem is a two dimensional low pass filter. The high frequency cutoff could be adapted to expected object size and shape and implementation is straightforward. The problem is that object shape is variable and such filtering tends to blur the object boundaries and edges which are perceptually important in distinguishing objects.

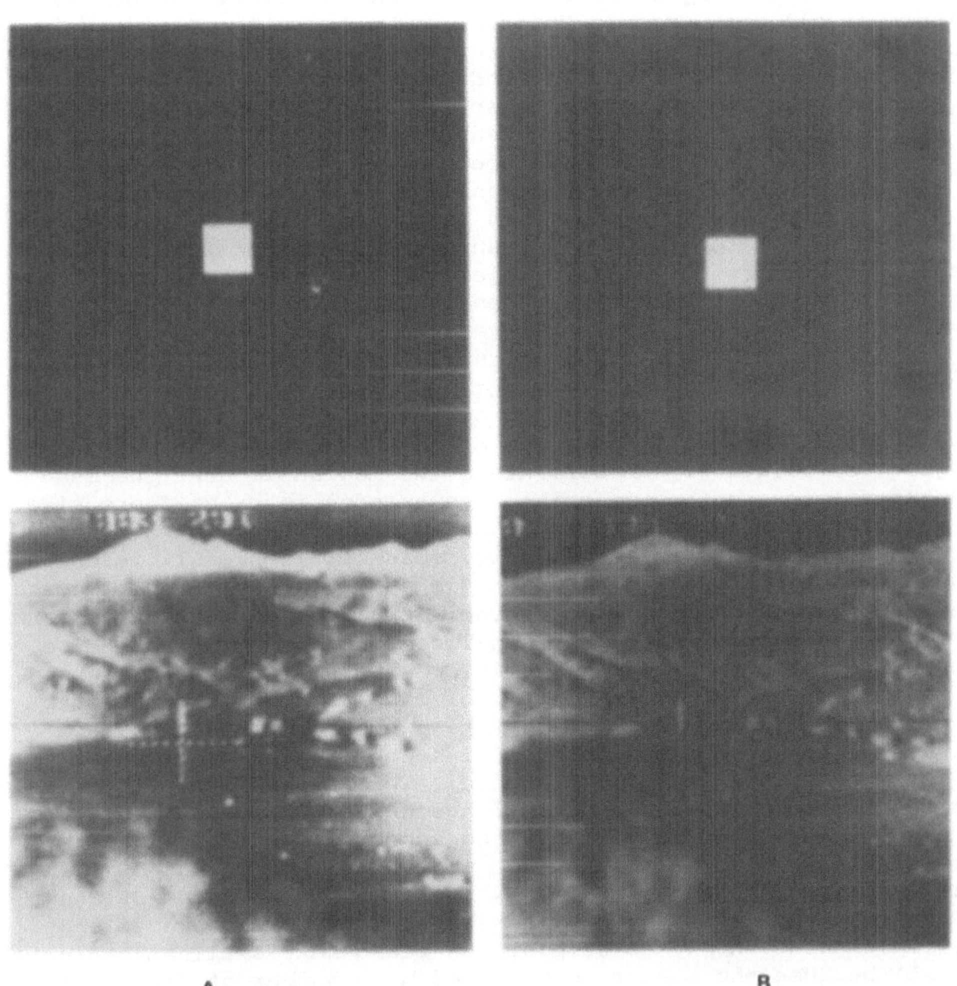

A
DC RESTORATION PROBLEM

B
PSEUDO-DC RESTORED

FIG 5. EFFECTS OF PSEUDO-DC RESTORATION

One attempt to overcome these objections is the use of an adaptive filter. Such a filter consists of several two dimensional linear low pass filters with different high frequency cutoffs operating simultaneously on the image with delays such that all are outputing their current estimate of the same filtered image point simultaneously. Some control function related to perceptually important image properties (e.g., intensity gradient in the input) is used to select the estimate to be used at each point. Thus, the filter is most aggressive in large uniform areas and hardly filters at all near edges. Unfortunately, noise too produces intensity gradients (and intensity curvature and any other property usable to denote edges) and so tends to preserve itself by reducing the aggressiveness. This effect is most prominent at low signal-to-noise ratios where it is least desirable. In any case, the signal-to-noise gain is signal dependent and non-predictable due to the non-linearity of the process. It is always, however, less (typically far less) than the gain of the most aggressive low pass section used. Each of the components (e.g., low pass filters and gradient detectors) is easily implemented but the total hardware complexity may be substantial depending on the number of filters used.

Another approach is the median filtering process. This is accomplished by sliding a window over the image. The filtered image is composed by placing the median of the values in the input window at the location of the center of that window in the output image. For relatively uniform areas the median is near the mean and the filter smooths out noise. As an edge is crossed one side or the other dominates the window and the output switches sharply between these values as the edge is crossed. The result is that edges are not blurred and gradient slopes are not altered. However, in two dimensions object corners are rounded. The algorithm is signal dependent and highly non-linear so the signal to noise gain is not predictable although it is always less than or equal to that of an average value filter of equal size. Like the adaptive filter, low signal-to-noise ratios breakup image edges and produce false noise edges which confuse the filter and reduce its effectiveness when it is most needed. This type of filter can be constructed from CCD taped delay lines and a CCD intensity sorter or histogrammer [2]. A variant, the separable median filter is also available. This computes the median of the medians of the line segments in the moving window instead of the median of the elements in the entire window. The results seem comparable and the implementation may be simpler for reasonably small sizes (up to 5x5) because the sorter is replaced by a number of diode networks [3].

Clearly, these techniques are of little use at the moment as they either destroy perceptually relevant information or work poorly in the low signal to noise environment in which they are most needed. Improvements on this unsatisfactory state await a better understanding of imagery as a signal and of the psychophysics of human perception.

6. AUTOMATIC TARGET DETECTION ALGORITHMS

Although many different approaches to target cueing exist, they may broadly be said to depend on the application of a few types of equivalent processes generally applied in sequence; segmentation, topological analysis, feature measurement, and classification. Images are segmented into those points which may be part of a target image and those points which are not. This segmentation procedure may be very simple (e.g., intensity threshold) or quite complex (e.g., based on multi-spectral content, gradient intensity, connectivity, etc.). The segmentation process defines target-like points and is critical because it sets a ceiling on the probability of target detection. Further, a poor or inconsistent segmentation may prevent target recognition by destroying object shape or produce a high false alarm rate by generating small target-like regions in the background.

Next the segmented image elements (pixels) undergo a topological analysis to establish disjoint regions of mutually connected pixels. This process, which may be embodied as an edge follower, an edge surroundness criteria, or an analysis of pixel to pixel or line to line connectivity, is responsible for defining objects to the system. It is only after such an analysis that the features (properties) of the objects in the image may be measured (e.g., object length, width, area).

A suitable set of features is then measured on each object. Features commonly used include length/width, area, perimeter^2/area, and edge direction histograms which are measured on the segmented image and average intensity, texture, and intensity moments which are measured on the corresponding portions of the original image. Features are generally selected in a heuristic manner and refined based on experience gained in training the classifier. It is important to select features which are simple enough to be computed on many objects in real-time and yet are complex enough to reliably differentiate between the targets and background objects.

Finally, the feature values of each object are compared to those of a known and limited set of target types. If the object resembles a known target class sufficiently well, it is so classified, otherwise it is assumed to be a nuisance and is ignored. Such classifiers are generally designed using statistical classification approaches and so may be quite dependent on the training imagery provided. It is the classifier which determines false alarm rate and target recognition probability.

6.1 Early approaches

The earliest workable automatic target detection algorithms began emerging in the early 1970's [4,5]. These approaches depended on measurement of a few relatively high level image properties to segment the entire image at a rapid rate. The segmented objects were then handled at a much reduced data rate. One approach measured edge intensity and closure [4] to produce this segmentation. The other approach looked for edges and areas significantly brighter or darker than that of the norm [5]. Required thresholds were preset based on experience with training imagery and were constant across the frame. No attempt was made to miniaturize the required hardware which weighed several hundred pounds, and operated at the equivalent of about 1-full image frame in 2-3 seconds. This speed required hardware handling approximately $5x10^5$ operations (equivalent additions) per second which was achieved using special purpose hardwired digital pipeline [4] and analog parallel [5] architecture preprocessors.

In 1975, computer simulations on a common digital imagery data base provided a crude test of these algorithms and produce the results shown in Table 2 [6,7]. These results are probably biased toward lower than normal performance by the limited nature of the tests (which allowed only a small amount of algorithm adjustment to the unfamiliar imagery) and by the relatively poor quality of the imagery used. Nevertheless, the results are strongly indicative. In general, the stability of results between test and training sets was less than desirable which indicates a lack of robustness in the algorithms. In this regard, the Westinghouse extraction procedures were more robust, while the Honeywell classification procedures were apparently better. This can be understood, in part, by realizing that the edge based methods used by Westinghouse are allowing a great deal of local adaptation of the algorithm to show variations in image brightness over the image. This was not available to the Honeywell method due to incorporation of fixed intensity thresholds. The edge information is, however, much less robust than the boundary information in the presence of noise, a fact which derives from the higher redundancy of interconnection through an image region as compared to the redundancy of interconnection along a narrow band of edge surrounding it. Clearly, the boundary will give a more stable indication of object shape under such conditions. In any case, the tests also indicated that robust segmentation and classification of small, degraded targets was particularly difficult.

6.2 Current approaches

Several automatic target detection systems are currently under development for completion in the 1979 timeframe. In general, they are extensions to the earlier work. The size and power of the devices has decreased and the operating speed has increased as shown in Table 3 due to the use of the improved LSI circuitry which is now available and attention to optimizing the design for these factors. Aside from the increase in processing rate to 10-full frames per second (an increase of up to 30 times in processing requirements), the major improvement has been the inclusion of various forms of self adaptive thresholds.

	Honeywell		Westinghouse	
	Training	Test	Training	Test
Prob. of Extraction	79%	79%	88%	91%
Prob. of Detection	65%	44%	59%	57%
Prob. of Recognition	61%	34%	47%	38%
False Alarm Rate	1.2/frame	2.0/frame	> 10/frame*	> 10/frame *

*Estimated number of false alarms/frame based on false alarm rate per window and relative sizes of window and frame.

Extracted from references 6 and 7.

TABLE 2. SIMULATION RESULTS OF EARLY CUEING APPROACHES

	1974	1979[8]	1981[7]
Image Enhancement			
Gain/Brightness Control	No	Yes	Yes
Auto Focus	No	No	Yes[8]
DC Restore	No	Yes	Yes
SNR Enhancement	No	No	Yes[9]
Operations/Sec*	0	1×10^{10}	2×10^{10}
Target Cueing			
Prob. of Detection	57% †	85%	92%
False Alarm Rate	1.5/frame †	1.0/frame	1.0/frame
Full Frame Rate	.33/sec	10/sec	10/sec
Operations/Sec*	5×10^{5}	5×10^{8}	1×10^{9}
Nominal Physical Characteristics			
Volume	5-10 $ft.^3$	1.5 $ft.^3$	<.5 $ft.^3$
Weight	200 lbs.	50 lbs.	<5 lbs.
Power	100 W	50 W	10 W

*Equivalent additions (1 multiply = 4 additions) if done digitally.
† Honeywell results test + training set.

TABLE 3. PROGRESSION OF APPROACHES

In approaches which extract high level features to achieve segmentation over the entire image [1], the edge and intensity thresholds involved have become adaptive to local variations in the image. This implies an understanding that most images are composed of large areas of significantly varying intensity and complexity which contain small regions. Detection of the small regions is therefore improved by accounting for the effects of the large area variations in which they are imbedded. This can be accomplished by adapting the threshold to the local mean and edge content in a manner analogous to the adaptation of display characteristics in LABGC algorithms. The result is an improvement in the extraction probability of low contrast, noisy objects, a concurrent improvement in the robustness of the shape of all objects extracted, and a reduced production of nuisances due to non-optimal thresholding of the background.

In approaches which rely on low level features to denote areas of interest for further processing [8], the local adaptation is a natural result of processing each interest window separately. In current approaches of this type, adaptation is achieved by using a crude edge surroundedness criterion to denote interior and exterior points. Intensity histograms of these two regions can then be used to generate a maximum likelihood classification of each point in the local window on which the final segmentation is based. Clearly, this approach is very computationally expensive and can only be applied for a small percentage of the total image area due to hardware constraints. Indeed, this is just as well, because intensity variations from object to object are likely to be sufficient to prevent successful maximum likelihood classification of intensities on a full frame basis. In any case, this approach also results in an improvement in the quality of segmentations produced and a reduction of nuisance production.

Both the above approaches still tend to produce spurious segmentations in highly textured areas where local variability is not related to small objects. Similar effects occur with very noisy target images where the result is an undesirably large variation in the shape of the segmented target regions. Nevertheless, a simulation of both approaches on the same data base used in the 1975 studies results in a considerable improvement in detection performance and some reduction in false alarm rate as shown in Table 3.

6.3 Advanced approaches

The already high computational complexity of the algorithms just described implies that further improvements will require simultaneous and interactive design of the hardware implementation concept with the algorithm. An attempt to accomplish this goal has just been successfully completed [2]. The result is an algorithm which optimizes the detection of small, faint, noisy targets and which can be implemented in special purpose CCD structures for minimum size and power consumption.

The algorithm begins by applying a median filter to the image. This reduces both noise and texture and emphasizes the bimodality of the intensity histograms in local areas containing objects. Next several intensity thresholds are applied to the entirety of the filtered image in parallel and connected component analysis is applied to each of these segmented images. The result is a series of exemplars for each object in the scene. The best exemplar of each object is then chosen based on maximal agreement between the border of the connected component and the maximal strength edges which surround the object. The implicit assumption is that the maximal edge information provides the best (but least robust) indication of object shape while the object border is more robust (especially after median filtering) but potentially less accurate (due to improper threshold selection). The choice of exemplar with the best match of the two insures the benefits of both while minimizing the problems due to either.

The result is an algorithm which chooses the best threshold for each object in the image. The segmentation of small, noisy targets is optimized first by the median filter (which tends to make the segmented objects more compact and their edges more pronounced) and second by the incorporation of connectivity in the threshold selection process. Further, the object boundaries extracted are unusually clean and robust even at high noise level when compared to gradient detection schemes.

Test results of computer simulation of this algorithm are shown in Table 3. These results were achieved on the same imagery used to produce the other results shown. The improvement is obvious and significant. The hardware estimates shown for this algorithm are based on a detailed study which included construction of one of the key new CCD components. Only a few square inches of silicon is required for all the needed circuitry (the redundant thresholding and connected component analysis accounts for most of the space). Of course, implementation awaits sufficient demand to amortize the development of the special purpose CCD devices and is thus several years in the future.

6.4 Detection algorithm limitations

It is interesting to note that the performance noted for the best algorithm in Table 3 is essentially equal to that expected of a human observer on the same imagery except for the false alarm rate (the man's is generally much lower). Thus, we are on the verge of lacking even an "existence" proof that further improvements are possible within the limitations of the imagery used. It seems likely that further algorithm improvements will require use of additional data (e.g., multi-spectral properties or frame sequences for detection of relative target motion) and improvements to the classifier to extract and use higher level information. Likely candidates in this latter category include attempts to detect major scene components so that spatial context may be used and attempts to generate syntactical and other non-statistical classification schemes so that targets of variable shape can be detected.

It is also worth noting the continuing reduction of the hardware in the face of a concurrent massive increase in algorithm complexity. This implies that the major limitations are due to lack of optimality in the algorithms and that continued efforts in improved image modeling and algorithm development can be expected to produce major improvements.

REFERENCES

1. Prototype Automatic Target Screener, Quarterly Progress Report 21 September to 31 December 1977 to US Army, Night Vision and Electro-Optics Laboratory on Contract DAAK70-77-C-0248, January 15, 1978.
2. Algorithms and Hardware Technology for Image Recognition, Final Report to US Army, Night Vision and Electro-Optics Laboratory on Contract DAAG53-76C-0138, March 31, 1978.
3. Final Report on Automated Image Enhancement Techniques for Second Generation FLIR, Final Report to US Army, Night Vision and Electro-Optics Laboratory on Contract DAAG53-76-C-0195, December 1977.
4. Automatic Cueing Study for Helicopter Fire Control, Final Report, Phase II to US Army, Frankford Arsenal on Contract DAAA-25-72-C-0154, January 22, 1975.
5. Augmented Target Screener Subsystem (ATSS), Final Report (AFAL-TR-74-184) to US Air Force Avionics Laboratory, October 1974.
6. Demonstration of Westinghouse Automatic Cueing Techniques Using NV&EOL Imagery, Final Report to US Army, Night Vision and Electro-Optics Laboratory on Contract DAAG53-75-C-0225, May 23, 1976.
7. Flir Image Analysis with the Autoscreener Computer Simulation, Final Report to US Army, Night Vision and Electro-Optics Laboratory on Contract DAAG53-75-C-0264, February 1976.
8. Rubin, L. M., D. Y. Tseng, Automatic Target Cueing, to be presented at the 4th International Joint Pattern Recognition Conference, Kyoto, Japan, November 1978.

STOCHASTIC MODELS FOR IMAGES AND APPLICATIONS*

J. W. Modestino and R. W. Fries

Electrical and Systems Engineering Department
Rensselaer Polytechnic Institute
Troy, New York 12181

ABSTRACT. A useful class of two-dimensional (2-D) random fields is described and several of its more important properties discussed. This class of random fields is modeled as a marked point process evolving according to a spatial parameter. The autocorrelation and power spectral density properties are investigated for some special cases. Several applications are discussed including the modeling of real-world imagery possessing inherent edge structure. A class of efficient edge detection algorithms is described based upon 2-D least mean-square Wiener filtering concepts where the stochastic model for edge structure is chosen from this class of random fields. Additional applications are discussed.

1. INTRODUCTION

In a number of image processing applications it is important to have available a realistic and conveniently parameterized stochastic model for the class of images of interest. Examples include; image enhancement, image coding, texture discrimination and edge detection in noisy digitized images. This latter application provided the initial motivation for the work described here. It is clearly important in this case to have available a stochastic model for edge structure in two-dimensional (2-D) imagery data. This is also the case in the other applications cited as we shall see.

The ubiquitous 2-D Gaussian random field [1],[4], has often been proposed as an appropriate model for real-world imagery. This has been particularly the case in image coding applications (cf. [5],[6]). Unfortunately, the 2-D Gaussian random field, except

* This work was supported by ONR under Contract N00014-75-C-0281.

under pathological assumptions on the covariance, cannot account for the predominant and pronounced edge structure present in real-world imagery. Here we use the intuitive idea of an edge element as a point in the image field which lies on the boundary between two regions of distinctly different gray levels or intensities. It is of some interest then to develop alternative and more appropriate 2-D stochastic models for image data exhibiting pronounced edge structure.

In this paper a useful class of 2-D random fields appropriate for this purpose are described and several of their more important properties discussed. This class of random fields is modeled as a marked point process [7] evolving according to a spatial parameter. According to this model the plane is randomly partitioned into a number of disjoint geometric regions by an appropriately defined field of random lines which form the boundaries of these regions. The density of these random lines or edges is defined in terms of a rate parameter λ. Gray levels are then assigned within elementary regions to possess correlation coefficient ρ with gray levels in contiguous regions. Several schemes are described for partitioning the plane into elementary geometrical regions. Given a particular partitioning scheme, the random fields are completely defined in terms of the two parameters λ and ρ. The parameter λ represents the "edge business" associated with an image while ρ is indicative, at least on an ensemble basis, of the strength of an edge. For ρ large (in magnitude) and negative there is an abrupt almost black-to-white or white-to-black transition across an edge boundary. If $\rho>0$, on the other hand, the transition across an edge boundary is much more gradual. It is relatively easy to define these parameters for wide classes of images.

A number of applications of this model are discussed. These include; the development of a class of efficient edge detectors based upon 2-D least mean-square filtering concepts, image enhancement based upon 2-D homomorphic filtering concepts, stochastic source models for image coding and as models for texture in monochrome images.

2. PRELIMINARIES

We consider an image as a family of random variables $\{f_{\underline{x}}(\omega), \underline{x}\varepsilon R^2\}$, or a random field, defined on some fixed but unspecified probability space (Ω, a, P). It will prove particularly convenient to suppress the dependence upon the underlying probability space and consistently write $f(\underline{x})$ for $f_{\underline{x}}(\omega)$. The covariance function of the random field then becomes*

$$R_{ff}(\underline{x},\underline{y}) = E\{f(\underline{x})f(\underline{y})\} \ ; \ \underline{x},\underline{y}\varepsilon R^2 \tag{1}$$

* We assume the field is of second order (i.e., variances exist) and possesses zero mean.

where $E\{\cdot\}$ represents the expectation operator. If a random field $\{f(\underline{x}),\underline{x}\varepsilon R^2\}$ possesses a covariance function invariant under all Euclidean motions it will be called homogeneous and isotropic. (cf. [3] for definitions). In this case the covariance function of the field evaluated at two points can depend only upon the Euclidean distance between these two points so that

$$E\{f(\underline{x}+\underline{u})f(\underline{x})\} = R_{ff}(||\underline{u}||), \tag{2}$$

where $\underline{u}^T=(u_1,u_2)$ is an element of R^2 and $||\underline{u}||$ represents the ordinary Euclidean norm defined in terms of an inner product $<\cdot,\cdot>$ according to

$$||\underline{u}||^2 = <\underline{u},\underline{u}> = u_1^2+u_2^2 \ . \tag{3}$$

By construction, the 2-D random fields to be described here are of this category.

The corresponding power spectral density function is given by

$$S_{ff}(\underline{\omega}) = \int_{R^2} R_{ff}(||\underline{u}||)\exp\{-j<\underline{\omega}\ \underline{u}>\}d\underline{u}, \tag{4}$$

where $\underline{\omega}^T=(\omega_1,\omega_2)$ represents a 2-D spatial frequency vector and $d\underline{u}$ is the differential volume element in R^2. This expression can be evaluated up to functional form with the aid of a theorem of Buchner [7] with the result

$$S_{ff}(\underline{\omega}) = S_{ff}(r)=2\pi \int_0^\infty \lambda R_{ff}(\lambda)\ J_0(\lambda r)d\lambda \ , \tag{5}$$

where $r = ||\underline{\omega}|| = \sqrt{\omega_1^2+\omega_2^2}$ represents radial frequency. Here $J_0(\cdot)$ denotes the ordinary Bessel function of the first kind of order zero. Thus $S_{ff}(\cdot)$ and $R_{ff}(\cdot)$are related through a Hankel transform [8],[9].

3. CONSTRUCTION PROCEDURE

A fundamental role in the construction of this class of processes will be played by the integer-valued random field* $\{N(\underline{x}),\ \underline{x}\geq 0\}$ which provides a 2-D generalization of a counting process [11]. In particular, suppose the vector $\tilde{\underline{x}}$ is obtained from $\underline{x}$ according to $\tilde{\underline{x}}=\underline{\underline{A}}\ \underline{x}$ where $\underline{\underline{A}}$ is the unitary matrix

* By the notation $\underline{x}\geq 0$ we mean that $\underline{x}^T=(x_1,x_2)$ is such that $x_i\geq 0$, i=1,2.

$$\underline{\underline{A}}= \begin{bmatrix} \cos\theta & \sin\theta \\ -\sin\theta & \cos\theta \end{bmatrix} \tag{6}$$

defined for some $\theta\varepsilon[-\pi,\pi]$. This transformation results in a rotation of the Cartesian coordinate axes (x_1,x_2) by θ radians as illustrated in Fig. 1.

Consider now the integer-valued random field defined by

$$N(\underline{x}) = N_1(\tilde{x}_1) + N_2(\tilde{x}_2) \; ; \; \underline{x} \geq 0 \tag{7}$$

where $\theta\varepsilon[-\pi,\pi]$ is chosen according to some probability density function (p.d.f.) $P(\theta)$ and $\{N_i(\ell), \ell\geq 0\}$, i=1,2, are mutually independent 1-D counting processes. That is, $N_i(\ell)$ represents the number of events which have occurred in the interval $[0,\ell]$. We will be particularly concerned with the case where $\{N_i(\ell),\ell \geq 0\}$, i=1,2, are renewal point processes defined in terms of their interarrival distribution.

The random field $\{N(\underline{x}), \underline{x} \geq 0\}$ in (7) then assumes constant integer values on non-overlapping rectangles whose sides are parallel to the transformed axes $(\tilde{x}_1,\tilde{x}_2)$ and whose locations are determined by the event times of the corresponding point processes $\{N_i(\ell), \ell\geq 0\}$, i=1,2. Consider now the random field $\{f(\underline{x}),\underline{x} \geq 0\}$ which undergoes transitions at the boundaries of these elementary rectangles. The gray level assumed throughout any elementary rectangle is zero-mean Gaussian* with variance σ^2 and correlation coefficient ρ with the gray levels in contiguous rectangles. More specifically, let $X_{i,j}$ represent the amplitude or gray level assumed by the random field after i transitions in the $\tilde{x}_1$ direction and j transitions in the $\tilde{x}_2$ direction. The sequence $\{X_{i,j}\}$ is assumed generated recursively according to

$$X_{i,j} = \rho\, X_{i-1,j}+\rho\, X_{i,j-1}-\rho^2 X_{i-1,j-1}+W_{i,j} \; ; \; i,j\geq 1 \; , \tag{8}$$

where $|\rho|\leq 1$ and $\{W_{i,j}\}$ is a 2-D sequence of independent and identically distributed (i.i.d.) zero-mean Gaussian variates with common variance $\sigma_w^2=\sigma^2(1-\rho^2)^2$. The initial values $X_{k,0}$, $X_{0,\ell}$, $k,\ell\geq 0$ are jointly distributed zero-mean Gaussian variates with common variance σ^2 and covariance properties chosen to result in stationary conditions. An alternative interpretation of the sequence $\{X_{i,j}\}$ is as the output of a separable 2-D recursive filter excited by a white noise field. It is easily seen that

$$E\{X_{i,j}X_{i+k_1,j+k_2}\} = \sigma^2\rho^{k_1+k_2} \; ; \; k_1,k_2 \geq 0. \tag{9}$$

---

* For definiteness we assume Gaussian statistics. This assumption is not critical to the development which follows and is easily removed.

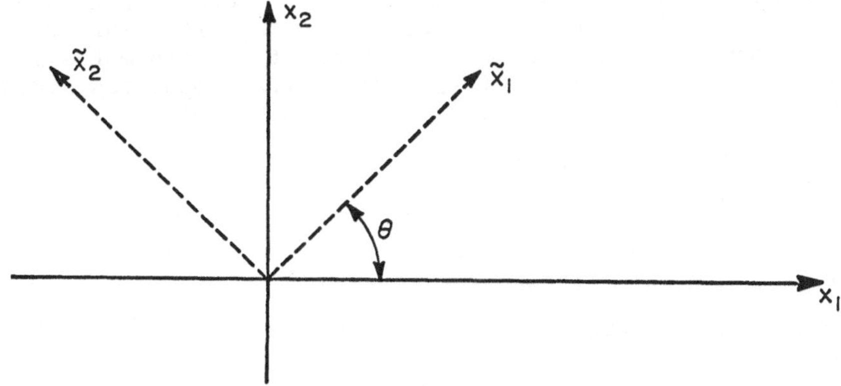

Figure 1

Rotation of Cartesian Coordinate Axes

Typical computer-generated realization of the resulting random field are illustrated in Fig. 2 for selected values of ρ when $p(\theta)$ is uniform over $[-\pi,\pi]$ and $\{N_i(\ell),\ell\geq 0\}$, $i=1,2$ are Poisson with intensities $\lambda_1=\lambda_2=\lambda$. The displayed images here and throughout the remainder of this paper are square arrays consisting of 256 elements or samples on a side. In Fig. 2, λ is measured in normalized units of events per sample distance so that there are an average 256 λ transitions along each of the orthogonal axes. Similarly in Fig. 3 we illustrate realizations of the resulting random field when the point processes $\{N_i(\ell), \ell\geq 0\}$, $i=1,2$, undergo jumps of unit height at equally spaced intervals $\ell=1/\lambda$. The starting positions ε_i, $i=1,2$, will be assumed uniformly distributed over the interval $[0,\ell]$.

The preceding two examples are special cases of the situation where the point processes $\{N_i(\ell), \ell\geq 0\}$, $i=1,2$, are delayed renewal processes [11], [12] with Gamma distributed interarrival times. This class of random fields represents a 2-D generalization of the class of 1-D processes described in [13]. In particular, we assume the common interarrival distribution of the two mutually independent point processes $\{N_i(\ell), \ell\geq 0\}$, $i=1,2$, possesses p.d.f.

$$f(x) = \frac{x^{\nu-1}}{\Gamma(\nu)\beta^{\nu}} \exp\{-x/\beta\} \tag{10}$$

where $\nu=1,2...$ and $\beta=1/\lambda\nu$ for fixed $\lambda>0$. For example, if $\nu=1$ we have the exponential distribution

$$f(x) = \lambda e^{-\lambda x} \quad ; \quad x\geq 0 \tag{11}$$

associated with the Poisson process, while in the limit $\nu\to\infty$ we have

$$f(x) = \delta(x-1/\lambda) \quad ; \quad x\geq 0 \tag{12}$$

corresponding to the case of periodic partitions as illustrated in Fig. 3.

In Fig. 4 we illustrate selected realizations of the resulting random field for several values of ν all with $\lambda=0.05$ and $\rho=0.0$. Clearly the parameter ν provides a measure of the degree of randomness or homogeneity of the edge structure. For small ν the random field $\{f(\underline{x}), \underline{x}\varepsilon R^2\}$ appears as a random rectangular mosaic. As ν increases individual realizations rapidly approach a more periodic mosaic in appearance. The parameters λ, ρ and ν then completely describe this class of 2-D random fields.

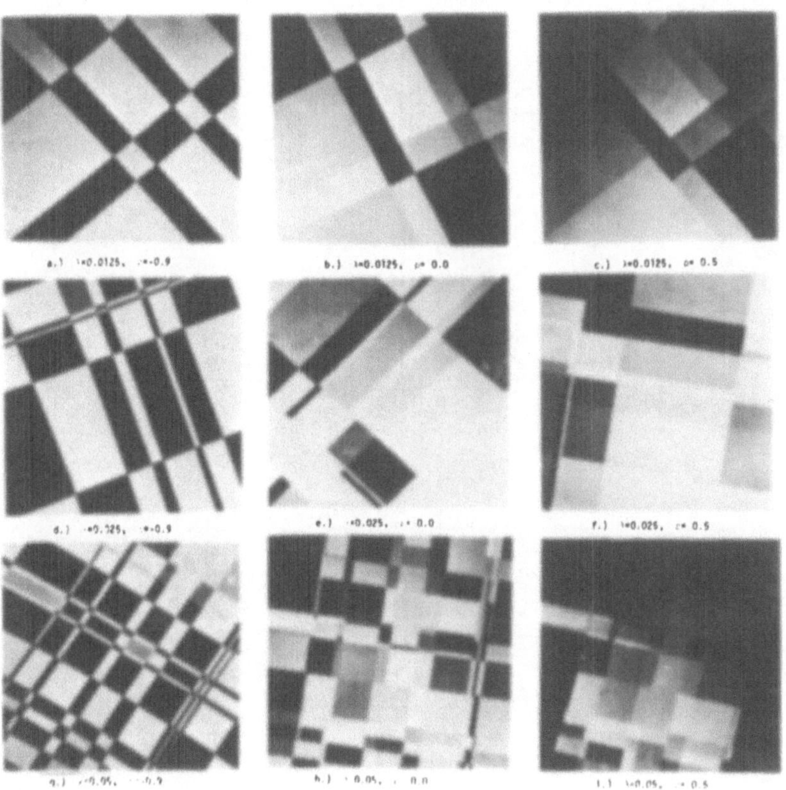

Figure 2

Selected Realizations of Random Field Generated by Poisson Partitions

Figure 3

Selected Realizations of Random Field Generated by Periodic Partitions

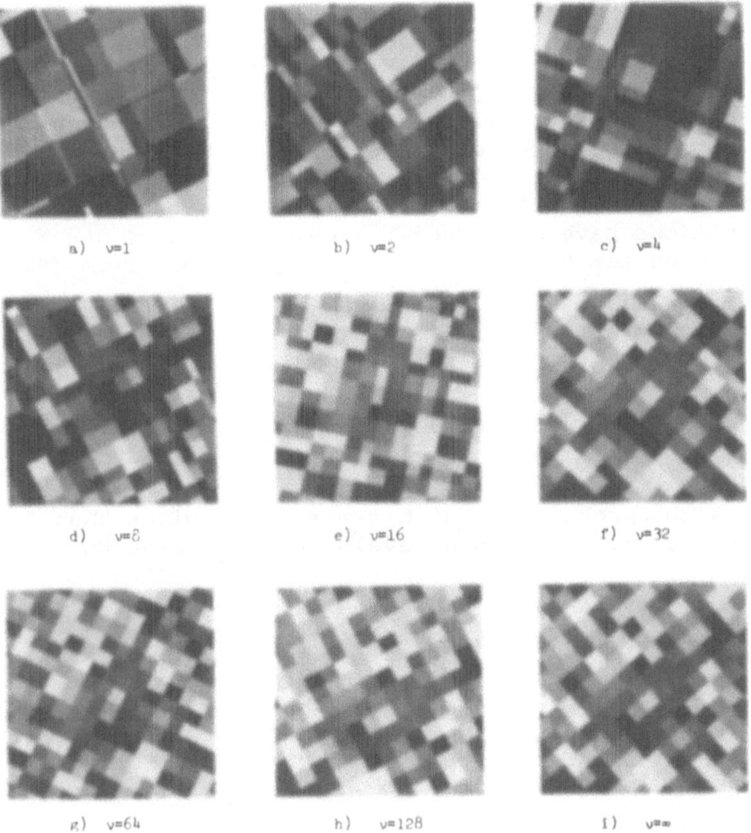

Figure 4

Selected Realizations of Random Field Generated by Delayed Renewal Point Process Possessing Gamma Distributed Interarrival Distribution and with $\lambda=0.05$ and $\rho=0.0$

4. SECOND-ORDER PROPERTIES

We turn now to the second-order properties of the class of 2-D random fields described in the preceding section. In the interests of brevity the treatment will be condensed and make extensive use of results reported elsewhere.

As a first step in the development of the covariance function, assume that the random orientation $\theta\varepsilon[-\pi,\pi]$ has been chosen and that k transitions have occurred* between the two points $\underline{x}$ and $\underline{x}+\underline{u}$ where we assume for the moment $\underline{u}>0$. It follows from (9) that

$$E\{f(\underline{x}+\underline{u})f(\underline{x})|\theta,k\} = \sigma^2\rho^k \; ; \quad k=0,1,2.. \tag{13}$$

The conditioning upon k is easily removed according to

$$E\{f(\underline{x}+\underline{u})f(\underline{x})|\theta\} = \sum_{k=0}^{\infty} E\{f(\underline{x}+\underline{u})f(\underline{x})|\theta,k\}\, p_{k|\theta}, \tag{14}$$

where $p_{k|\theta}$ is the probability of k transitions given that θ is acting. In particular, $p_{k|\theta}$ can be evaluated according to

$$p_{k|\theta} = \sum_{j=0}^{k} q^{(1)}_{k-j|\theta}\, q^{(2)}_{j|\theta} \; ; \quad 0,1,2..., \tag{15}$$

where $q^{(i)}_{j|\theta}$ is the probability that $\{N_i(\ell),\ \ell\geq 0\}$ has undergone j transitions in the interval $\tilde{u}_i$, i=1,2, which depends upon $\underline{u}^T = (u_1,u_2)$ and θ according to

$$\tilde{u}_1 = u_1\cos\theta+u_2\sin\theta, \tag{16a}$$

and

$$\tilde{u}_2 = u_2\cos\theta-u_1\sin\theta. \tag{16b}$$

Substituting (13) and (15) into (14) we obtain

$$E\{f(\underline{x}+\underline{u})f(\underline{x})|\theta\}=\sigma^2 \sum_{k=0}^{\infty} \rho^k \sum_{j=0}^{k} q^{(1)}_{k-j|\theta}\, q^{(2)}_{j|\theta} \;, \tag{17}$$

and by simple rearrangement of the double summation in this last expression we find

---

* By this we mean that $k=k_1+k_2$ where k_i, i=1,2, represents the number of transitions along each of the orthogonal axes which have now been rotated by θ radians.

$$E\{f(\underline{x}+\underline{u})f(\underline{x})|\theta\} = \sigma^2 \sum_{j=0}^{\infty} \sum_{k=j}^{\infty} \rho^{k-j} q^{(1)}_{k-j|\theta} \rho^{j} q^{(2)}_{j|\theta} \tag{18}$$

$$= \sigma^2 \left[\sum_{m=0}^{\infty} \rho^m q^{(1)}_{m|\theta} \right] \cdot \left[\sum_{n=0}^{\infty} \rho^n q^{(2)}_{n|\theta} \right]$$

Assuming a uniform distribution for θ, it follows that the covariance function becomes

$$R_{ff}(\underline{x}+\underline{u},\underline{x}) = \frac{1}{2\pi} \int_{-\pi}^{\pi} E\{f(\underline{x}+\underline{u})f(\underline{x})|\theta\}d\theta \tag{19}$$

with the integrand given by (18). While not immediately apparent, it is easily shown that this last expression depends only upon $||\underline{u}||$ so that the resulting random field is indeed homogeneous and isotropic.

While explicit evaluation of (19) is in general quite cumbersome, it can be evaluated in special cases. For example, in the Poisson case $\nu=1$, it can be shown [14] that

$$R_{ff}(||\underline{u}||) = \frac{2\sigma^2}{\pi} \int_{0}^{\pi/2} \exp\{-\sqrt{2}(1-\rho)\lambda||\underline{u}||\cos(\theta-\pi/4)\}d\theta \tag{20}$$

while the corresponding power spectral density computed according to (5) becomes

$$S_{ff}(r) = \frac{8(1-\rho)\lambda\sigma^2}{r^2+2(1-\rho)^2\lambda^2} \left[\frac{1}{r^2+(1-\rho)^2\lambda^2} \right]^{1/2} \tag{21}$$

Typical covariance surfaces together with intensity plots of the corresponding power spectral density in the case of periodic partitions (i.e., $\nu=\infty$) are illustrated in Fig. 5. The autocorrelation function is plotted as a function of the normalized spatial variable $||\underline{u}||\lambda$ over the range $0 \leq ||\underline{u}||\lambda \leq 3$, while the power spectral density is plotted as a function of the normalized spatial frequency variable $||\underline{\omega}||/2\pi\lambda$ over the range $0 \leq ||\underline{\omega}||/2\pi\lambda \leq 5$. Additional details can be found in [15]. Explicit evaluation of these quantities for the general case of Gamma distributed interarrival times is provided in [16].

5. NON-RECTANGULAR PARTITIONS

Although this class of 2-D random fields has proven useful in a number of image processing applications, the rectangular pattern

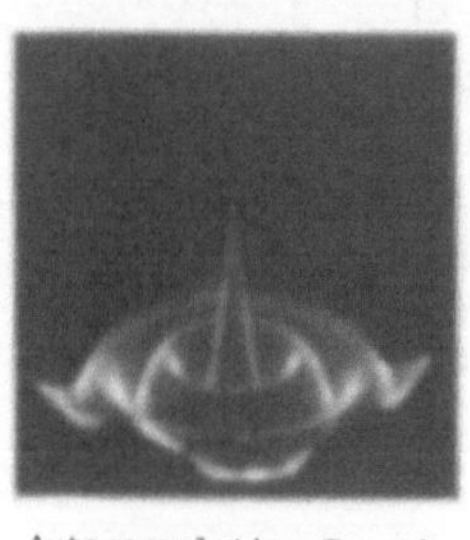

a) Autocorrelation Function
$\rho = -0.9$

b) Power Spectral Density
$\rho = -0.9$

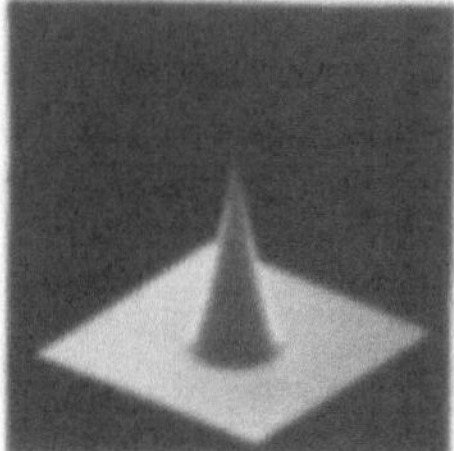

c) Autocorrelation Function
$\rho = 0.0$

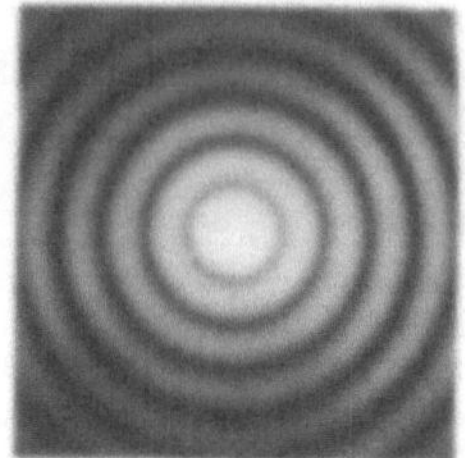

d) Power Spectral Density
$\rho = 0.0$

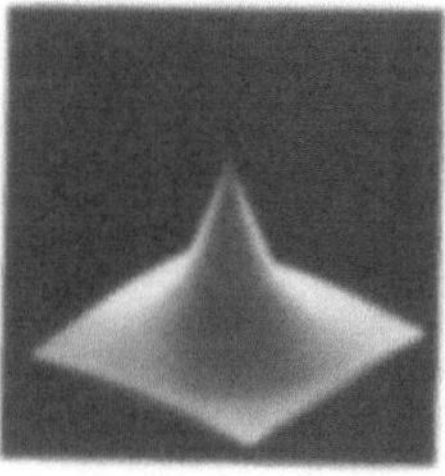

e) Autocorrelation Function
$\rho = 0.5$

f) Power Spectral Density
$\rho = 0.5$

Figure 5
Autocorrelation Function and Power Spectral Density of 2-D Random Checkerboard Process

exhibited by individual realizations is somewhat disturbing. That is, edge structure in real-world imagery does not generally exhibit this rectangular mosaic but a much more random edge orientation. Here we describe two distinct approaches to removing this deficiency.

Parallelogrammical Partitions: Again the plane is partitioned by two mutually independent renewal point processes evolving along appropriately defined coordinate axes. In this case, however, these coordinate axes are determined by a non-unitary or nonisometric transformation of the Cartesian coordinate frame. More specifically, we suppose the vector $\underline{\tilde{x}}$ is obtained from $\underline{x}$ according to the linear transformation $\underline{\tilde{x}} = \underline{\underline{A}}\underline{x}$ where now $\underline{\underline{A}}$ is the 2x2 matrix defined for $\phi,\theta\varepsilon[-\pi,\pi]$ according to

$$\underline{\underline{A}} = \frac{1}{\sin\phi}\begin{bmatrix} \sin(\theta+\phi) & -\cos(\theta+\phi) \\ -\sin\theta & \cos\theta \end{bmatrix} \tag{22}$$

This transformation has an interpretation as a distance preserving rotation of the Cartesian coordinate frame by θ radians followed by a non-distance preserving scaling. The new coordinate axes $(\tilde{x}_1,\tilde{x}_2)$ are illustrated in Fig. 6. In what follows we assume that the angle ϕ is fixed while θ is chosen uniformly on $[-\pi,\pi]$. The point processes $\{N_i(\ell),\ \ell\geq 0\}$ now evolving along the respective coordinate axes $\tilde{x}_i$, $i=1,2$, result in a partition of the plane into elementary regions comprised of disjoint parallelograms whose sides are parallel to the new coordinate axes. Gray levels are then assigned within these elementary regions as described previously.

In Fig. 7 we illustrate typical realizations of the resulting 2-D random field for selected values of ϕ all with $\lambda=0.025$ and $\rho=0.0$. The point processes generating the random field in this case were poisson corresponding to Gamma distributed interarrival times with parameter $\nu=1$. The second-order properties of this class of random fields for arbitrary ν are described in [16]. In particular, they are shown to be homogeneous and isotropic.

Polygonal Partitions: As a second approach to the construction of 2-D random fields with random edge orientation, consider the partition of the plane R^2 by a field of random sensed lines. More specifically, an arbitrary sensed line can be described in terms of the 3-tuple (r,θ,d). Here r represents the perpendicular or radial distance to the line in question, $\theta\varepsilon[-\pi,\pi]$ represents the orientation of this radial vector, and finally d is a binary random variable assuming values ± 1 which specifies the sense or direction imparted to this line segment. The pertinent geometry is illustrated in Fig. 8 for the case d=1. By virtue of the direction imposed on this line segment the plane is partitioned into two disjoint regions, R(right of line) and L(left of line) such that $R\cup L=R^2$.

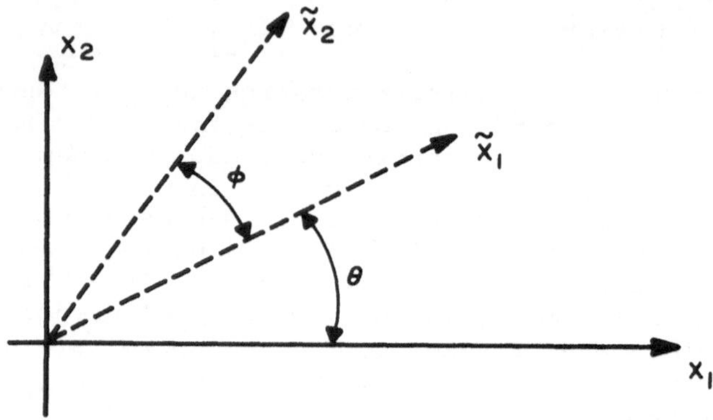

Figure 6

Nonisometric Transformation of Cartesian Coordinate Frame

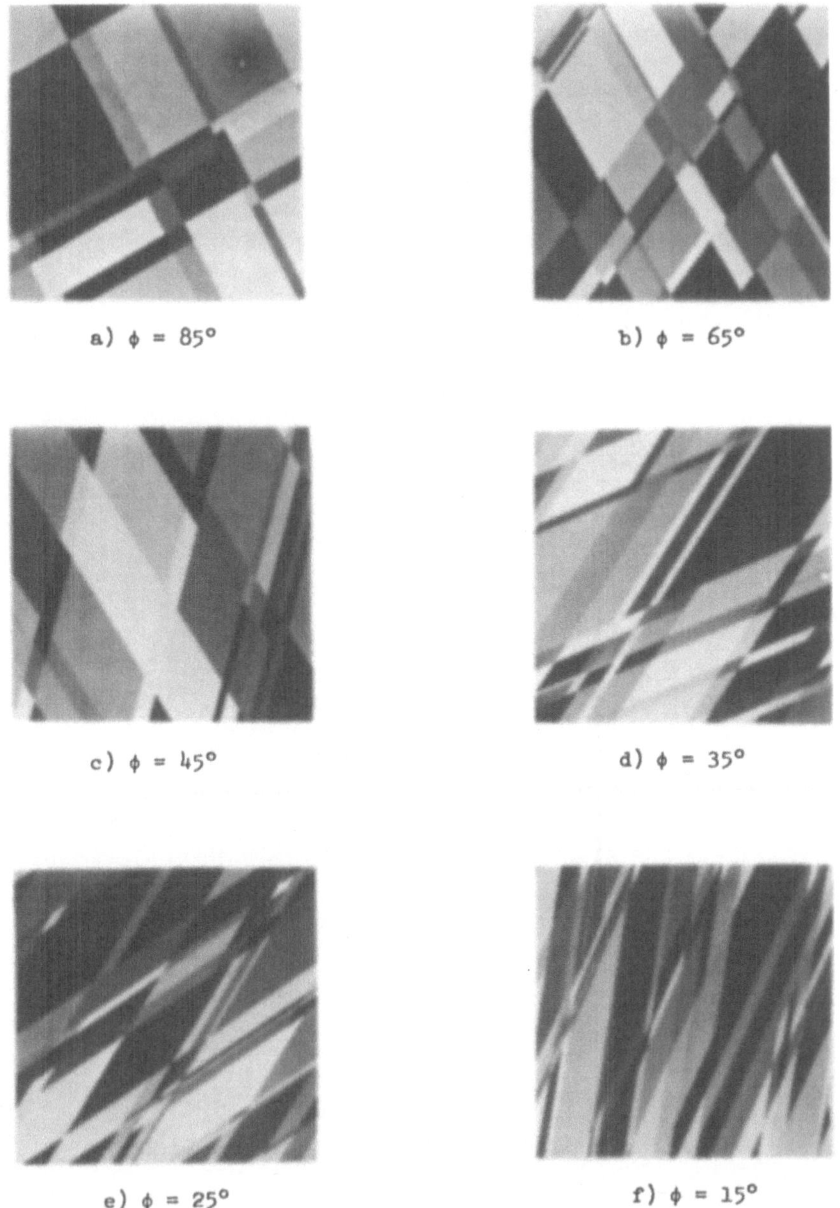

a) $\phi = 85°$ b) $\phi = 65°$

c) $\phi = 45°$ d) $\phi = 35°$

e) $\phi = 25°$ f) $\phi = 15°$

Figure 7

Selected Realizations of Non-Rectangular Random Field Generated by Poisson Point Processes with $\lambda=0.025$ and $\rho=0.0$

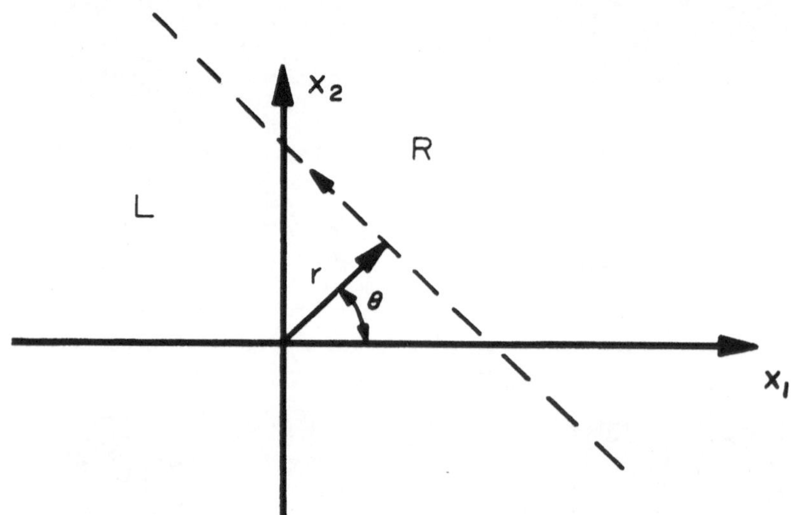

Figure 8

Geometry Defining Random Line on the Plane

Now consider the field of line generated by the sequence $\{r_i, \theta_i, d_i\}$. Here the sequence $\{r_i\}$ represents the "event times" associated with a homogeneous Poisson process $\{N(r), r \geq 0\}$ with intensity λ events/unit distance evolving according to the radial parameter r. The sequence $\{\theta_i\}$ is i.i.d. and uniform on $[-\pi, \pi]$ while $\{d_i\}$ is also i.i.d. assuming the values ± 1 with equal probability.

The field of random lines so generated results in a partition of the plane into disjoint polygonal regions. Gray levels are assigned as described in [17] to result in correlation coefficient ρ with gray levels in contiguous regions. Typical realizations of the resulting random field are illustrated in Fig. 9 for selected values of $\lambda_e = \lambda/\pi$ and ρ. This class of random fields is shown in [17] to be homogeneous and isotropic with covariance function

$$R_{ff}(||\underline{u}||) = \sigma^2 \exp\{-\lambda_e(1-\rho)||\underline{u}||\} \tag{23}$$

and power spectral density

$$S_{ff}(r) = \frac{2\pi\sigma^2(1-\rho)\lambda_e}{[r^2+(1-\rho)^2\lambda_e^2]^{3/2}} \tag{24}$$

6. APPLICATIONS

We consider now some selected applications of the 2-D random fields described in the preceding.

Edge Detection: This problem is treated in some detail in [14]. We assume that the true edge structure in an image is described by the random field $f(\underline{x})$ modeled as one of the previously developed 2-D random fields. In many applications, the observed image is a noise corrupted version of $f(\underline{x})$ described by

$$g(\underline{x}) = f(\underline{x}) + n(\underline{x}) \tag{25}$$

where $n(\underline{x})$ is a zero-mean homogeneous and isotropic noise field possessing noise spectral density $S_{nn}(r) = \sigma_n^2$, i.e., a white noise field. Here one assumes that the noise field $n(\underline{x})$ represents any additive noise or spurious detail not considered part of the essential contours or edges represented by the random field $f(\underline{x})$.

In [14] the problem of edge detection was posed as a 2-D Wiener filtering problem. More specifically, if $\ell(\underline{x})$ represents the output of some desired operation on f(x) then design the imaging system with optical transfer function (OTF) $H_0(\underline{\omega})$ whose output $\hat{\ell}(\underline{x})$ in response to $g(\underline{x})$ at its input minimizes the mean-square error

a) λ_e=.0125, ρ=-0.9 b) λ_e=.0125, ρ=0.0 c) λ_e=.0125, ρ=0.5

d) λ_e=.025, ρ=-0.9 e) λ_e=.025, ρ=0.0 f) λ_e=.025, ρ=0.5

g) λ_e=0.05, ρ=-0.9 h) λ_e=0.05, ρ=0.0 i) λ_e=0.05, ρ=0.5

Figure 9

Selected Realizations of Polygonal Partition Process Generated by Poisson Line Process

$$I_e = E\{[\ell(\underline{x})-\hat{\ell}(\underline{x})]^2\} \tag{26}$$

Assuming the desired operation possesses OTF $H_d(\underline{\omega})=||\underline{\omega}||^2\exp\{-\frac{1}{2}||\underline{\omega}||^2\}$ (cf.[14] for justification) the optimum Wiener filter is isotropic with OTF.

$$H_0(r) = \frac{r^2 e^{-r^2/2} S_{ff}(r)}{S_{ff}(r)+\sigma_n^2} \quad ; \quad r \geq 0 \tag{27}$$

For example, if $\{f(\underline{x}),\ x\varepsilon R^2\}$ is the rectangular 2-D random field with Poisson partitions then the power spectral density $S_{ff}(r)$ is given by (21). The resulting Wiener filter is then completely defined in terms of the three parameters λ,ρ and $\zeta \triangleq \sigma^2/\sigma_n^2$ which represents the signal-to-noise ratio (SNR). Typical results are illustrated in Fig. 10 employing a digital implementation of the optimum Weiner filter. Additional results appear in [18].

Image Enhancement: This class of 2-D random fields has found application as a stochastic model for spatially varying illumination in homomorphic filtering of images. More specifically, consider the formation of an image as indicated in Fig. 11. Here the observed image $s_i(\underline{x})$ is the product of the true image $s(\underline{x})$ and the illumination function $i(\underline{x})=\exp\{f_i(\underline{x})\}$ where $f_i(\underline{x})$ is a 2-D random field with polygonal partitions as described in the preceding section. The true image $s(\underline{x})$ will be similarly modeled although the edge density will be assumed much higher than that of the illumination process. Finally, we assume that the observational or recording process introduces the multiplicative white noise field $n(\underline{x})$. The image available for processing is then

$$s_0(\underline{x}) = s(\underline{x})\cdot i(\underline{x})\cdot n(\underline{x}) \tag{28}$$

As indicated in Fig. 11, by a simple application of homomorphic filtering concepts [19],[20] it is possible to design a linear least mean-square spatial filter to provide an estimate $\hat{s}(\underline{x})$ of the true image field. This approach attempts to minimize the effects of both the nonconstant illumination and the multiplicative noise. Typical results are indicated in Fig. 12 for a transaxial tomography image. Additional results can be found in [16].

Texture Discrimination: This class of 2-D random fields has also been used as a model for texture discrimination studies. For example, in Fig. 13 we illustrate a mixture of several textures all with $\lambda=0.05$ and $\rho=0.5$. In the upper left-hand corner we have the polygonal partition process, in the lower right-hand corner we have the rectangular partition process with $\nu=\infty$ (periodic) while in the lower left-hand corner we have a sample with $\nu=1$ (Poisson). The problem is to distinguish the boundaries of these regions. This is typical of the image segmentation problem on the basis of local texture. An approach to this problem is described in [16].

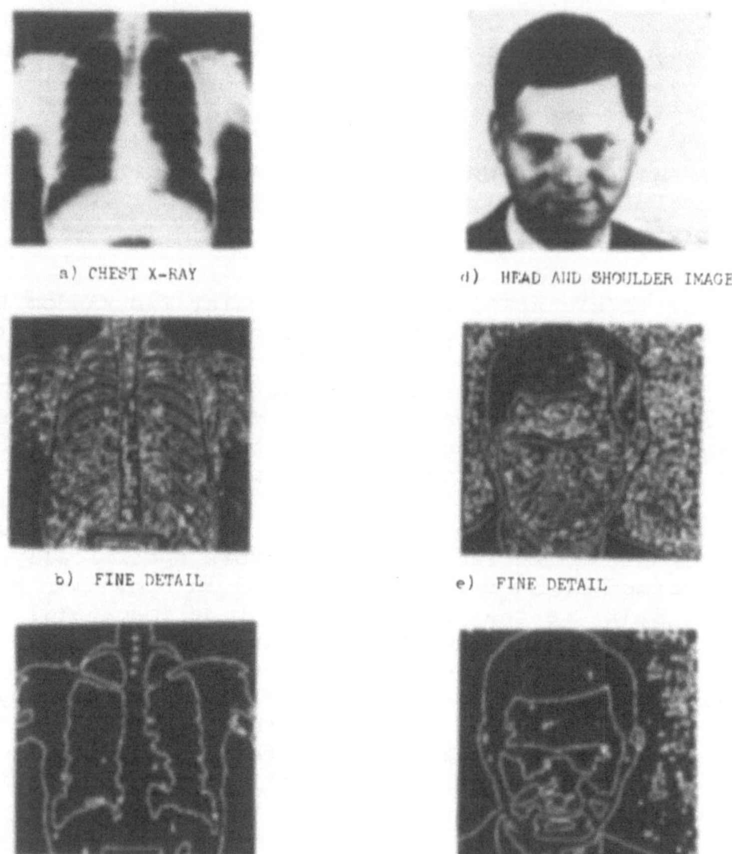

a) CHEST X-RAY

b) FINE DETAIL

c) COARSE DETAIL

d) HEAD AND SHOULDER IMAGE

e) FINE DETAIL

f) COARSE DETAIL

Figure 10

Typical Results of Wiener Edge Detector
Applied to Real-World Images

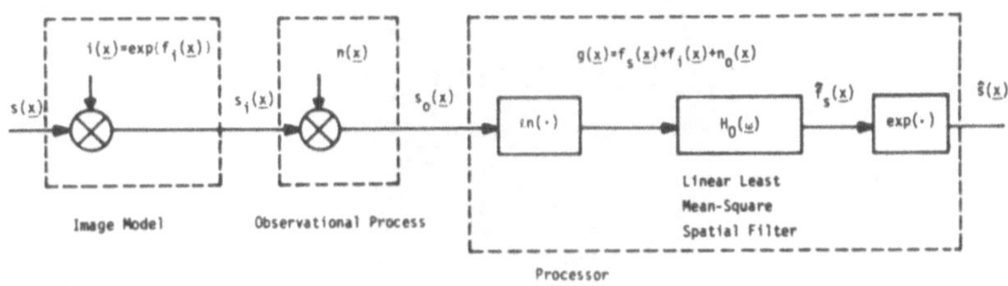

Figure 11

Homomorphic Filtering of Degraded Images

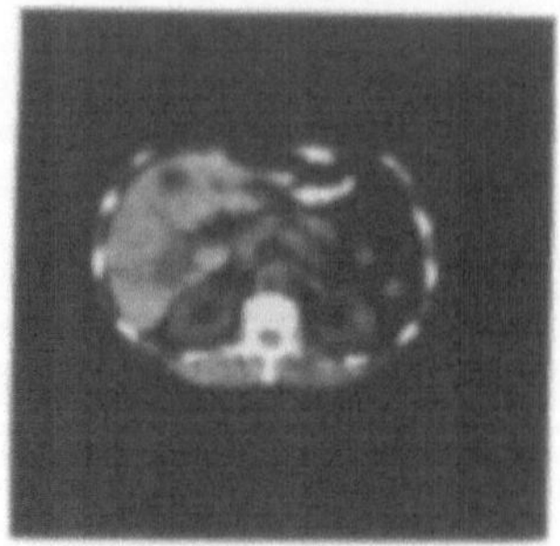

a) ORIGINAL

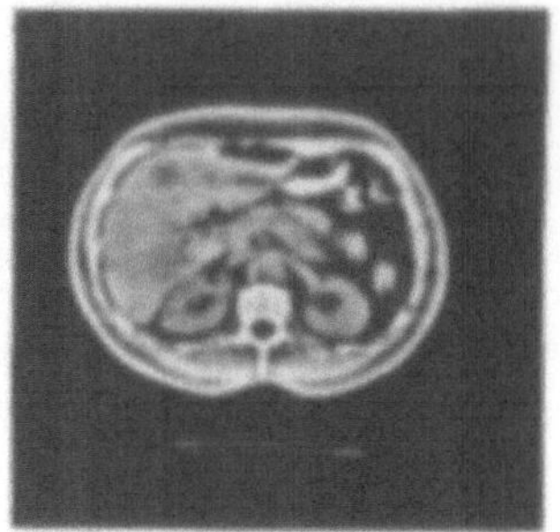

b) PROCESSED VERSION 1

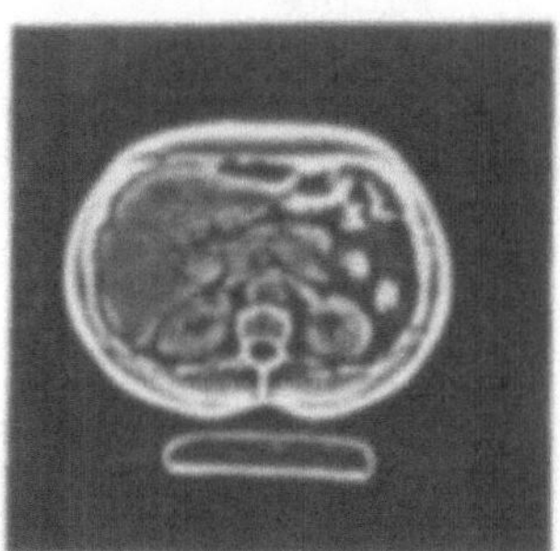

c) PROCESSED VERSION 2

Figure 12
Typical Results of Homomorphic Filtering of Transaxial Tomography Image

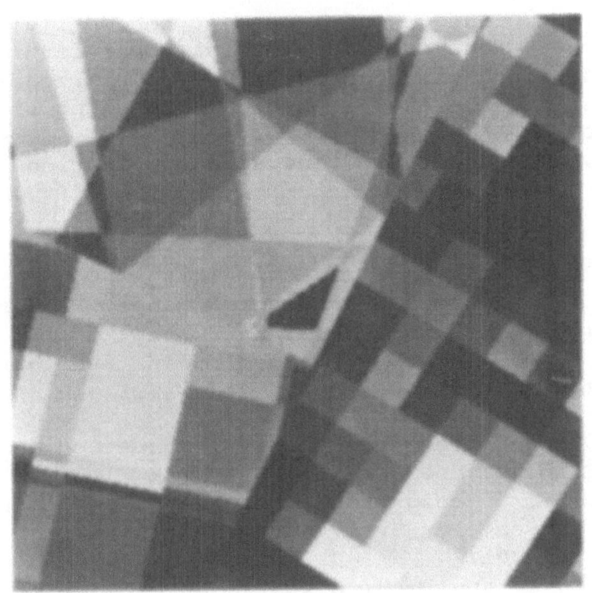

Figure 13

Illustration of Different Textured Regions Described by Random Fields with $\lambda=0.05$, $\rho=0.5$

Image Coding: Studies have been made of the performance of 2-D DPCM encoders operating on this class of random fields [21]. Work is in progress to determine optimum encoding schemes.

7. SUMMARY AND CONCLUSIONS

A class of 2-D homogeneous and isotropic random fields has been described which we feel provide a useful model for real-world imagery possessing pronounced edge structure. This model is conveniently parameterized by several physically meaningful quantities. Several of the more important properties of this class of 2-D random fields have been discussed and some applications described.

REFERENCES

1. Wong, E., "Homogeneous Gauss-Markov Random Fields", Ann. Math. Stat., vol. 40, pp. 1625-1634, 1969.
2. Wong, E., "Two-Dimensional Random Fields and the Representation of Images", SIAM J. Appl. Math., vol. 16, pp. 756-770, 1968.
3. Wong, E., Stochastic Processes in Information and Dynamical Systems, Chap. 7, McGraw-Hill, New York, 1971.
4. Yaglom, A. M., "Second-Order Homogeneous Random Fields", In Proc. Fourth Berkeley Symp. Math. Stat. and Prob., vol. 2, pp. 593-620, 1961.
5. Sakrison, D. J. and V. R. Algazi, "Comparison of Line-by-Line and Two-Dimensional Encoding of Random Images", IEEE Trans. Inform. Theory, vol. IT-17, pp. 386-398, July 1971.
6. O'Neal, J. B. Jr., and T. Raj Natarajan, "Coding Isotropic Images", IEEE Trans. Inform. Theory, vol. IT-23, pp. 697-707, Nov. 1977.
7. Snyder, D. L., Random Point Processes, Wiley, New York, 1975.
8. Bochner, S., Lectures on Fourier Integrals, Annals of Math. Studies, Nov. 42, Princeton Univ. Press, Princeton, N. J., pp. 235-238, 1959.
9. Papoulis, A., "Optical Systems, Singularity Functions, Complex Hankel Transforms", J. Opt. Soc. Amer., vol. 57, pp. 207-213, 1967.
10. Papoulis, A., Systems and Transforms with Applications in Optics, McGraw-Hill, New York, 1968.
11. Parzen, E., Stochastic Processes, Holden-Day, San Francisco, 1962.
12. Cinlar, E., Introduction to Stochastic Processes, Prentice-Hall, Englewood Cliffs, N. J., 1975.
13. Modestino, J. W., and R. W. Fries, "A Generalization of the Random Telegraph Wave", submitted to IEEE Trans. on Inform. Theory.

14. Modestino, J. W., and R. W. Fries, "Edge Detection in Noisy Images Using Recursive Digital Filtering", Computer Graphics and Image Processing 6, pp. 409-433, Oct. 1977.
15. Modestino, J. W., R. W. Fries and D. G. Daut, "A Generalization of the Two-Dimensional Random Checkerboard Process", submitted to J. Opt. Soc. Amer.
16. Modestino, J. W., and R. W. Fries, "Construction and Properties of a Useful Two-Dimensional Random Field", submitted to IEEE Trans. on Inform. Theory.
17. Fries, R. W., "Theory and Application of a Class of Two-Dimensional Random Fields", Ph. D. Dissertation, Electrical and Systems Engineering Department RPI, Troy, New York, 1978.
18. Fries, R. W., and J. W. Modestino, "An Empirical Study of Selected Approaches to the Detection of Edges in Noisy Digitized Images", Proc. of 1977 IEEE Computer Society Conf. on Pattern Recognition and Image Processing, Troy, New York, June 1977.
19. Oppenheim, A. V., R. W. Schafer, and T. G. Stockhom, Jr., "Nonlinear Filtering of Multiplied and Convolved Signals", Proc. IEEE, vol. 56, pp. 1264-1291, Aug. 1968.
20. Oppenheim, A. V. and R. W. Schafer, Digital Signal Processing, Prentice-Hall Englewood Cliffs, N. J., Chap. 10, 1975.
21. Daut, D. G., "An Empirical Study of Two-Dimensional Differential Pulse Code Modulation Encoding of Images", M. S. Thesis, Electrical and Systems Engineering Department, RPI, Troy, New York, July, 1977.

FEATURE-BASED SCENE ANALYSIS AND MODEL MATCHING

Carol S. Clark, Anthony L. Luk, and Charles A. McNary

Hughes Research Laboratories, 3011 Malibu Canyon Road, Malibu, California 90265

ABSTRACT. This paper describes a feature-based scene-matching technique that offers a viable alternative to more conventional correlation-based techniques. It effectively combines top-down (artificial-intelligence) and bottom-up (pattern-recognition) strategies for real-time image analysis, and it is applicable to a variety of image-analysis problems, including change detection, damage assessment, and image-based guidance. The basis for this type of analysis is the identification, characterization, and extraction of key scene features, which become the elements of unique scene models. The specific application of feature-based scene modeling discussed in this paper is automatic, image-based guidance for terminal homing and aimpoint maintenance.

Two fundamental feature-extraction techniques, region bounding and region growing, are exploited to model natural terrain and complex structural imagery. Region bounding is the subject of this paper. It is based on contrast-edge extraction, edge-element filtering, line-segment generation, and line linking. The resultant lines are the basis for a composite line and vertex scene model. This model is stored as a data structure including the feature identifier, its priority, and parameters such as position, orientation, and contextual relationships with other model features.

Scene model matching for match-point identification involves the sequential matching of local features in the reference and sensed models, calculation of sensed-to-reference model transformations based on the local matches, and verification of the matching transformation either globally or piecewise locally over the matched models. The resultant transformation provides the platform position information necessary for the trajectory updating and precision match-point maintenance.

Scene model generation and matching results are included to show the status of our algorithm development and the robustness of these techniques for multispectral, complex structural scenes.

1. INTRODUCTION

Feature-based scene analysis of outdoor scenes requires the identification, characterization, and extraction of key scene features. Since these candidate features are both environmentally and sensor dependent, an additional requirement of feature invariance as a function of the imaging sensor, atmospheric conditions, scene content, and surface conditions must be satisfied to guarantee robust scene models. The requirements for feature invariance and easily implementable feature extraction techniques have restricted our choice of region bounding features to lines (straight and curvilinear), vertices, and simple geometric boundaries (circles and ellipses).

The low-level feature for line generation and boundary formation is the contrast edge. Following a global edge extraction process, edge filtering algorithms are implemented to reject edge elements that do not contribute to line segment generation. Candidate edge element groups for line segment generation are formed on the bases of edge element orientation and local image space grouping of adjacent, similarly oriented edge elements. Line segments are fitted to the resultant edge element groups and tested for goodness-of-fit. Two processes follow the generation of straight line segments: linking of locally collinear line segments and vertex location at line intersections. The resultant straight lines, curvilinear contours, and vertices provide the principal high-level features for scene modeling.

Scene models are then generated from the line and vertex features. These models consist of data structures listing the feature type, priority, position, orientation, and contextual links to other features. These models are matched to prestored, prioritized reference models of the image area, with the resulting matching transformation providing the scale, rotation, displacement, and perspective change measures necessary to estimate the imaging sensor position. Reference model feature priorities are assigned on the basis of feature invariance to sensor, atmospheric, and scene conditions, thus permitting dynamic reference model revision based on known imaging conditions. The final step in this image-based guidance scenario is the derivation of guidance commands to the platform control surfaces to maintain the required trajectory and to minimize the match-point spherical error probability.

2. MODEL BUILDING

2.1 Edge-element extraction

In developing feature-based models of structural and natural scenes, our most important low-level feature is the edge element. It is the basic building block of our line and vertex scene models and is being incorporated in our region-based scene models as a measure of texture.

Two approaches have been taken to the extraction of edge elements. First, the complex Heuckel [1] edge operator was implemented with a large extraction window to provide high tolerance to noise. This same operator, however, degrades in performance with complex industrial and structural scenes having high edge densities. Corners are poorly defined because of the large extraction window, and the presence of multiple edge elements within the window results in inaccurate edge element location.

Our second approach to edge-element extraction uses the Sobel [2] operator with a small 9-pixel extraction window. This operator effectively accommodates the high edge densities of structural scenes and establishes boundary and corner locations accurately, but at the expense of reduced noise tolerance. Numerous noise edges are extracted and must be separated from those contributing to straight-line segments in the scene.

The Sobel edge operator is applied to each pixel position in an image (with the exception of a 1-pixel band at the image boundary) and extracts gradients in two principal directions. This process is shown schematically in Figure 1. The edge magnitude, the sum of the magnitudes of the two gradients, is tested against a preset (or dynamically adjustable) threshold to establish the presence of an edge. Each identified edge is further parameterized by its orientation, $\theta \equiv \tan^{-1} \nabla y/\nabla x$, and principal direction. The edge element principal direction is defined as the direction of its principal gradient, thus dividing the extracted edge elements into four directional groups separated by the 45° diagonals in the x-y image plane (see Figure 2). This principal direction parameter is used to simplify the composite edge filtering process described next.

7610-1

A	B	C
D	E	F
G	H	J

$$\nabla x = (C + 2F + J) - (A + 2D + G)$$

$$\nabla y = (G + 2H + J) - (A + 2B + C)$$

- SOBEL EDGE GRADIENTS, ∇x AND ∇y, ARE EXTRACTED WITH 9-ELEMENT WINDOW CENTERED AT PIXEL POSITION E.

- EDGES ARE ESTABLISHED BY THRESHOLD TEST OF EDGE MAGNITUDE, $|\nabla x| + |\nabla y|$

  IF $\dfrac{|\nabla x| + |\nabla y|}{T} > 1$, EDGE IS ESTABLISHED AT PIXEL POSITION E.

- EDGE ORIENTATION, $\theta = \tan^{-1} [\nabla y/\nabla x]$.

Figure 1. Sobel edge extraction.

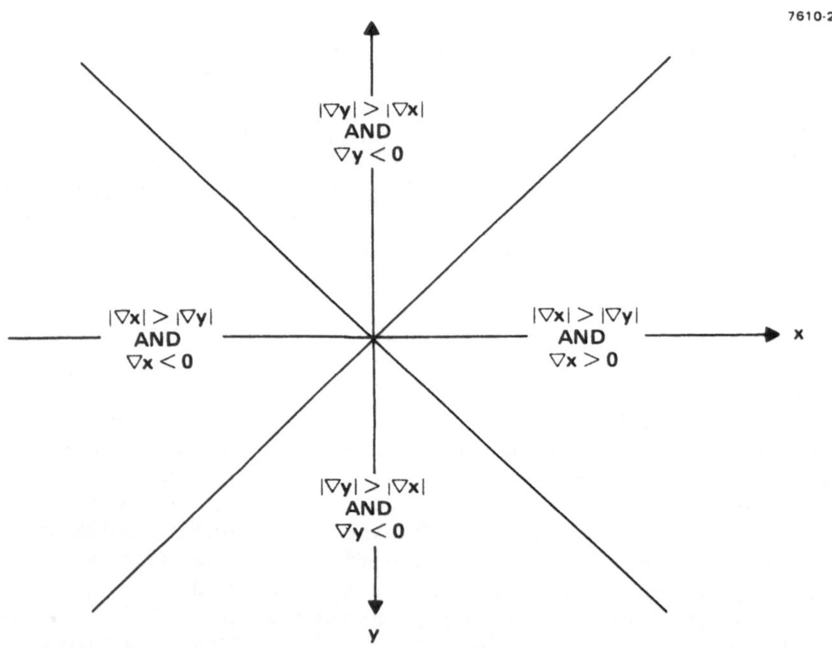

Figure 2. Edge element principal directions.

The result of edge extraction is an edge picture, which, for complex industrial scenes, may include as many as 30,000 edges in a 225 x 256 pixel array. A typical visible-band image of an industrial site is shown in Figure 3(a), and the corresponding edge picture is shown in Figure 3(b). The high edge densities, multiple edge elements along the significant scene lines, shadows, and "noise" edge elements do not contribute to important scene lines. These corrupting elements must be removed to permit identification of key scene lines, both straight and curvilinear.

2.2 Composite edge filtering

Several steps of edge element filtering are implemented to reduce the image edge density and to keep only those edge elements contributing to significant scene line structure. These filters are local windowing operations, which eliminate adjacent edge elements with isotropic orientations, and thin contiguous edge element clusters in each of the four principal Sobel directions.

The first filter step tests each extracted edge element for common orientation with adjacent edge elements. This straight-line edge filter [3] retains an edge element at pixel position $P_{i,j}$ with orientation θ if and only if a minimum of n edge elements with angles in the range $\theta \pm \Delta\theta$ are found within a window of length ℓ and width w centered at pixel $P_{i,j}$ and oriented in the θ-direction (see Figure 4). The filter parameters $\Delta\theta$, n, ℓ, and w are scene, sensor, and edge-operator dependent. For the imagery shown here with scene ranges of 200 to 400 and 3 mrad angular resolution, $\Delta\theta = 15°$, $n = 2$, $\ell = 5$ pixels, and $w = 1$ pixel. The results of this filtering step are shown in Figure 3(c), where the number of edge elements has been reduced by more than 15%. At this stage, however, the resultant edge-element density is still too high to accurately locate the dominant scene line structure.

The next step of the filtering process is to thin the edge element clusters to eliminate multiple edges and shadow effects. This thinning filter [4] first groups the edge elements by the principal gradient directions and thins each group in its respective direction. The operation of the filter is simple, testing triplets of pixels oriented parallel to the selected principal gradient direction to determine the pixel position with the maximum gradient. As an example, consider edge elements with principal gradient direction ∇x and $\nabla x > 0$ (see Figure 2).

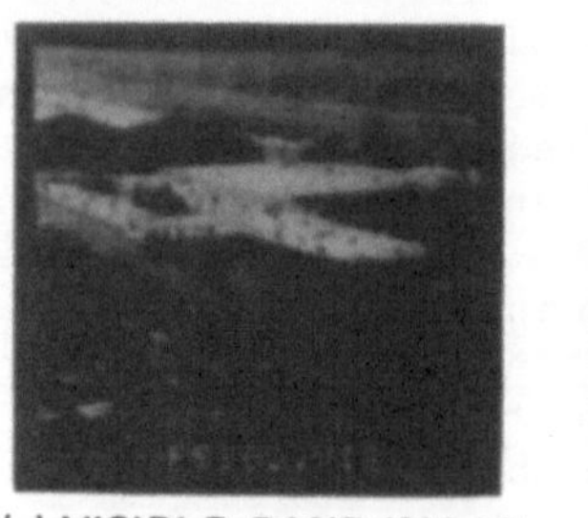

(a) VISIBLE-BAND IMAGE

(b) SOBEL EDGE PICTURE

(c) FILTERED EDGE PICTURE FIRST APPLICATION OF STRAIGHT LINE EDGE FILTER

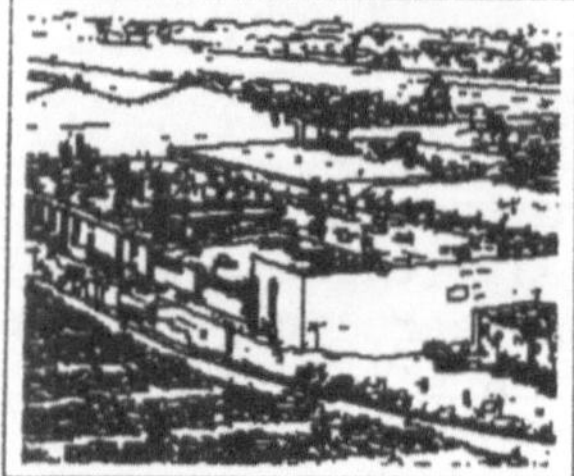

(d) THINNED EDGE PICTURE

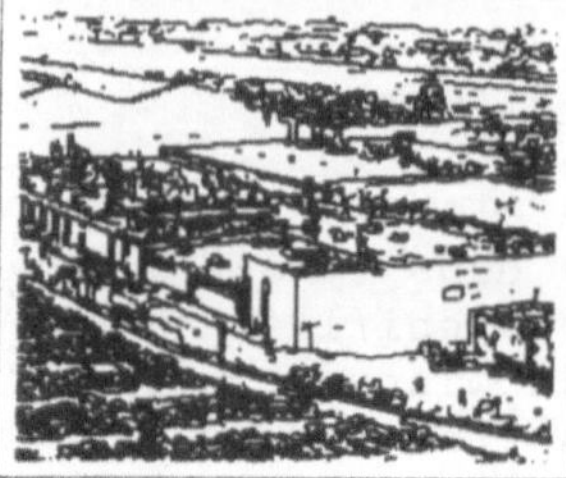

(e) FILTERED EDGE PICTURE SECOND APPLICATION OF STRAIGHT LINE EDGE FILTER TO THINNED EDGE PICTURE

Figure 3. Sobel edge extration and edge element filtering.

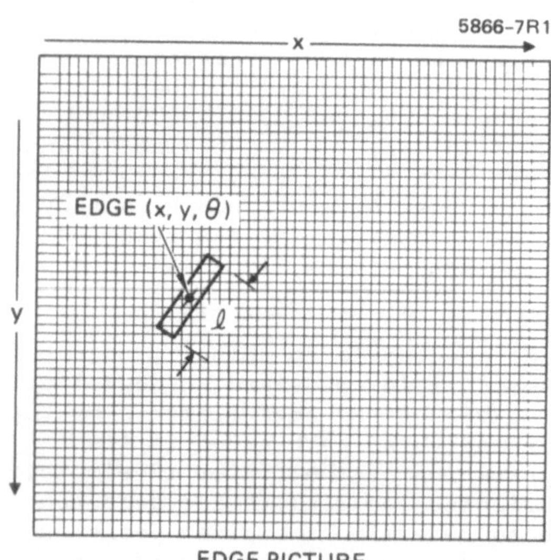

Figure 4. Straight line edge filter.

Moving a 3-pixel horizontal window from left-to-right and top-to-bottom through the image, the center pixel $P_{i,j}$ of the triplet is tested for the presence of an edge element. If an edge element is present, the adjacent pixels $P_{i-1,j}$ and $P_{i+1,j}$ are tested for edge element content. If no edge elements are found, the edge element in pixel $P_{i,j}$ is maintained. If adjacent edge elements are found, the central edge element in pixel $P_{i,j}$ is kept only if its gradient, $\nabla x_{i,j}$, is greater than those of the adjacent edge elements ($\nabla x_{i,j} > \nabla x_{i-1,j}$ and $\nabla x_{i,j} > \nabla x_{i+1,j}$). This thinning filter (shown schematically in Figure 5) and a second application of the straight-line edge filter using the filter parameters described previously are implemented separately in each of the four principal directions. The result of the thinning filter is a skeletonization of the multiple and shadowed edges, while the second application of the straight-line edge filter eliminates non-contiguous edge elements revealed by the thinning process. Figure 3(d) shows the results of the thinning operation with a reduction in the image edge element content to less than 50% of the original edge elements. Figure 3(e) shows the result of the second application of the straight-line edge filter; the number of edge elements is reduced to about 25% of those extracted initially. At this stage of the processing, multiple and non-collinear edge elements have been effectively removed, and the eye can now easily group and integrate the resultant edge elements to define the prominent scene lines. The computer implementation of these processes is described in Section 2.3.

2.3 Edge element grouping and line segment generation

At this stage of our processing, the remaining edge elements are grouped by the Sobel gradient direction into the four principal direction groups. These groups are the basis for our next grouping operation, spatial (or x-y) grouping, which establishes edge element groups based on spatial continuity of edge elements with similar orientations. The continuity criterion is a gap tolerance between adjacent edge elements of no more than 2 pixels. Edge elements within a principal direction group that satisfy this criterion form a single spatial group; edge elements that fail the criterion are candidates for new spatial groups. The result of this grouping operation is the formation of several spatial groups for each principal direction. These spatial groups are the initial candidates for line generation. Figure 6(a) shows filtered edge elements in the principal direction, $\nabla y < 0$; a typical spatial group extracted from this principal direction group is shown in Figure 6(b).

7610 4

THINNING FILTER FOR ∇x DIRECTIONS

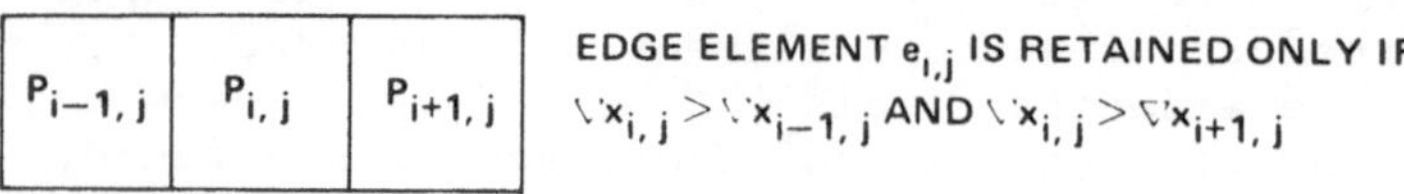

$P_{i-1,j}$
$P_{i,j}$
$P_{i+1,j}$
EDGE ELEMENT $e_{i,j}$ IS RETAINED ONLY IF
$\nabla x_{i,j} > \nabla x_{i-1,j}$ AND $\nabla x_{i,j} > \nabla x_{i+1,j}$

THINNING FILTER FOR ∇y DIRECTIONS

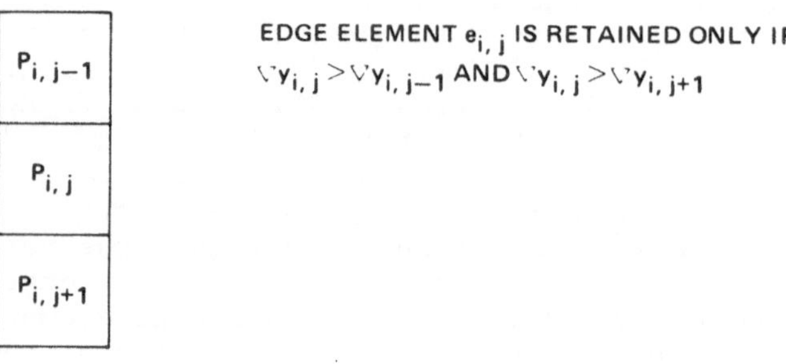

Figure 5. Edge element thinning.

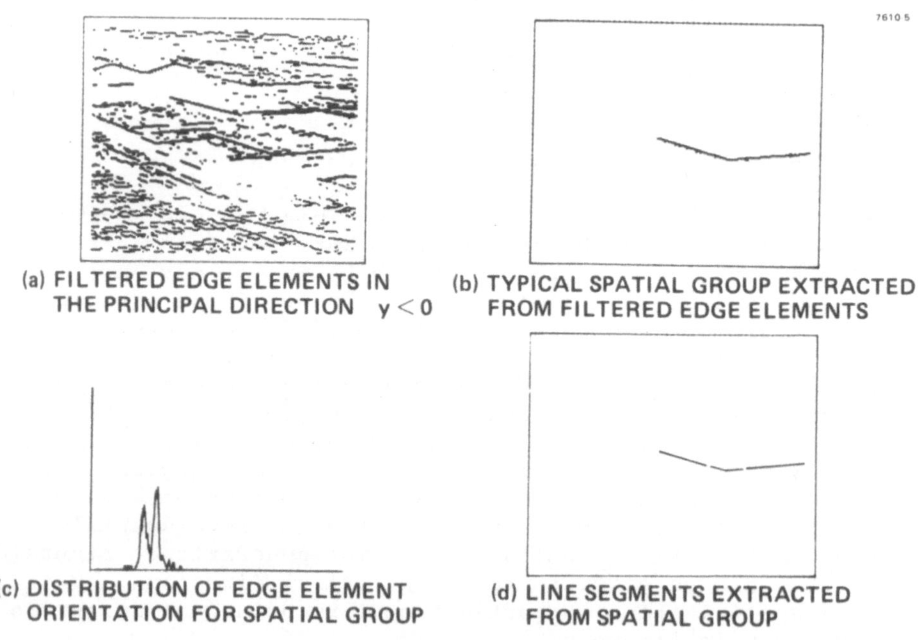

Figure 6. Edge element grouping and line segment filtering.

Line segment generation is basically a process of fitting a line to the edge elements of a spatial group. The best-fit straight line is the line that minimizes the sum of the squares of the perpendicular distances of the individual edge elements from the candidate line segment. This process is indicated in Figure 7. The line segment end points are determined from the perpendicular projection of each edge element position onto the candidate line segment (see Figure 8). The extreme values of p_i establish the end points of the line segment.

Two goodness-of-fit tests (see Figure 9) are applied to the resultant line segments. The first requires the rms perpendicular distance of the edge elements from the line segment to be less than 2 pixels:

$$D_{rms} = \sqrt{\frac{1}{N}\sum_{i=1}^{N} (\rho - x_i \cos\theta - y_i \sin\theta)^2} < 2 \text{ pixels} .$$

A perfectly fit line segment yields $D_{rms} = 0$. The second test requires the ratio of edge element density per unit line length, μ, to fall within a specified range: $0.8 < \mu \leq 1.2$. This ratio is calculated as

$$\mu = \frac{N}{Lc} ,$$

where N is the number of edge elements in the candidate x-y group, L is the best-fit line segment length, and c is a geometric factor depending on the line segment orientation:

$$c = \cos\theta \text{ for } |\tan\theta| \leq 1$$

$$c = \sin\theta \text{ for } |\tan\theta| > 1 .$$

Failure of either of these tests requires additional edge element grouping on the basis of more precise orientation information than that provided by the principal direction groups.

The required operation is θ-grouping [5] or grouping based on the specific orientations of edge elements in an x-y group that failed the goodness-of-fit tests. A histogram of the edge element orientations within the spatial group is generated, and θ-groups are specified on the basis of local minima in the distribution. Figure 6(c) is the histogram for the spatial group in Figure 6(b) that failed the goodness-of-fit tests. As a result of θ-grouping and a second spatial grouping operation, two line segments are generated (see Figure 6(d)), each of which

6247 11R1

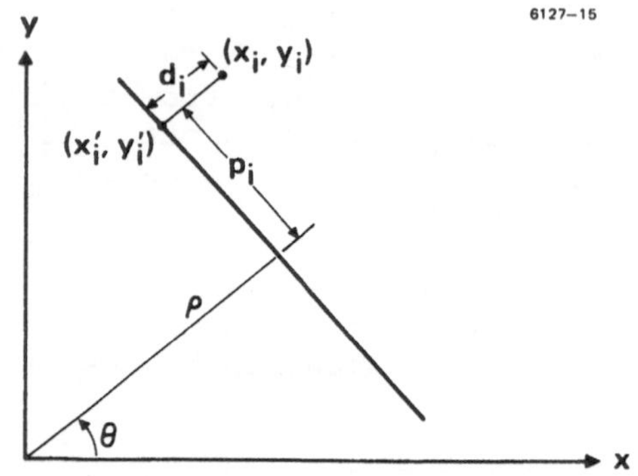

D^2 = SUM OF SQUARES OF ⊥ DISTANCES FROM EDGE ELEMENTS TO LINE

$$= \Sigma \, d_i^2$$

$$= \sum_{i=1}^{N} (\rho - X_i \cos\theta - Y_i \sin\theta)^2$$

$$\frac{\partial D^2}{\partial \theta} = 0 \quad \text{AND} \quad \frac{\partial D^2}{\partial \rho} = 0$$

GIVING (ρ_1, θ_1) AND (ρ_2, θ_2)

CHOOSE BEST-FIT STRAIGHT LINE PARAMETERS (ρ, θ)

WHERE $D^2 (\rho, \theta)$ = MINIMUM

Figure 7. Straight-line fit to grouped edge elements.

6247–10

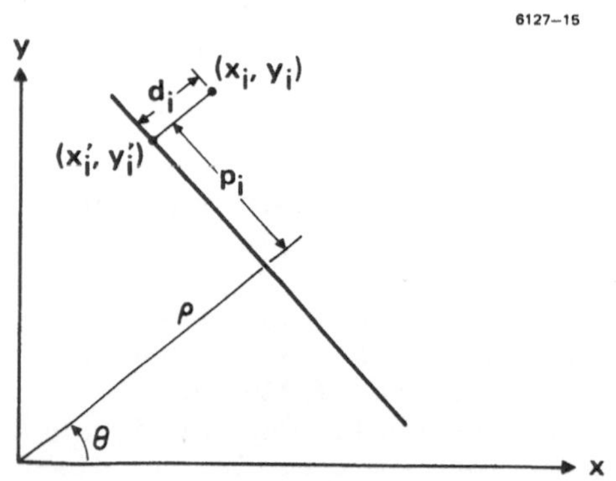

$p_i = - x_i \sin\theta + y_i \cos\theta$

WHERE θ IS THE BEST ESTIMATED PARAMETRIC VALUE

SORT p_i's TO OBTAIN THE TWO EXTREMA (x_o, y_o) AND (x_∞, y_∞)

END-POINTS PROJECTED ON THE BEST-FIT LINE

$$x' = x + \rho \cos\theta - x \cos^2\theta - y \sin\theta \cos\theta$$

$$y' = y + \rho \sin\theta - x \cos\theta \sin\theta - y \sin^2\theta$$

Figure 8. End-point calculation.

6497-1R1

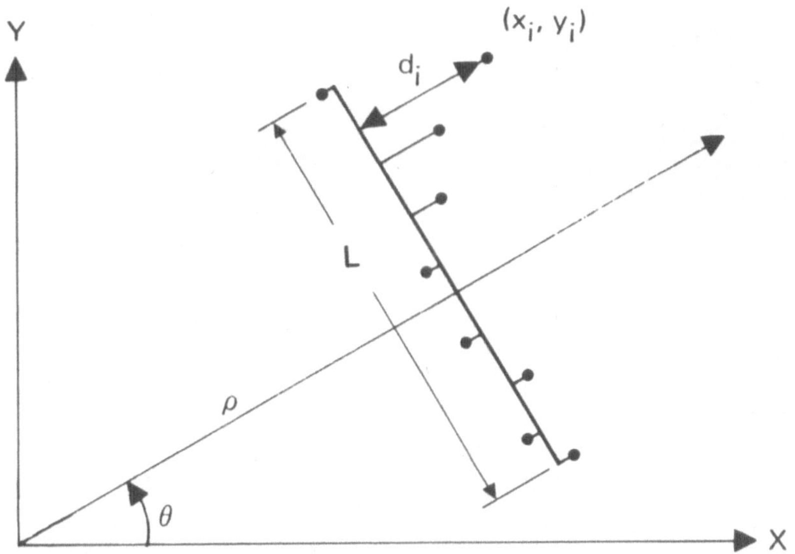

● RMS PERPENDICULAR DISTANCE TEST

$$d_i = x_i \cos\theta + y_i \sin\theta - \rho$$

$$D_{rms} = \sqrt{\frac{1}{N}\sum_{i=1}^{N}(\rho - x_i \cos\theta - y_i \sin\theta)^2}$$

BEST ESTIMATED LINE SHOULD GIVE $D_{rms} \approx 0$

● EDGE ELEMENT DENSITY TEST

$$\eta = \frac{N}{Lc}$$

WHERE $c = \cos\theta$ IF $|\tan\theta| \leqslant 1$

$= \sin\theta$ IF $|\tan\theta| > 1$

N = NUMBER OF POINTS

L = LENGTH OF LINE

BEST ESTIMATED LINE SHOULD GIVE $\eta \approx 1$

Figure 9. Line segment goodness-of-fit tests.

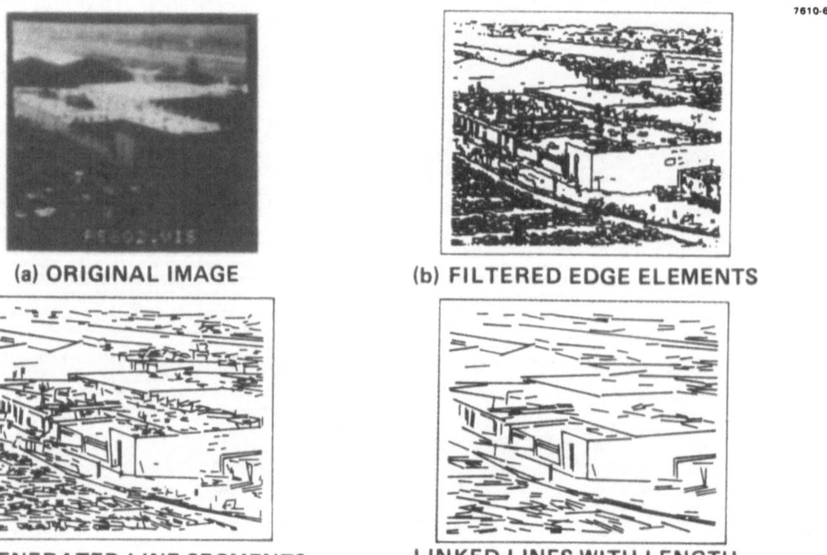

Figure 10. Line segment linking for line and vertex models.

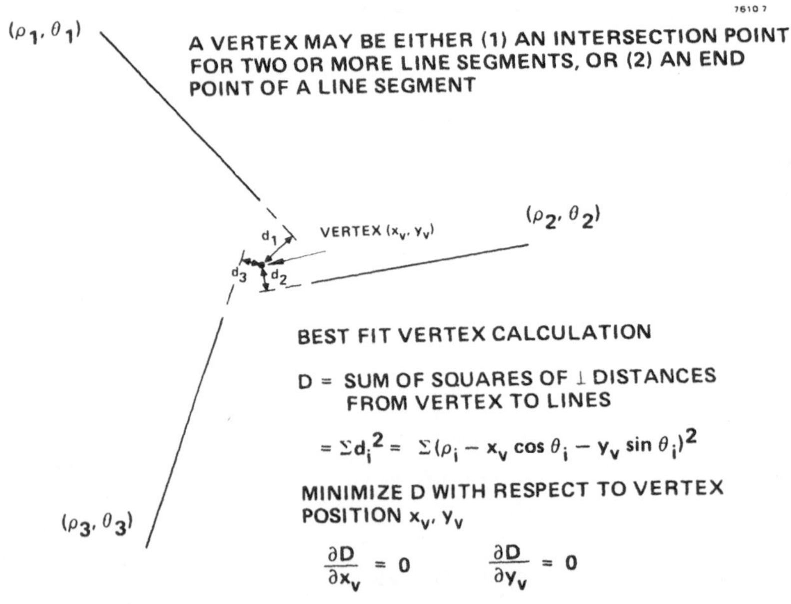

Figure 11. Location of scene vertices.

satisfies the goodness-of-fit criteria. The second spatial grouping operation within each of the θ groups is required to identify isolated spatial groups within a θ-group.

At this stage of the processing, the new line segments are tested with the goodness-of-fit criteria and those failing the test are processed further with a split-and-merge routine [6]. Computationally, this is a branching process in which the unsatisfactory line segment is replaced by several shorter line segments, each of which must ultimately satisfy the goodness-of-fit criteria. This routine is used infrequently, but is occasionally required to fit short-radius-of-curvature edge contours [7].

2.4 Line linking and model generation

Numerous line segments of varying lengths and orientations result from the edge-element-grouping and line-segment-generation processes. A typical example of these line segments is shown in Figure 10(c). Line segment orientation errors result from edge element position errors produced in the composite filtering operations, particularly the edge element thinning process. Line segment end-point errors are contributed both by the filtering operations and the low threshold of the spatial grouping step. Numerous other error sources remain to be identified, but their net effect is to break up long lines into shorter line segments. Thus, an important step prior to model generation is the linking of line segments contributing to long straight and curvilinear edges. This process must consider error-induced variations in collinear line segment properties. These include line segment orientation, overlap, and lateral and longitudinal gaps between candidate line segments. Tolerances are set for the variation of each of these parameters based on sensor resolution, scene content and range, and algorithm performance. Candidate line segments are linked if they meet these criteria. Figure 10(d) shows the result of this linking process for the visible-band image of this example.

At this point, two steps remain to complete the model-building process. First, the remaining short line segments (typically those less than 10 pixels long) are rejected from the model, and vertices are located at the intersections of the remaining model lines. A vertex is defined as the intersection of two or more lines; its degree specifies the number of intersecting lines. Locating the vertex is a process of minimizing the squares of the perpendicular distances from the lines to the candidate vertex position (see Figure 11).

The feature-based scene model is the combination of model vertices, dominant lines, and connected line groups. Vertices are prioritized by their degree and parameterized by their position, the lines intersecting at the vertex, the orientation and length of these intersecting lines, and adjacent vertices linked by these lines. Lines are prioritized on the basis of length and parameterized by their position and orientation. Connected line groups are prioritized based on their planar orientation with respect to the sensor line-of-sight, the number and degrees of vertices, and the number, orientations, and lengths of lines included in the group. The scene model is reduced to a data structure containing the model feature information.

3. MODEL MATCHING

Model matching is the process of identifying common features in scene models and, based on several such identifications, deriving the photogrammetric transformation between the two models [8]. With this transformation, the relative positions of the imaging sensors for the two models can be calculated. If the real (or simulated) position of one sensor is known, the position of the other sensor may be derived in map (or inertial) coordinates.

For the image-based guidance applications, one of the models (referred to as the sensed model) is generated autonomously in real time on the imaging platform, while the other model (the reference model) is prepared off-line with analyst interaction. The purpose of analyst interaction is to prepare the most probable reference model for the anticipated imaging and scene conditions. The degree of interaction depends on several factors: sensor commonality for the sensed and reference images, differences in perspective and range to the imaged scene, and diurnal and seasonal changes in weather and scene surface conditions. Under the most difficult conditions of different sensors, grossly different imaging geometries, and significant differences in the propagation medium and scene surface conditions, unique reference models may have to be synthesized from image data, cartographic information, and analytic and experimental results for weather, diurnal, and surface variations. We have not undertaken this complex reference preparation task; instead, we have made several assumptions to reduce the reference preparation task to a model editing function. Only images from a limited set of short-wavelength sensors are used to generate scene models. Perspective, scale, rotation, and displacement changes between sensed and reference models are constrained by platform trajectory variation limits. Propagation medium variations primarily change the image signal-to-noise ratio (S/N). Scene surface effects similarly manifest themselves as S/N changes and contrast reversals. S/N changes are accommodated by feature prioritization, and our selected features are invariant to contrast reversals.

In summary, reference models are prepared using the identical algorithms described in the previous section for feature extraction and model building. These models are edited to reflect the feature changes anticipated in the sensed model, and the features are prioritized to reflect their importance in the model and their expected degradation due to propagation medium and surface conditions. This edited and prioritized model is then matched to the autonomously derived sensed model. The details of this matching process are described in Sections 3.1 and 3.2, and examples of successful matches and false fixes are presented in Section 3.3.

3.1 Description of the feature-matching algorithm

To fully utilize and integrate all available information obtained by feature extraction in the ensuing model-matching algorithm, a composite scene model has been introduced. This model may include all of the feature types obtained in the feature-extraction processes, such as vertices, lines, circles, and regions. Where appropriate, pointers are introduced to establish relations between different feature types, such as the lines that meet at a vertex. The model information includes feature labels, priorities, parameters, and links; sensor type; and indexing information. Additional information, such as contextual relations and match priorities, may be inserted in the reference preparation and editing processes.

The basic features of our composite scene model are singly and multiply connected straight lines and vertices. These features are prioritized on the basis of their invariance properties and ordered based on their importance in the matching process. The ordering is (1) open or closed groups of connected line segments with a connectivity index equal to the number of connected lines, (2) vertices with a degree index corresponding to the number of lines joined at the vertex, and (3) single lines indexed by length. Line connectivity and vertex degree are considered more important than line length since a group of connected lines possesses a useful local structure that is more significant and easily identified than isolated long lines.

The matching process is implemented on the ordered lists of reference and sensed model features. Features in these reference models are matched based on the predetermined order of feature types, information, and priorities assigned to the features in the reference model. Features are considered according to their priorities using known constraints on rotation, perspective, and scale change to narrow the search space during matching.

Feature matching proceeds as shown in Figure 12 in an effort to match a minimum of three noncollinear and nominally coplanar lines in the reference model to the equivalent lines in the sensed model. These matches are first sought in the connectivity groups with index greater than three; then as three or more lines connecting context-related vertices; and finally as individual, context-related scene lines. When an acceptable number of reference and sensed features have been matched, these features, particularly the subset in the immediate neighborhood of the desired match point, are used to compute an estimated local transformation between the reference and sensed models. Rotation between the reference and sensed models is estimated from the line-angle correlation of matched lines least sensitive to perspective-induced distortion (see Figure 13). If a clear correlation peak is not found, an average rotation of the

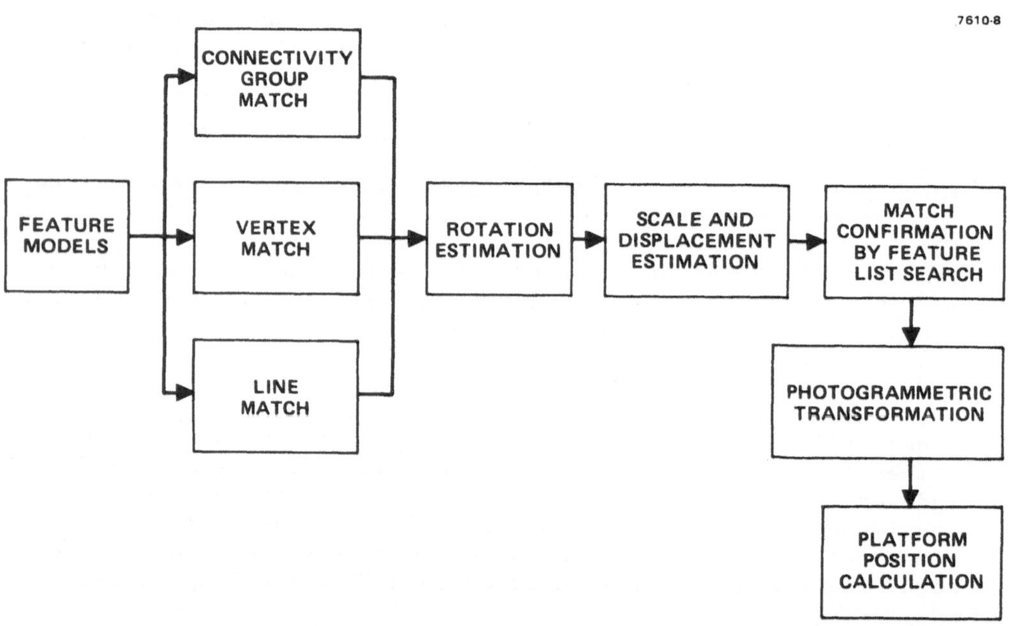

Figure 12. Flow diagram for feature matching.

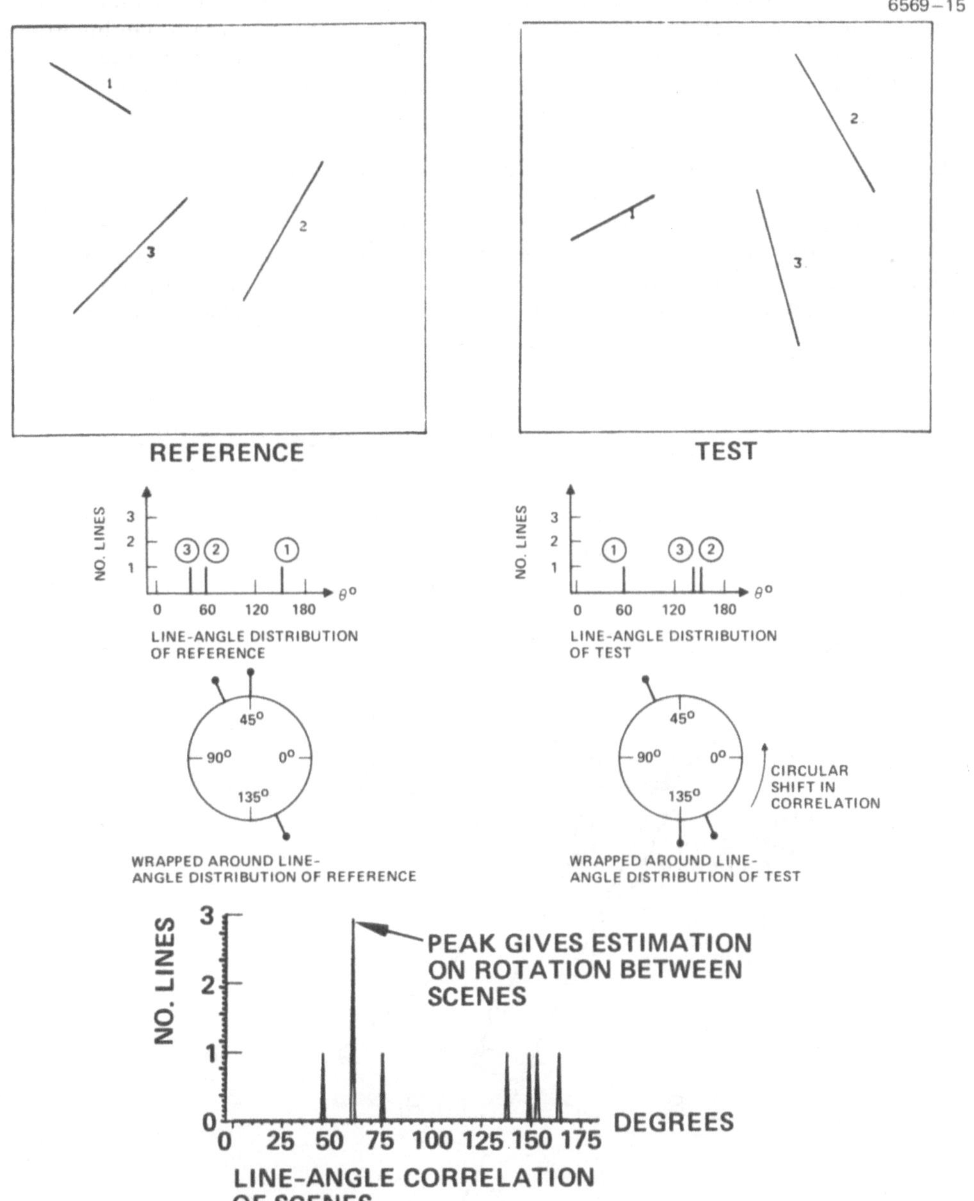

Figure 13. Correlation of line-angles to estimate rotation.

tentatively matched reference and sensed lines is computed. Scale and displacement is estimated by a least-squares technique which minimizes the distances between matched reference and sensed lines. This procedure is described in the following paragraph.

Let $\vec{d}_i$ be the distance vector from the midpoint of the reference line, R_i, to the corresponding sensed lines, S_i, in the direction perpendicular to the sensed line, as shown in Figure 14. Let $\vec{r}_i$ be the vector from the origin to the midpoint of the reference line. If the scale factor between the reference and sensed coordinate frames is expressed as $(1 + \alpha_o)$, then the distance

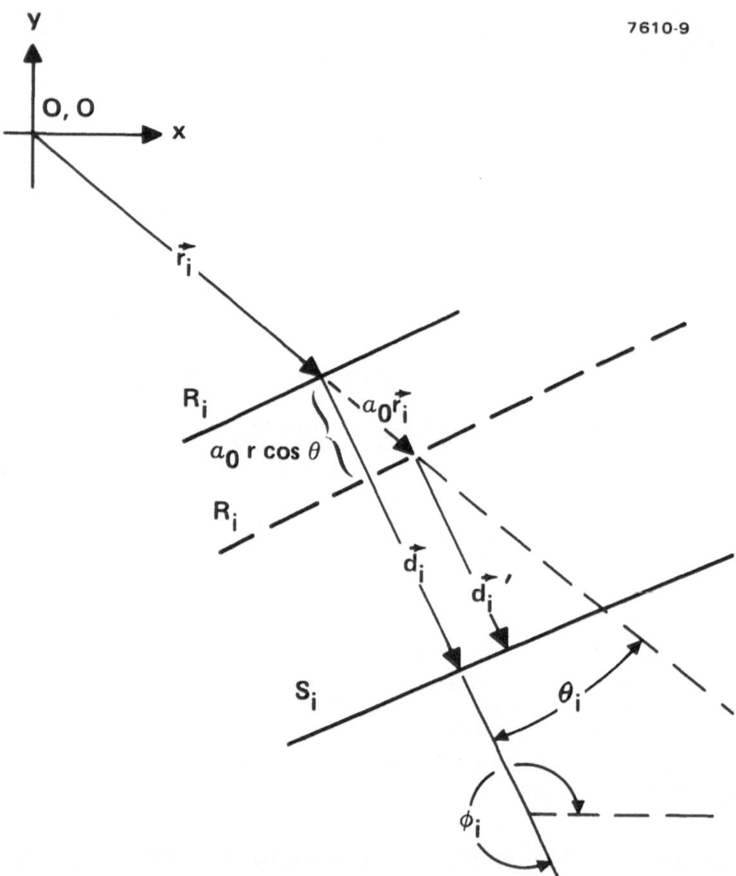

Figure 14. Line geometry for least-squares scale and rotation estimation.

between the scaled reference line R_i' and the sensed line can be written as $d_i' = d_i - \alpha_o r_i \cos\theta_i$, where θ_i is the angle from $\vec{d}_i$ to $\vec{r}_i$. If x_o and y_o are the translational displacement values, and ϕ_i is the angle from the x axis to d_i, then the quantity $[\Sigma_i (d_i'')^2]^{1/2}$, where $d_i'' = d_i' - x_o \cos\phi_i - y_o \sin\phi_i$, can be minimized to obtain α_o, x_o, and y_o with the following system of linear equations:

$$\begin{bmatrix} \Sigma\cos^2\phi_i & \Sigma\sin\phi_i\cos\phi_i & \Sigma r_i \cos\theta_i\cos\phi_i \\ \Sigma\sin\phi_i\cos\phi_i & \Sigma\sin^2\phi_i & \Sigma r_i \cos\theta_i\sin\phi_i \\ \Sigma r_i\cos\phi_i\cos\theta_i & \Sigma r_i\cos\theta_i\sin\phi_i & \Sigma r_i^2 \cos^2\phi_i \end{bmatrix} \begin{bmatrix} x_o \\ y_o \\ \alpha_o \end{bmatrix} = \begin{bmatrix} \Sigma d_i\cos\phi_i \\ \Sigma d_i\sin\phi_i \\ \Sigma d_i r_i\cos\theta_i \end{bmatrix}.$$

With these estimates of scale change and displacement between the sensed and reference models and the relative rotation derived earlier, the list of reference features is searched further for more correspondences with the sensed model features. Each reference feature is appropriately rotated, scaled, and translated, and a search is made in the sensed model for the corresponding feature. Several match criteria must be satisfied, depending on the feature type. Lines, multiply or singly connected, must satisfy two scene- and scenario-dependent criteria. The perpendicular distances from the end points of the transformed reference line to the sensed line and the degree of overlap between sensed and transformed reference lines must be within established thresholds. Vertices must also satisfy two criteria. They may differ in degree by no more than one, and the transformed reference vertex position must satisfy a distance threshold requirement similar to that for matched lines. Any additional feature correspondences found serve to strengthen confidence that the initial match is correct. Values of rms error for line perpendicular distance and overlap and vertex position add a quantitative measure of match correctness.

3.2 Photogrammatic transformation for aimpoint estimation

The coordinates of matched control points obtained from the model-matching step are used to derive the match point in the sensed scene. In the most general case of three-dimensional scenes, the coordinates of a sensed-scene point corresponding to an arbitrary reference-scene point are found by bringing the two scenes into registration via three translations and three rotations in the three-dimensional (3-D) reference coordinate frame (Figure 15(a)). The effect of these changes, in

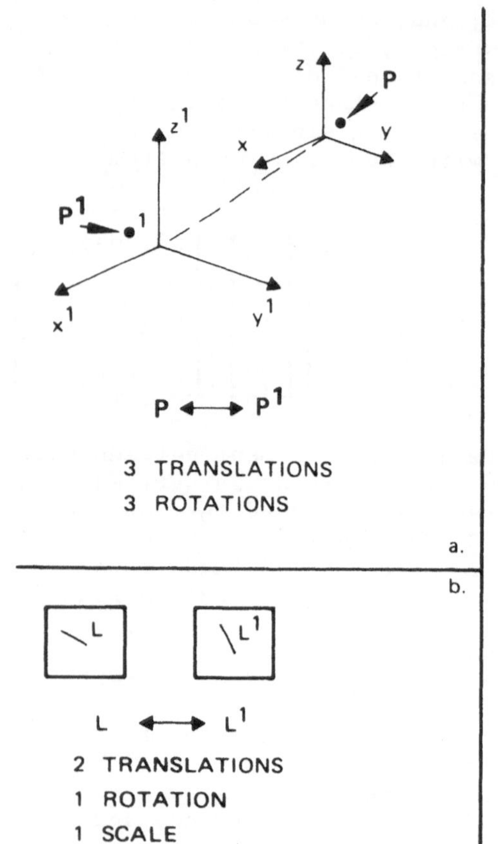

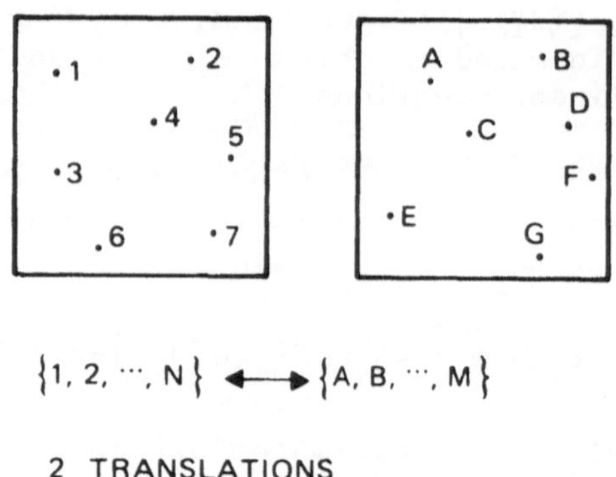

Figure 15. Match point computation (a) point registration in a three-dimensional scene; (b) linear two-dimensional transformation; (c) nonlinear two-dimensional transformation.

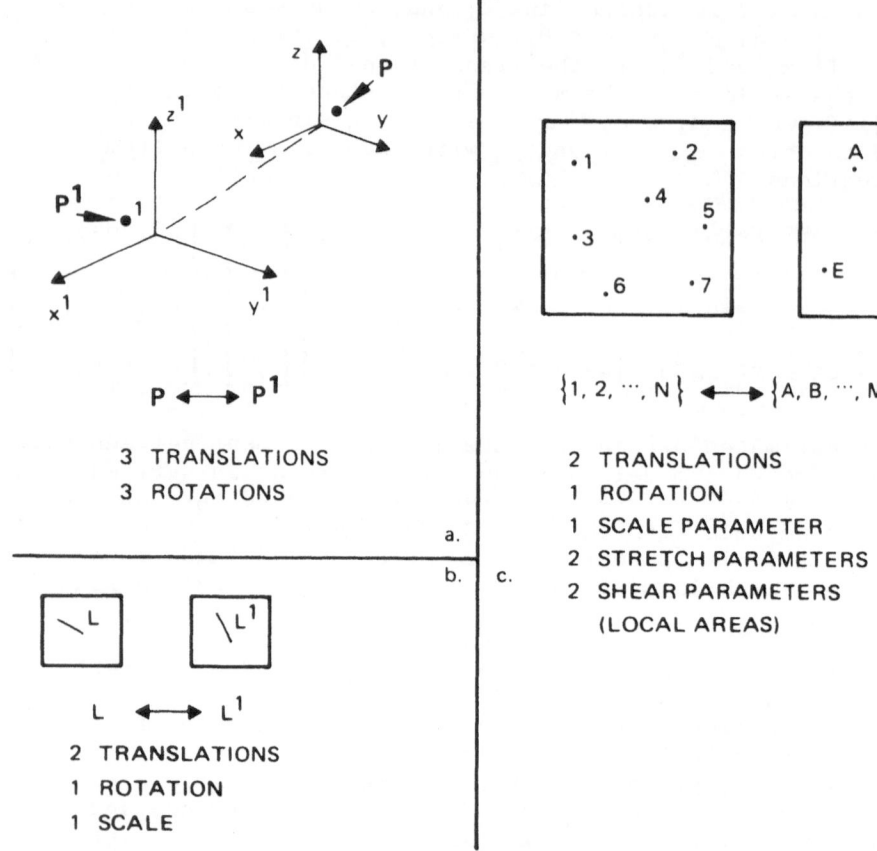

general, is to introduce nonlinear transformations between corresponding points in the 2-D reference and sensed images. This is the well-known "camera-model" problem of computer vision and terrain imaging. Under appropriate conditions, a linear 2-D transformation is possible. Then, two translations, one rotation, and one scale parameter are adequate to define the geometry precisely. This linear transformation is valid only for coaxial coordinate frames and a planar scene (constant range). In the more general (and more usual) case, varying sensor locations and scene range extent lead to a nonlinear 2-D transformation.

In general, to counter the effects of a nonlinear 2-D transformation, a 3-D model is required because parameters describing the nonlinear 2-D transformation vary across the image plane. This is indicated in Figure 15(c), where small areas around corresponding control points follow a nonlinear transformation specified by several parameters, and these parameters vary across the image.

A restrictive assumption can be made that will permit calculating a match point without requiring 3-D images. It is that, for the cases encountered, the perspective change is sufficiently small that the nonlinear 2-D transformation has constant distortion parameters. A polynomial approximation with constant coefficients can then be used to map the reference control points to the corresponding sensed control points using the following equations [9,10]:

$$x = \sum_{i=0}^{N} \sum_{j=0}^{N-i} a_{ij} u^i v^j$$

$$y = \sum_{i=0}^{N} \sum_{j=0}^{N-i} b_{ij} u^i v^j ,$$

where (u,v) is the reference control point, (x,y) is the sensed control point, and a_{ij} and b_{ij} are the constant polynomial coefficients. For most cases, a second degree (N = 2) approximation is adequate [11], and at least six pairs of matching control points are then needed to determine the 12 polynomial coefficients. For M points, the M relations for the x coordinate may be combined as

$$\begin{bmatrix} x_i \\ x_2 \\ x_3 \\ \cdot \\ \cdot \\ \cdot \\ \\ x_M \end{bmatrix} = \begin{bmatrix} 1 & u_i & v_i & u_1^2 & v_1^2 & u_1 \cdot v_1 \\ 1 & u_2 & v_2 & u_2^2 & v_2^2 & u_2 \cdot v_2 \\ 1 & u_3 & v_3 & u_3^2 & v_3^3 & u_3 \cdot v_3 \\ \cdot & \cdot & \cdot & \cdot & \cdot & \cdot \\ \cdot & \cdot & \cdot & \cdot & \cdot & \cdot \\ \cdot & \cdot & \cdot & \cdot & \cdot & \cdot \\ \\ 1 & u_M & v_M & u_M^2 & v_M^2 & u_M \cdot v_M \end{bmatrix} \times \begin{bmatrix} a_{00} \\ a_{10} \\ a_{01} \\ a_{20} \\ a_{02} \\ a_{11} \end{bmatrix} ,$$

where $M \geq 6$. A corresponding equation exists for y and b_{ij}. Both may be written in matrix notation as

$$X = UA$$

$$Y = UB \; .$$

Using linear least-squares regression, the best estimate for A and B in the over-determined case ($M > 6$) is given by:

$$A = (U^T U)^{-1} \, U^T X$$

$$B = (U^T U)^{-1} \, U^T Y \; ,$$

where $(U^T U)^{-1} \, U^T$ is the generalized inverse of U. The polynomial transformation represents an approximation to the general spatial warp transformation. The procedure is relatively simple, computationally efficient, and does not require 3-D information, instead relying only on data contained entirely in the 2-D images. This is an important advantage over the perspective transformation. For these reasons, we have chosen the second-degree polynomial transformation as a method to accommodate perspective changes in our match-point estimation.

3.3 Results

Numerous structural scene models were matched during the development of our feature-based model-matching technique. The most objective sample of these matching demonstrations is the recent independent test of our model-building and matching algorithms. Our algorithms were compared with those of other image-processing organizations on a common set of test images. Reference images were provided for model preparation, and corresponding sensed images were then delivered for non-interactive model-building and matching. This scene-matching evaluation resulted in correct matches for 7 of 12 cases with a one-sigma match accuracy of less than 1.5 pixels. Samples of these matches are shown and discussed in the following paragraphs. Model-building problems were the primary cause of the five matching failures. Limited scene content in several of the small reference images resulted in non-unique reference models and erroneous matches. The Hueckel edge detector was ineffective with low-resolution imagery. The single, step-edge approximation of the Hueckel operator failed when pixel intensities varied widely within the relatively large extraction window. This was the case with low-resolution structural imagery with high spatial density of features. These problems can be alleviated by using larger reference images (including key scene features) and the Sobel edge detector with its small extraction window.

Two examples of successful scene matches are shown in Figures 16 and 17. In each case, the reference and sensed images are shown with the resulting scene models. The various lines matched are numbered in each of the models and their correspondences shown in the accompanying table. The reference-to-sensed model coordinate transformation parameters and the resulting match point accuracy are both included. Figure 16 shows a match of 10.6 μm active IR images, and Figure 17 shows a match of a visible-band reference image with an 8 to 12 μm passive IR sensed image. In the second example, several contrast reversals on key scene features were successfully handled in our model-building and matching process.

Figure 18 shows an attempted match that resulted in a false fix. In this case, the reduced resolution of the reference image resulted in poor edge extraction with the Hueckel operator. Thus, the reference model built from these low-level features does not uniquely represent the reference image. The corresponding match is incorrect with a match-point error of 21 pixels.

7223-10

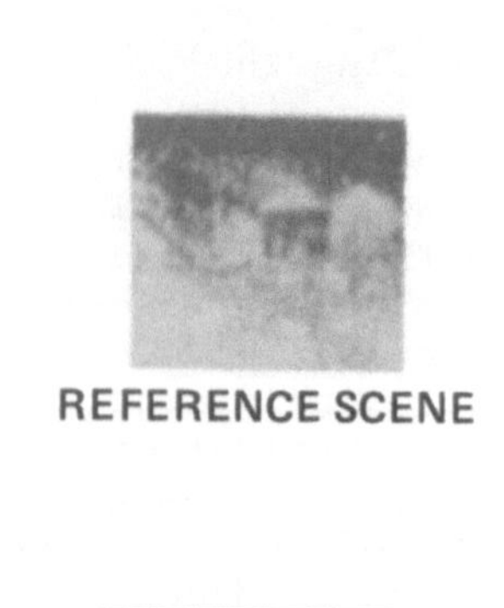

REFERENCE SCENE

SENSED SCENE

REFERENCE MODEL

SENSED MODEL

MATCHING LINES	
REF	SENSED
1	8
2	20
3	22
4	18
5	27
6	12
7	21
10	29
11	34
12	24
13	31
ROTATION	0.7°
X-TRANSLATION	29.6
Y-TRANSLATION	21.3
SCALE	0.97

Figure 16. Line-matching results with same sensor type for reference and sensed scene.

7223-9

REFERENCE SCENE

SENSED SCENE

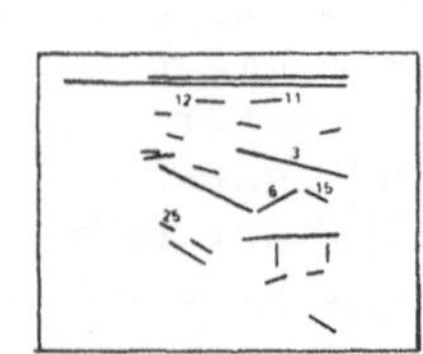

REFERENCE MODEL

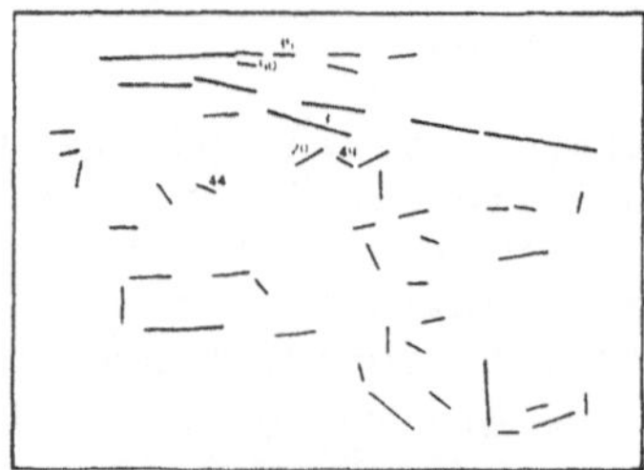

SENSED MODEL

MATCHING LINES	
REF	SENSED
3	3
6	20
11	35
12	50
15	49
25	44
ROTATION	0.4°
X-TRANSLATION	24.2
Y-TRANSLATION	0.8
SCALE	1.02

Figure 17. Line-matching results with different sensor type for reference and sensed scene.

7223-22

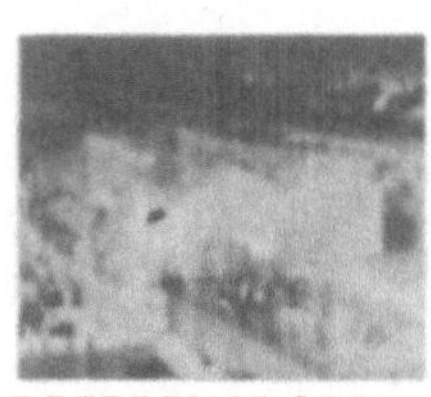

REFERENCE SCENE

SENSED SCENE

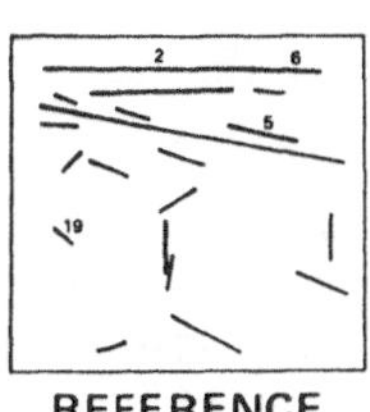

REFERENCE MODEL

SENSED MODEL

MATCHING LINE	
REF	TEST
2	1
5	8
6	14
19	33
ROTATION	0°
X-TRANSLATION	31.4
Y-TRANSLATION	9.2
SCALE	0.98

Figure 18. Example of false fix.

In general, six well-defined model lines in one or more connectivity groups or as separate but contextually related lines are sufficient to establish a successful model match. Additional line and vertex correspondences serve to confirm and improve the match confidence. In the sample imagery and model matches shown here, perspective changes were small (less than 5° in azimuth). As a result, a linear, two-dimensional transformation applied globally was adequate to achieve the required match-point accuracies.

The performance of our model-building and matching techniques on the evaluation image pairs has demonstrated the robustness and versatility of this approach to image analysis. But more importantly, it has identified weak elements in our processing chain and helped to specify image requirements. The algorithms described in this paper are responsive to the identified weaknesses and, when executed on imagery with sufficient resolution and feature content, result in significantly improved models compared to those shown in Figures 16 through 18.

ACKNOWLEDGEMENT. The line matching and least-squares techniques for minimizing distance between sensed and reference lines described here were orginated by R. Nevatia, a consultant from the University of Southern California; they will be described in more detail in a forthcoming paper by R. Nevatia and C. Clark.

REFERENCES

1. M.H. Hueckel, "A Local Visual Operator which Recognizes Edges and Lines," Journal ACM 20, 4, October 1973, pp. 634-647.
2. R.O. Duda and P.E. Hart, Pattern Classification and Scene Analysis (John Wiley & Sons, New York, 1973).
3. S.A. Dudani, A.L. Luk, J.P. Stafsudd, C.S. Clark, and B.L. Bullock, "Model-based Scene Matching," Hughes Research Laboratories, Research Report No. 509, July 1977.
4. R.B. Eberlein, "An Iterative Gradient Edge Detection Algorithm," Computer Graphics and Image Processing 5, 1976, pp. 245-253.
5. S.A. Dudani and A.L. Luk, "Locating Straight-Line Edge Segments on Outdoor Scenes," Pattern Recognition, June 1978.
6. U. Ramer, "Extraction of Line Structures from Photographs of Curved Objects," Computer Graphics and Image Processing, 4, 1975, pp. 81-103.
7. A.L. Luk and J.P. Stafsudd, "Circle Extraction in Outdoor Scenes," submitted to be published.
8. S.A. Dudani and C. Clark, "Vertex-Based Model Matching," Proc. Symp. on Current Mathematical Problems in Image Science, November 1976, Monterey, California.
9. Robert Y. Wong, "Sensor Transformation," IEEE Transactions on Systems, Man and Cybernetics, Vol. SMC-7, No. 12, Dec. 1977.
10. E.G. Johnston and A. Rosenfeld, "Geometric Operations on Digitized Pictures," Picture Processing and Psycopictories, B.S. Lipkin and A. Rosenfeld, Eds. New York: Academic, pp. 217-240, 1970.
11. Y.G. Kawamura, "Automatic Recognition of Changes in Urban Development from Aerial Photographs," IEEE Trans. Syst., Man and Cybern. Vol. SMC-1, No. 3, 230 (July 71).

COMPUTER INTERPRETATION OF SPEECH PATTERNS BASED ON FORMANT TRAJECTORIES

Renato De Mori
Istituto di Scienza dell'Informazione
Università - Corso Massimo d'Azeglio 42
10125 Torino, Italy

Pietro Laface and Elio Piccolo
CENS-IENGF - Istituto di Elettrotecnica Generale
Politencnico
Corso Duca degli Abruzzi 24 - 10129 Torino, Italy

ABSTRACT. A method for the interpretation of speech patterns based on fuzzy graphs of formant trajectories is described. Experimental results are presented. They refer to the use of such features for automatic recognition of vowels, liquid and nasal consonants in continuous speech.

1. INTRODUCTION AND PRECATEGORICAL CLASSIFICATION

Large efforts have been devoted in the last years for developing systems capable of understanding connected speech. Recent reviews of the problems involved in designing such systems and of the proposed solutions have been done by Reddy [1], [2], Martin [3], Jelinek [4], Klatt [5], and Wolf [6].

The purpose of this paper is to describe the extraction of speech patterns from the acoustic signal and to show the possibility of their interpretation in terms of phonetic and phonemic labels.

Automatic interpretation of speech patterns is often affected by a degree of vagueness because sentences are not always pronounced perfectly in all their details. Furthermore, we are only able to make doubtful assumptions about the structure of the process involved in speech perception and the organization of the sources of knowledge used for grasping the meaning of a spoken sentence.

Not only the whole process of speech understanding involves a complex hierarchy of operations, but even the lowest levels, consisting in labelling the speech waveform by phoneme symbols, are controlled by a deeply structured source of knowledge. Such source of knowledge can be thought as a grammar generating a language describing the possible spectral patterns of a verbal message belonging to a given protocol.

Phonetic and phonemic features are syntactic categories of such a grammar. Features like vocalic-nonvocalic, sonorant-nonsonorant, etc., that can be extracted by a precategorical classification, with a procedure that does not require the knowledge of the context, are considered as the deepest syntactic categories.

Other categories, like liquid-nasal, are introduced for immediately generating phonetic transcriptions with procedures that are context-dependent.

Phoneme labels are expressed by rules on the basis of such categories and new features extracted from the acoustic data.

In practice there is not always enough evidence in the acoustic data about the presence of a feature and more candidates have to be postulated. This leads to a number of competing hypotheses the evaluation of which is made by a set of concurrent processes. Consequently, the emission of hypotheses about the phonemic labelling of acoustic data is performed by an algorithm that is substantially a fuzzy one: in such an algorithm membership functions are related to the evidence of phonetic and phonemic features in the acoustic patterns.

In a paper by Stevens [7] a number of findings relating to speech acoustics have been recalled to state the reason for a precategorical classification of phonetic units in speech perception.

On the basis of experiments about the acquisition of phonetic distinctions by young children, it has been possible to conclude that a few simple property detectors can provide a basis for classifying consonants into broad categories with no essential modifications in multi-talker situations.

The possibility of precategorical classification is confirmed by an analysis of speech spectrograms, after which some property detectors operating on the time evolutions of spectral energies have been proposed by Stevens [7] for establishing the place of coarticulation of certain consonants independently from their context.

The tree of fig. 1 summarizes a scheme for precategorical classification that has been implemented giving satisfactory results with little ambiguities; it requires simple operations on the time evolutions of broad-band spectral energies. The "father-sons" relations represented in the tree have to be interpreted as conditions on which the features representing the sons can be extracted. Thus, for example, the feature "affricate" can be recognized only for nonsonorant consonants.

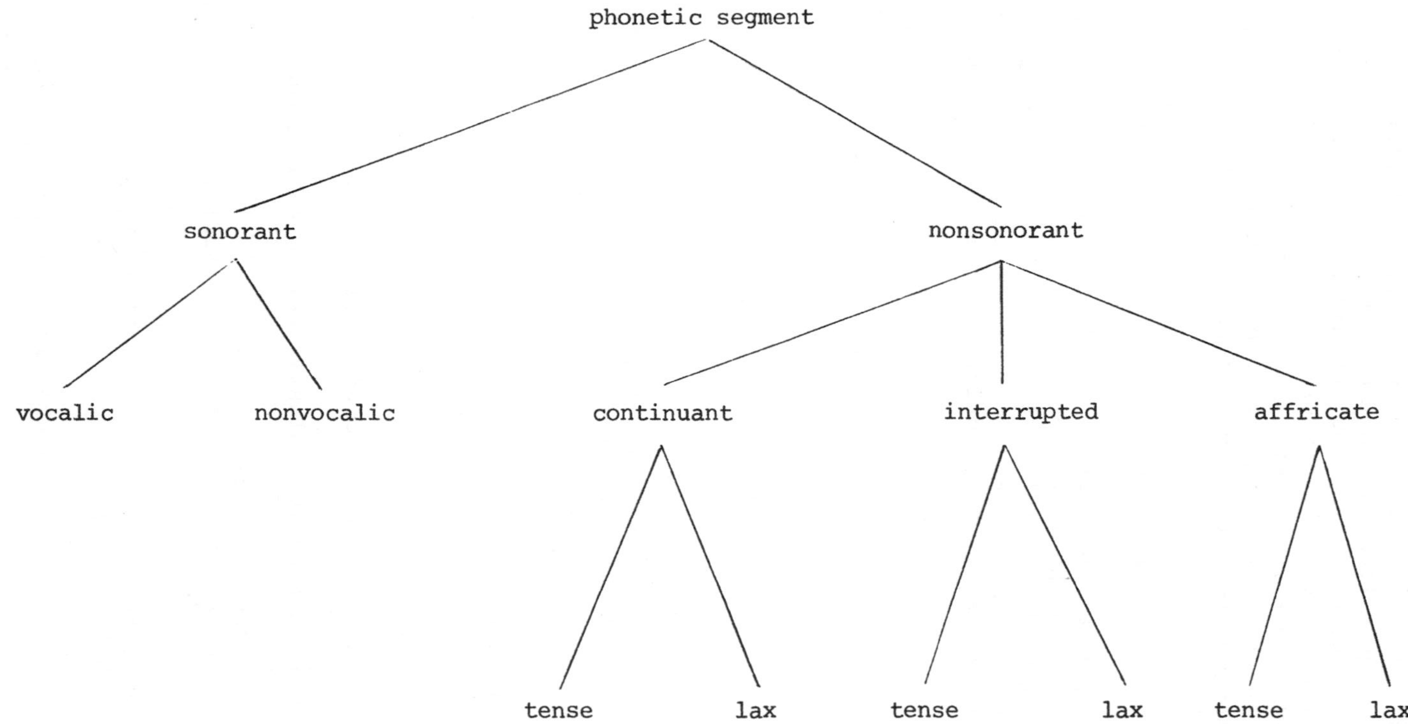

Figure 1. Tree of the features detectable at the level of precategorical classification using context-independent algorithms.

The classification of context-independent attributes provides a basis for the acquisition of a set of context-dependent cues. This is done in practice by looking for different spectral features like formants (i.e. time evolution of energy concentrations), burst or frication noise parameters, etc., whose extraction is decided after the precategorical classification of each speech segment.

More details on this item can be found in [8].

2. EXTRACTION OF SPEECH PATTERNS FROM SPECTRAL FEATURES

The extraction of detailed spectral features depends on precategorical classification and is performed independently on each syllable nucleus. Syllable nuclei, called PSS (pseudo syllable segments) are delimited by a procedure described in [9] and controlled by a grammar GSEG whose rules are briefly recalled in the following:

GSEG; (V_T, V_N, PSEG, PSS);

where V_T = { V,UT,SL,VC} is the terminal alphabet,

V_N = { VLK, UN, PSS }

is the nonterminal alphabet, PSS is the starting symbol and PSEG is the following set of rewriting rules:

PSEG: (VLK) → (V) / (VC)(V)
(UN) → (SL) / (UT) / (SL) (UT) / (UT) (SL)
(PSS) (VLK) → (V) (VLK) (VLK) / (UN) (VLK) (VLK)
(PSS) (UN) → (V) (VLK) (UN) / (UN) (VLK) (UN) / (V) (VLK) (VC)
(UN) / (UN) (VLK) (VC) (UN)

The following primitive symbols:

V : vowels,
UT : unvoiced tracts,
SL : silences,
VC : tracts of one or more voiced consonants,

are obtained from precategorical classification as follows:

〈nonsonorant-tense-continuant〉 → (UT)
〈nonsonorant-tense-interrupted〉 → (SL)/(SL)(UT)
〈nonvocalic sonorant〉 → (NV)
〈nonsonorant-lax〉 → (NV)
〈nonsonorant-tense affricate〉 → (SL)/(SL)(UT)
(NV) (NV)* → (VC).

Feature extraction for the nonsonorant intervals, except silences, consists in extracting average spectra in an unambiguous manner.

For sonorant intervals formants are tracked; by such operation, a fuzzy graph instead of the usual pattern of three formants is extracted from a segment of spectrogram. The arcs of the graph are possible tracts of formants and the weight associated with an arc expresses the possibility that the arc is a part of a formant. The arcs are directed concurrently with the time axis and the procedure for representing a piece of spectrogram by a fuzzy graph is an algorithm because it performs a finite number of steps on a finite number of possibilities. The algorithm is made of three parts, namely, extraction of a multilinked data structure from a spectrogram, elimination of links that would generate arcs in the fuzzy graph that cannot be part of formants and assignment of weights.

Formants have been widely used in the recent speech understanding projects for emitting phonemic hypotheses (see recent reviews by Reddy [10] and De Mori [11]).

2.1 Extraction of a multilinked data structure from a spectrogram

Let G(n,f) be the n-th short-time spectrum. Let I(n,j) be the frequency intervals of energy concentration of G(n,f); $j = 1,2,\dots,N_n$ where N_n is the number of intervals detected in the n-th spectrum.

A binary variable LP is associated to each interval. LP = 1 means that the interval contains a local maximum of the spectrum computed from linear prediction coefficients; LP = 0 otherwise.

Algorithm ARGEN

Step 1 - The intervals I(n,j) are obtained for $n_1 \leq n \leq n_2$ where n_1 is the time index of the first spectrum and n_2 is the time index of the last spectrum of the spectrogram segment.

The value of LP is computed for each interval and appended to it.

Each interval is represented with a cell in a data structure. Each cell has five fields containing respectively:

- the initial frequency of the interval.
- the final frequency of the interval,
- the average energy density of the interval,
- a label identifying the arc which the interval belongs to,
- LP.

Links are created between nodes corresponding to intervals of successive spectra having partially overlapped projections along the frequency axis.

Intervals are represented by points in the time-frequency plane and labels are assigned to points in a way such that points belonging to the same arc have the same label except for the graph nodes for which two types are distinguished according with the definition shown in fig. 2.

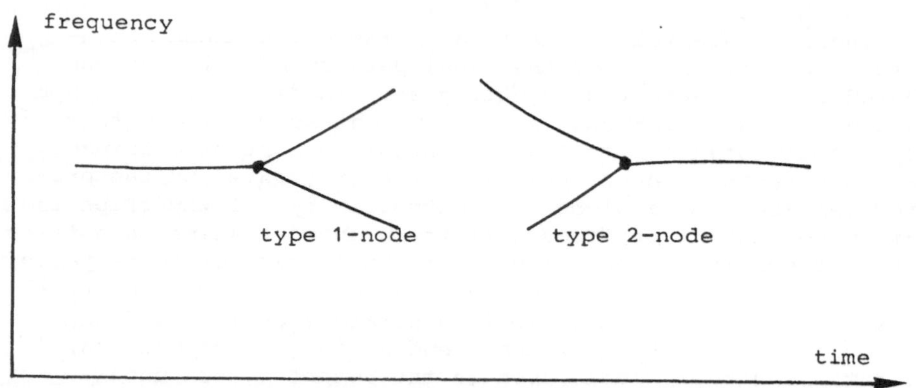

Figure 2. Types of graph nodes that may appear in the extraction of formant patterns.

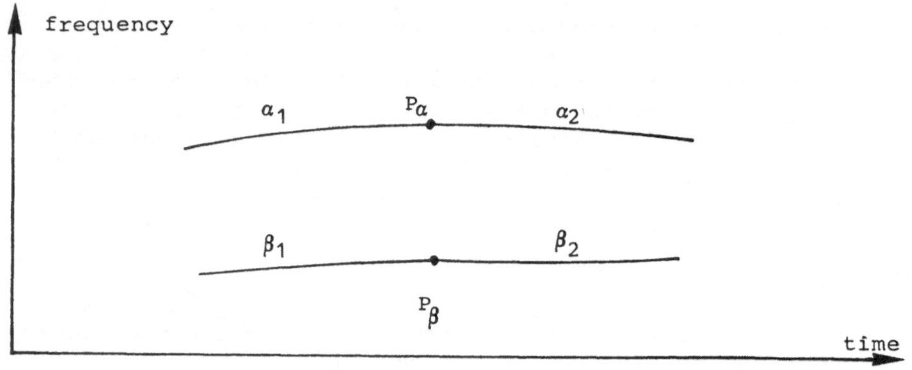

Figure 3. Establishment of a long link between two formant arcs.

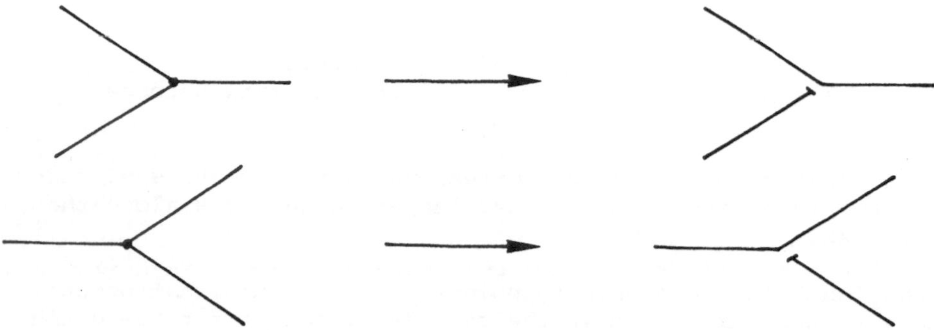

Figure 4. Sample of transformational rules used for pruning the formant graph.

Descriptions of type 1 nodes are stored in a data structure T_1; descriptions of type 2 nodes are stored in a data structure T_2 and a stack Σ_1 is introduced to store informations about the intervals that have to be considered for possible forward links.

All the intervals that do not exhibit possible left-connections with other intervals are listed as "initial intervals"; the addresses of the corresponding cells are pushed into stack Σ_1.

Step 2 - A type -1 node "generating" all the initial intervals is created in T_1.

The first element of Σ_1 is popped, i.e.:

$$\Sigma_1 \rightarrow P \tag{1}$$

The first available label of a label list Λ is assigned to P: this operation is represented as follows:

$$l(P) = \lambda_1 . \tag{2}$$

The point P of the h-th spectrum may be identified by its label as follows:

$$P = \lambda_1(h) \quad . \tag{3}$$

Let $\lambda_k(h)$ be the interval that has been labelled λ_k by the algorithm processing the h-th spectrum.

Step 3 - Possible "sons" of $\lambda_k(h)$, are searched looking for intervals belonging to spectra successive to the spectrum containing $\lambda_k(h)$.

These sons correspond to a possible continuation of the arc containing $\lambda_k(h)$.

The following cases are possible:

3a. $\lambda_k(h)$ has only one son.
The algorithm goes to step 4.

3b. $\lambda_k(h)$ has two or more sons.
The algorithm goes to step 5.

3c. $\lambda_k(h)$ has no sons.
The algorithm goes to step 6.

Step 4 - The following cases are possible:

4a. The son was not previously labelled.
In this case the son is assigned the same label as "father", i.e.: $l(\mathrm{son}(\lambda_k(h)) = \lambda_k(h)$; thus the son becomes $\lambda_k(h+1)$, and the algorithm goes back to step 3.

4b. The son was already labelled, but not as a node. The interval is considered a type-2 node and the data structure T_2 is updated with the definition of the new node.

A new label λ_{k+1} is derived and assigned to the line going forward from the new node to the first previously defined node. The algorithm goes to step 7.

4c. The son was already labelled as a node. In this case the only operation performed is the updating of the node description. The algorithm goes to step 7.

A node which two or more lines end in, and several lines go out from is defined as a type-2 node followed immediately by a type-1 node.

Step 5 - A type-1 generating as many lines as the number of sons is defined in the data structure T_1.

All the sons generated by the node are pushed into the stack Σ_1 with reference to the node stored in T_1.

The first element of Σ_1 is popped, a new label is assigned to it and used for updating the node description in T_1.

The algorithm goes back to step 3.

Step 6 - A "long linkage" to intervals far in frequency or in time is tried. The research of such linkages is controlled by rules that allow only situations of rapid shifts of the formant frequencies and brief gaps in the formant lines.

Only a single long linkage is allowed and, if it is found, the algorithm goes to step 4.

If no linkage is found the algorithm goes to step 7.

Step 7 - The stack Σ_1 is now popped and two cases are possible.

7a. Σ_1 is empty. The algorithm stops.

7b. Σ_1 is not empty. The element popped from Σ_1 is assigned the next available label and the definition of the node generating such interval is updated with the label of the new path.

Step 6 is controlled by a rule that may be formulated in general form considering two arcs as in fig. 3. If point P_α of arc α is linked with point P_β of arc β, each one of the old arcs is split into two arcs leading to four new arcs: α_1, α_2, β_1 and β_2 (see fig. 3). The new link is established and the new arcs are generated if some conditions hold. These conditions are complex and may depend on the frequency range; they involve the length of α_1, α_2,

β_1 and β_2, their average amplitudes, the number of points for which LP = 1, the amplitude and slope discontinuities.

2.2 Deletion of unsuitable links

The graph obtained with the algorithm ARGEN contains paths that are to be eliminated because there are other paths that are much more likely to be formants than the formers. A formant graph simplification is performed by the following algorithm FGS.

Algorithm FGS

Step 1 - A pointer points to the first type-1 node in the data structure T_1.

Step 2 - A research is performed to establish if there is a pair of paths not previously considered starting from the selected generating node in T_1 and going to the same type-2 node contained in the data structure T_2.

Two cases are possible:

2a. A pair of paths is found. The algorithm goes to step 3.

2b. No pair of paths are found. The algorithm goes to step 4.

Step 3 - The uncommon parts of the two paths are considered for possible reduction to one. The decision of reduction is based on local transformational rules summarized in fig. 4; the rules apply if some conditions hold in the line amplitudes, number of gaps, the number of points with LP = 1 and the amplitude and slope discontinuities.

If a decision of reduction is made, some cuts are performed in the graph in order to eliminate the redundant path.

The algorithm goes back to step 2.

Step 4 - A new type-1 node is extracted from the structure T_1. If T_1 is empty, the algorithm stops, otherwise it goes back to step 2.

Finally, a pruning is performed and all the arcs that do not belong to any path going approximately from the beginning to the end of the spectrogram are eliminated.

2.3 Assignment of weights to the arcs

Let Ω_1 be a segment of spectrogram that is supposed to have G formants, namely $F_1,F_2,\ldots,F_G$. If different formant choices are allowed there will be different descriptions for the piece of spectrogram Ω_1 and each description will be assigned a membership function expressing how well the description represents a piece of spectrogram Ω_1. Let:

$$\mu(d_{1\ell}, \Omega_\ell);\ \mu(d_{2\ell},\Omega_\ell);\ldots\ldots\ldots\ldots;\ \mu(d_{i\ell},\Omega_\ell);$$

$$\ldots\ldots\ldots;\ \mu(d_{E_\ell}, \Omega_\ell)$$

be such membership functions and E_1 and the number of the descriptions.

Let d_{i1} correspond to the selection of the paths π_{1i} for the first formant, π_{2i} for the second formant, π_{3i} for the third formant. Because d_{i1} is the description of the three paths π_1,π_2,π_3 the membership $\mu(d_{i1},\Omega_1)$ refers to the fact that, given Ω_1 the three paths $\pi_{1i},\pi_{2i},\pi_{3i}$ are the three formants.

The total membership is computed as follows:

$$\mu(d_{i\ell},\Omega_\ell) = (\pi_{1i},\Omega_\ell) \wedge (\pi_{2i},\Omega_\ell) \wedge (\pi_{3i},\Omega_\ell) \tag{4}$$

Let us assume that π_1 is the concatenation of the following J_1 paths: $\pi_{11i}, \pi_{12i}, \pi_{1J_1i}$, thus:

$$\mu(\pi_{1i},\Omega_\ell) = \mu(\pi_{11i},\Omega_\ell) \wedge \mu(\pi_{12i}, \pi_{11i}\Omega_\ell) \wedge \mu(\pi_{13i}, \pi_{11i}\pi_{12i}\Omega_\ell)\ldots\ldots$$

$$\ldots\ldots\mu(\pi_{1J_1i}, \pi_{11i}\pi_{12i}\ldots\ldots\pi_{1(J_1-1)i}) \tag{5}$$

A simplification is now introduced for computing (5),

$$\mu(\pi_{1ki}, \pi_{11i}\ldots\ldots\ldots \pi_{1(k-1)i}\Omega_1) = \mu(\pi_{1ki}, \pi_{1(k-1)i}\Omega_1) \tag{6}$$

This is justified by the fact that the behavior of a formant in a certain instant does depend only on the immediate past values.

The second member in (6) is computed considering all the possible paths that can follow $\pi_{1(k-1)i}$. Let us assume, for the sake of simplicity that there are two paths, namely π_{1ki} and π_{1ji} then the following definition is used:

$$\mu(\pi_{1ki}, \pi_{1(k-1)i}) = \frac{\xi c(\pi_{1ki}) + \eta g(\pi_{1ki})}{\xi\{c(\pi_{1ki}) + c(\pi_{1ji})\} + \eta\{g(\pi_{1ki}) + g(\pi_{1ji})\}} \tag{7}$$

where ξ and η are weighting coefficients; $c(\pi_{1ki})$ is a continuity function between $\pi_{1(k-1)i}$ and the argument of the function, and $g(\pi_{1ki})$ is a function equal to the sum of the amplitudes of the points of π_{1kj} and of all the points that can be reached going forward from the last point of π_{1kj}. The weight assigned to a line leading type-1 node is thus proportional to how close is the direction of the new line with respect to the previous one and how much energy can be associated with the new line and the paths that can be reached through it.

Minor modifications can be introduced for particular and well-recognizable situations.

Fig. 5 shows an example of extraction of a fuzzy graph from a spectrogram segment corresponding to the syllable ⟨ere⟩. Fig. 5a shows all the intervals of energy concentration extracted from the spectrogram: each interval is represented by a vertical segment made by repeating an alphanumeric character, the intervals marked with an asterix are those for which LP = 1. The $\log_2$ of the energy associated to an interval is proportional to the ASCII value of the alphanumeric characters, the sequence of digits being placed after the sequence of literals by changing the coefficient of proportionality.

Fig. 5b shows the links established by the algorithm ARGEN (points of the same arc are represented by the same label); arc labels do not have any relation with spectral energies.

Fig. 5c shows the results of the deletion of unsuitable links performed by algorithm FGS. Such operation extends some arcs the most relevant of which are marked by a black contour.

Fig. 5d contains the fuzzy graph extracted from the spectrogram segment. The true formants are made of arcs with highest membership.

3. THE PROBLEM OF VOWEL RECOGNITION

Formant values are considered for the spectral segments for which a consistent hypothesis about the presence of a vowel has been emitted. Formant evolutions are described by a language defined in [9]. The primitives of the language are basically stable zones (SZ) and lines (LN). Stable zones correspond to the stationary portions of the speech waveform. Nonreduced vowels are almost always represented by stable zones.

Vowel classification is based on the detection of stable zones in speech intervals that have been previously hypothesized as vocalic segments. Some SZs belonging to segments that have not been previously classified as "vocalic" may cause the assignment of such a feature provided that some conditions involving relations with gross spectral features are verified. This assignment is controlled by rules that are omitted for the sake of brevity.

The vocalic SZ are then named by one or more vowel symbols and a membership function is assigned to each label. This operation uses a definition of vowel loci as fuzzy sets in the F1-F2 plane. Fig. 6 shows an example of such loci. Each SZ has the coordinates of its center of gravity as attributes. The coordinates identify a point in the F1-F2 plane that may belong to one or more vowel loci with certain membership functions.

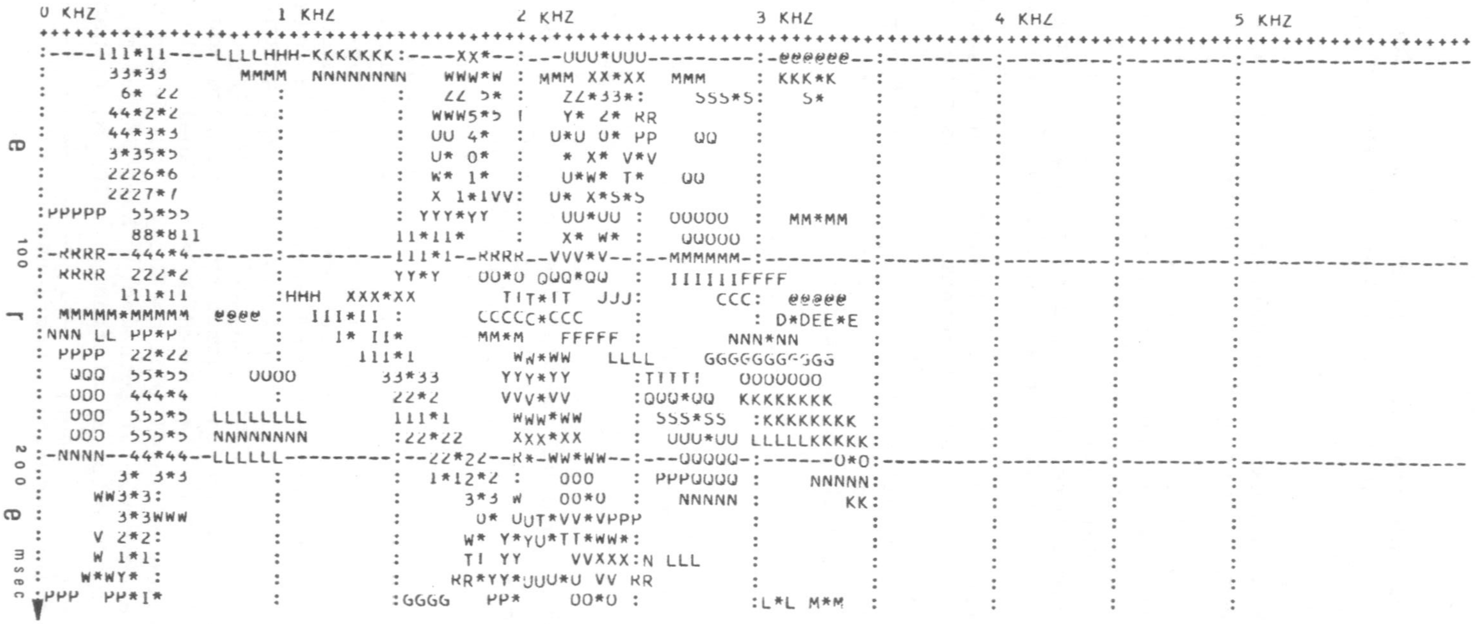

```
     0 KHZ               1 KHZ               2 KHZ               3 KHZ               4 KHZ               5 KHZ
     ++++++++++++++++++++++++++++++++++++++++++++++++++++++++++++++++++++++++++++++++++++++++++++++++++++++++++++++++++++++++
     :----111*11----LLLLHHH-KKKKKKK:----XX*--:---UUU*UUU---------:-@@@@@@--:---------:---------:---------:-----------------
     :     33*33      MMMM  NNNNNNNN   WWW*W : MMM XX*XX  MMM    : KKK*K   :         :         :         :
     :      6* 22        :         :   22 5* :   22*33*:    SSS*S:   S*    :         :         :         :
     :     44*2*2        :         :  WWW5*5 I   Y* 2* RR        :         :         :         :         :
     :     44*3*3        :         :  UU 4*  :  U*U U* PP    QQ  :         :         :         :         :
e    :     3*35*5        :         :  U* 0*  :    * X* V*V       :         :         :         :         :
     :     2226*6        :         :  W* 1*  :   U*W* T*   QQ    :         :         :         :         :
     :     2227*7        :         :  X 1*1VV: U* X*S*S          :         :         :         :         :
     :PPPPP  55*55       :         : YYY*YY  :   UU*UU :  00000  :  MM*MM  :         :         :         :
     :       88*811      :         11*11*    :   X* W* :   QQQQQ :         :         :         :         :
100  :-RRRR--444*4-------:---------111*1--RRRR--VVV*V--:--MMMMMM-:---------:---------:---------:---------:-----------------
     : RRRR  222*2       :         YY*Y   00*0 QQQ*QQ  :  IIIIIIFFFF       :         :         :         :
     :      111*11       :HHH  XXX*XX       TIT*IT  JJJ:      CCC: @@@@@   :         :         :         :
r    : MMMMM*MMMMM  @@@@ :  III*II :      CCCCC*CCC    :         : D*DEE*E :         :         :         :
     :NNN LL PP*P        :    I* II*      MM*M   FFFFF :       NNN*NN      :         :         :         :
     : PPPP  22*22       :      111*1        WW*WW   LLLL    GGGGGGGGGG    :         :         :         :
     :  QQQ  55*55     0000       33*33     YYY*YY     :TTTT!   0000000    :         :         :         :
     :  000  444*4       :         22*2     VVV*VV     :QQQ*QQ  KKKKKKKK   :         :         :         :
     :  000  555*5  LLLLLLLL       111*1     WWW*WW    : SSS*SS  :KKKKKKKK :         :         :         :
     :  000  555*5  NNNNNNNN      :22*22    XXX*XX     :  UUU*UU LLLLLKKKKK:         :         :         :
200  :-NNNN--44*44--LLLLLL---------:--22*22--R*-WW*WW--:---QQQQQ-:------0*0:---------:---------:---------:-----------------
     :      3* 3*3       :         :  1*12*2 :   000   : PPPQQQQ :    NNNNN:         :         :         :
     :    WW3*3:         :         :     3*3 W   00*0  :   NNNNN :       KK:         :         :         :
e    :      3*3WWW       :         :      0* UUT*VV*VPPP         :         :         :         :         :
     :    V 2*2:         :         :     W* Y*YU*TT*WW*:         :         :         :         :         :
     :    W 1*1:         :         :     TI YY    VVXXX:N LLL    :         :         :         :         :
msec :  W*WY* :          :         :    RR*YY*UUU*U VV RR        :         :         :         :         :
     :PPP  PP*1*         :         :GGGG   PP*    00*0 :         :L*L M*M  :         :         :         :
```

Fig. 5a

Figure 5. Example of extraction of a fuzzy graph representing a formant pattern from the spectrogram segment of the syllable ⟨ere⟩ .

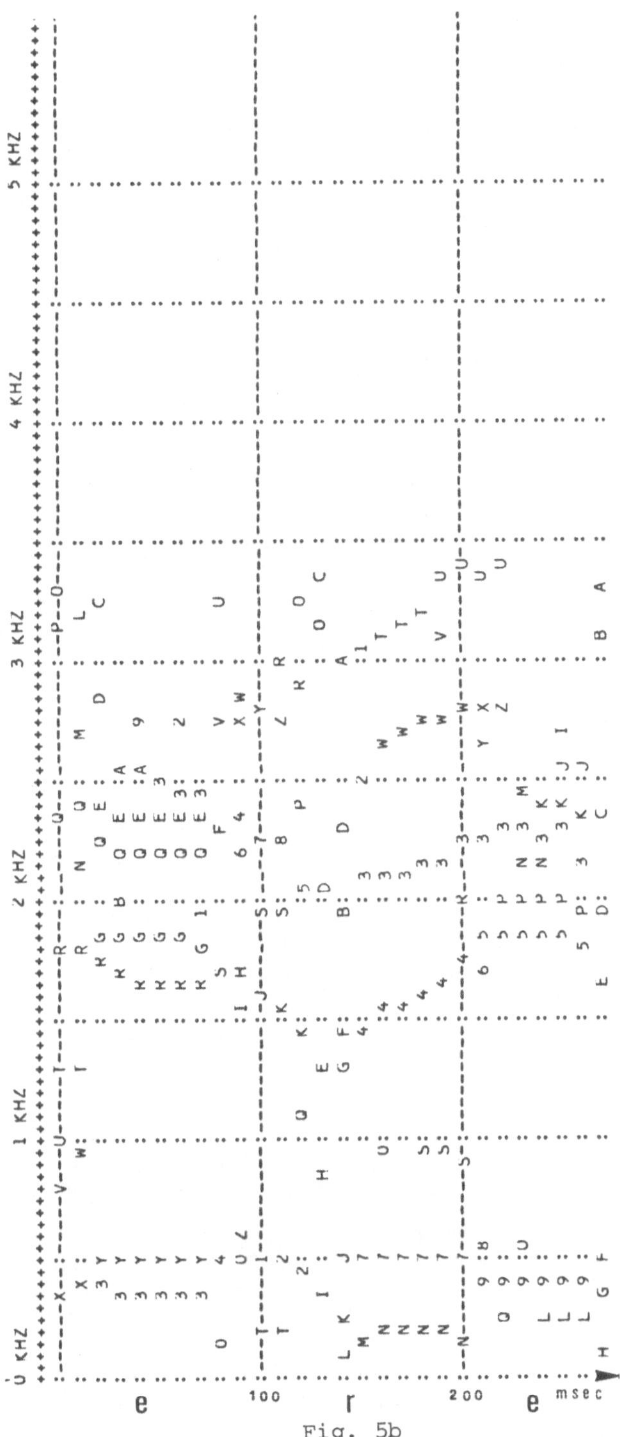

Fig. 5b

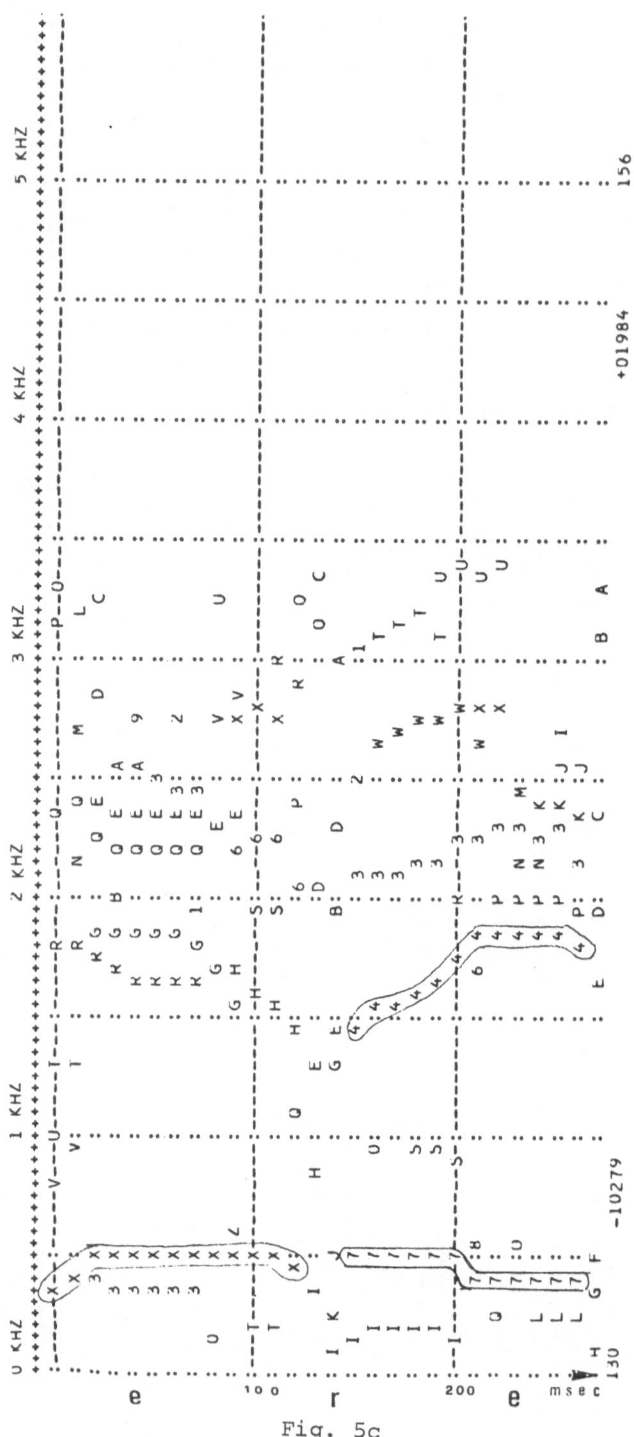

Fig. 5c

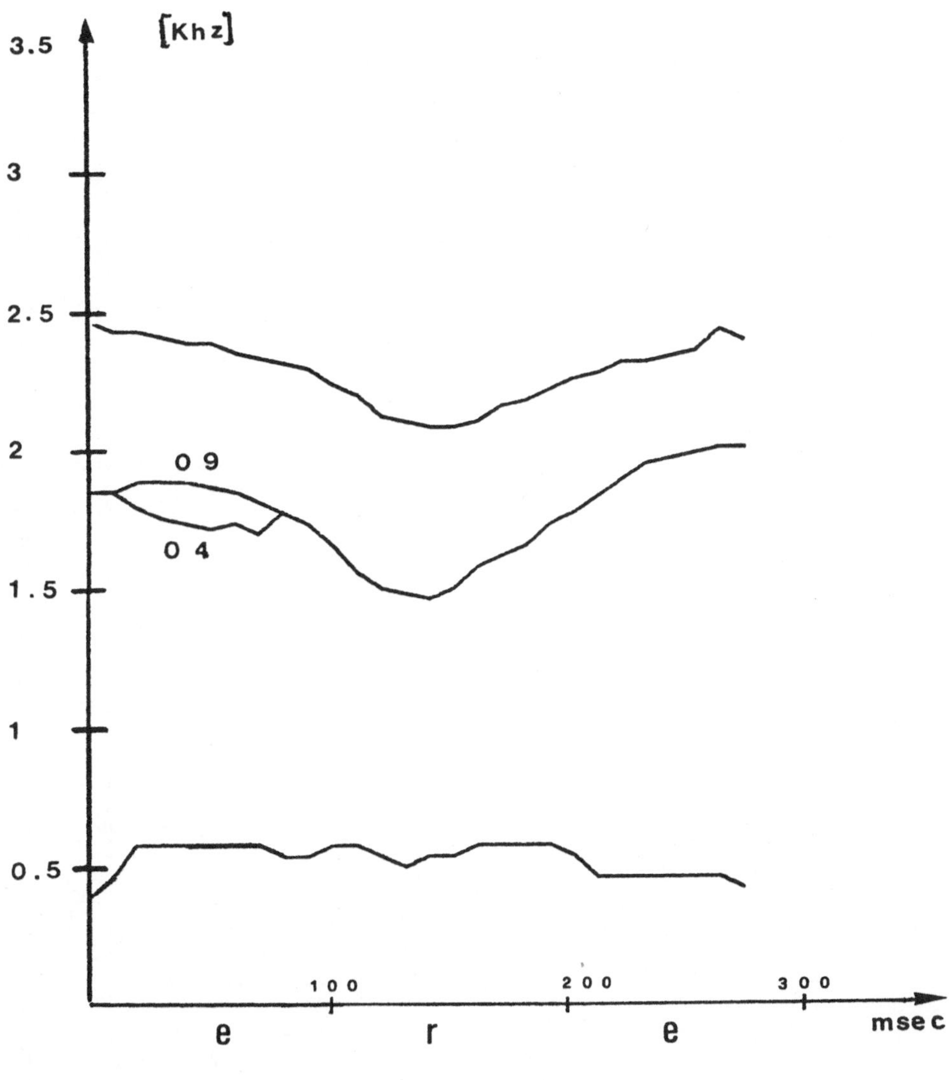

Fig. 5d

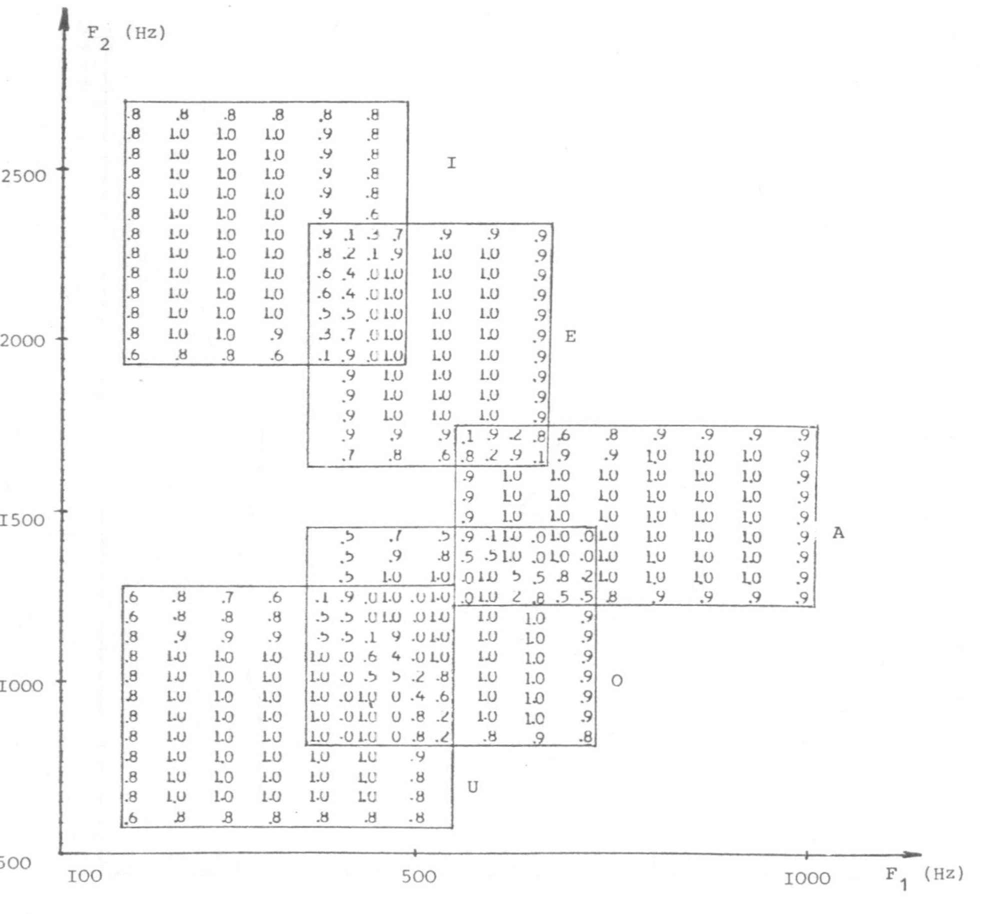

Figure 6. Definition of vowel loci as fuzzy sets in the F_1F_2 plane.

Vowel loci, defined as fuzzy sets, have an assignment of membership functions that can be made by a subjective adjustment of probability densities estimated after a long period of learning. The resulting assignment is speaker dependent.

A vocalic hypothesis v_i is defined by the following rule:

$$\langle v_i \rangle ::= \langle \text{vocalic} \rangle \quad \langle SZ_v \rangle \quad \langle v_i(F_1,F_2) \rangle \tag{8}$$

where ⟨vocalic⟩ is a phonetic feature attached during precategorical classification, $\langle SZ_v \rangle$ represents the fact that a vocalic stable zone has been found inside the vocalic segment and $v_1(F_1,F_2)$ represents the fact that the center of gravity of SZ_v is in the locus of the vowel v_i.

A semantic rule is associated with (8) for evaluating the vocalic hypothesis v_i:

$$\mu_{\langle v_i \rangle} = \mu_{\langle \text{vocalic} \rangle} \wedge \mu_{\langle SZ_v \rangle} \wedge \mu_{\langle v_i(F_1,F_2) \rangle} \tag{9}$$

$\mu_{\langle \text{vocalic} \rangle}$ is a result of the precategorical classification,

$\mu_{\langle v_i(F_1,F_2) \rangle}$ is the degree of membership of the center of gravity of SZ_v in the fuzzy set which is the locus of v_i in the F1-F2 plane and $\mu_{\langle SZ_v \rangle}$ is a degree, defined by heuristic rules, of the stationarity of the formant pattern.

<u>Example</u>

Fig. 7 shows the formant pattern of the syllable ⟨ilu⟩ . Dashed rectangles contain stable zones. Both of them are recognized as SZ_v with membership equal to 1. The centers of gravity of the two stable zones are respectively:

	F1	F2 (Hz)
SZ_1	292	2068
SZ_2	310	955

Such centers of gravity belong respectively to the fuzzy sets of ⟨i⟩ and ⟨u⟩ , as defined in fig. 6, with membership 1.

The experimental results related to 200 vowel samples are reported in Table 1 and confirm the effectiveness of the method.

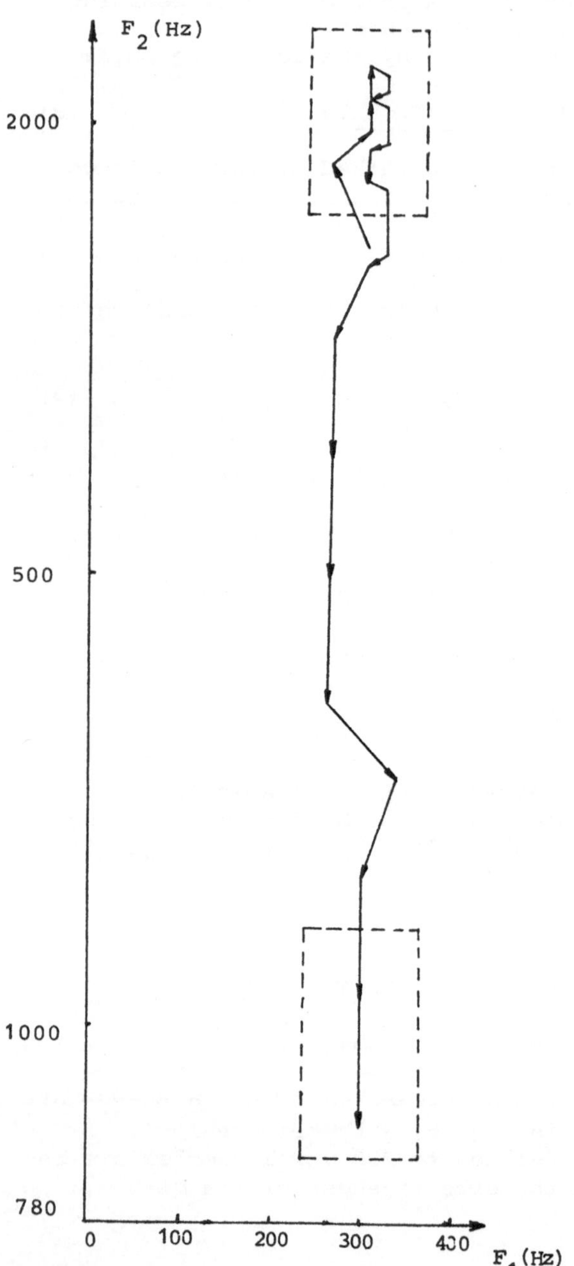

Figure 7. Formant pattern of the syllable <ilu> .

TABLE 1

Percentage of missed vowels	: 0.6%
Percentage of vowels recognized with highest membership	: 98%
Percentage of vowels recognized within two choices	: 99%

4. USE OF FORMANTS FOR THE RECOGNITION OF LIQUIDS AND NASALS

After experiments on liquid and nasal intervocalic consonants, it has been found that, if the algorithm used for vowels finds consistent paths with no discontinuities joining the formant lines of the two vowels, it generally tracks the right formants.

This result, together with the constraint that the first formant is allowed to have only a downward shift in its frequency interval, made it possible to track correctly the formants for all the liquid and for some of the nasal consonants. Such a correctness was confirmed in many cases by synthesis experiments [12].

For some nasal consonants, particularly in the context of two front vowels, formant tracking appears to be very difficult (especially for the second formant of / m /) even if FFT and up to 16-pole LPC spectra are used together. This is due to the complex interaction of the anti-resonances of the vocal tract and the resonances of the pharingal and nasal tract.

Such difficulties have been successfully avoided by controlling formant tracking with rules for those cases for which there is not a single, consistent path in the consonant region that joins, with no discontinuities, the formant lines of the vowels.

The special rule inferred after experiments is briefly introduced in the following. After tracking the first formant, the most consistent tract of energy concentration in the consonant segment and in the frequency range from 2 kHz to 2.8 kHz is assigned to the third formant. Consistency is based on stationarity in time, spectral energy and number of LPC spectra peaks along the candidate tract. Then, the first consistent tract of energy concentration below this third formant is assigned to the second formant of the consonant. Such a rule is consistent with synthesis experiments. Fig. 8 shows an example of formant tracking of a pseudosyllable / emi /; dashed lines represent the connections established by the rule just described.

4.2 Generation of hypotheses based on formant patterns

An hypothesis is the answer to a composite classificational question $Q \triangleq B$, where the bodies of the component questions are fuzzy linguistic variables $B_1, B_2, \ldots, B_n$ involved in an analytic representation of B [13]. Learning in this framework consists in a

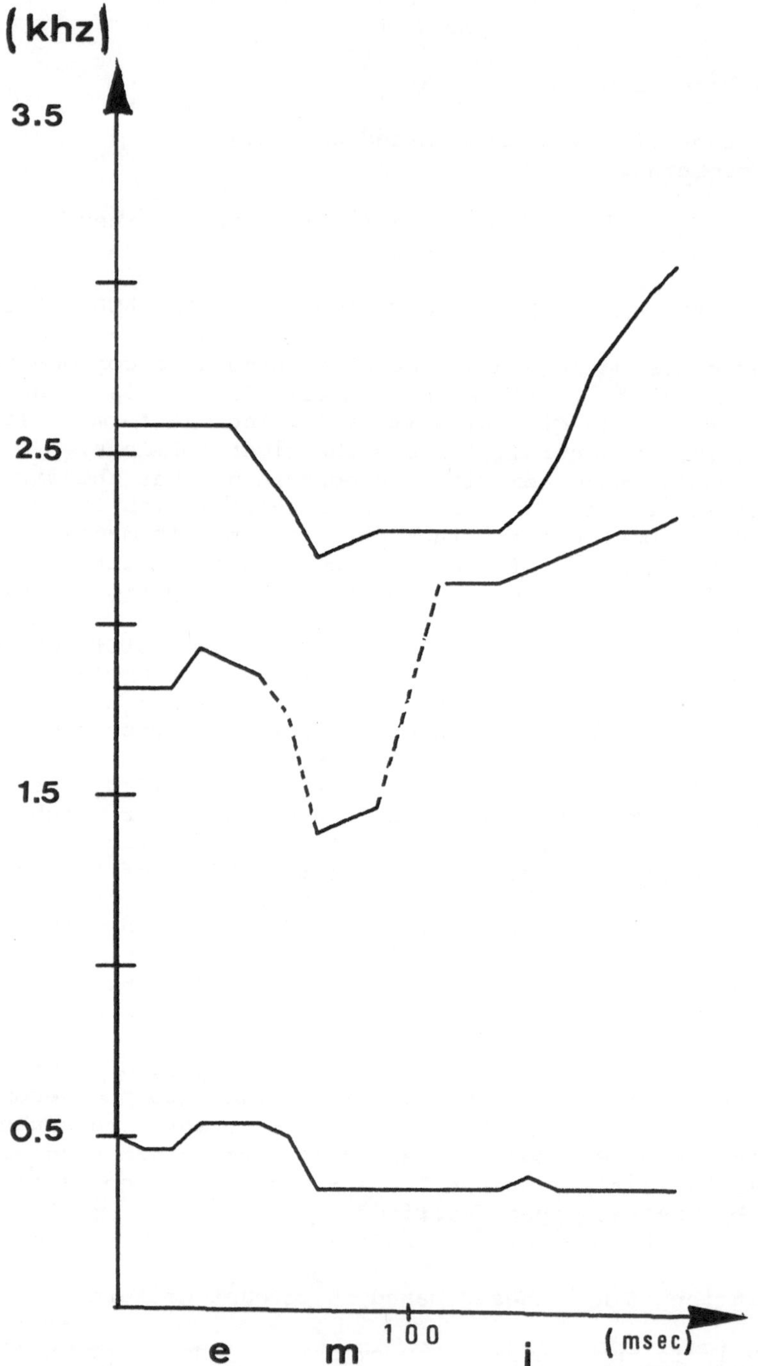

Figure 8. Example of formant tracking for the syllable <emi> .

subjective assignment of a membership function to each answer to every atomic question and to the composition of the answers to the atomic questions in order to obtain the answer to the composite question. The atomic questions are asked on the time evolutions of acoustic parameters like total energy, energy in specified bands, formant frequencies and amplitudes. The fuzzy sets that are the answers to the atomic questions, are defined over the ranges of acoustic parameters.

The assignment of the membership functions to each linguistic variable is made considering the distribution of the answers to the composite question over the range of the parameter for which the component question is asked. For example, let Q_j be the question: does a given piece of spectrogram correspond to a nasal /m/? Let us assume that Q_{ji} is the i-th component question of Q_j and that it refers to the parameter p_{ji} that exhibits high values for /m/ and low values for other nasals. Let us assume that the fuzzy variables h_{ji} and l_{ji} (high, low) are the answers to Q_{ji}. In order to assign a membership function, for example to h_{ji}, an interval of p_{ji} where only samples of /m/ are contained is isolated; a membership function $\mu_{hji} = 1$ is assigned to the interval; analogously, a membership function $\mu_{hji} = 0$ is assigned to the interval of p_{ji}, where no /m/ samples have been found. The values of μ_{hji} for the points not included in the above mentioned intervals are assigned subjectively using simple rational functions of p_{ji}. μ_{lji} is defined as follows:

$$\mu_{lji} = 1 - \mu_{hji}$$

As the learning proceeds, the interval in which μ_{lji} and μ_{hji} assume value 0 or 1 may only decrease; continuing indefinitely such learning, the interval where both the membership are less than one may cover the entire range allowed for a given parameter. This expansion may be stopped after acquiring a satisfactory amount of data and membership functions may be assigned to the rules introduced for obtaining the answer to the composite question from the answers to the atomic questions. These rules may be learned automatically with some adjustments introduced by the experimenter [14]. Rules with fuzzy grammaticality have been proven useful [14] for improving the classification sonorant/nonsonorant, but, fortunately, they do not seem to be necessary for nasal classification.

Furthermore, it is not reliable to have a small distance between the break points delimiting the intervals where μ_{lji} is 0 or 1. This distance can grow if new atomic questions on new parameters are introduced.

The relation between a speech pattern and its possible phonemic transcriptions becomes, with the approach proposed in this section, a fuzzy composite relation.

4.3 Application to liquid-nasal classification

The liquid/nasal classification is a process of emission of hypotheses about the consonant sounds classified as sonorants with high membership function value in the precategorical classification. The decision within the sonorant consonant class is the answer to the composite question:

$$Q_1 \triangleq B_1$$

The answer to Q_1 is obtained by a relation on the answers to the component questions $Q_{1i}\cdot(i = 1,2,\ldots,K_1)$ the answers to the component questions Q_{1i} are fuzzy linguistic variables h_{1i}, m_{1i}, l_{1i} defined over the continuous intervals of the following acoustic parameters:

- S the total energy of a short-time spectrum,
- G the energy of a short-time spectrum in the 5-10 KHz band,
- B the energy of a short-time spectrum in the 200-900 Hz band,
- R = B/G, (10)
- dur(dip(S)) the duration of a dip of S defined as the time interval for which the values of S are below a value obtained decreasing by a fixed threshold the minor relative maximum at the boundaries of the dip,
- dur (dip(R)) is the duration of the dip of R,
- $F_i(nT)$: the n-th sample of the i-th formant frequency (T is the sampling period = 10 msec)
- $A_i(nT)$: the n-th sample of the energy associated with the i-th formant,
- $D_{12} = A_1(n^*T) - \text{Max } A_2(n^*T), A_3(n^*T)$; where n T is the time interval where $A_2(nT)$ has the absolute minimum in the syllable,
- $D_{13} = A_2(n^*T) - A_3(n^*T)$.

m_{1i} is a fuzzy set having membership equal to 1 when both $\mu_{h_{1i}}$ and $\mu_{\ell_{1i}}$ are less than 0.7, equal to zero when $\mu_{h_{1i}}$ or $\mu_{\ell_{1i}}$ are equal to 1; it assumes values between zero and one in the other intervals.

After experiments on hundreds of consonants in VCV syllables extracted from sentences pronounced by a single male talker, a set of rules has been derived. This set of rules is considered as a starting assumption for an inference procedure that will be developed in a recognition stage. The inference will have the purpose of refining the definition of the fuzzy sets associated to each component question. Furthermore it will better specify the relations between atomic answers and the answers to the composite question and the semantic (relations between membership functions) associated with such relations.

For the sake of simplicity, the intervals for which the membership functions assume value 1 and 0 are identified by the boundary values ℓ_{ji} and h_{ji} that are listed in Tables 2,3,4 together with their associated acoustic variable x_{ji}. The parameters h_{1i}, ℓ_{1i} and x_{1i} for the composite question Q_1 are summarized in Table 2; $\ell_{18} = \ell_{19} = 0$ means that the variables ℓ_{18} and ℓ_{19} have membership function equal to 1 only when the corresponding durations are 0.

It has been found possible to classify correctly with membership equal to 1 all the training set using regular expressions as relations between the answers to the atomic questions and the answers to the composite question. Some of the terms of this relations are valid only if some associated predicates are true. These predicates refer to the vocalic context of the sonorant sound and are represented by brackets containing the vowel before the consonant followed by the vowel after the consonant; an asterix represents a "don't care" condition. The fact that the relation derived shows a context dependency in the distinction between liquids and nasals means that context dependency is relative to the particular acoustic parameters used. Nevertheless, the experiments carried out for deriving rules in general cases make appear unlikely that the distinction liquid/nasal is feasible with context-independent procedure, even if this is possible for limited protocols [9].

The expressions obtained for the answers to the question Q_1 are the following:

$$\begin{aligned}\text{LIQUID} = {} & \ell_{11} + m_{11}\ ([u,*])\ell_{12} + [e,*](\ell_{13} + m_{13}h_{17}) + \\ & + [i,*](m_{18} + h_{18}(\ell_{14}+m_{14}h_{17})) + [o,*](\ell_{15} + \\ & + m_{15}h_{17}) + [a,*](\ell_{16}+m_{16}(\ell_{19} + m_{19}h_{17})))\end{aligned} \tag{11}$$

$$\begin{aligned}\text{NASAL} = {} & h_{11} + m_{11}\ ([e,*](h_{13} + m_{13}\ell_{17})+[i,*]h_{18}(h_{14} + \\ & + m_{14}\ell_{17})+[o,*](h_{15}+m_{15}\ell_{17})+[a,*]\ h_{19}(h_{16}+ \\ & + m_{16}\ell_{17})\end{aligned} \tag{12}$$

The membership functions can be computed with min-max operations [15].

4.4 Applications to the classification of liquids

The classification of liquids mainly concerns "l" and "r" because the liquid "gl" (like in the Spanish word Sevilla) is easily distinguished by a syntax controlled procedure that recognize typical evolutions in the F1-F2 graph. The acoustic parameters involved in this classification are dur (dip (S)), D_{23}, dur (cons), i.e.

the duration of the consonant tract, F_{1c} and F_{2c} that are respectively the first and the second formant frequencies corresponding to the minimum of the second formant energy.

The question Q_L for the classification of liquids has component questions and fuzzy set definitions according with Table 3; the expressions derived are the following:

$$\text{"LIQUID l"} = h_{L1} + m_{L1}\ (l_{L2} + h_{L3} + m_{L3}(l_{L4}\ h_{L5})) \tag{13}$$

$$\text{"LIQUID r"} = l_{L1} + m_{L1}\ (l_{L3} + h_{L4}) \tag{14}$$

TABLE 2

QUESTION	x_{li}	h_{li}	ℓ_{li}	DIMENSION
Q_{11}	D_{12}	15	4	dB
Q_{12}	D_{12}	15	14	dB
Q_{13}	D_{12}	12	10	dB
Q_{14}	D_{12}	7	6	dB
Q_{15}	D_{12}	12	9	dB
Q_{16}	D_{12}	10	5	dB
Q_{17}	D_{23}	1	-6	dB
Q_{18}	Dur(dip(S)) (Dur(dip(R)=0)	20	0	msec
Q_{19}	Dur(dip(S))	20	0	msec

TABLE 3

QUESTION	x_{Li}	h_{Li}	ℓ_{Li}	DIMENSION
Q_{Li}	Dur(dip(S))	40	20	msec
Q_{L2}	D_{23}	2	0	dB
Q_{L3}	Dur(conson.)	50	30	msec
Q_{L4}	F_{1c}	560	400	Hz
Q_{L5}	F_{2c}	1300	800	Hz

TABLE 4

QUESTION	x_{Ni}	h_{Ni}	ℓ_{Ni}	DIMENSION
Q_{N1}	D_{12}	23	17	dB
Q_{N2}	D_{12}	23	15	dB
Q_{N3}	F_{12}	1100	950	Hz
Q_{N4}	F_{22}	2000	1500	Hz
Q_{N5}	F_{21}	1700	1400	Hz
Q_{N6}	D_{12}	20	13	dB
Q_{N7}	D_{12}	20	15	dB
Q_{N8}	F_{12}	1700	1600	Hz
Q_{N9}	D_{12}	26	13	dB
Q_{N10}	F_{22}	2120	2000	Hz
Q_{N11}	F_{21}	1360	1160	Hz
Q_{N12}	F_{22}	1700	1500	Hz
Q_{N13}	F_{21}	1600	1400	Hz
Q_{N14}	F_{22}	2000	1600	Hz
Q_{N15}	F_{21}	900	800	Hz
Q_{N16}	F_{22}	1900	1400	Hz
Q_{N17}	F_{21}	1200	1000	Hz
Q_{N18}	F_{22}	1320	1280	Hz
Q_{N19}	F_{22}	1460	1260	Hz
Q_{N20}	F_{22}	1200	1000	Hz
Q_{N21}	$Max(F_2)$	1200	800	Hz
Q_{N22}	$Max(F_2)$	1240	900	Hz
Q_{N23}	$Max(F_2)$	1520	1160	Hz
Q_{N24}	D_{12}	26	15	dB
Q_{N25}	D_{12}	15	9	dB
Q_{N26}	F_{22}	1700	1400	Hz

5. APPLICATION TO THE DETAILED CLASSIFICATION OF NASALS

5.1 Parameters of the atomic questions

The classification of nasals is obtained as an answer to the composite question Q_N. This answer is obtained by a relation on the answers to the component questions Q_{Ni} $(i = 1,2,\ldots,K_N)$; the answers to the component questions Q_{Ni} are fuzzy linguistic variables h_{Ni}, ℓ_{Ni} defined by Table 4 on the basis of the acoustic parameters defined in (10) and the following ones:

- F_{21}: the lowest value of $F_2(nT)$ at the time corresponding to the beginning of the downward shift of the first formant frequency of the nasal consonant,
- F_{22}: the value of $F_2(nT)$ at the end of the downward shift of the first formant frequency of the nasal consonant.

Detection of F_{21} and F_{22} is very easy because the formant evolutions in the F_1-F_2 plane are described by a language [9] that contains lines as primitives. Fig. 9a shows, as an example, the values of F_{21} and F_{22} in a plot representing the time evolution of the formant of a VNV utterance. Fig. 9b shows the same points in a representation of the same utterance in the F_1-F_2 plane.

5.2 The recognition rules

The recognition rules for the classification of nasals were inferred subjectively with the goals of avoiding a wrong answer with degree of worthiness equal to one and trying to minimize the number of times the wrong answer assumes a degree of worthiness higher than the right answer [16].

The vocalic contexts are represented by places of articulation according to Table 5 and the break points at the end of the intervals where $(\ell_{Ni} = 1) \wedge (h_{Ni} = 0)$ or $(h_{Ni} = 1) \wedge (\ell_{Ni} = 0)$ are reported in Table 6 with column heading h_{Ni} and ℓ_{Ni} respectively. In Table 6, the type of parameters on which the questions are asked is reported with column heading p_{Ni}.

The places of articulation have been determined on the basis of the values assumed by the second formant on the stationary portion of the vocalic segment according to the following definitions:

$$
\begin{array}{lll}
1900\ \text{Hz} < F_2 & \text{front} & \\
1300\ \text{Hz} < F_2 < 1900\ \text{Hz} & \text{central} & (15) \\
F_2 < 1300\ \text{Hz} & \text{back.} &
\end{array}
$$

An analytic representation of the inferred rules is the following:

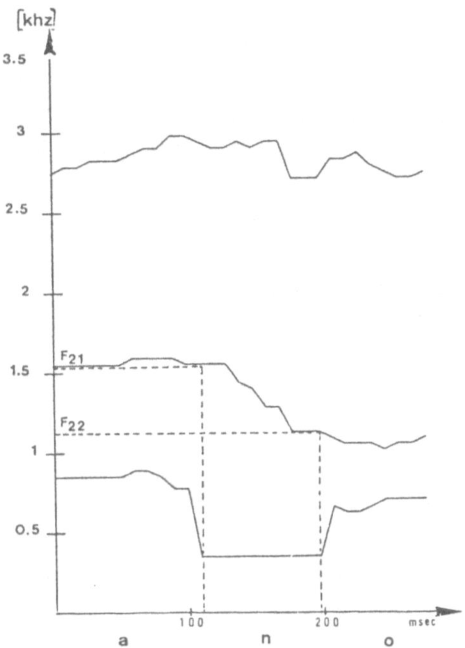

Figure 9a. Example of points F_{21} and F_{22} in the time evolution of formants for the syllable ⟨ano⟩ .

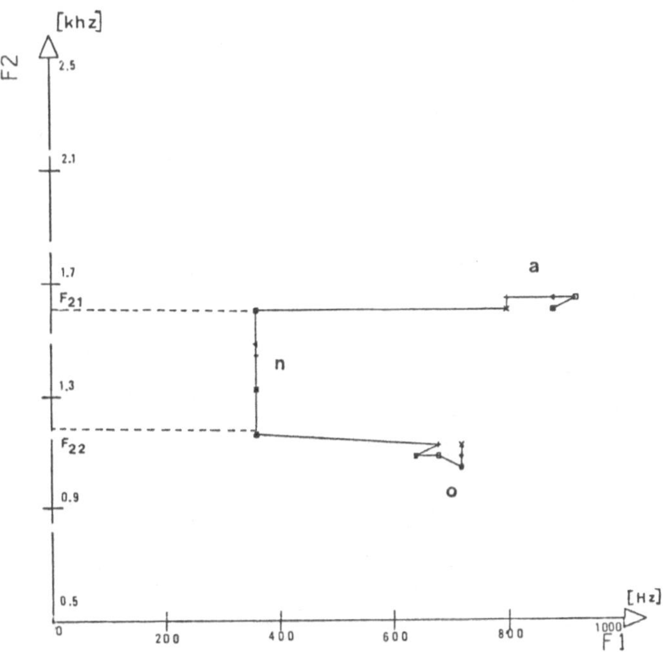

Figure 9b. The same as in fig. 9a but in F_1 F_2 plane.

TABLE 5

Assignment of symbols to coarticulation instances

Place of articulation of the preceding and following vowel in VNV utterance	Symbol assigned
front-front	ff
front-central	fc
front-back	fb
central-front	cf
central-central	cc
central-back	cb
back - front	bf
back - central	bc
back - back	bb

TABLE 6

Definition of the component questions of QN

Question	p_{Ni}	L_{Ni}	H_{Ni}	Dimension
Q_{N1}	F_{21}	1880	2320	Hz
Q_{N2}	D_{12}	0	15	dB
Q_{N3}	F_{21}	1750	2150	Hz
Q_{N4}	F_{22}	1500	1700	Hz
Q_{N5}	D_{12}	6	17	dB
Q_{N6}	F_{21}	1400	2250	Hz
Q_{N7}	F_{22}	900	1400	Hz
Q_{N8}	D_{12}	9	23.5	dB
Q_{N9}	F_{21}	1500	1700	Hz
Q_{N10}	D_{12}	8	17	dB
Q_{N11}	F_{21}	1300	1500	Hz
Q_{N12}	F_{22}	1300	1500	Hz
Q_{N13}	F_{21}	1300	1480	Hz
Q_{N14}	F_{22}	960	1100	Hz
Q_{N15}	F_{21}	940	1200	Hz
Q_{N16}	F_{22}	900	1100	Hz
Q_{N17}	D_{12}	6	20.5	dB
Q_{N18}	F_{21}	900	1000	Hz
Q_{N19}	F_{22}	980	1400	Hz
Q_{N20}	D_{12}	10	19	dB
Q_{N21}	F_{21}	980	1050	Hz
Q_{N22}	F_{22}	900	1000	Hz

$$\langle m\rangle = [ff](\ell_{N1}+\ell_{N2})+[fc](\ell_{N3}+\ell_{N4}+\ell_{N5})+[fb](\ell_{N6}+\ell_{N7}+\ell_{N8})$$
$$+ [cf](\ell_{N9}+\ell_{N10})+[cc](\ell_{N11}+\ell_{N12}) +[cb](\ell_{N13}+\ell_{N14})$$
$$+ [bf](\ell_{N15}+\ell_{N16}+\ell_{N17})+[bc](\ell_{N18}+\ell_{N19}+\ell_{N20}) +$$
$$+ [bb](\ell_{N21}+\ell_{N22}) \qquad (16)$$

$$\langle n\rangle = [ff](h_{N1}+h_{N2})+[fc](h_{N3}+h_{N4}+h_{N5})+[fb](h_{N6}+h_{N7}+h_{N8})$$
$$+ [cf](h_{N9}+h_{N10})+[cc](h_{N11}+h_{N12})+[cb](h_{N13}+h_{N14})$$
$$+ [bf](h_{N15}+h_{N16}+h_{N17})+[bc](h_{N18}+h_{N19}+h_{N20})+$$
$$+ [bb](h_{N21}+h_{N22}) \qquad (17)$$

Table 7 reports the error rates for four speakers.

8. CONCLUSIONS

Extraction of formants from speech spectrograms and a method for taking into account the possible ambiguities that may arise in extracting such features have been presented.

Good experimental results have been achieved using such features in the classification of sonorant sounds. The reason is that significant context-dependent cues are very often clearly detectable. This allows to infer powerful context-dependent rules for phoneme recognition. The use of formant patterns together with burst spectra has been found also promising for plosives recognition.

9. ACKNOWLEDGEMENTS

The work described in this paper has been carried out at the Centro per l'Elaborazione Numerale dei Segnali (CENS) Turin Italy and was supported by Consiglio Nazionale delle Ricerche (CNR) of Italy.

The authors wish to thank all the colleagues and the guests of the CENS which contributed to the development of the researches described in this paper. Their contributions are referenced in the bibliography.

TABLE 7

Four speakers detailed distribution of the error rates for each coarticulation instance

Coasticulation Instance	Error Rate
front-front	10%
front-central	9%
front-back	16%
central-front	9%
central-central	-
central-back	9%
back-front	-
back-central	-
back-back	-
average	6%

REFERENCES

1. Reddy, D. Raj, ed., "Speech Recognition: Invited Papers of the IEEE Symp.", New York: Academic Press, 1975.
2. Reddy, D. Raj, "Speech Recognition by Machine: A Review", Proc. IEEE, vol. 64, No. 4, Apr. 1976, pp. 501-531.
3. Martin, T. B., "Practical Applications of Voice Input to Machines", Proc. IEEE, vol. 64, No. 4, Apr. 1976.
4. Jelinek, F., "Continuous Speech Recognition by Statistical Methods", Proc. IEEE, vol. 64, No. 4, Apr. 1976, pp. 532-556.
5. Klatt, D. H., "Review of the ARPA Speech Understanding Project", Journ. Acoust. Soc. of America, vol. 60, Fall. 1976, S 10.
6. Wolf, J. J., "Speech Recognition and Understanding" in Pattern Recognition, K. S. Fu ed., New York: Springer, 1975.
7. Stevens, K. N., "Potential Role of Property Detectors in the Perception of Consonant", M.I.T.-R.L.E., Quarterly Progress Report No. 110, July 1973, pp. 155-168.
8. De Mori, R., P. Laface, "Automatic Detection of Distinctive Features in Continuous Speech", Proc. 3rd Int. Joint Conference on Pattern Recognition, Coronado, Nov. 1976, pp. 609-616.
9. De Mori, R., P. Laface, E. Piccolo, "Automatic Detection and Description of Syllabic Features in Continuous Speech", IEEE Transactions, vol. ASSP-24, Oct. 1976, pp. 365-379.
10. Reddy, D. R., "Hearsay Speech Understanding Systems: Summary of Results of the Five-Year Research Effort at Carnagie-Mellon University", Department of Computer Science, Carnagie-Mellon University, Aug. 1977.
11. De Mori, R., "Recent Advances in Automatic Speech Recognition", Invited Paper 4th Int. Joint Conference on Pattern Recognition, Kyoto (Japan), nov. 1978.
12. Mezzalama, M., E. Rusconi, "General System for Synthetizing Speech", Proc. of the Speech Communication Seminar, vol. 2, Almqvist and Wikselt, Uppsala, 1975, pp. 307-314.
13. Zadeh, L. A., "Fuzzy Languages and Their Relation to Human and Machine Intelligence", Proc. Int. Conference on Man and Computer, Bordeaux (France), ed. by S. Kargle Basel, Munchen, New York 1972, pp. 130-165.
14. De Mori, R., L. Saitta, "Automatic Learning of Fuzzy Relation" (in preparation).
15. Zadeh, L. A., "A Fuzzy Algorithmic Approach to the Definition of Complex or Imprecise Concepts", Memo ERL-M474-Oct. 1974- Electronic Research Laboratory, College of Engineering, University of California, Berkeley.
16. Gubrynowicz, R., R. De Mori, P. Laface, "La Description au Niveau Acoustique des Consonnes Nasales Prononcees dans un Discours Continu", Actes des Nemes Journees d'Etude sur la Parole, Lannion, Juin 1978, pp. 287-296.

A NEW ALGORITHM FOR PITCH EXTRACTION AND VOICED/UNVOICED SEGMENTATION OF SPEECH USING ONLY ZERO CROSSING INTERVAL SEQUENCES (ZCIS) AND VOICED PHONEME RECOGNITION WITH ZCIS.

Davras Yavuz
Department of Electrical Engineering
Middle East Technical University
Ankara, Turkey

Nezih C. Geçkinli
Electronic Research Unit
Marmara Scientific and Industrial Research Institute
Gebze, Kocaeli, Turkey

ABSTRACT. A novel digital computer algorithm for pitch period analysis of connected speech is presented. The basic idea is to find, through an adaptive search, the "periodic-like" portions of the speech from its zero-crossing interval (ZCI) sequence. The algorithm requires only additions and comparisons, no multiplications. A PDP-8 computer implementation performs the analysis in real-time, as accurately as manually performed pitch measurement, from a plot of the speech waveform, low-pass filtered to a band-width of 900 Hz.

The same basic idea may also be used for voiced/unvoiced segmentation of speech. Again, segmentation performance as accurate as any reported in the literature is possible in real-time.

As a further extension of ZCI usage, techniques for recognition of voiced phonemes, in particular Turkish vowels, from ZCI patterns and ZCI distributions is discussed. Using these techniques, it is possible to obtain very fast voiced phoneme recognition with relatively good accuracy.

A unique feature of the pure Turkish language is the so-called "vowel harmony" rule, which may be summarized as two state diagrams. Using this, vowel recognition results on connected speech may be tested and corrected with software.

Implementation results related to each of the above individually and in combination will be presented.

1. INTRODUCTION

Recognition of connected speech has been and is still the subject of extensive research. The presupposition of the fifties that, given 'better' computers, the problem of speech recognition would easily be solved, has obviously not been materialized. In fact at this point in time, it might be possible to state that the rapid development of computing power, where we use the word power to describe the combined attributes memory size, speed, efficient architecture, might even have hindered the appearance of 'clever' and efficient approaches.

One characteristic property of the research related to automatic speech recognition (OSR) in the last decade, has been, that, the speed of operation of the components of OSR systems has generally been considered of secondary relevance. Although it is clear that in research related to such complex topics as OSR systems, one has to explore every avenue for possible bits of clues, we feel that the approach of relegating speed of operation to secondary roles has been detrimental. It should also be mentioned that, the very rapid development of digital as opposed to analog signal processing hardware has probably had some misguiding effects. In our opinion, any approach to the realization of the components of OSR systems, which do not have the potential of realization of operation in at most a few times real-time with standard computers of today is probably more an exercise in computer programming than OSR research.

Having thus fired up the expectations of the reader we will now proceed to give some information related to an OSR system that appears to have some of the properties we advocate. The system we will describe will be a dedicated system, in the sense that it will operate with Turkish speech only. However, some components of the system can be modified to operate on other languages readily.

The system we propose is the following. The speech is passed through two band-pass filters with non-overlapping pass-bands which have been determined through experimental observations. In our work we have used 30 - 900 Hz and 1100 - 4000 Hz which appeared suitable for Turkish. In fact, as we will mention later, a third band-pass filter around 4 - 8 KHz might improve performance with unvoiced phonemes.

The output of the band-pass filters is passed through a suitable computer interface such as the one proposed by Niederjohn and Stick [1] and the zero crossing interval sequences (ZCIS) are obtained. At the outset, we are thus constraining ourselves to use only the ZCIS of the signals at the outputs of the band-pass filters. With the exception of recognition of unvoiced phonemes, this constraining does not appear to give too much trouble.

We then perform voiced/unvoiced segmentation of the speech signal. It turns out that only ZCIS of the output of the first 30 - 900 Hz filter are sufficient for acceptable v/u segmentation in real time [2]. Although we use the word acceptable, we must state the segmentation performance of the ZCIS based algorithm we propose is comparable to any published in the open literature.

Once the segmentation is performed, an attempt is made to determine if each voiced or unvoiced interval consists of a single or multiple phoneme, and then to classify each of the segments or partitioned segments into phoneme classes. For this part of the OSR system ZCIS from the outputs of all the filters are necessary. In fact for the recognition and partitioning of unvoiced segments that contain more than a single unvoiced phoneme, ZCIS do not appear to be sufficient. As we will mention later, due to certain unique properties of the Turkish language, these problems can be avoided.

At this point we will have obtained a sequence of characters, with possibly ordered multiple characters for some of the segments. This sequence of characters may be put into a final form using some form of linguistic information. At the moment the simplest approach to this appears to be the use of statistical information such as character occurrance probabilities, digrams, trigrams etc., simple syntax rules. It is this last component of an OSR system that we feel is the determining factor for successful speech recognition. We will give some information and preliminary results related to the decision of correct voiced phoneme sequences from the character sequences, that make use of a unique feature of the Turkish language, the so-called vowel harmony rule and some other related simple rules. All these relate to the character sequences forming individual words, so that assuming word separation has been performed perfectly, individual word recognition can be completed with relatively simple and fast software.

2. VOICED/UNVOICED SEGMENTATION ALGORITHM

The algorithm we have developed for voiced/unvoiced speech segmentation operates with only the ZCIS from the speech low-pass filtered to 900 Hz. Basically, the algorithm performs a generalized quasi-periodicity test on the ZCIS. Most of the recently proposed techniques for segmentation or pitch extraction require the sampling of speech at a rate around 7-20 KHz and only a few of these have the potential of performing the segmentation analysis in real-time with very fast computers. The algorithm to be presented below briefly, obtains comparable results in faster than real time with a moderate speed computer using only ZCIS of the low-pass filtered speech.

Let the ZCI's of the speech be denoted by $z(1)$, $z(2)$, ..., $z(p)$, ... These time intervals will normally be simply positive integers, as a result of time quantization. The possibility of the existance of two successive periods of ZCIS, whose ZCI's satisfy the properties summarized below as Test 1, is searched starting from the first ZCI, $z(1)$.

Test 1: The k+m successive ZCI's

$$z(p),\ z(p+1),\ \ldots,\ z(p+k-1),$$
$$z(r),\ z(r+1),\ \ldots,\ z(r+m-1)$$

with

$$S_1 = \sum_{i=p}^{r-1} z(i)\ ,\ S_2 = \sum_{i=r}^{s-1} z(i)\ ,\ r=p+k\ ,\ s=r+m$$

are said to comprise the first two periods of a periodic-like section with pitch periods S_1 and S_2, or a portion of a voiced segment, if

$z(p) \geq T_{10}$ (Stable ZCI's usually correspond to large amplitude so they cannot be too small. This threshold not only speeds up the procedure, but it also helps avoiding incorrect results due to unvoiced fricative conconants which have small ZCI's)

k , m are positive, even integers smaller than or equal to $T_9 = 16$ (Since the speech signal is low-pass filtered to 900 Hz, the number of ZCI's in a period of human speech very rarely reaches 16)

$|z(r) - z(p)| \leq T_1$ (This threshold is to verify the stability of z(p))

$T_3 \leq S_1 < T_4$ where $T_4 = 2\ T_3$ (This constraint on the speaker dependent threshold values T_3 and T_4 is realistic because the pitch frequency range of a speaker very rarely spans more than an octave. These constraints speed up the search process, make the results more dependable and help avoiding the possibility of accepting two or more periods as one period)

$|S_1 - S_2| \leq T_5$ (Avoids the acceptance of large changes in pitch periods)

$$d(p,r-1;r,s-1) \leq \begin{cases} T_6 & \text{if any of } k,m = 2 \\ T_7 & \text{if none of } k,m = 2 \end{cases}$$

where d( ) is the 'distance' between the ZCI's of the first period and the ZCI's of the second period [2].

The existance of a third periodic ZCIS is similarly tested when the periodic properties of the first two ZÇIS is verified. At all steps the algorithm performs only comparisons and additions with integers.

This very brief description of the algorithm is given here only to give an idea of the operations involved and not for a full exposition. The algorithm is described fully with the relevant thresholds T_{ij} and the flow-chart in [2].

The segmentation performance of the algorithm tested on six males and females with Turkish speech, is comparable to the best in the literature and always in less than real-time with a relatively slow computer PDP-8 [2]. In passing we should state that the algorithm, in a way, implements the visual pitch extraction portion of the man interacting pitch detection technique described in [3] and used as a standard in the evaluation of seven pitch extraction algorithms in [4].

3. RECOGNITION OF PHONEMES WITH ZCIS

We will discuss briefly some techniques that appear to be suitable for voiced phoneme recognition with ZCIS.

Single voiced phoneme segments can be classified with good average accuracy using a simple and very fast ZCIS template matching technique that we will call 'ZCIS periodic pattern matching', which is readily implementable in conjunction with the segmentation algorithm.

The segmentation or the pitch extraction algorithm removes the 'unstable' ZCI's, Fig. 1, and determines the ZCIS that constitute the pitch periods for the voiced segments of speech. The ZCIS periodic pattern matching, simply compares these ZCIS patterns with appropriate normalizations and modifications, for the voiced segments. The ZCIS of 2,4,6,8,10,12,14, or 16 ZCI's are obtained from the training operation for each of the voiced phonemes and stored as prototypes. The phoneme to be classified is reduced to its periodic ZCIS in the same way and compared with the stored patterns using standard Euclidian distance measure. This simple technique was found to be 70 - 75% accurate for classifying Turkish voiced phonemes uttered by the same speaker. As the number of different speakers is increased the results deteriorate quickly as perhaps expected from such a simple scheme [5,6]. If this procedure is implemented not on the natural speech signal but on its differentiated version, there is considerable improvement for single person voiced phoneme classification performance (approximately 90% overall accuracies are attainable) however multiple person performance does not improve [6].

We have also performed some other tests on classification of single phoneme segments obtained after the application of the segmentation routine. In these experiments, the ZCI's of the signals at the output of the two filters (30 - 900 Hz and 1.1 - 4 KHz) are measured stored for each phoneme quantized to 42 μs. The positive integers thus obtained are sorted into N_L (for low frequency signal ZCI's) and N_H (for high frequency signal ZCI's) channels, each of which has a width a multiple of 42 μs, except the last one which

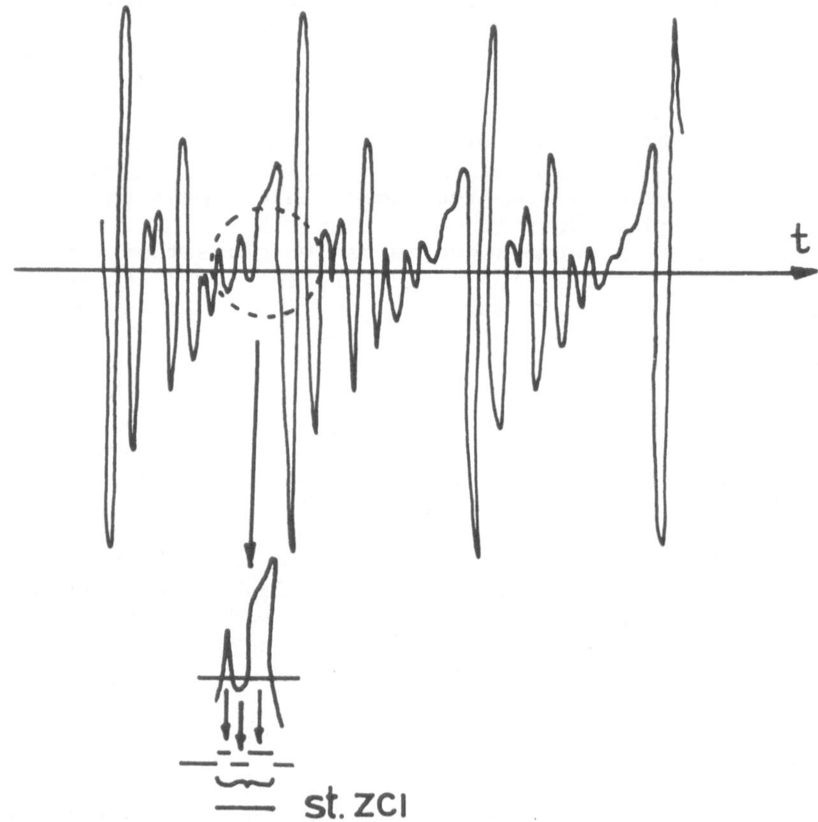

Figure 1. A typical portion of a voiced segment of speech indicating how small changes of the waveform near zero can change the ZCIS structure. The algorithm tries to determine ZCI's due to this type of changes and replace the three ZCI by a stable ZCI whose length is the sum of the three unstable ZCI's.

is infinitely wide. The channel values are then normalized to segment length. What is thus obtained is nothing but the probability density histograms of the ZCI's of a segment or a phoneme. It was found more convenient to use these low frequency and high frequency histograms independently but in an identical way, rather than together in $(N_L \times N_H)$ space.

Let a general definition of the distance between two R dimensional patterns X, Y be D(X,Y). With

$$X=(x_1,x_2,\ldots,x_r)^T$$

$$Y=(y_1,y_2,\ldots,y_r)^T$$

$$D(X,Y) = \sum_{r=1}^{R} d^2(x_r,y_r)w_r^2.$$

where d(,) is some elementary signed distance defined along a coordinate, and w_r is the weight of the rth coordinate. A convenient elementary distance definition was determined as [7],

$$d(x_r,y_r) = \frac{(x_r - y_r)^2}{x_r + y_r} \operatorname{sgn}(y_r - x_r)$$

which appears not to possess some of the undesirable properties of the usual Euclidian metric in this particular application. The determination of the weights w_r is quite critical and effects the success of the classification to a large extent. It is possible to show that a possible optimum choice for the weights w_{kr}, the weight of the kth phoneme class along the rth coordinate is

$$w_{kr}^2 = \left(\sum_{s=1}^{R} v_{ks}\right)^{1/R} \Big/ v_{kr},$$

where v_{kr} is the sample variance of the kth class along the rth coordinate [7].

Since the training or learning prototypes and the unknown test prototypes are high dimensional and distance measure is non-linear, it is difficult to visualize the clusters of prototypes within a phoneme class and the relationships of the different phoneme classes with each other. An interesting way to get around this problem is to map non-linearly the N points of an M-space into N points in 2-space while preserving interdistances. An algorithm suggested by Sammon [8], which starts with randomly selected N points in 2-space and adjusts these through a steepest descent procedure to minimize an error function of the distances in M-space and 2-space, is available. Using this approach it is possible to observe the allophones in feature space and the relationships between pattern clusters [7].

The remaining point is how to combine the individual decisions in the low frequency and high frequency feature spaces. For this we have used what might be termed 'best position + worst position' rule as exemplified in Fig. 2, [7], and we are investigating other possibilities.

Our preliminary results indicate that using the above techniques it is possible to obtain 80-90% overall accuracies with voiced phonemes and 30-40% with unvoiced phonemes for a single person uttering words from a list of 200 most used roots in Turkish with the removal of 34 that do not have alternate v/con. structure. At this point, it appears that unvoiced phoneme recognition could be improved if the ZCI's at the output of a third non-overlapping band-pass filter (4.4 - 8 KHz) were similarly used as a third feature space. However we feel that this improvement will probably not be sufficient and other techniques not necessarily based solely on ZCIS will have to be investigated or developed.

4. SOFTWARE FINALIZATION OF THE DECISIONS

Turkish language has certain properties that make it very convenient for OSR research. The most important of these is that, it is a 'phonetic' lauguage, i.e., a character or a sequence of characters corresponds to the same phoneme in whichever word it appears and wherever in a given word it appears. This allows for example, a complete stranger to the language to learn the alphabet (which is the same as the English alphabet with the removal of Q, X, W and addition of Ç, I/I, G, O, Ş, Ü) and be able to read or pronounce Turkish words completely intelligibly to a Turk within a matter of few hours of concentrated phonetic exercises. In addition to this phonetic structure the Turkish language has the following properties which were verified by an extensive statistical computer study of a large corpus of present day Turkish, composed of portions of novels, short stories, newspaper articles etc. [9]. The corpus size was 1.6×10^5 characters, 6×10^4 syllables or 2.2×10^3 words.

1. The probability of occurrance of all the words is determined. If the probability of occurrance of all the words with alternating vowel/consonant or consonant/vowel structure are summed we obtain a probability slightly larger than 50% (52.1%). That is, if '1' denotes a vowel and '0' a consonant, the sum of the probabilities of occurrance of the words with the patterns 1, 01, 10, 101, 010, 1010, ... , 01010101010101 is slightly more than 0.5. Another way to state the same property is the following. If the probability of occurrance or more precisely frequency of occurrance of the roots in the Turkish language is determined, the cumulative frequency of occurrance of the most common 200 roots is 55.2%, of these 164 obey the alternate vowel consonant structure and these have a cumulative frequency of occurrance of 49.2%. In other words the more common words have the alternate vowel/consonant structure.

LIST 1

Position	Name of the class	Distance to the unknown pattern in the low frequency space
1	C_4	369
2	C_3	685
3	C_6	852
4	C_2	904
5	C_1	973
6	C_5	999

LIST 2

Position	Name of the class	Distance to the unknown pattern in the high frequency space
1	C_6	97
2	C_4	289
3	C_3	355
4	C_2	390
5	C_5	1365
6	C_1	1535

Two lists of the voiced phoneme classes, $C_1, C_2, \ldots, C_6$, ordered according to the distance to an unknown segment, in the low and high frequency spaces.

Name of the class	Worst position	Best position
C_4	2	1
C_6	3	1
C_3	3	2
C_2	4	4
C_1	6	5
C_5	6	5

The ordering obtained from the above lists

Figure 2. An example showing how the distances in the low frequency and high frequency feature spaces are used i.e. the best 'best position+worst position' rule.

2. 99.65% of the words do not have adjacent vowels.

3. 99.67% of all the words beginning with a consonant, begin with a single consonant.

4. 99.11% of all words have at most two adjacent consonants within a word

5. 99.25% of all the words ending with a consonant, end with a single consonant.

   (0.35%, 0.33%, 0.89%, 0.75% that violate constraints 2, 3, 4, 5, above are mostly of foreign origin)

6. Vowels of each word satisfy the vowel harmony rules fully described by the two unconnected Markov state diagrams given in Fig. 3. If the words of foreign origin, a few compound words, and some words with invariable suffixes are not included, 99% of the words in the corpus obey the harmony rules. Even if all the violations are included, vowel harmony rules are satisfied in 91% of all cases.

If the words for OSR tests are chosen among those that satisfy 1, 2, 3, 5, 6 above which for the Turkish language is clearly non-restrictive, then, it is possible to finalize the character sequences at the output of the waveform recognition operation, by invoking the vowel harmony rules as given in Fig. 3. Our results indicate that with a vocabulary of 200 two or more vowel words, selected as described, it is possible to increase character recognition accuracies of 50-60% to 80-90% <u>word</u> recognition accuracies with the application of these rules for a single person or for two persons whose voices are perceived not too different by a normal 'ear'.

The application of the vowel harmony rule is as follows. From the sequence of first and second choice characters for a given word, the vowels are extracted. If the first choices of the characters of each segment satisfy one of the vowel harmony rules, no operation is performed. Otherwise a duration weighted procedure is employed. For example, if the word has two vowels and the vowel harmony rules are not satisfied, the first choice of the segment with the larger duration is taken with the second choice of the shorter segment. If this too does not satisfy the harmony rules, the first choice of the shorter segment and the second choice of the longer segment are checked. If this too is negative, second choices of both segments are checked. If this is negative too, the first and second choices are printed with an error message. Three or more vowel words are treated similarly.

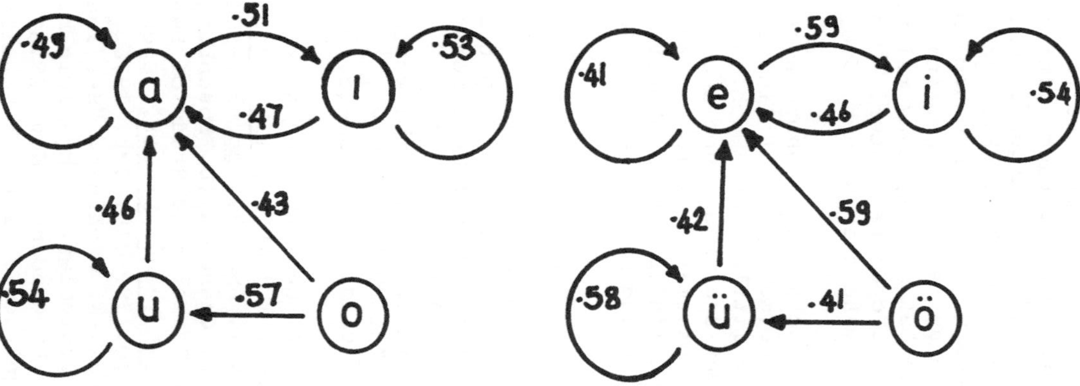

Figure 3. The two vowel harmony rules of the Turkish language summarized in the form of two unconnected Markov state diagrams.

5. CONCLUSION

An outline and some preliminary results of an OSR system for recognition of Turkish speech has been given. All components of the system have the potential of operating in real-time. The voiced/ unvoiced segmentation algorithm operates only on the ZCIS and gives satisfactory results in real-time for multiple speakers.

The recognition of voiced segments appears possible using only ZCIS with either ZCI periodic pattern search and matching or ZCI histogram derived feature-space distance minimization. Consonant classification with the latter technique is only 30-40% accurate.

Some unique properties of the Turkish language may be used to perform word based software correction based on vowel harmony rules. This approach has the potential of correcting the errors of the initial components of an OSR system, thus relaxing their requirements and allowing them to operate faster. Other languages that may have properties that make them suitable for OSR research are Hungarian and Finnish.

It is hoped that these studies, where a successful attempt has been made to introduce some form of linguistic information, although admittedly very little, into machine recognition of speech, will give further insight and experience about the techniques of introducing linguistic information into OSR research. Unless we find ways of introducing linguistic information into machine recognition, we are bound to spend many more fruitless years.

REFERENCES

1. Niederjohn, R. J., P. P. Stick, "A Computer Interface for efficient Zero-crossing Interval Measurement", IEEE Trans. Comput., vol. C-24, pp. 329-331, Mar. 1975.
2. Geçkinli, N. C., D. Yavuz, "Algorithm for Pitch Extraction Using Zero-Crossing Interval Sequence", IEEE Trans. Acoust., Speech, Signal Processing, vol. ASSP-25, pp. 559-564, Dec. 1977.
3. McGonegal, C. A., L. R. Rabiner, and A. E. Rosenberg, "A Semi-Automatic Pitch Detector (SAPD)", IEEE Trans. Acoust., Speech, Signal Processing, Vol. ASSP-23, pp. 570-574, Dec. 1975.
4. Rabiner, L. R., M. J. Cheng, A. E. Rosenberg, and C. A. McGonegal, "A Comparative Performance Study of Several Pitch Detection Algorithms", IEEE Trans. Acoust., Speech, Signal Processing, vol. ASSP-24, pp. 399-418, Oct. 1976.
5. Yavuz, D., T. Şenocak, N. Geçkinli, "Ses İşaretlerinin Taninmasinda kullanilan Yöntemler, Türkçe için bir Uygulama" TBTAK IV Science Congress Proceedings, Nov. 1973.
6. Şenocak, T., "Recognition of Turkish Vowels Using Zero-Crossing Information", M. Sc. Thesis, Middle East Technical University, E. E. Department, Ankara, 1973.

7. Geçkinli, N., "Some Novel, Optimal Spectral Windows, Speech Segmentation, Recognition and Pitch Extraction Algorithms", Ph. D. Thesis, E. E. Department, Middle East Technical University, Ankara, 1976.
8. Sammon, J. W. Jr., "A Nonlinear Mapping for Data Structure Analysis", IEEE Trans. Comput., vol. C-18, pp. 401-409, May, 1969.
9. Töreci, E., "Statistical Investigations on the Turkish Language Using Digital Computers", M. Sc. Thesis, E. E. Department, Middle East Technical University, Ankara, 1974.

SPECTRAL CLASSIFICATION OF RADAR CLUTTER USING THE MAXIMUM ENTROPY METHOD

Simon Haykin

Communications Research Laboratory
Faculty of Engineering, McMaster University
Hamilton, Ontario, Canada

1. INTRODUCTION

In the characterization of a wide-sense stationary random process, use of the spectral density is often preferred to the autocorrelation function, because a spectral representation may reveal such useful information as hidden periodicities or close spectral peaks. Until 1967, most of the procedures used for estimating the spectral density of a random process were based on the classical work by Blackman and Tukey [1]. In this method, the available data are first used to estimate the sample autocorrelation function for a number of lags, and then the estimate is multiplied by a window function that goes to zero beyond the largest available lag. Next, the Fourier transform of this product is determined to obtain an estimate of the spectral density. The expectation of such an estimate is equivalent to the convolution of the true spectral density of the random process with the spectral window. The statistical stability and resolution of the spectral density estimate using the Blackman-Tukey procedure are highly dependent on the choice of the window function, and a considerable amount of effort has been expended in order to determine a good window function.

An alternate procedure for estimating the spectral density is based on the so-called periodogram [2], which is defined as the squared magnitude of the Fourier transform of the available data. This approach has become rather popular with the introduction of the fast Fourier transform (FFT) algorithm for performing discrete Fourier transformation. In particular, the procedure described by Welch [3], based on time averaging over several short modified periodograms, results in a significant reduction in the number of computations and in the amount of core storage required for long data records. As with the Blackman-Tuckey approach, spectral density estimation procedures based on the periodogram involve the use of a window function. Since the window function is completely independent of the data being analyzed, we find that misleading or false conclusions may sometimes be drawn by using such an approach.

The disadvantage of procedures based on the use of a window function may be overcome by using the maximum entropy method proposed by Burg [4, 5]. The basic idea of this approach is to extrapolate the autocorrelation function of a random process by maximizing the entropy of the corresponding probability density function [6]. It turns out that this maximization is equivalent to the least-squares fitting of an all-pole (autoregressive) model to the random process. Thus, the methods proposed by Burg and Parzen [7] are equivalent, differing only in approach. The maximum entropy method has been successfully applied to truncated sinusoidal waves as test signals [8, 9, 10], to geophysical data [11, 12, 13], and radar data [14, 15, 16]. The applicability of the method in the spatial processing of array data has also been indicated [17].

This paper is intended as a tutorial review of the maximum entropy method, with some experimental results included to illustrate application of the method to the spectral analysis of recorded clutter data obtained from a coherent air traffic control radar. Spectral analysis of clutter returns in a scanning radar may be helpful in forming a "clutter map" of the surrounding environment. For example, the spectral characteristics of the clutter signal can be used as a means of controlling the adjustable filter coefficients in an adaptive moving target indicator filtering scheme [18, 19]. The speed of adaptation depends on how fast the power spectral density can be computed from the observed clutter data. For a typical air traffic control radar, the number of data samples collected from a range-azimuth cell in one scan is between 10 and 20. To get a reliable estimate of the spectral density of the clutter in this cell with conventional methods, we need to use data gathered from several antenna scans in the analysis. If, however, the statistical properties of the surrounding environment change continuously and somewhat rapidly because of prevailing stormy weather conditions or bird migrations and the like, it may be necessary to use a short-term estimate of the pertinent clutter signal. The use of the maximum entropy method is rather well-suited for such an application.

2. PREDICTION - ERROR FILTER

Consider a wide-sense stationary Gaussian process $X(t)$ of zero mean and duration T seconds. The $X(t)$ may be complex-valued. Let $x(t)$ denote a sample function of this process, which is sampled uniformly at the rate $1/T_s$ to produce a time series consisting of N samples, as shown by:

$$\{x(n)\} = \{x(1),\ x(2),\ \ldots,\ x(N)\} \tag{1}$$

The x(n) denotes the nth sample of the time series. For the sake of simplicity, we have omitted dependence on the sampling duration T_s.

The z-transform of this time series is defined by the polynomial

$$X(z) = \sum_{n=1}^{N} x(n)\ z^{-n} \tag{2}$$

where z^{-1} is the unit-delay operator. Expressing X(z) in its factored form, we may write

$$X(z) = x(N) \prod_{i=1}^{N} (z^{-1} - z_i^{-1}) \tag{3}$$

where the z_i denote the zeros of X(z). Each factor $z^{-1} - z_i^{-1}$ represents the z-transform of a dipole consisting of 2 samples. The given time series {x(n)} may thus be considered as the multiple convolution of N such dipoles. Furthermore, if the dipole components of the time series {x(n)} are all arranged so that the coefficient of greatest magnitude occurs first in each dipole, we find that all the zeros of the z-transform of this time series lie inside the unit circle of the z-plane. Such a time series is said to be "minimum phase" or "minimum delay".

The autocorrelation function of the time series {x(n)} for a time lag of mT_s seconds and time index n is defined by:

$$R_x(m,n) = E\ [x(n+m)x^*(n)] \tag{4}$$

where E[ ] denotes the expectation operator, and the asterisk denotes the complex conjugate operation. With the random process x(t) assumed wide-sense stationary, we find that the autocorrelation function $R_x(m,n)$ is independent of the time index n, as shown by:

$$R_x(m,n) = R_x(m) \tag{5}$$

Also, the autocorrelation function exhibits conjugate symmetry, that is, $R_x(m)$ is a Hermitian function:

$$R_x(-m) = R_x^*(m).$$

If we further assume that X(t) is an ergodic process, then we may compute the autocorrelation function by using the time-average formula:

$$R_x(m) = \frac{1}{N} \sum_{n=1}^{N} x(n+m)x^*(n), \qquad -N \le m \le N \tag{7}$$

Suppose that the time series $\{x(n)\}$ is applied to a linear digital filter of impulse response $\{h_M(n)\}$ of order M. The filter produces an output time series $\{y(n)\}$ that is designed to approximate a desired time series $\{d(n)\}$, as shown in Fig. 1, according to the least mean-square error criterion. The resulting error time series is:

$$\varepsilon(n) = d(n) - y(n) \tag{8}$$

By minimizing the mean-square value of the error, that is, $E[\varepsilon(n)\, \varepsilon^*(n)]$, we find that impulse response of the filter is defined by the Wiener-Hopf equation in matrix form as follows [20]:

$$\begin{bmatrix} R_x(0) & R_x(-1) & \cdots & R_x(1-M) \\ R_x(1) & R_x(0) & \cdots & R_x(2-M) \\ \cdot & \cdot & \cdot & \cdot \\ \cdot & \cdot & \cdot & \cdot \\ \cdot & \cdot & \cdot & \cdot \\ R_x(M-1) & R_x(M-2) & \cdots & R_x(0) \end{bmatrix} \begin{bmatrix} h_M(1) \\ h_M(2) \\ \cdot \\ \cdot \\ \cdot \\ h_M(M) \end{bmatrix} = \begin{bmatrix} R_{dx}(0) \\ R_{dx}(1) \\ \cdot \\ \cdot \\ \cdot \\ R_{dx}(M-1) \end{bmatrix} \tag{9}$$

where $R_{dx}(M)$ is the cross-correlation function of $\{d(n)\}$ and $\{x(n)\}$, as shown by:

$$R_{dx}(m) = E[d(n+m)x^*(n)] \tag{10}$$

For the special case of a filter designed to predict the sample value of a random process $X(t)$, one time unit ahead, by using the present and past values of the time series $\{x(n)\}$, we have:

$$d(n) = x(n+1) \tag{11}$$

and

$$\begin{aligned} R_{dx}(m) &= E[x(n+1+m)x^*(n)] \\ &= R_x(m+1), \qquad m = 0, 1, \ldots, M-1 \end{aligned} \tag{12}$$

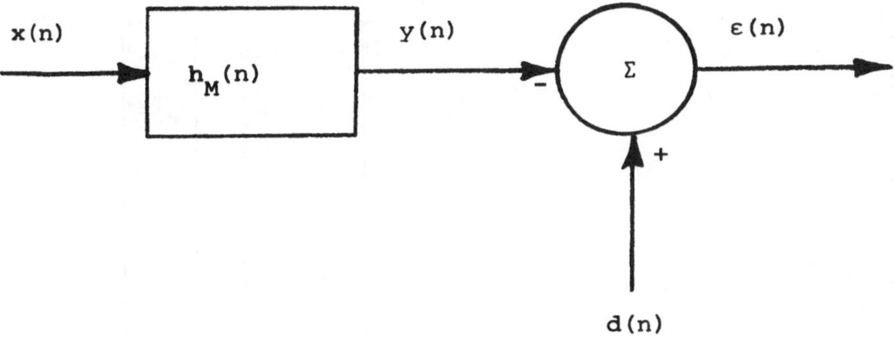

Fig. 1 Linear predictive filter

Accordingly, we may rewrite Eq. (9) as follows:

$$\begin{bmatrix} R_x(0) & R_x(-1) & \cdots & R_x(1-M) \\ R_x(1) & R_x(0) & \cdots & R_x(2-M) \\ \cdot & \cdot & \cdot & \cdot \\ \cdot & \cdot & \cdot & \cdot \\ \cdot & \cdot & \cdot & \cdot \\ \cdot & \cdot & \cdot & \cdot \\ R_x(M-1) & R_x(M-2) & \cdots & R_x(0) \end{bmatrix} \begin{bmatrix} h_M(1) \\ h_M(2) \\ \cdot \\ \cdot \\ \cdot \\ \cdot \\ h_M(M) \end{bmatrix} = \begin{bmatrix} R_x(1) \\ R_x(2) \\ \cdot \\ \cdot \\ \cdot \\ \cdot \\ R_x(M) \end{bmatrix} \tag{13}$$

That is:

$$\sum_{m=1}^{M} h_M(m) R_x(k-m) = R_x(k), \qquad k = 1, 2, \ldots, M \tag{14}$$

The deviation of the actual time series $\{y(n)\}$ from the desired time series $\{d(n)\}$ is defined by the error time series:

$$\begin{aligned} \varepsilon(n) &= d(n) - y(n) \\ &= x(n) - \sum_{k=1}^{M} h_M(k) x(n-k) \\ &= \sum_{k=0}^{M} w_M(k) x(n-k) \end{aligned} \tag{15}$$

where the coefficient $w_M(k)$ is defined by:

$$w_M(k) = \begin{cases} 1, & k = 0 \\ -h_M(k), & k = 1, 2, \ldots, M \end{cases} \tag{16}$$

The digital filter with impulse response $\{w_M(n)\}$ is called the "prediction-error filter" of order M+1, with the order referring to the number of coefficients needed to uniquely define $\{w_M(n)\}$. According to Eq. (15), the error time series $\{\varepsilon(n)\}$ may be viewed as the response produced by passing the original time series $\{x(n)\}$ through this filter. The functional relationship between the linear predictive filter, characterized by the impulse response

$\{h_M(n)\}$, and the prediction-error filter, characterized by the impulse response $\{w_M(n)\}$, is illustrated in Fig. 2. Note that the order of the prediction-error filter is greater than that of the predictive filter by one.

The output power of the prediction-error filter is given by:

$$
\begin{aligned}
P_{M+1} &= E[\varepsilon(n)\varepsilon^*(n)] \\
&= E\{[x(n) - \sum_{k=1}^{M} h_M(k)x(n-k)][x^*(n) - \sum_{k=1}^{M} h_M^*(k)x^*(n-k)]\} \\
&= E[x(n)x^*(n)] - \sum_{k=1}^{M} h_M(k)E[x(n-k)x^*(n)] \\
&\quad - \sum_{k=1}^{M} h_M^*(k)E[x(n)x^*(n-k)] \\
&\quad + \sum_{k=1}^{M} h_M^*(k) \sum_{i=1}^{M} h_M(i)E[x(n-i)x^*(n-k)] \\
&= R_x(0) - \sum_{k=1}^{M} h_M(k)R_x(-k) - \sum_{k=1}^{M} h_M^*(k)R_x(k) \\
&\quad + \sum_{k=1}^{M} h_M^*(k) \sum_{i=1}^{M} h_M(i)R_x(k-i)
\end{aligned}
\tag{17}
$$

Therefore, substituting Eq. (14) in (17), we get

$$
P_{M+1} = R_x(0) - \sum_{k=1}^{M} h_M(k)R_x(-k) \tag{18}
$$

Finally, using the relationship between $\{h_M(k)\}$, as defined by Eq. (16), we get the following expression for the prediction-error power:

$$
P_{M+1} = \sum_{k=0}^{M} w_M(k)R_x(-k) \tag{19}
$$

Augmenting Eq. (14) by this expression for the prediction-error power, we obtain the so-called "prediction-error filter equations":

$$
\sum_{k=0}^{M} w_M(k)R_x(m-k) = \begin{cases} P_{M+1}, & m = 0 \\ 0, & m = 1,2,\ldots,M \end{cases} \tag{20}
$$

which in matrix form may be expressed as follows:

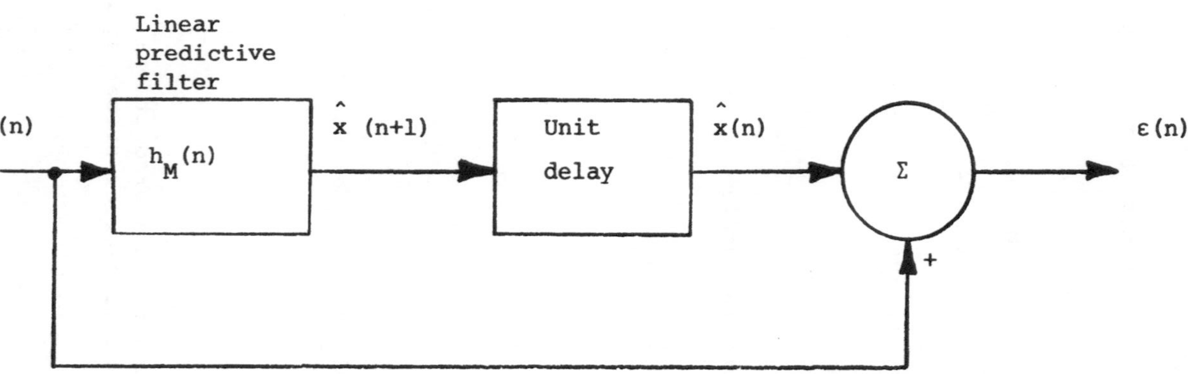

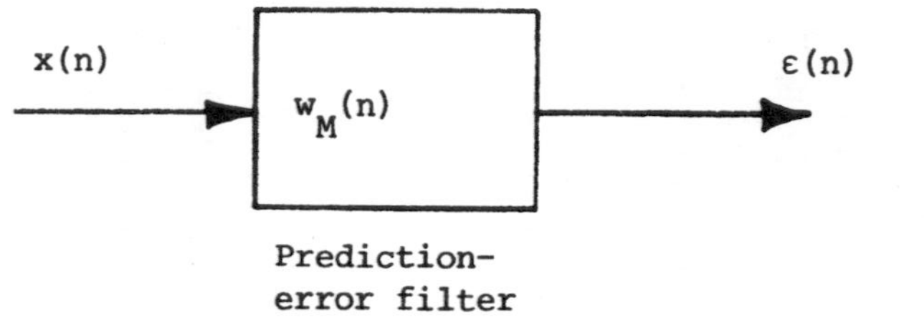

Fig. 2 Illustrating the relationship between the predictive filter and prediction-error filter

$$\begin{bmatrix} R_x(0) & R_x(-1) & \cdots & R_x(-M) \\ R_x(1) & R_x(0) & \cdots & R_x(1-M) \\ \vdots & \vdots & & \vdots \\ R_x(M) & R_x(M-1) & \cdots & R_x(0) \end{bmatrix} \begin{bmatrix} 1 \\ w_M(1) \\ \vdots \\ w_M(M) \end{bmatrix} = \begin{bmatrix} P_{M+1} \\ 0 \\ \vdots \\ 0 \end{bmatrix} \tag{21}$$

The equidiagonal autocorrelation matrix of order M+1, defined by:

$$\underline{R}_x(M+1) = \begin{bmatrix} R_x(0) & R_x(-1) & \cdots & R_x(-M) \\ R_x(1) & R_x(0) & \cdots & R_x(1-M) \\ \vdots & \vdots & & \vdots \\ R_x(M) & R_x(M-1) & \cdots & R_x(0) \end{bmatrix} \tag{22}$$

is called a "Toeplitz matrix". In our case, this is also a Hermitian matrix. A solution to Eq. (21) always exists if the Toeplitz matrix $\underline{R}_x(M+1)$ is non-negative definite.

It is of interest to note that, as the order of the prediction-error filter approaches infinity, the limiting form of the determinant of the Toeplitz matrix $\underline{R}_x(M+1)$ is related to the spectral density $S_x(f)$ of the input random process X(t) as follows [13]:

$$\lim_{M \to \infty} \left\{ \det[\underline{R}_x(M+1)] \right\}^{1/M+1} = 2B \exp\left\{ \frac{1}{2B} \int_{-B}^{B} \ln[S_x(f)]df \right\} \tag{23}$$

where it is assumed that X(t) is limited to the frequency band $-B \le f \le B$.

3. BURG'S RECURSIVE ALGORITHM

We now describe a recursive algorithm developed by Burg [4, 5] for solving the set of prediction-error filter equations, that is, Eq. (21).

Suppose that we know the solution to the set of M equations pertaining to a prediction-error filter of order M, as shown by:

$$\begin{bmatrix} R_x(0) & R_x(-1) & \cdot & \cdot & R_x(1-M) \\ R_x(1) & R_x(0) & \cdot & \cdot & R_x(2-M) \\ \cdot & \cdot & \cdot & & \cdot \\ \cdot & \cdot & & \cdot & \cdot \\ R_x(M-1) & R_x(M-2) & \cdot & \cdot & R_x(0) \end{bmatrix} \begin{bmatrix} 1 \\ w_{M-1}(1) \\ \cdot \\ \cdot \\ w_{M-1}(M-1) \end{bmatrix} = \begin{bmatrix} P_M \\ 0 \\ \cdot \\ \cdot \\ 0 \end{bmatrix} \qquad (24)$$

Using the solution to this set of equations, we will develop an algorithm for solving the corresponding set of M+1 equations for the prediction-error filter of order M+1. In order to do this, we take the complex conjugate of both sides of Eq. (24) and recognize that: (1) the prediction-error power P_M is a real quantity, and (2) the autocorrelation matrix is Hermitian, that is, $R_x^*(m) = R_x(-m)$. We thus obtain:

$$\begin{bmatrix} R_x(0) & R_x(1) & \cdot & \cdot & R_x(M-1) \\ R_x(-1) & R_x(0) & \cdot & \cdot & R_x(M-2) \\ \cdot & \cdot & \cdot & & \cdot \\ \cdot & \cdot & & \cdot & \cdot \\ R_x(1-M) & R_x(2-M) & \cdot & \cdot & R_x(0) \end{bmatrix} \begin{bmatrix} 1 \\ w^*_{M-1}(1) \\ \cdot \\ \cdot \\ w^*_{M-1}(M-1) \end{bmatrix} = \begin{bmatrix} P_M \\ 0 \\ \cdot \\ \cdot \\ 0 \end{bmatrix} \qquad (25)$$

Rearranging this set of equations so that the autocorrelation matrix is of exactly the same form as in Eq. (24), we have:

$$\begin{bmatrix} R_x(0) & R_x(-1) & \cdot & \cdot & R_x(1-M) \\ R_x(1) & R_x(0) & \cdot & \cdot & R_x(2-M) \\ \cdot & \cdot & \cdot & & \cdot \\ \cdot & \cdot & & \cdot & \cdot \\ R_x(M-1) & R_x(M-2) & \cdot & \cdot & R_x(0) \end{bmatrix} \begin{bmatrix} w^*_{M-1}(M-1) \\ w^*_{M-1}(M-2) \\ \cdot \\ \cdot \\ 1 \end{bmatrix} = \begin{bmatrix} 0 \\ 0 \\ \cdot \\ \cdot \\ P_M \end{bmatrix} \qquad (26)$$

We may distinguish between the 2 sets of relations in Eqs. (24) and (26) by noting that the first set of relations pertains to a prediction-error filter of order M, operated in the forward direction.

On the other hand, the second set of relations defined by Eq. (26) pertains to a prediction-error filter of the same order, except that it is operated in the backward direction. This is illustrated in Fig. 3.

We may combine these two sets of relations to expand the number of prediction-error filter equations by one as follows:

$$\begin{bmatrix} R_x(0) & R_x(-1) & \cdot & \cdot & R_x(1-M) & R_x(-M) \\ R_x(1) & R_x(0) & \cdot & \cdot & R_x(2-M) & R_x(1-M) \\ \cdot & \cdot & \cdot & & \cdot & \cdot \\ \cdot & \cdot & & \cdot & \cdot & \cdot \\ R_x(M-1) & R_x(M-2) & \cdot & \cdot & R_x(0) & R_x(-1) \\ R_x(M) & R_x(M-1) & \cdot & \cdot & R_x(1) & R_x(0) \end{bmatrix} \left[\begin{bmatrix} 1 \\ w_{M-1}(1) \\ \cdot \\ \cdot \\ w_{M-1}(M-1) \\ 0 \end{bmatrix} + w_M(M) \begin{bmatrix} 0 \\ w^*_{M-1}(M-1) \\ \cdot \\ \cdot \\ w^*_{M-1}(1) \\ 1 \end{bmatrix} \right]$$

$$= \begin{bmatrix} P_M \\ 0 \\ \cdot \\ \cdot \\ 0 \\ \Delta_M \end{bmatrix} + w_M(M) \begin{bmatrix} \Delta^*_M \\ 0 \\ \cdot \\ \cdot \\ 0 \\ P_M \end{bmatrix} \qquad (27)$$

However, for the corresponding prediction-error filter of order M+1 we have:

$$\begin{bmatrix} R_x(0) & R_x(-1) & \cdot & \cdot & R_x(1-M) & R_x(-M) \\ R_x(1) & R_x(0) & \cdot & \cdot & R_x(2-M) & R_x(1-M) \\ \cdot & \cdot & \cdot & & \cdot & \cdot \\ \cdot & \cdot & & & \cdot & \cdot \\ R_x(M-1) & R_x(M-2) & \cdot & \cdot & R_x(0) & R_x(-1) \\ R_x(M) & R_x(M-1) & \cdot & \cdot & R_x(1) & R_x(0) \end{bmatrix} \begin{bmatrix} 1 \\ w_M(1) \\ \cdot \\ \cdot \\ w_M(M-1) \\ w_M(M) \end{bmatrix} = \begin{bmatrix} P_{M+1} \\ 0 \\ \cdot \\ \cdot \\ 0 \\ 0 \end{bmatrix} \qquad (28)$$

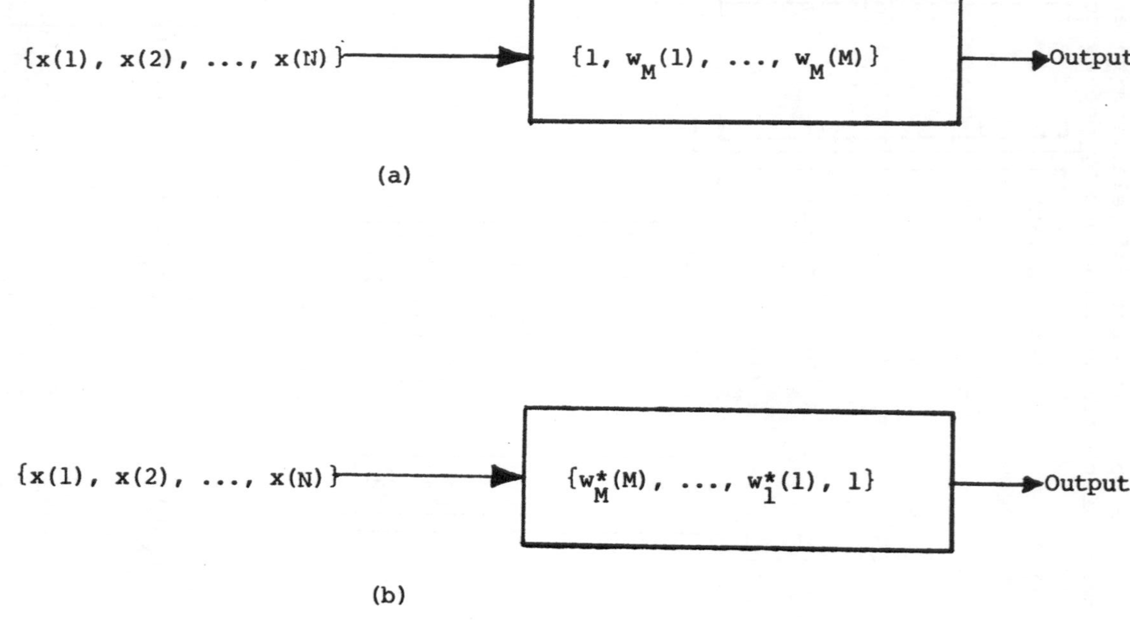

Fig. 3. Illustrating: (a) forward, and (b) backward prediction-error filtering.

Therefore, comparing Eqs. (27) and (28), we deduce that:

$$w_M(k) = w_{M-1}(k) + w_M(M)\, w^*_{M-1}(M-k), \quad k = 0, 1, \ldots, M \tag{29}$$

$$P_{M+1} = P_M + w_M(M)\Delta^*_M \tag{30}$$

and

$$0 = \Delta_M + w_M(M)P_M \tag{31}$$

In the recursive formula of Eq. (29), note that for all values of M:

$$w_M(k) = \begin{cases} 1, & \text{for } k = 0 \\ 0, & \text{for } k > M \end{cases} \tag{32}$$

Equation (31) will always have a solution provided that $P_M > 0$. Thus, using Eq. (31) to eliminate Δ^*_M from Eq. (30), we get:

$$P_{M+1} = P_M[1-|w_M(M)|^2] \tag{33}$$

With the prediction-error power decreasing with increasing filter order, we find that $P_{M+1} < P_M$. Hence, from Eq. (33), we deduce that $|w_M(m)| < 1$. By analogy with the transmission of power through a terminated two-port network, we may view $w_M(M)$ as a "reflection coefficient".

Note that Eq. (29) is reversible provided that $|w_M(M)| \neq 1$. That is, given the coefficients of a prediction-error filter of order M+1, we can obtain the corresponding coefficients of the prediction-error filter of order M.

The algorithm is first considered for the case of a prediction-error filter of order one, that is, M=0. For this case, Eq. (19) reduces to:

$$P_1 = R_x(0) \tag{34}$$

where $R_x(0)$ is the zero-lag value of the autocorrelation function, that is,

$$R_x(0) = \frac{1}{N} \sum_{n=1}^{N} x(n)x^*(n) \tag{35}$$

Next, for M = 1 we use Eq. (33) to write:

$$P_2 = P_1[1-|w_1(1)|^2] \tag{36}$$

The prediction-error power P_2 is taken as the average value of the error powers calculated for forward and backward prediction. For forward prediction, the error power is (see Fig. 3a):

$$P_{f,2} = \frac{1}{N-1} \sum_{n=1}^{N-1} |x(n+1) + w_1(1)x(n)|^2 \tag{37}$$

On the other hand, for backward prediction, the error power is (see Fig. 3b):

$$P_{b,2} = \frac{1}{N-1} \sum_{n=1}^{N-1} |x(n) + w_1{}^*(1)x(n+1)|^2 \tag{38}$$

Averaging $P_{f,2}$ and $P_{b,2}$, we thus get:

$$\begin{aligned} P_2 &= \frac{1}{2}(P_{f,2} + P_{b,2}) \\ &= \frac{1}{2(N-1)} \sum_{n=1}^{N-1} [\,|x(n+1)+w_1(1)x(n)|^2 + |x(n)+w_1^*(1)x(n+1)|^2] \end{aligned} \tag{39}$$

The value of $w_1(1)$ for which P_2 is a minimum is obtained by solving:

$$\frac{\partial P_2}{\partial w_1(1)} = 0$$

Therefore, differentiating Eq. (39) with respect to $w_1(1)$ and setting the result equal to zero, we get:

$$w_1(1) = \frac{-2 \sum_{n=1}^{N-1} x^*(n)x(n+1)}{\sum_{n=1}^{N-1} [\,|x(n)|^2 + |x(n+1)|^2]} \tag{40}$$

For this optimum value of $w_1(1)$, we can use Eq. (36) to compute the corresponding value of the prediction-error power P_2. The solution for M = 1 is thereby completed, and we may thus proceed on to the next value of M.

For M=2, we use Eqs. (29) and (33) to obtain, respectively,

$$w_2(1) = w_1(1)+w_2(2)w_1^*(1) \tag{41}$$

and

$$P_3 = P_2[1-|w_2(2)|^2] \tag{42}$$

Once again, we express the error power P_3 as the average value of the error power for forward prediction and the error power for backward prediction. We may thus write (see Fig. 3):

$$P_3 = \frac{1}{2(N-2)} \sum_{n=1}^{N-2} [|x(n+2)+w_2(1)x(n+1)+w_2(2)x(n)|^2 + |x(n)+w_2^*(1)x(n+1)+w_2^*(2)x(n+2)|^2] \tag{43}$$

We choose $w_2(2)$ so as to minimize the value of P_3, thereby obtaining the following solution:

$$w_2(2) = \frac{-2 \sum_{n=1}^{N-2} [x(n)+w_1^*(1)x(n+1)]^* [x(n+2)+w_1(1)x(n+1)]}{\sum_{n=1}^{N-2} [|x(n)+w_1^*(1)x(n+1)|^2 + |x(n+2)+w_1(1)x(n+1)|^2]} \tag{44}$$

Using this optimum value, we complete the solution for M=2 by computing the remaining filter coefficient $w_2(1)$ and the prediction-error power P_3 by using Eqs. (41) and (42), respectively.

We may generalize the above recursive procedure in a way that after the Mth recursion, the predictions-error filter of order M+1 is completely defined. Specifically, we may write:

$$w_M(M) = \frac{-2 \sum_{n=1}^{N-M} p_{M-1}^*(n) q_{M-1}(n)}{\sum_{n=1}^{N-M} [|p_{M-1}(n)|^2 + |q_{M-1}(n)|^2]} \tag{45}$$

where

$$p_M(n) = \sum_{k=0}^{M} w_M^*(k)x(n+k) \tag{46}$$

and

$$q_M(n) = \sum_{k=0}^{M} w_M(k)x(n+M+1-k) \tag{47}$$

The filter coefficient $w_M(k)$ is defined by Eqs. (29) and (32).

We may develop recursive relations for the computation of the polynomials $p_M(n)$ and $q_M(n)$ as follows. Substitution of Eq. (29) in (46) yields:

$$p_M(n) = \sum_{k=0}^{M-1} w_{M-1}^{*}(k)\,x(n+k) + w_M^{*}(M) \sum_{k=0}^{M-1} w_{M-1}(k)\,x(n+M-k)$$

$$= p_{M-1}(n) + w_M^{*}(M)\,q_{M-1}(n) \qquad (49)$$

Similarly, by substituting Eq. (29) in (47) and using the fact that $w_{M-1}(M)=0$, we obtain:

$$q_M(n) = \sum_{k=0}^{M=1} w_{M-1}(k)\,x(n+M+1-k) + w_M(M) \sum_{k=1}^{M} w_{M-1}^{*}(M-k)\,x(n+M+1-k)$$

$$= q_{M-1}(n+1) + w_M(M)\,p_{M-1}(n+1) \qquad (50)$$

A signal-flow graph representation of the recursive pair of relations in Eqs. (49) and (50), in the form of a lattice structure, is presented in Fig. 4. Note that for M=0, Eqs. (46) and (47) reduce to:

$$p_0(n) = x(n) \qquad (51)$$

and

$$q_0(n) = x(n+1) \qquad (52)$$

Having computed $w_M(M)$, and knowing the set of filter coefficients for the prediction-error filter of order M, we are in a position to completely define the prediction-error filter of order M+1. Specifically, we use Eq. (33) to determine the prediction-error power P_{M+1}, and Eq. (29) to determine the remaining set of filter coefficients $\{w_M(k)\}$, where k=1,2,...,M-1.

Thus, Burg's recursive algorithm generates a Toeplitz matrix $\underline{R}_x(M+1)$, prediction-error filter, and prediction-error power which satisfy the prediction-error equations (21). A key step in the development of the algorithm is the minimization of the prediction-error power taken as the average for forward and backward prediction. This averaging procedure ensures that the Toeplitz matrix $\underline{R}_x(M+1)$ is non-negative definite.

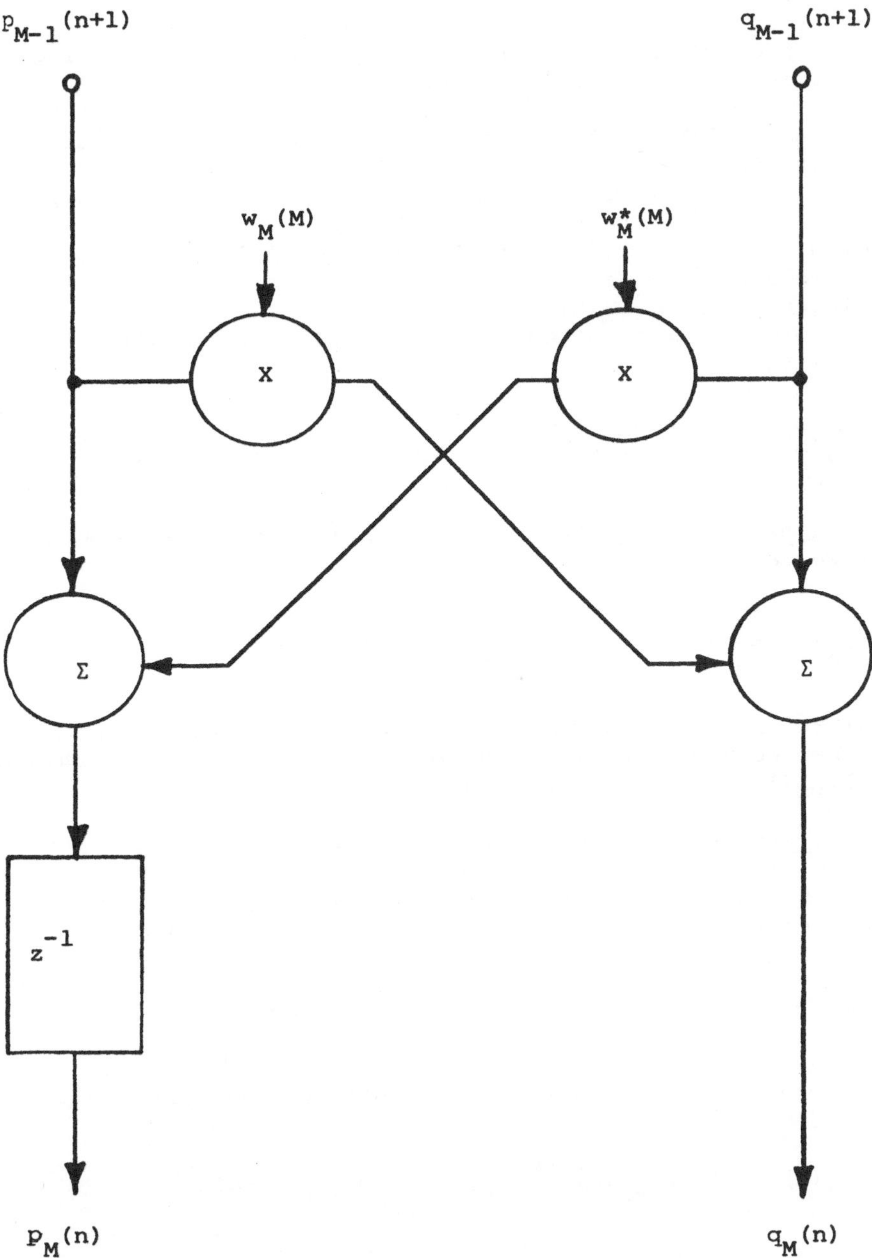

Fig. 4. Lattice structure for the recursive computation of $p_M(n)$ and $q_M(n)$.

4. PROPERTIES OF THE PREDICTION-ERROR FILTER

The prediction-error filter has a number of interesting properties which we shall discuss next.

Property 1: Relation among the autocorrelation function and the reflection coefficients.

The set of autocorrelation values $\{R_x(0), R_x(1), R_x(2), \ldots\}$ is uniquely determined by specifying the corresponding set of numbers $\{R_x(0), w_1(1), w_2(2), \ldots\}$, where $R_x(0)$ is the zero-lag value of the autocorrelation function and the $w_M(M)$, $M=1,2,\ldots$, are reflection coefficients, with $|w_M(M)|<1$.

To prove this relationship, we deduce from the last line of Eq. (27) that:

$$R_x(M) + \sum_{k=1}^{M-1} w_{M-1}(k) R_x(M-k) = \Delta_M \tag{53}$$

Next, eliminating Δ_M between Eqs. (31) and (53), and solving for $R_x(M)$ we get:

$$R_x(M) = -w_M(M) P_M - \sum_{k=1}^{M-1} w_{M-1}(k) R_x(M-k) \tag{54}$$

Therefore, if we are given the set of numbers $\{R_x(0), w_1(1), w_2(2), \ldots\}$ then we may recursively generate the unique set of numbers $\{R_x(0), R_x(1), R_x(2), \ldots\}$ by using Eq. (54).

Property 2: Minimum-Phase Theorem.

The prediction-error filter is minimum phase.

To prove this property, let $W_M(z)$ denote the transfer function of a prediction-error filter of order M+1, as shown by:

$$W_M(z) = \sum_{k=0}^{M} w_M(k) z^{-k} \tag{55}$$

where $w_M(0)=1$. Substitution of Eq. (29) in (55) yields

$$W_M(z) = \sum_{k=0}^{M-1} W_{M-1}(k) z^{-k} + w_M(M) \sum_{k=1}^{M} w^*_{M-1}(M-k) z^{-k} \tag{56}$$

where we have made use of the fact that $w_{M-1}(M)=0$. The first summation in the right-hand side of Eq. (56) is recognized as the transfer function of the corresponding error-prediction filter of order M, that is,

$$W_{M-1}(z) = \sum_{k=0}^{M-1} w_{M-1}(z) z^{-k} \tag{57}$$

As for the second summation term, we substitute k for M-k, obtaining

$$\sum_{k=1}^{M} w_{M-1}^{*}(M-k) z^{-k} = \sum_{k=0}^{M-1} w_{M-1}^{*}(k) z^{-M+k}$$

$$= z^{-M} W_{M-1}^{*}(1/z^{*}) \tag{58}$$

where

$$W_{M-1}^{*}(1/z^{*}) = \sum_{k=0}^{M-1} w_{M-1}^{*}(k) z^{k} \tag{59}$$

We may therefore rewrite Eq. (56) as follows

$$W_M(z) = W_{M-1}(z) + w_M(M) z^{-M} W_{M-1}^{*}(1/z^{*}) \tag{60}$$

Since $|w_M(M)|<1$, we find that on the unit circle in the z-plane (i.e., for $|z|=1$),

$$|w_M(M) z^{-M} W_{M-1}^{*}(1/z^{*})| < |z^{-M} W_{M-1}^{*}(1/z^{*})| = |W_{M-1}(z)| \tag{61}$$

Now suppose that we know that $W_{M-1}(z)$ has no zeros outside the unit circle in the z-plane. Then, from Rouche's theorem* we deduce that $W_M(z)$ also has no zeros outside the unit circle. This proof by induction is completed by noting that the prediction-error filter of order one (i.e., M=0) has no zeros at all and is thus minimum phase. It follows therefore that a prediction-error filter of any order is minimum phase.

<u>Property 3</u>: Maximum Entropy Spectral Estimate.

Burg's recursive algorithm generates a spectral density characterized by an entropy rate that is both stationary with respect to the unknown autocorrelation values and consistent with the known

---

* Rouche's theorem [21] states that if the function F(z) is analytic upon a contour C in the z-plane and within the region enclosed by this contour, and if a second function G(z), in addition to satisfying the same analyticity conditions, fulfills the relation, $|G(z)|<|F(z)|$ on the contour C. then the function F(z)+G(z) has the same number of zeros within the region enclosed by C as does the function F(z).

autocorrelation values. Specifically, this spectral density estimate of the input random process X(t) is defined by

$$\hat{S}_x(f) = \frac{P_{M+1}}{2B\left|1+\sum_{k=1}^{M} w_M(k)\exp(-j2\pi kfT_s)\right|^2} \tag{62}$$

where T_s is the sampling period, and X(t) is band-limited to -B, B. When the process X(t) is sampled at the Nyquist rate, we have $T_s=1/2B$.

To derive this relation, we note that X(t) is a zero-mean, Gaussian process (by assumption). Therefore, the "entropy" (which provides a measure of the average information content) of the set of random variables obtained by uniformly sampling X(t) is given by, except for a scaling factor, [22]

$$H = \frac{1}{2}\ln\{\det[\underline{R}_x(M+1)]\} \tag{63}$$

where $\underline{R}_x(M+1)$ is a Toeplitz matrix defining values of the autocorrelation function of the process. When the process is of infinite duration, however, the entropy H diverges and so we cannot use it as a measure of information content. In such a case we may use the "entropy rate" defined by [23]

$$\begin{aligned} h &= \lim_{M\to\infty} \frac{H}{M+1} \\ &= \lim_{M\to\infty} \frac{1}{2}\ln\{\det[\underline{R}_x(M+1)]\}^{1/(M+1)} \\ &= \frac{1}{2}\ln(2B) + \frac{1}{4B}\int_{-B}^{B}\ln[S_x(f)]df \end{aligned} \tag{64}$$

where we have made use of Eq. (23). The spectral density $S_x(f)$ is related to the autocorrelation function $R_x(m)$ by

$$S_x(f) = T_s \sum_{m=-\infty}^{\infty} R_x(m)\exp(-j2\pi mfT_s) \tag{65}$$

Therefore, substituting Eq. (65) in (64), and assuming sampling at the Nyquist rate (that is, $T_s=1/2B$), we get

$$h = \frac{1}{4B}\int_{-B}^{B}\ln\left[\sum_{m=-\infty}^{\infty} R_x(m)\exp(-j2\pi mfT_s)\right]df \tag{66}$$

The most reasonable choice for the unknown values of the autocorrelation function is one that adds no information or entropy. That is,

$$\frac{\partial h}{\partial R_x(m)} = 0, \qquad |m| \geq M+1 \tag{67}$$

Hence, carrying out this differentiation, we find that the conditions for an extremum are as follows:

$$\int_{-B}^{B} \frac{\exp(-j2\ mfT_s)}{\hat{S}_x(f)}\, df = 0, \qquad |m| \geq M+1 \tag{68}$$

where $\hat{S}_x(f)$ is the spectral density estimate corresponding to Eq. (67). Equation (68) implies that the reciprocal of $\hat{S}_x(f)$ is expressible in the form of a truncated Fourier series as follows:

$$\frac{1}{\hat{S}_x(f)} = \sum_{n=-M}^{M} c(n)\exp(-j\pi nfT_s) \tag{69}$$

where $c(-n)=c^*(n)$ so as to ensure that $\hat{S}_x(f)$ is a real quantity.

The next step in the derivation is to determine values for the set of coefficients $\{c(n)\}$ such that the spectral density estimate $\hat{S}_x(f)$ is consistent with the known values of the autocorrelation function, that is,

$$R_x(m) = \int_{-B}^{B} \hat{S}_x(f)\exp(j2\pi mfT_s)\,df, \qquad 0 \leq m \leq M \tag{70}$$

Therefore, substituting Eq. (69) in (70), we get

$$R_x(m) = \int_{-B}^{B} \frac{\exp(j2\pi mfT_s)}{\sum_{n=-M}^{M} c(n)\exp(-j2\pi nfT_s)}\, df, \quad 0 \leq m \leq M \tag{71}$$

At this point, it is convenient to use z-transform notation. Thus, with the unit-delay operator z defined by

$$z = \exp(j2\pi fT_s) \tag{72}$$

and

$$df = \frac{1}{j2\pi T_s}\left(\frac{dz}{z}\right) = \frac{B}{j\pi}\left(\frac{dz}{z}\right) \tag{73}$$

we may rewrite Eq. (71) in the form

$$R_x(m) = \frac{B}{j\pi} \oint \frac{z^{m-1}}{\sum_{n=-M}^{M} c(n) z^{-n}} dz, \qquad 0 \leq m \leq M \tag{74}$$

where the contour integration is carried out around the unit circle in the z-plane in a counter-clockwise direction. Because of the Hermitian property of the set of coefficients $\{c(n)\}$, we may express the summation in the denominator of the integrand in Eq.(74) as the product of two polynomials as follows:

$$\sum_{n=-M}^{M} c(n) z^{-n} = G_M(z) G_M^*(1/z^*) \tag{75}$$

where

$$G_M(z) = \sum_{n=0}^{M} g_M(n) z^{-n} \tag{76}$$

and

$$G_M^*(1/z^*) = \sum_{n=0}^{M} g_M^*(n) z^{n} \tag{77}$$

The first polynomial $G_M(z)$ is chosen to be minimum-phase (i.e., with its zeros all located inside the unit circle), whereas the second polynomial $G_M^*(1/z^*)$ is chosen to be maximum-phase (i.e., with its zeros all located outside the unit circle). Furthermore, the zeros of the two polynomials are the inverse of each other with respect to the unit circle. Thus, substituting Eq. (75) in (74), we get

$$R_x(m) = \frac{B}{j\pi} \oint \frac{z^{m-1}}{G_M(z) G_M^*(1/z^*)} dz, \qquad 0 \leq m \leq M \tag{78}$$

Using this relation, we next form the summation

$$\begin{aligned} \sum_{k=0}^{M} g_M(k) R_x(m-k) &= \frac{B}{j\pi} \oint \frac{z^{m-1} \sum_{k=0}^{M} g_M(k) z^{-k}}{G_M(z) G_M^*(1/z^*)} dz, \\ &= \frac{B}{j\pi} \oint \frac{z^{m-1}}{G_M^*(1/z^*)} dz, \qquad 0 \leq m \leq M \end{aligned} \tag{79}$$

Since the polynomial $G_M^*(1/z^*)$ has no zeros inside the unit circle, it follows that the integrand in Eq. (79) is analytic on and inside the unit circle for $m \geq 1$, in which case we find from Cauchy's residue theorem [21] that the contour integral in Eq. (79) is zero. However, for m=0 the integrand has a simple pole at z=0 with a residue equal to $1/g_M^*(0)$. We thus conclude that

$$\sum_{k=0}^{M} g_M(k) R_x(m-k) = \begin{cases} \dfrac{2B}{g_M^*(0)}, & m=0 \\ 0, & m=1,2,\ldots,M \end{cases} \tag{80}$$

This set of equations is similar in form to the prediction-error filter equations. In particular, by comparing Eqs. (20) and (80) we deduce that

$$g_M(k) = \frac{2B}{g_M^*(0) P_{M+1}} w_M(k), \qquad 0 \leq k \leq M \tag{81}$$

Thus, noting that $w_M(0) = 1$, we find that for k=0,

$$|g_M(0)|^2 = \frac{2B}{P_{M+1}} \tag{82}$$

Hence, using Eqs. (81) and (82), we may rewrite Eq. (75) in the form

$$\sum_{n=-M}^{M} c(n) z^{-n} = \frac{2B}{P_{M+1}} W_M(z) W_M^*(1/z^*) \tag{83}$$

where $W_M(z)$ is the transfer function of the prediction-error filter (see Eq. 55). Finally, substituting this result, with $z=\exp(j2\pi fT_s)$, in Eq. (69), we get the desired expression for the spectral density estimate $\hat{S}_x(f)$, namely, Eq. (62).

<u>Property 4</u>: Whitening of the Time Series.

The error time series produced by a prediction-error filter of order M+1, in response to an autoregressive process of order M+1, is a white noise process.

The error time series $\{\varepsilon(n)\}$, given by Eq. (15), is reproduced here for convenience:

$$\varepsilon(n) = \sum_{k=0}^{M} w_M(k) x(n-k) \tag{84}$$

In order for this difference equation to have a wide-sense stationary solution satisfying the condition

$$E[x(m)\varepsilon(n)] = 0, \quad \text{for all } m \leq n-1, \qquad n=0,\pm 1,\ldots \tag{85}$$

for the case when $\varepsilon(n)$, $n=0,\pm 1,\ldots$, is a white noise process, it is sufficient that all of the zeros of the transfer function of the prediction-error filter, represented by the set of coefficients $\{w_M(n)\}$, lie inside the unit circle [24]. Since it is known that the prediction-error filter is minimum phase (see Property 2), we therefore deduce that with an autoregressive process of order M+1 as input, the error time series produced by the prediction-error filter of order M+1 is a white noise process.

Indeed, virtually every time series encountered in practice (after appropriate preprocessing) can be approximated to any desired accuracy by a finite autoregressive model of sufficiently high degree [24]. It follows therefore that, in the limit, with an arbitrary time series as input, the error time series produced by a prediction-error filter of infinite length is a white noise process.

5. CHOICE OF FILTER ORDER

A major issue in the design of the maximum entropy spectral estimator is the choice of a suitable order for the prediction-error filter. To resolve this problem, we may use the so-called final prediction error (FPE), which is defined as the mean-square value of the prediction error [25].

For a prediction-error filter of order M+1, the final prediction error is

$$(\mathrm{FPE})_M = \frac{N+M+1}{N-M-1} P_{M+1} \tag{86}$$

where P_{M+1} is the prediction-error power. Equation (86) is evaluated for a successively higher order of prediction-error filter at every recursion step. The optimum value M_{opt} turns out to be the particular value of M for which the $(\mathrm{FPE})_M$, expressed as a function of M, is a minimum. Empirical investigations have shown that if the filter order is too low, the corresponding spectral estimate is broadened; while if it is too high, spurious details appear in the spectrum.

To illustrate the applicability of minimum FPE criterion, consider a second-order all pole process defined by the difference equation:

$$x(n) = -\alpha_1 \cdot x(n-1) - \alpha_2 \cdot x(n-2) + z(n) \tag{87}$$

where z(n) is a white noise process with zero mean and variance σ_n^2, and α_1 and α_2 are constants. The theoretical spectral density of x(n) is given by

$$S_x(f) = \frac{\sigma_n^2}{|1+\alpha_1 \exp(-j2\pi f)+\alpha_2 \exp(-j4\pi f)|^2} \quad (88)$$

where, for convenience, we have put the sampling period T_s=1 second. The curve labelled "theoretical" in Fig. 5 is a plot of Eq. (88) for α_1=-1, α_2=0.8 and σ_n^2=2.56. The remaining curves represent the maximum entropy spectral density estimates obtained from 128 samples of x(n), with z(n) obtained from a Gaussian-distributed random number generator. The plot for M=9 corresponds to the optimum prediction-error filter, for which the FPE is minimum. In Fig. 5 we see that for both M=5 and M=14, the maximum entropy spectral density estimator exhibits minor peaks which do not appear in the plot for M_{opt}. The number of these peaks and their amplitudes increase with increasing filter order. Also, the location of the mean peak shifts to the right, and its amplitude increases by increasing M. We may thus consider the curve corresponding to M_{opt} as an optimum estimate in the sense that it is a compromise between fidelity of the estimate and the computation time required to obtain the filter coefficients. Note that since the curve of FPE as a function of M has a relatively broad minimum, the resulting estimate is somewhat insensitive to the choice of M in the immediate vicinity of M_{opt}.

For a second example, consider the case of a complex second-order all-pole process involving data generated in a manner similar to that in the previous example. In this case, however, two different sets of random numbers are generated according to Eq.(88), which form the real and imaginary parts of the process. The parameters of the process are α_1=-1 and α_2=0.8-j0.2, while σ_n^2 is chosen to be

$$\sigma_n^2 = |1+\alpha_1+\alpha_2|^2 = 0.68,$$

so that the normalized spectral density at zero frequency is equal to unity.

The theoretical spectral density of such a process, and its maximum entropy estimates for three different values of M are shown in Figures 6a, 6b, and 6c, with each part of the figure corresponding to a specific set of record lengths. In part (a) of the figure, the theoretical spectrum exhibits asymmetry in that the peak in the negative frequency region is broader, further away from the origin, and has a significantly smaller magnitude than the peak in the positive frequency region. The spectral density estimates in Fig. 6a are obtained with 64 data samples of the process, with the first 50 samples ignored in order to reduce transient effects. It can be

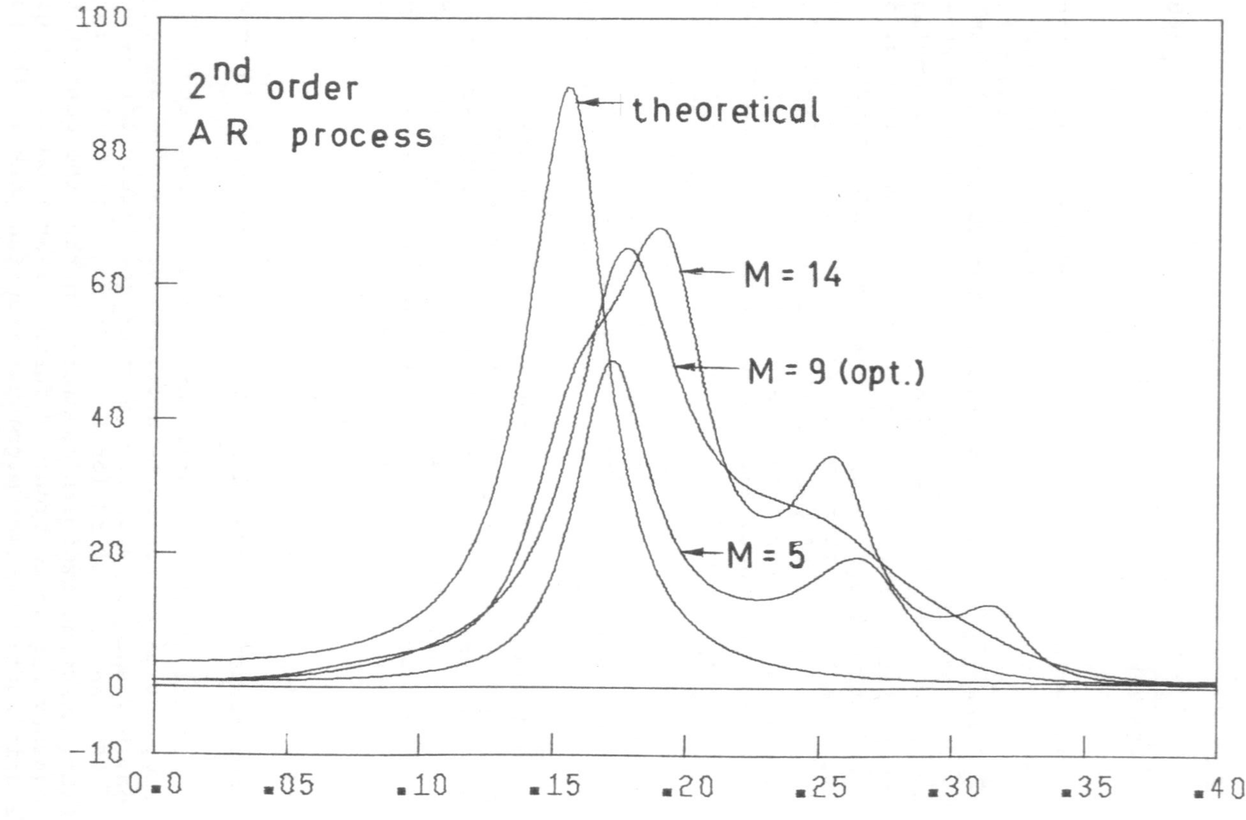

Fig. 5. Theoretical power spectral density for the all-pole process $x(n) = x(n-1) - 0.8x(n-2) + z(n)$; $\sigma_w^2 = 2.56$, and its ME estimates for $M = 5$, $M = M_{opt} = 9$, and $M = 14$.

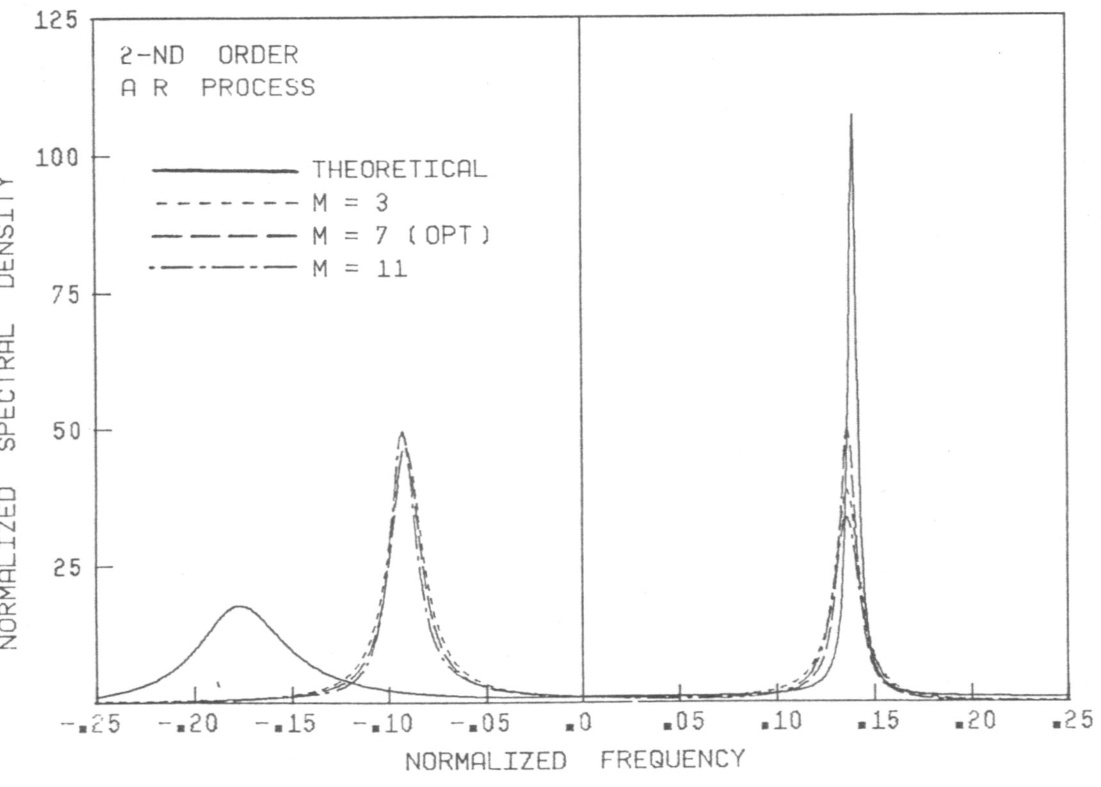

Fig. 6(a). The theoretical power spectral density for the complex second-order all-pole process with $a_1 = 1.$, $a_2 = 0.8\text{-}j0.2$, and $\sigma_w^2 = 0.68$, and its ME estimates with the first 50 samples ignored and N = 64.

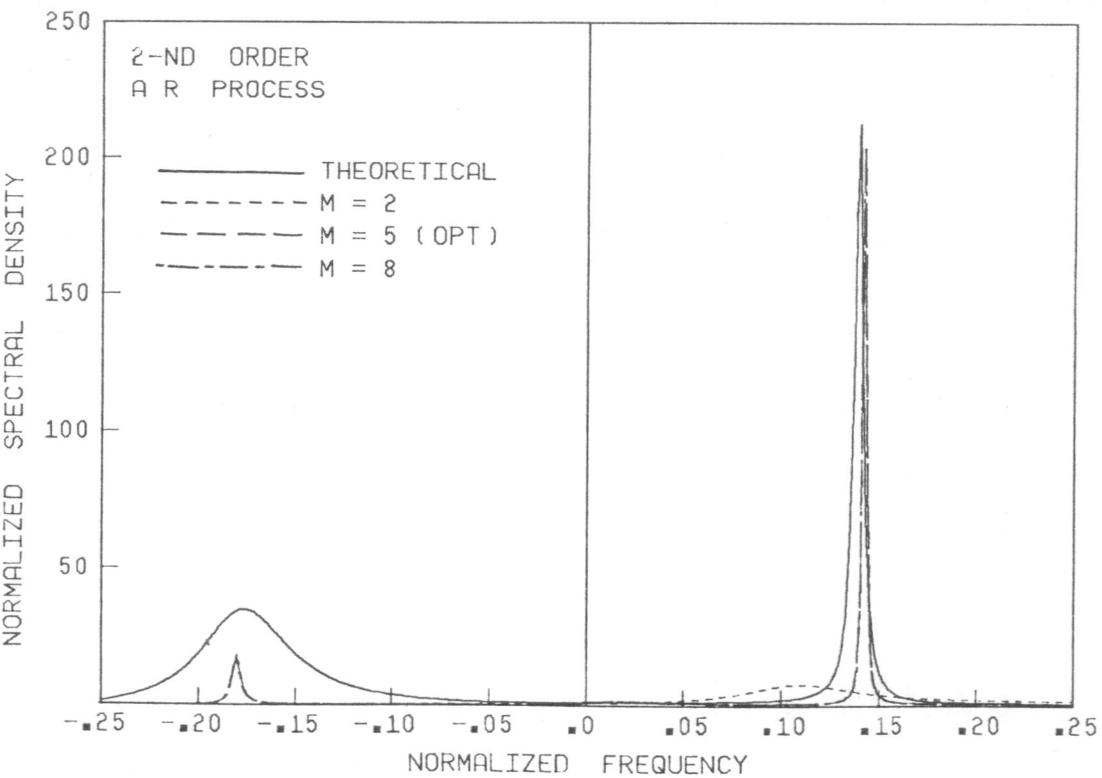

Fig. 6(b). The same spectra as in Fig. 3.1 - 2(a) except that the first 114 samples are ignored and N = 64.

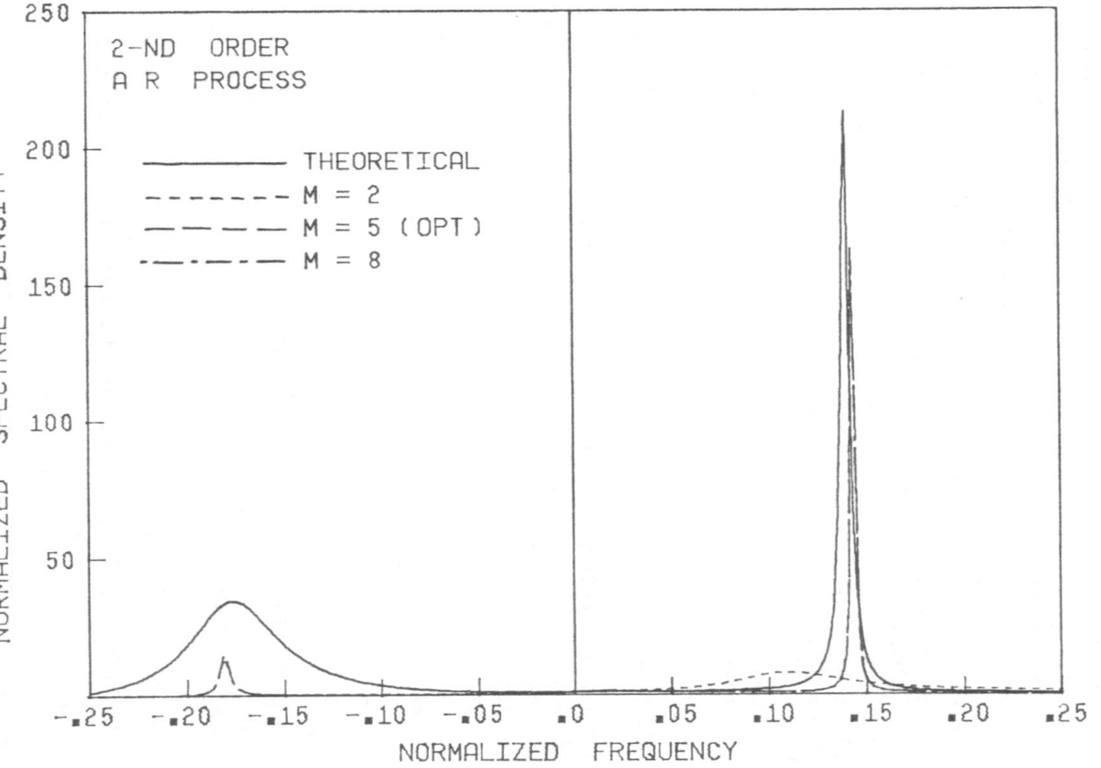

Fig. 6(c). The same spectra with the first 114 samples ignored and N = 192.

seen from Fig. 6a that the spectral estimates give a good representation of the theoretical spectrum in the positive frequency region, again with the closest approximation obtained for the optimal value of M; in the negative frequency region, however, the estimated peaks are shifted away from their actual positions. The reason for this phenomenon is that the number of front-end data samples of the process that were ignored were not adequate to ensure that all transient effects have died away completely. This observation is confirmed in Fig. 6b where 114 data samples were ignored as compared to 50 in the case of Fig. 6a. Note, however, that in Fig. 6b, the optimum length of the filter is equal to 5. We thus find that in Fig. 6b the maximum entropy estimates provide a closer approximation to the actual spectrum position of the peak for negative frequencies, for both the optimal value of M and for M=8, than the previous estimates in Fig. 6a. Note also that the filter of order two gives a poor estimate of the peak in the positive frequency region, and that the peak in the negative frequency region is not resolved at all.

Provided that an adequate number of front-end data samples are ignored so as to make transient effects negligible, the maximum entropy estimates are relatively insensitive to the number of data samples used for driving the estimators. This is confirmed in Fig. 6c where the number of data samples used in the spectral analysis is three times as many.

6. COMPUTATIONAL PROCEDURE

We are now in a position to formulate the steps in the recursive procedure involved in calculating the prediction-error filter coefficients and related values of the spectral density and autocorrelation function:

(1) For the given time series $\{x(n)\}$, $n=1,2,\ldots,N$, and $M=0$, use Eqs. (34) and (35) to calculate the power

$$P_1 = \frac{1}{N}\sum_{n=1}^{N} x(n)x^*(n)$$

and use Eqs. (51) and (52) to calculate

$$p_0(n) = x(n)$$

$$q_0(n) = x(n+1)$$

(2) For M=1, use Eq. (40) to calculate the reflection coefficient

$$w_1(1) = \frac{-2\sum_{n=1}^{N-1} x^*(n)x(n+1)}{\sum_{n=1}^{N-1}\left[\,|x(n)|^2+|x(n+1)|^2\right]}$$

and use Eq. (36) to calculate the prediction-error power

$$P_2 = P_1[1-|w_1(1)|^2]$$

(3) Increment M by 1, and use Eqs. (49) and (50) to calculate the next values

$$p_M(n) = p_{M-1}(n)+w_M^*(M)q_{M-1}(n)$$

$$q_M(n) = q_{M-1}(n+1)+w_M(M)p_{M-1}(n+1)$$

(4) For the current value of M, use Eq. (45) to calculate the reflection coefficient

$$w_M(M) = \frac{-2\sum_{n=1}^{N-M} p_{M-1}^*(n)\, q_{M-1}(n)}{\sum_{n=1}^{N-M} [\,|p_{M-1}(n)|^2+|q_{M-1}(n)|^2]}$$

and use Eq. (33) to calculate the corresponding value of the error-prediction power

$$P_{M+1} = P_M[1-|w_M(M)|^2]$$

(5) Use Eq. (29) to calculate the remaining coefficients of the prediction-error filter

$$w_M(k) = w_{M-1}(k)+w_M(M)w_{M-1}^*(M-k), \qquad k=1,2,\ldots,M-1$$

(6) Repeat steps (3) to (5) for each value of M, up to the optimum value M_{opt} for which the final prediction-error

$$(FPE)_M = \frac{N+M+1}{N-M-1} P_{M+1}$$

is a minimum.

(7) We now have all the quantities required for using Eq. (54) to calculate the extrapolated value of the autocorrelation function

$$R_x(M) = -w_M(M)P_M - \sum_{k=1}^{M-1} w_{M-1}(k)R_x(M-k)$$

and for using Eq. (62) to calculate the spectral density estimate

$$\hat{S}_x(f) = \frac{P_{M+1}}{2B\left|1+\sum_{k=1}^{M} w_M(k)\exp(-j2\pi kfT_s)\right|^2}$$

The computation is thereby completed.

7. MODELING OF RADAR CLUTTER AS AN AUTOREGRESSIVE PROCESS

The maximum entropy method attempts to fit, in a least squares sense, an autoregressive model to the input time series. Accordingly, use of the maximum entropy method is usually favored for the spectral analysis of a stochastic process that can be modeled as an autoregressive process [26].

In the case of clutter generated by reflections of the transmitted wave from various objects located in a radar environment, there exists no unique way of defining the spectral properties of such signals. However, a considerable amount of effort has been expended in order to find a clutter model which best fits the particular situation. According to Barlow [27], the clutter spectrum may be assumed to have a Gaussian shape, as shown by

$$S(f) = S(0)\exp\left(-\frac{f^2}{2\sigma_f^2}\right) \tag{89}$$

where S(0) is the value of the spectral density S(f) at zero frequency, and σ_f is the spectral spread. Other investigators have found that a better fit to experimental data is achieved when the spectral density S(f) is represented by an all-pole spectrum of order three [28]. It is therefore apparent that there is no unique representation of the clutter spectrum which will be valid for all circumstances.

The clutter generated in a radar environment may, however, be modeled as an autoregressive (all-pole) process of a finite order. The representation in [28] gives the model order as three. On the other hand, the Gaussian-shaped model may be well-approximated by an all-pole model of a relatively low order. For example, the absolute error in approximating the Gaussian function

$$S(f) = \frac{1}{\sqrt{2\pi}}\exp\left(-\frac{f^2}{2}\right)$$

by an autoregressive model of order 6 is less than 2.7×10^{-3}.

We conclude therefore, that the application of the maximum entropy method to the spectral analysis of radar clutter is theoretically justified. Furthermore, the experimental results given in section 9 confirm the practical usefulness of the method.

8. RECORDING AND COMPUTER PROCESSING OF CLUTTER DATA

A block diagram of the experimental system used for the processing of recorded clutter data is shown in Fig. 7. The system consists of an RCA wideband video recorder/reproducer, an interface unit, a CDC 1700 digital computer, and an oscilloscope.

The recorder stores the incoming video signal onto two wide-band video channels, which have the following characteristics:

Frequency response:	15 Hz to 5 MHz, with ± 3 dB points at 10 Hz and 6 MHz
Signal-to-noise ratio:	36 dB peak-to-peak rms signal in reference to 1V peak-to-peak signal between 10 Hz and 6 MHz
Amplitude stability:	± 0.5 dB at 1V peak-to-peak
Time-base stability:	75 ns rms absolute
Head switching transients:	6 dB minimum below peak noise
Crosstalk:	Not greater than 35 dB below reference signal of 1V peak-to-peak.

For the analysis of coherently detected video signals, recordings of the in-phase and quadrature components of the video signal, at the phase detector output of the radar receiver, are made.

In addition to the video signal, timing information is needed to determine the range and azimuth of a particular video segment. The range of a target is obtained from a measurement of the time delay between transmission of a radar pulse and reception of the target echo. Hence, the time of pulse transmission must be known. This timing is given by the system trigger. Azimuth's information is given by the position of the antenna. This information is provided by three synchro signals which are sinusoidal waves with a frequency equal to the scanning rate of the antenna, 120° apart in phase, and which modulate a 60 Hz carrier. These three signals can be resolved to provide the azimuth position of the antenna. These synchro signals are used to drive an azimuth digitizer, producing a series of azimuth change pulses (ACP), usually 4096 per revolution, as well as an azimuth reference pulse (ARP) once every revolution, usually at north reference. Thus, there are three signals needed for positional information: system triggers, ACP's and ARP's. These signals are recorded onto two auxiliary audio channels (with an upper curoff frequency at 15 kHz) available on

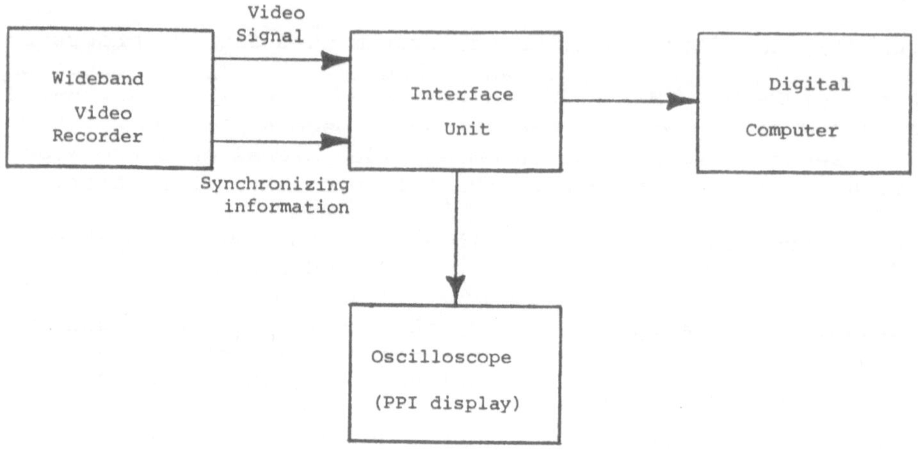

Fig. 7. Experimental set-up.

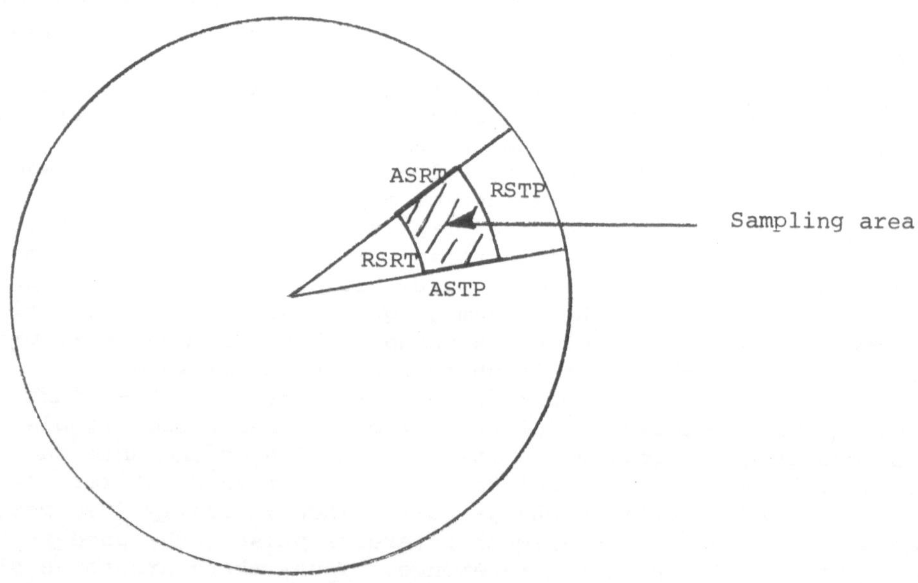

Fig. 8. Definition of sampling area.

the recorder. The pulses become rounded because of the limited bandwidth, but remain sufficiently well-defined to allow their recovery upon replay.

The interface unit enables coupling of the recorder to the computer. This unit is logically divided into two parts: display circuitry and sampler/digitizer circuitry. The display circuitry is designed to take the video and timing signals from the recorded video tape and configure them to produce a pseudo plan-position indicator (PPI) display suitable for use on an oscilloscope with X, Y, Z input capabilities. The sampler/digitizer part of the interface unit, on the other hand, is designed to perform two functions: (1) to properly sample and digitize the recorded video signal corresponding to a preselected sampling area of the PPI display, and store the digitized data thus obtained in a random access memory (RAM), and (2) to allow the computer to access the data stored in the RAM, and control the parameters that are used to define the sampling area. The sampling area is bounded in range by range start (RSRT) and range stop (RSTP) and in azimuth by azimuth start (ASRT) and azimuth stop (ASTP), as shown in Fig. 8.

The recorded radar signal is two-dimensional in nature in that it is a function of both range and azimuth coordinates. On the other hand, the processing has to be done in one dimension. A choice must therefore be made about the number of clutter samples in range and those in azimuth, and the order in which the samples are processed. Suppose that we wish to analyze an area that contains N_R bins in range and N_A bins in azimuth, giving a total number of range-azimuth cells equal to $N_R N_A$. The azimuth bin or resolution is determined by the 3-dB beamwidth of the antenna. The number of ACP's within a specified azimuth bin determines the number of hits per beamwidth. For the radar used in the analysis, the 3-dB beamwidth is 1.35°, so that the number of hits per beamwidth is

$$N_H = \frac{4096}{360^o} \times 1.35^o = 15.4$$

which, for the purpose of analysis, may be rounded to 16. On the other hand, range bin or resolution is determined by the transmitted pulse duration. Thus, one procedure for ordering the clutter samples is to take 16 samples from the first range-azimuth cell, then the next 16 samples from the neighbouring cell in azimuth and so on, up to the last azimuth bin in the same range ring. Then, we shift to the next range ring and repeat the same procedure, until we reach the N_R-th ring. For example, if we choose $N_R = 2$, the arrangement is as shown in Fig. 9(a).

A second procedure is to select the successive samples along range in one azimuth bin, then shift to the next azimuth bin, and so on. This order of data arrangement is exactly the same as that existing in computer memory, since the incoming data appear in that order. The arrangement for 2 successive range rings is shown in Fig. 9(b).

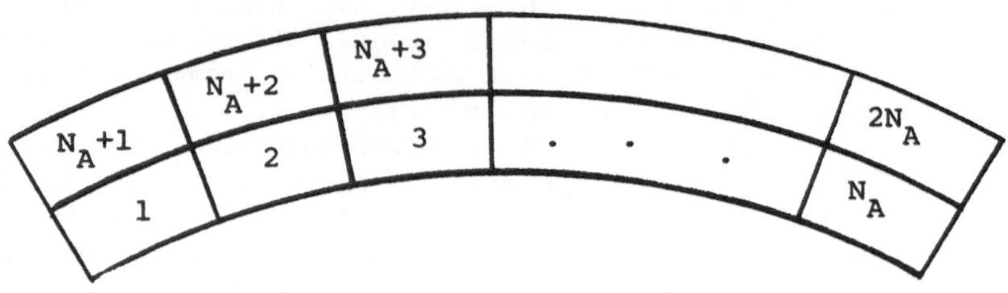

(a)

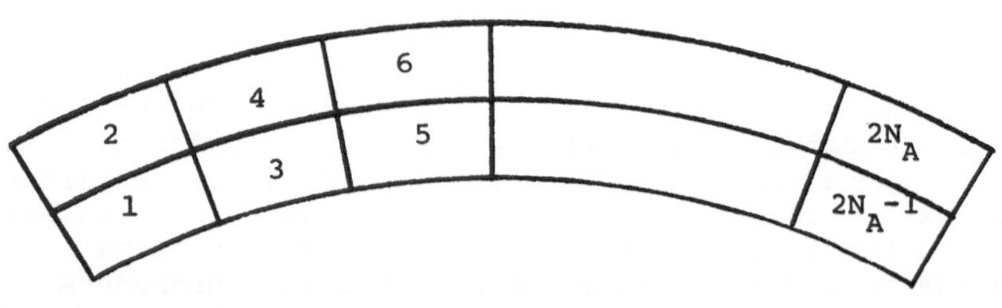

(b)

Fig. 9. Illustrating 2 different arrangements for ordering of clutter data for processing.

9. EXPERIMENTAL RESULTS

The results reported in this section are based on clutter data obtained by using an airport surveillance radar ASR-8 located at Bagotville, Quebec. This radar is coherent, generating in-phase and quadration bipolar signals.

Figure 10 shows the spectral density estimates of ground clutter obtained with the maximum entropy method, using the minimum FPE criterion, for different numbers of data samples. The frequency values are normalized with respect to the average pulse-repetition frequency of the radar (equal to 1.04 kHz). Spectral density values are also normalized with respect to the maximum value, so that they extend from 0 dB down to the limiting value which is selected in advance. These normalizations are applied in all the figures that follow. A feature that is immediately obvious from these curves is the presence of a very narrow central peak at zero frequency. Furthermore, there is not much difference between the curves for N = 256, 100, and 14. It is only when the number of data samples is low (N = 8) that we find the spectral spread is relatively wide and sidelobe level high. It is of interest that the spectral analysis of the data ordered as indicated in Fig. 9(a) produced practically the same results as that obtained using the data arrangement of Fig. 9(b). This suggests that, in the case of ground clutter, the spatial correlations of samples in range and azimuth are of the same order of magnitude.

Figure 11 shows the maximum entropy spectral density estimates of weather clutter for different number of data samples. Here, again, we see that the spectral change for different values of data samples is rather small. We also see that for weather clutter the spectral spread is larger and the sidelobe level is higher than the corresponding values for ground clutter.

Figure 12 shows the maximum entropy spectral density estimates of clutter produced by migrating birds. Comparing this set of curves with those for ground and weather clutter, we see that: (1) the spectral spread is significantly wider, (2) the sidelobe levels are higher, and (3) some of the main peaks are shifted away from zero frequency, thereby resulting in a pronounced asymmetry.

The features which distinguish the maximum entropy spectral density estimates of clutter due to ground, weather, and migrating flocks of birds are further illustrated in Fig. 13 for the case of 16 data samples. Figure 14 shows the corresponding spectral density estimates obtained by using the modified periodogram method based on the fast Fourier transforms [3]. These two sets of curves clearly illustrate the superiority of the maximum entropy method over the modified periodogram method in the spectral classification of different forms of radar clutter.

The qualitative results described above indicate that it is possible to provide a practical means for the on-line classification of the various types of clutter, based on the maximum entropy analysis of spectral density. In order to provide a feeling for

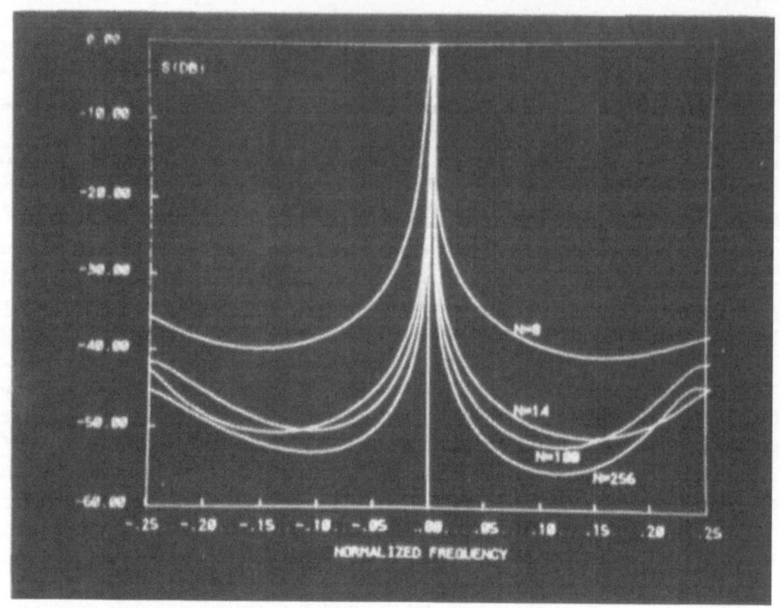

Fig. 10. ME spectral estimates of ground clutter, with N = 256, 100, 14, and 8; frequency range from -0.25 to 0.25.

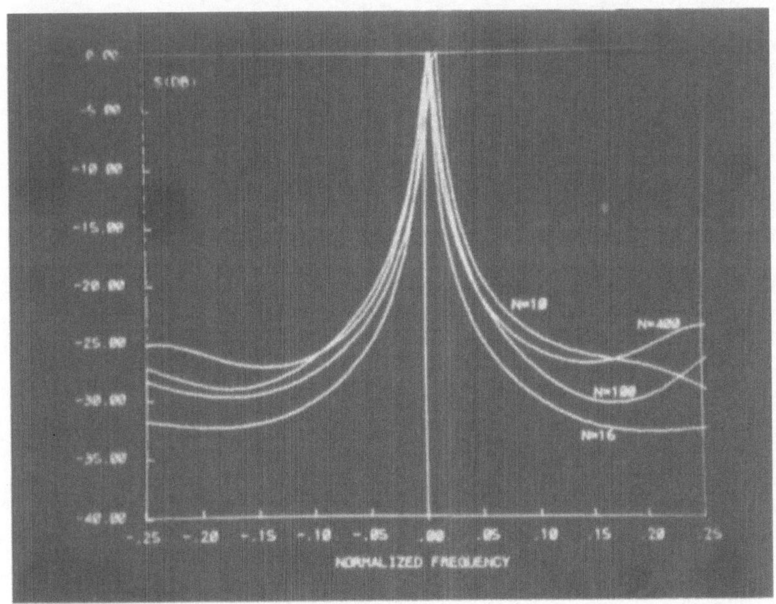

Fig. 11. ME spectral estimates of weather clutter, with N = 400, 100, 16, and 10; frequency range from -0.25 to 0.25.

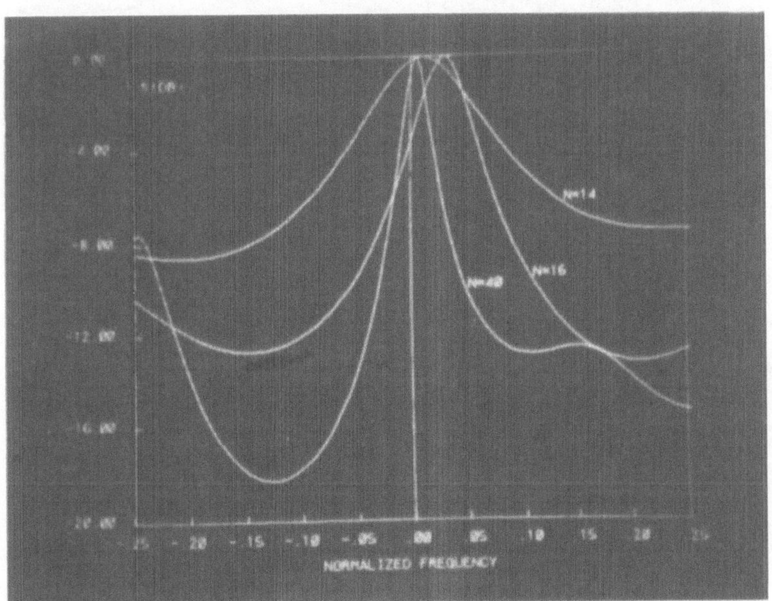

Fig. 12. ME spectral estimates of birds clutter, with N = 40, 16, and 14; frequency range from -0.25 to 0.25.

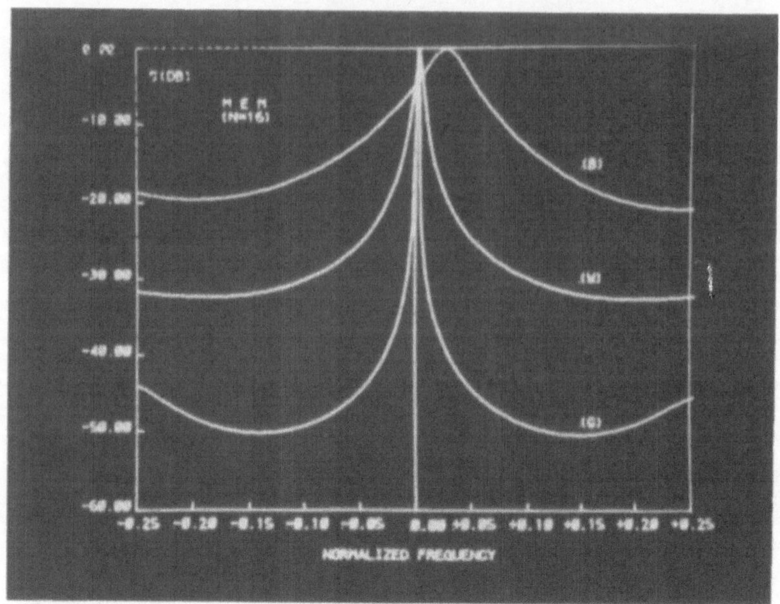

Fig. 13. M estimates of typical ground, weather, and bird clutter signals, with N = 16.

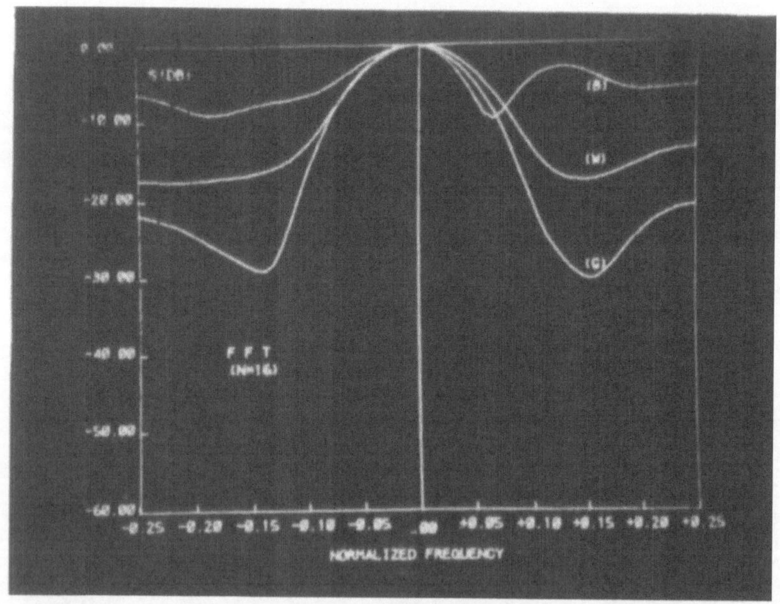

Fig. 14. Welch's estimates of typical ground, weather, and bird clutter signals, with N = 16.

the statistical variability of spectra spread for different clutter conditions and record lengths, a large number of spectra were calculated and the spectral spread, corresponding to the -10 dB points, were extracted. Figure 15 illustrates the differences in the statistical averages of spectral spread for the three types of clutter for 16 data samples. The solid vertical lines in the figure represent the sample mean, while the dotted lines represent the boundary values determined by standard deviations.

10. CONCLUSIONS

The experimental results described in Section 9 confirm that it is possible to classify the various types of clutter in an air traffic control environment by using the maximum entropy method. The separation between radar echoes due to ground, weather and migrating birds is made possible by a prediction-error filtering technique which exhibits excellent frequency resolution capabilities. A statistical analysis of recorded coherent radar signals has confirmed the ability of the method to resolve the spectra of echoes due to ground, weather, and birds with relatively short data record lengths as encountered in a typical air traffic control environment. This method has been shown to be superior to the conventional methods of spectral analysis using the periodogram.

11. ACKNOWLEDGEMENTS

The author is indebted to Dr. S. B. Kesler for numerous discussions covering the maximum entropy method and its signal processing applications. The experimental set-up described in Section 8 is based on an M. Eng. Thesis by Mr. B. W. Currie, and the experimental results described in Section 9 are based on a Ph. D. Thesis by Dr. S. R. Kesler, both of which were written at McMaster University.

He is also grateful to the National Research Council of Canada for supporting this research program.

REFERENCES

1. Blackman, R. B. and J. W. Tukey, "The Measurements of Power Spectra, From the Point of View of Communications Engineering," New York: Dover, Inc., 1959.
2. Jones, R. H., "A Reappraisal of the Periodogram in Spectral Analysis," Technometrics, vol. 7, pp. 531-542, Nov. 1965.
3. Welch, P. D., "The Use of FFT for Estimation of Power Spectra: a Method Based on Averaging Over Short Modified Periodograms," IEEE Trans. Audio Electroacoust., vol. AU-15, pp. 70-73, June 1967.

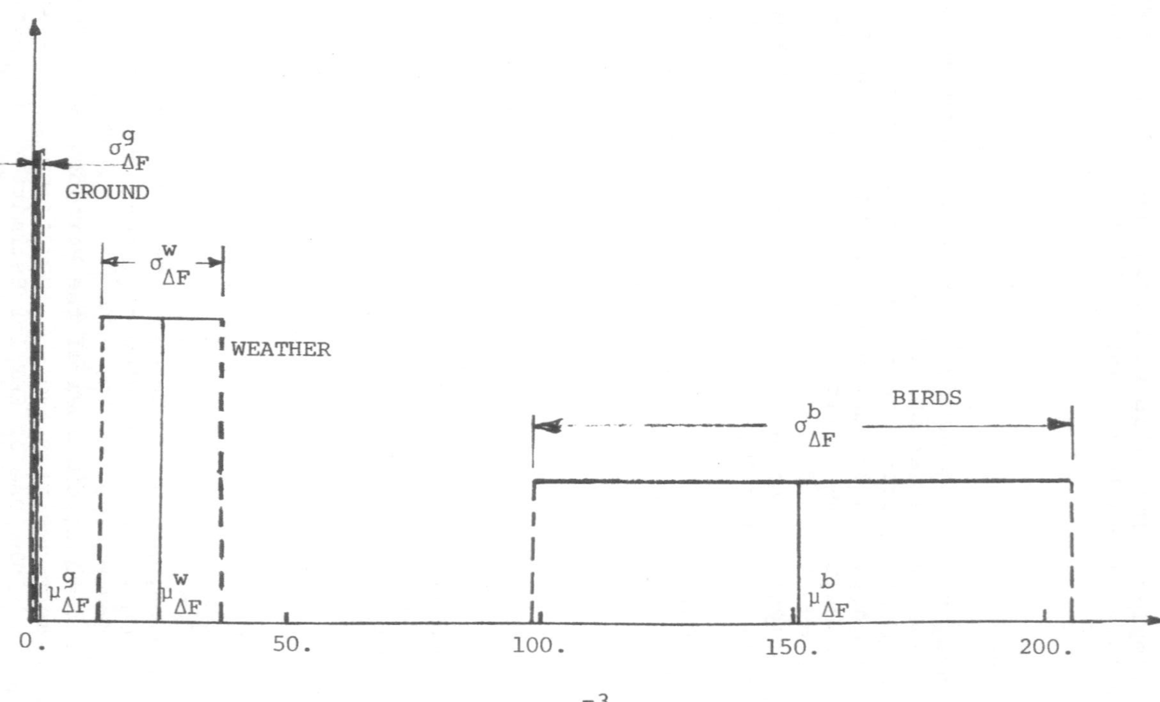

Fig. 15. Comparison of the statistical averages of ΔF for ground, weather, and birds; N = 16.

4. Burg, J. P., "Maximum Entropy Spectral Analysis," Paper presented at the 37th Annual SEG Meeting, October 31, 1967, Oklahoma City, Okla.
5. Burg, J. P., "A New Analysis Technique for Time Series Data," Paper presented at NATO Advanced Study Institute on Signal Processing, 1968.
6. Van den Bos, A., "Alternative Interpretation of Maximum Entropy Spectral Analysis," IEEE Trans. Information Theory, vol. IT-17, pp. 493-494, July 1971.
7. Parzen, E., "Statistical Spectral Analysis (single channel case) in 1968." Tech. Rep. 11 on Contract Nonr-225-(80), Stanford University, 1968.
8. Lacoss, R. T., "Data Adaptive Spectral Analysis Methods," Geophysics, vol. 36, pp. 661-675, Aug. 1971.
9. Ulrych, T. J., "Maximum Entropy Power Spectrum of Truncated Sinusoids," J. Geophys, Res., vol. 77, pp. 1396-1400, March 10, 1972.
10. Cehn, W. Y., and G. R. Stegen, "Experiments With Maximum Entropy Power Spectra of Sinusoids," J. Geophys, Res., vol. 79, pp. 3019-3022, July 10, 1974.
11. Radoski, H. R., P. F. Fougere, and E. J. Zawalick, "A Comparison of Power Spectral Estimates and Applications of the Maximum Entropy Method," J. Geophys, Res., vol. 80, pp. 619-625, Feb. 1, 1975.
12. Ulrych, T. J., and T. N. Bishop, "Maximum Entropy Spectral Analysis and Autoregressive Decomposition," Reviews in Geophysic and Space Physics, vol. 13, pp. 183-200, Feb. 1975.
13. Smylie, D. E., G. K. C. Clarke, and T. J. Ulrych, "Analysis of Irregularities in the Earth's Rotation," in Methods in Computational Physics, vol. 13, B. A. Bolt, ed., New York, Academic Press, 1973.
14. Kaveh, M., and G. R. Cooper, "An Empirical Investigation of the Properties of the Autoregressive Spectral Estimator, "IEEE Trans, Information Theory, vol. IT-22, pp. 313-323, May 1976.
15. Kesler, S. B., and S. Haykin, "The Maximum Entropy Method Applied to the Spectral Analysis of Radar Clutter," IEEE Trans. Information Theory, vol. II-24, pp. 269-272, March 1978.
16. Kesler, S. B., and S. Haykin, "A Comparison of the Maximum Entropy Method and the Periodogram Method Applied to the Spectral Analysis of Computer-Simulated Radar Clutter," Canadian Electrical Engineering Journal, vol. 3, pp. 11-16, 1978.
17. McDonough, R. N., "Maximum Entropy Spatial Processing of Array Data," Geophysics, vol. 39, pp. 661-675, Dec. 1974.
18. Hawkes, C. D., and S. Haykin, "Adaptive Digital Filter for Coherent MTI Radar," IEEE Int. Radar Conf. Washington, DC. Apr. 1976, pp. 57-62.
19. Haykin, S., and C. Hawkes, "Adaptive Digital Filtering for Coherent MTI Radar," Information Sciences, vol. 11, pp. 335-359, 1976.
20. Haykin, S., "Communication Systems," p. 476 (John Wiley and Sons, 1978).

21. Guillemin, E. A., "The Mathematics of Circuit Analysis," (John Wiley and Sons, 1949).
22. Shannon, C. E., and W. Weaver, "The Mathematical Theory of Communication," p. 90, (The University of Illinois Press, 1959).
23. Middleton, D., "Introduction to Statistical Communication Theory p. 306, (McGraw-Hill Book Company, 1960).
24. Koopmans, L. H., "The Spectral Analysis of Time Series," (Academic Press, 1974).
25. Akaike, H., "Fitting Autoregressive Models for Prediction," Ann. Inst. Stat. Math., vol. 21, pp. 243-247, 1969).
26. Gutowski, P. R., E. A. Robinson, and S. Treitel, "Spectral Estimation: Fact or Fiction," IEEE Trans. Geoscience Electronics, vol. GE-16, pp. 80-84, April 1978.
27. Barlow, E. J., "Doppler Radar", Proc. IRE, vol. 37, pp. 340-355, April 1949.
28. Fishbein, W., S. W. Graveline, and O. E. Rittenbach, "Clutter Attenuation Analysis," Tech. Rep. ECOM-2808, March 1967.

PREPROCESSING OF SEISMIC SIGNALS FOR PATTERN RECOGNITION*

John E. Brolley

University of California
Los Alamos Scientific Laboratory
Los Alamos, New Mexico 87545, USA

ABSTRACT. This study describes the preparation of seismic signals in order to identify the best technique for input to a pattern recognition scheme. The signal is first filtered with zero phase shift high and low pass Butterworth filters. It is then subjected to adaptive filtering and finally moving average filtering. Spectral decomposition in terms of circular functions is done via conventional Fourier and log P maximum entropy analysis. Spectral decomposition in terms of sequency functions, Walsh, and Chebyshev, is also performed. The Walsh decomposition is done with the conventional fast operator. The Chebyshev decomposition is done with an optimization procedure. Results, based on these various operations, are presented for an underground nuclear explosion and several earthquakes.

1. INTRODUCTION

A vast corpus of seismic data is recorded daily, and presumably a significant portion of it is visually displayed for scrutiny by trained analysts. A goal of some analysts is to classify the source of the signal. In the present context two sources are considered; underground nuclear explosions and natural events. Essentially all of the events in the latter category will be earthquakes. In view of the impending Complete Test Ban (CTB), it is desirable to enhance the accuracy of the analyst's decision, and better still, to make a computer-generated decision. It may be possible to almost

---

*Work performed under the auspices of the U. S. Department of Energy, under Contract No. W7405-ENG-36.

completely remove the human analyst from the decision process. This report is concerned with initial steps taken by the Digital Image and Signal Processing Group of the Los Alamos Scientific Laboratory towards the computer-based decision process.

Basically a characteristic feature of the seismogram is sought which tags the event unambiguously. In general, this may be a very difficult objective to achieve. However, it may be possible to supplement the seismogram tag with other information to minimize the error of decision.

Some obvious characteristics of a seismogram are time of arrival, duration, peak energy, total energy, and wave shape. The analysis to follow will be concerned mainly with the last two, and somewhat loosely with the others. All of the examples will be based on data recorded at the Albuquerque (ABQ) Seismic Research Observatory (SRO) of the United States Geological Survey (USGS). SRO stations are distributed around the world and represent the most modern facilities in the field [1]. The examples to be discussed will utilize the short period data, 20 samples per second, and consideration of the short period gain curve will be given.

2. DATA

Three sequences of short period data will be considered. In the plots to be presented the horizontal axis is time and the vertical axis is a measure of ground motion at ABQ. The calibration of an SRO is such that 2×10^6 digital counts equals 1 micron of ground displacement at 1 s period. All three sets include 20 s of data prior to the onset of the signal.

The seismogram of an underground nuclear explosion at the Nevada Test Site (NTS) is displayed in Fig. 1. The first 2048 points are plotted. The second seismogram was produced by an earthquake in Utah. The coordinates were lat. 40.51 N, long. 110.28 W. The origin time was 30 September 1977, 10:19:19.6 GMT. This was a normal earthquake with the nodal plane of the focal mechanism solution for the near shock striking NNW and dipping to the NE. The epicenters were confined to an area 5 x 2 km. The hypocenters were in the depth range 2-8 km with most activity in the range 4-8 km. The magnitude was $M_L = 5.1$ (National Earthquake Information Service, Golden, Colorado). Aftershocks continued for several weeks; one aftershock on 11 October 1977, 0756 GMT, reached a magnitude of $M_L = 4.7$. The Utah seismogram will be referred to as UTAH. The first 2048 points are displayed in Fig. 2.

The second earthquake occurred near the California-Mexico border. Its coordinates were lat. 32.47°N, long. 115.18 W, and the origin time was 17 January 1977, 11:13:19.4 GMT. This earthquake has not been subjected to detailed analysis yet. Preliminary indications are that its magnitude was $M_L = 4.2$, with a depth of 5 km. It will be referred to as CALMEX. The first 2048 points of its seismogram are displayed in Fig. 3.

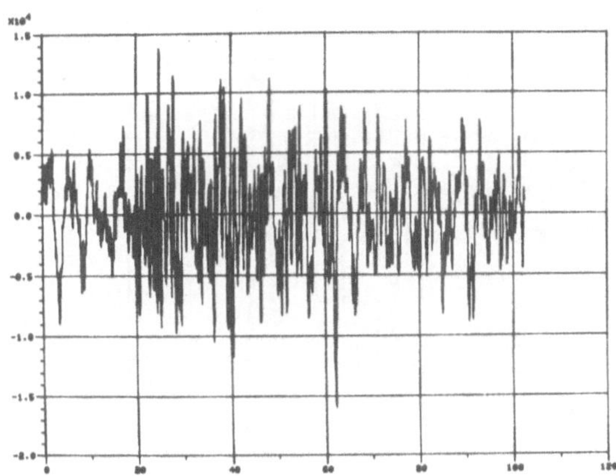

Fig. 1. First 2048 points of NEVADA data from ABQ/SRO.

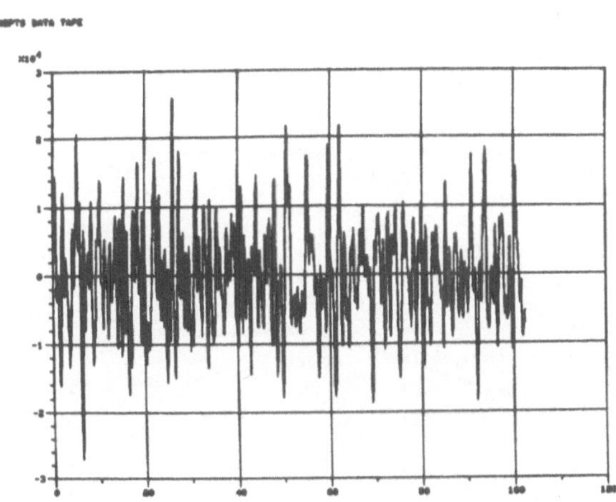

Fig. 2. First 2048 points of UTAH data from ABQ/SRO.

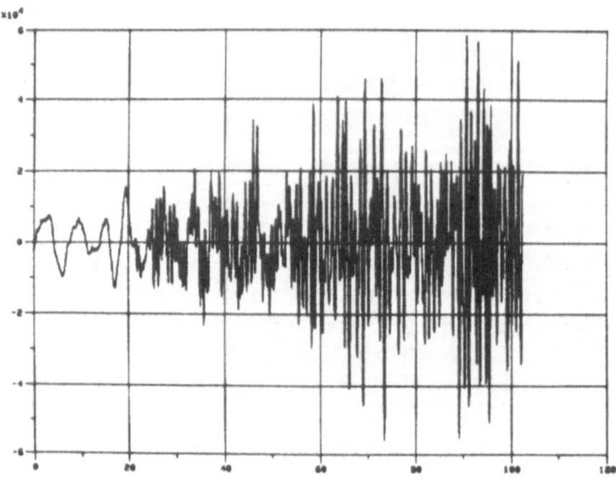

Fig. 3. First 2048 points of CALMEX data from ABQ/SRO.

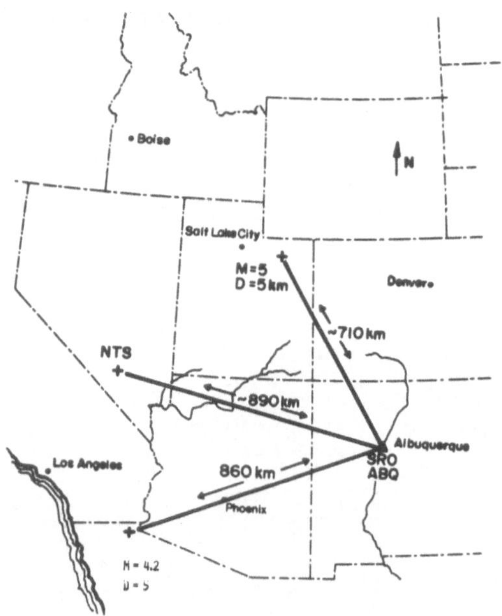

Fig. 4. Geographical disposition of sources and detector.

UTAH had the longest duration signal at ABQ/SRO. It lasted approximately four times longer than the other two.

The approximate positions of the three sources with respect to ABQSRO are shown in Fig. 4.

The three seismograms will now be used to illustrate various aspects of signal processing.

3. FILTERING

The simplest form of filtering is leveling in the amplitude domain. All three data sequences were scanned and the amplitudes proportionately reduced so that the maximum amplitude in any sequence was unity. This can also be viewed as a form of normalization. When all subsequent filtering operations have been performed, the total energy in the signal is computed. All amplitudes are then divided by the square root of the total energy. This is called total energy normalization [2]. It is not implied that this is the optimum normalization technique. The three sequences contain energy of frequencies that one may not wish to consider during a particular mode of analysis. Moreover, frequencies outside the band of interest may complicate the interpretation of frequencies of interest [3]. For these reasons high- and low-pass, three section Butterworth digital filters with zero phase shift have been constructed. These filters have smooth top and bottom roll-off with no ripple [3]. Bandpass filtering is available by operating the high- and low-pass filters in tandem. The result of bandpass filtering the NEVADA data is shown in Fig. 5. In this case the cutoff of the high-pass filter was 0.5 Hz, and the low pass, 5.0 Hz. It will be noted that the wide ranging excursions apparent in the first 20 s before onset have been removed.

It is possible that signals may come through the passband that has been used, which are not related to the seismic signal and may contribute some uncertainty in the analysis. In some cases this undesirable effect can be reduced. It is possible to construct a digital filter that is smart. If such a filter has the opportunity to observe, as it were, a signal, and if the signal is reasonably stationary, the filter can memorize the pattern and reject it. This is called adaptive filtering [4],[5]. The adaptive filters used in this work are nonrecursive. Recursive types will be studied shortly. Both types are suitable for real-time operation. The amount of training data in the three sequences is only 20 s long. This can be extended in future work. The 20 s period may not be long enough for the filter to generate as much reduction as possible, i.e., the training period may be too short. At the onset of the desired signal the filter configuration is frozen. The application of a 20 element filter, whose adaptive rate [5] was 0.04, to NEVADA is shown in Fig. 6. It is clear by comparison with Fig. 1 that in the short training interval the filter is

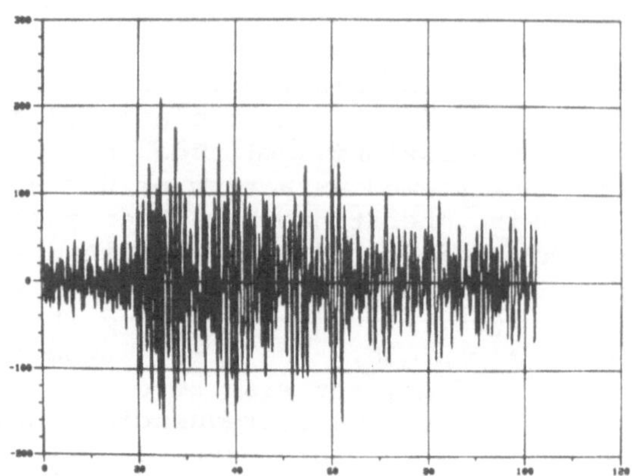

Fig. 5. Bandpass filtered NEVADA data.

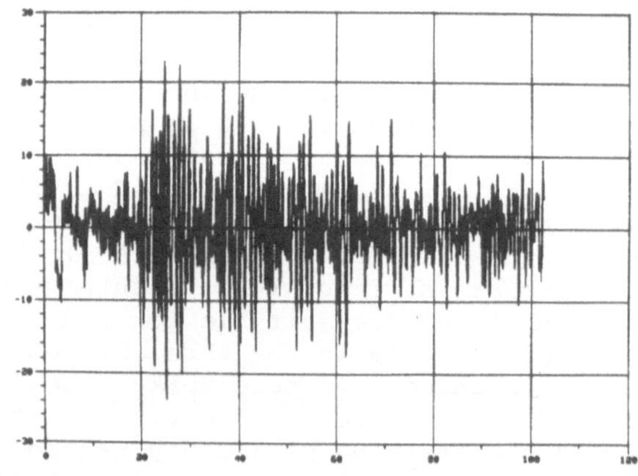

Fig. 6. Adaptive filtering of NEVADA data.

learning to remove some low frequency components.* In-depth studies of this type of filtering are planned. The result of operating on NEVADA with all the filters discussed so far is shown in Fig. 7.

Moving average filters is sometimes used advantageously after adaptive filtering [5]. The result of operating on the sequence shown in Fig. 7 with a moving average filter of length 5 is shown in Fig. 8. In what follows, CALMEX and UTAH were processed in precisely the same manner as NEVADA.

4. WAVE SHAPE

The description of a time series or wave in terms of an expansion utilizing circular functions has a hoary tradition and is ardently practiced by most students of the seismogram. However, for a given number of terms, it does not provide the most faithful description. The Karhuenen-Loeve (KL) (sometimes called KARLOV) achieves the best fit [6]. This achievement comes at the expense of great computing labor.

However, it has been conjectured that the most accurate representation may not provide the best differentiating capability [7]. It appears that there is an analytical basis for this conjecture [6]. In the present discussion, a pragmatic approach will be indicated, although a semi-rigorous investigation may be made later.

It will be assumed that the seismic signal can be built up by assembling various types of building blocks. For the present three types of blocks will be considered; circular functions, Walsh functions, and Chebyshev functions. Others could be added to the list. There are various ways of stacking these blocks to build seismic signals. Three that will be considered here are the fast projection operator, e.g., Fast Fourier Transform (FFT), log P maximum entropy, and optimization theory. Other approaches exist [8] and may be applied. In the following examples the appropriate operator acted on 64 consecutive data points and generated a power spectrum. Successive spectra were produced by displacing the operator 16 points in the direction of increasing time.

In the case of circular functions, two types of operators were employed. The first was the conventional FFT with a Hamming window encompassing the 64 data points. Thus, a three-dimensional plot of Fourier power can be generated. In what will follow, these surfaces have been smoothed somewhat with two-dimensional twice Tukey filtering. A prescription for this is sketched in Appendix B. For the present, only power in the higher frequencies will be indicated. In order to more clearly visualize this distribution, power corresponding to Fourier function indices 1 to 6 have been

---

* See Appendix A for an additional example.

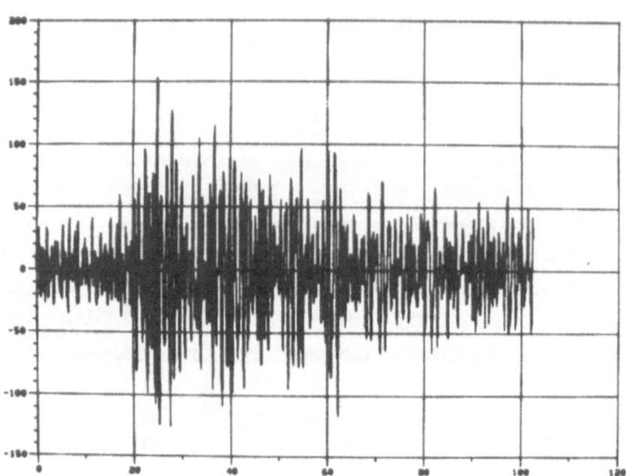

Fig. 7. Bandpass plus adaptive filtering of NEVADA data.

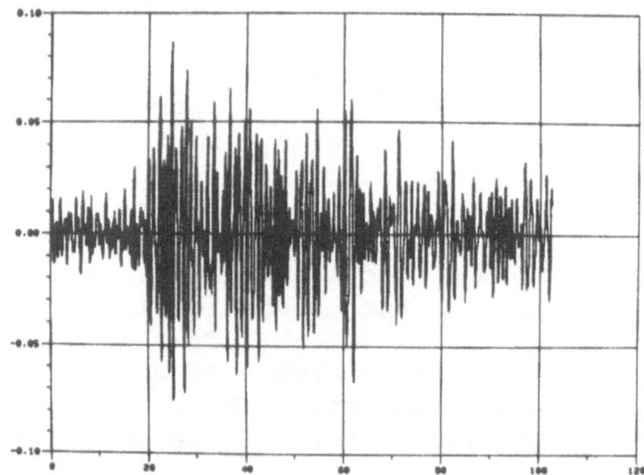

Fig. 8. Bandpass plus adaptive plus moving average filtering of NEVADA data.

set to zero. In addition, two-dimensional twice Tukey filtering, as done here, removes four border points all the way around the plot, a matter of no consequence for the present discussion. In Fig. 9 the Fourier power surface for the first 2048 points of NEVADA is displayed. The right hand corner is the origin of time and frequency. Time flows upward, and frequency flows to the left. The frequency index at which power is first plotted is 7. In the usual ordering, the index 1 corresponds to DC, 2 to one cycle, etc. Since each sequence is 64 points (63 x .05 = 3.15 sec) the first plotted frequency is 6/3.15=1.9 Hz, and the highest frequency shown on the plot is 4.44 Hz. The use of synthetic signals in the code is useful for checking performance. Thus, a damped sine wave starting at point 100 and running out to 2048 points is Fourier analyzed and is shown in Fig. 10. The damped sine had a frequency of 3.492 Hz, and therefore, should correspond to a function index of 12. The two-dimensional twice Tukey filter has removed the first 4 indices and the peak of the power occurs at 8 on the plot as it should. A wide range of synthetic signals is available in the code for studying various modes of analysis.

Burg's log P maximum entropy analysis was applied next to NEVADA. There appears in this operator a filter length parameter which will require some study. Richer [9] has given a prescription for evaluating it but there does not seem to be unanimity of views on this point [10]. With a filter length of 20 the result shown in Fig. 11 was obtained from NEVADA. The maximum entropy operator was not windowed.

There are many noncircular functions available for building blocks, and two examples will be considered here. Walsh functions constitute a complete orthonormal set having amplitudes +1, -1, and are, therefore, computationally relatively simple to deal with. A comparison of the first few Walsh functions with circular functions is given in Fig. 12. Corresponding to the concept of frequency as applied to circular functions, there exists the analogous concept of sequency for Walsh functions. In this case, sequency is defined as one-half the number of zero crossings over the time base. The first zero crossing at the beginning of the time base is not counted. Analogous to the FFT operator, there exists a fast operator for Walsh functions, FWT. This has been applied in manner similar to the Fourier case except that the data window was a triangle. The result of acting on the NEVADA data with the Walsh operator is shown in Fig. 13. Walsh analysis indicates more high-sequency power than Fourier analysis [7], a feature which may be useful in the pattern recognition problem.

Another set of building blocks is the orthonormal set of Chebyshev polynomials defined over the range of argument - 1 to + 1. A plot of the first five is shown in Fig. 14. The concept of sequency may also be used here. For this discussion Chebyshev sequency will be defined as the number of zero crossings over the argument range -1 to +1. Perhaps it should be called Chebquency. The operator used here for evaluating Chebyshev sequency power

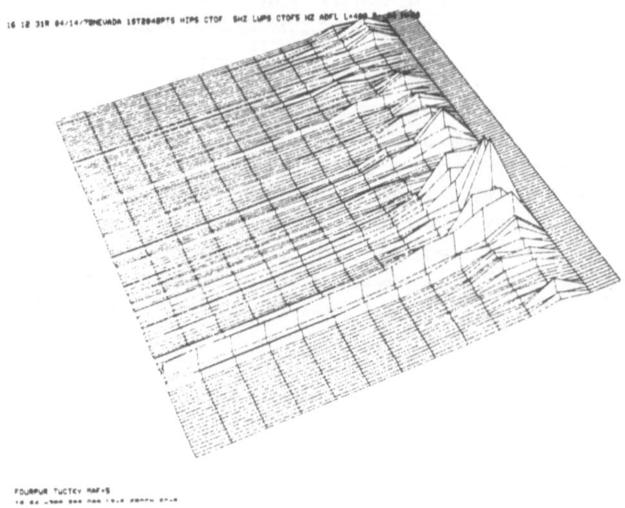

Fig. 9. Fourier power spectrum of NEVADA.

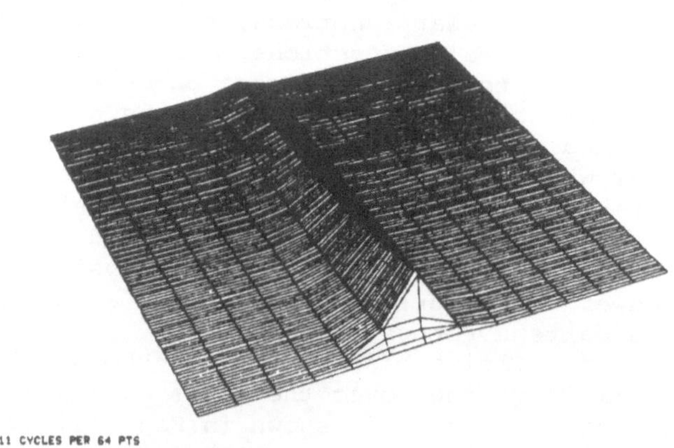

Fig. 10. Fourier power spectrum of a damped sine wave.

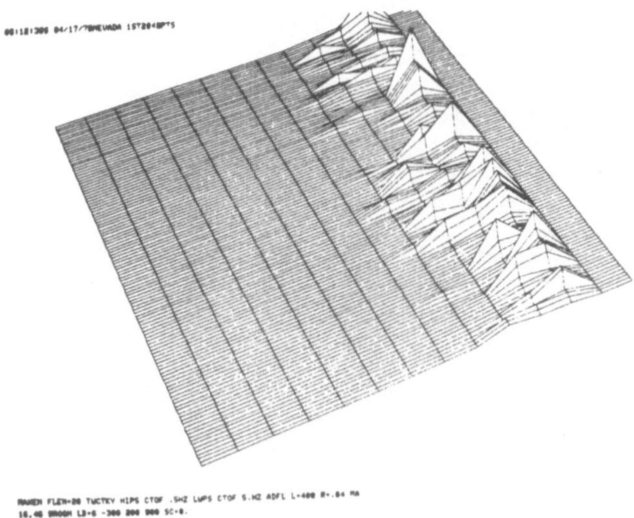

Fig. 11. Log P maximum entropy power spectrum of NEVADA.

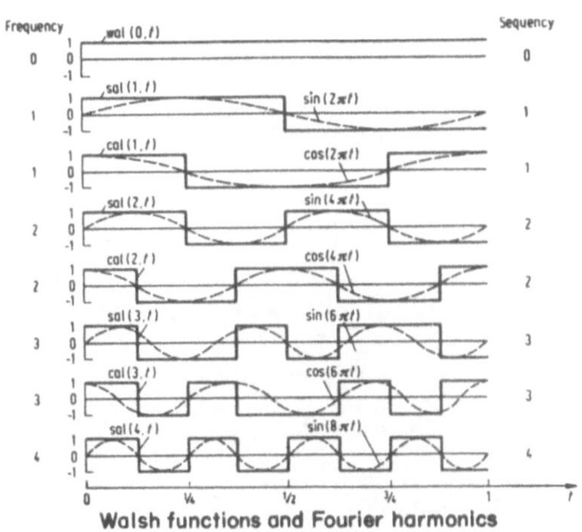

Fig. 12. Comparison of some Walsh and circular functions.

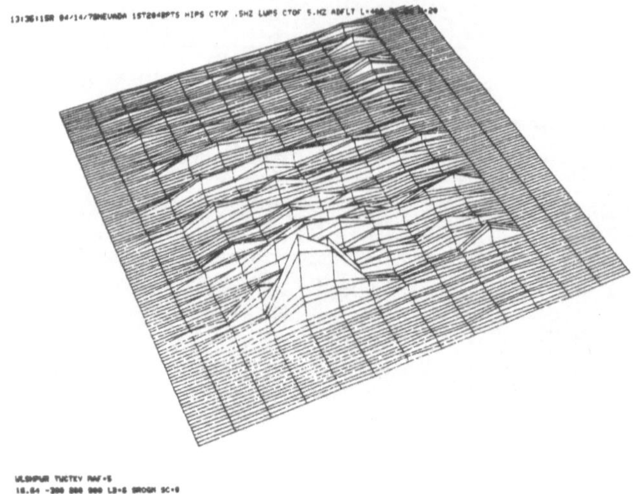

Fig. 13. Walsh power spectrum of NEVADA.

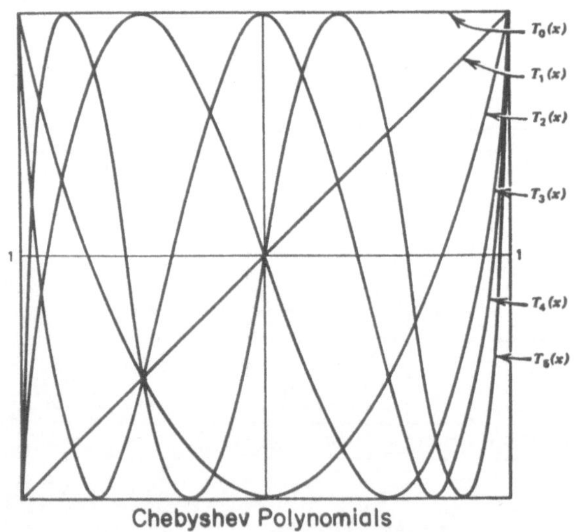

Fig. 14. First 5 Chebyshev polynomials.

differs from the fast operators previously alluded to. Basically, it performs an optimized fit of a given set of polynomials to a sequence of arbitrary length. For purposes of comparison with the other methods, 32 polynomials are used over a time base of 64 points. No windowing has been done. The result of acting on NEVADA with the Chebyshev operator is shown in Fig. 15. This plot and those that follow differ from the previous set in that only the DC term has been equated to zero. Also the vertical scaling factor is no longer automatic, but is fixed at a particular value. As in the other examples the sharp rise of high-sequency power occurs at the beginning of the seismic signal and then dies out.

It is interesting to compare this plot with the earthquake CALMEX (Fig. 2). Identical operations on the earthquake signal produced the result shown in Fig. 16. The plotting can be done quantitatively by passing planes through the peaks of power, i.e., contour plotting. The corresponding contour plot of NEVADA for three levels of power is shown in Fig. 17, and for CALMEX in Fig. 18. The same contour plot for UTAH (Fig. 3) is shown in Fig. 19.

5. CONCLUSIONS

Adaptive filtering, zero phase shift Butterworth filtering, moving average, and twice Tukey filtering have been incorporated in a seismic signal analysis code. Four modes of spectral decomposition are available for study: Fourier, Walsh, log P maximum entropy, and Chebyshev. Different power signatures have been noted for this set which may be augmented. A set of training data is on hand. A study can then be made to ascertain which combination of spectral decomposition and pattern recognition provides the best differentiating capability between seismic signals from natural events and underground nuclear explosions.

6. ACKNOWLEDGMENTS

I acknowledge, with pleasure, the vital and generous help received from four branches of the USGS. J. Peterson and J. Hoffman of ABQ/SRO have spent much time explaining SRO and providing all the data used in this paper. C. Langer of the Denver office provided the information on UTAH. G. Fuis of the Pasadena office provided information on CALMEX. J. F. Evernden, National Earthquake Research Center, has patiently explained some of the intricacies of geophysics and spectral problems.

N. Ahmed of Kansas State University, G. Elliot, S. Sterns, and W. Jacklin of Sandia Laboratory have given valuable guidance in various aspects of signal processing.

At Los Alamos, G. Wecksung guided the two-dimensional twice Tukey filtering, and J. House, R. Saunders, and N. Ginther helped

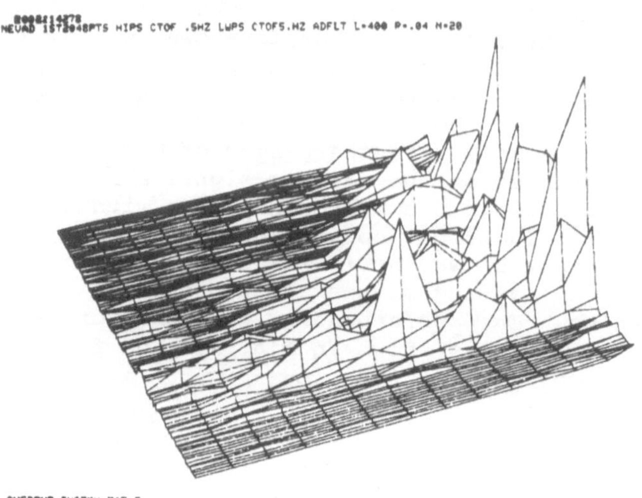

Fig. 15. Three-dimensional Chebyshev power spectrum of NEVADA.

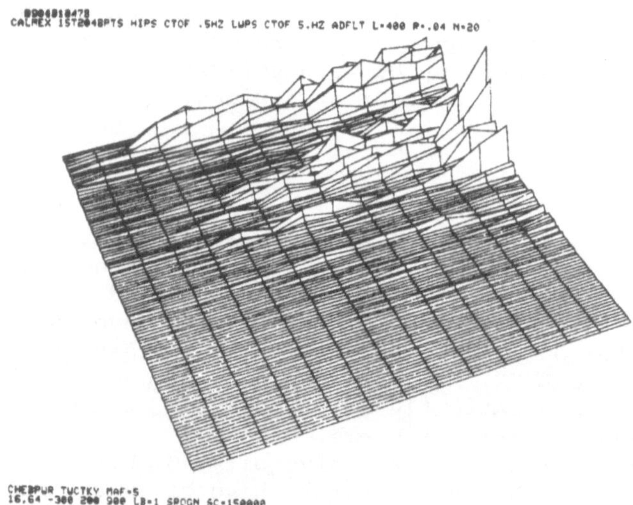

Fig. 16. Three-dimensional Chebyshev power spectrum of CALMEX.

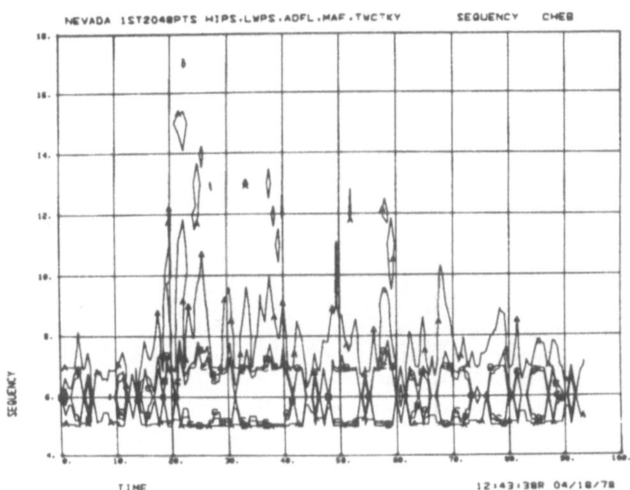

Fig. 17. Three-level contour plot of Chebyshev power spectrum of NEVADA.

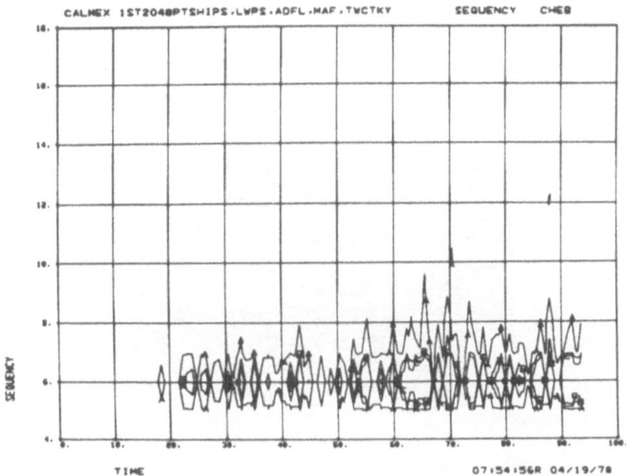

Fig. 18. Three-level contour plot of Chebyshev power spectrum of CALMEX.

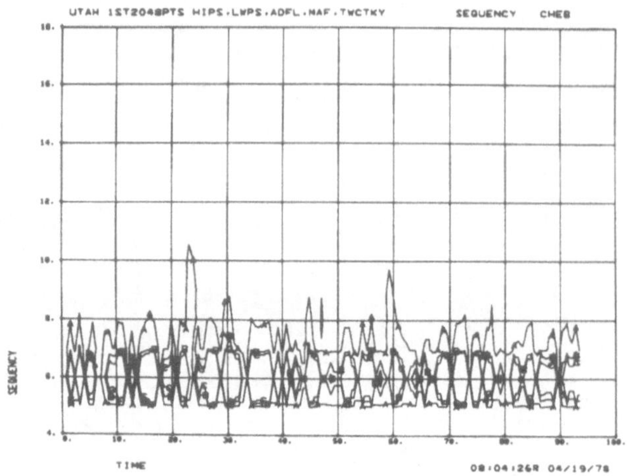

Fig. 19. Three-level contour plot of Chebyshev power spectrum of UTAH.

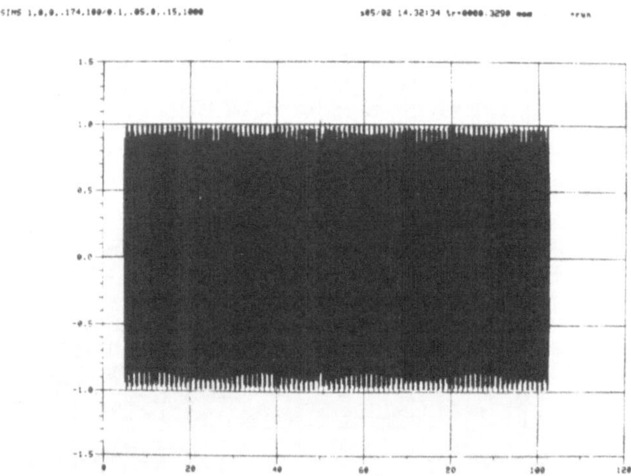

Fig. 20. Synthetic data: continuous sinewave plus damped sinewave.

in data acquisition. M. Cannon and R. Frank translated data tapes to local system files.

This work was made possible by the support and encouragement of W. E. Deal and D. H. Janney.

APPENDIX A

Example of Adaptive Filtering

The synthetic signal feature of the seismic code can be used to illustrate rather clearly the action of the adaptive filter. Figure 20 displays a synthetic composite data sequence 2048 points long. One sine wave of amplitude 1, frequency 3.48 Hz, starts at point 100 and runs without damping. A second sine wave starts at point 1000 with amplitude 0.1 and frequency 3.0 Hz, and is strongly damped. The filter was instructed to stop learning at point 995, and freeze its configuration after that. Its adaptive rate was 0.01. The results are shown in Fig. 21. Some appreciation of the signal to noise gain can be seen in the blown up plot of Fig. 22.

APPENDIX B

A Prescription for Two-Dimensional Twice Tukey Filtering

Given an n x m Matrix X.

Form a 5-point sampling cross

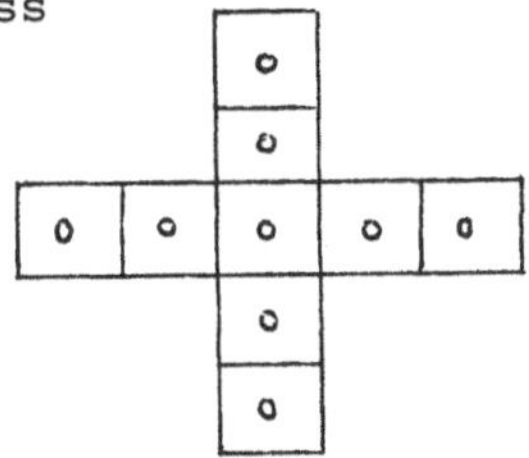

to operate on X. This operator starts for convenience in the upper left-hand corner of X such that the uppermost of the five vertical point samples the values in row one and the left-most in the horizontal set of five samples, the points in the first column. Nine values are obtained. The median value is extracted and loaded in the new matrix Y at $Y_{1,1}$. The operator is then stepped across X column by column until the right-most cell of the cross samples the last column, thereby filling up the first row of Y. The operator is displaced one row down and the operation repeated, etc. Y will then be (n-4)x(m-4). Form X-Y with Y placed symmetrically in X and the overlap thrown away. Repeat the cross operation on X-Y to form S. The final result is Y+S where overlap has been discarded. S is now (n-8)x(m-8). The reader may wish to consult Ref. 11.

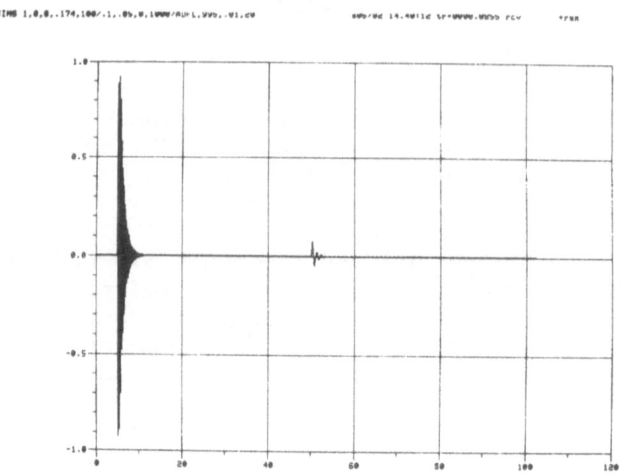

Fig. 21. Adaptive filter output for the input shown in Fig. 20.

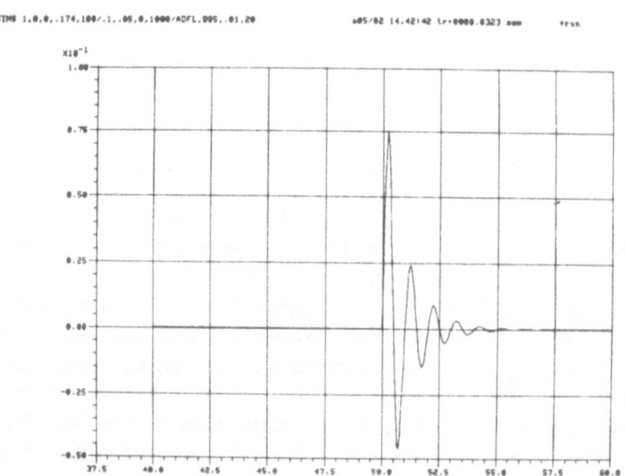

Fig. 22. Expanded view of the filter output around the damped wave.

REFERENCES

1. Peterson, J., H. M. Butler, L. G. Holcomb, and G. R. Hutt, "The Seismic Research Observatory," Bull. Seism. Soc. Amer. 66, 2049 (1976).
2. Youngblood, W. R., "Seismic Discrimination by Harmonic Analysis Techniques," AFIT-EN GE/EE/72-29.
3. Stearns, S. D., Digital Signal Analysis, Hayden, Rochelle Park, NY, 1975.
4. Elliott, G. R., "Improving the Performance of Perimeter Security Sensors Through Digital Signal Processing," Proc. Digital Processing Symposium, Sandia Lab. Albuquerque, NM, SAND 77-1845C, 1977.
5. Ahmed, N., "On Intrusion-Detection via Adaptive Filtering," SAND 77-1845C, 1977.
6. Pavlidis, T., Structural Pattern Recognition, Springer-Verlag, Berlin, 1977.
7. Brolley, J. E., "Preliminary Code Development for Seismic Signal Analysis Related to Test Ban Treaty Questions," SAND 77-1845C, 1977.
8. Brolley, J. E., R. B. Lazarus, B. R. Suydam, and H. J. Trussell, "Maximum Entropy Analysis for Some One- and Two-Dimensional Problems," SAND 77-1845C, 1977.
9. Richer, H. B., and T. J. Ulrych, "High Frequency Optical Variables," Ap. J. 192, 719 (1974).
10. Radoski, H. B., P. F. Fougere, and E. J. Zowalick, "A Comparison of Power Spectral Estimates and Applications of Maximum Entropy Method," J. Geophys. Resch., 80, 619 (1975).
11. Hamming, R. W., Digital Filters, Prentice-Hall, Englewood Cliffs, NJ, 1977.

BIOMEDICAL IMAGE PROCESSING*

A. Oosterlinck**, L. Vanderheydt, J. Van Daele***,
H. Van Den Berghe

Division of Human Genetics, University of Leuven
Minderbroedersstraat 12
B-3000 Leuven-Belgium

ABSTRACT. Image processing exists now as a separated discipline in computer science and electrical engineering. The development of computer techniques for picture analysis began 25 years ago. Most of the research in this field is problem-oriented. Some of the major applications fields are robot vision, high energy physics, remote sensing, visual inspection, etc... A very broad field, even so in number of applications as in the number of researchers, is the biomedical image processing.

The different sub-areas in the biomedicine with a high number of computerized image analysis are: cytology, chromosone analysis, computerized tomography (CT), radiology (with X-ray and radioisotopic picture), cancer, etc... The two drastic innovations in the mid 1970's with a high range of manpower saving and a substantial increase of accuracy are the white blood cells differentiation and the CT-scan.

PART I. IMAGE PROCESSING

1. Definitions

In a broad sense, picture processing is the manipulation of multivariate, multi-dimensional signals, with the purpose to transform the signals or extract a description. The goal of picture or scene analysis is given by Rosenfeld and Weszka [7] as: the extraction of a description from the given picture. This description may

---

* Supported by FGWO grant.
** Qualified research man National Science Foundation Belgium.
*** Bursaal I.W.O.N.L.

consist of a set of numerical data (feature measurements) or it may be some of data structure which represents relationships among significant parts (segments) of the picture, as well as properties of these parts.

Depending on the applications and goals of the picture processing and analysis operation, several topics are covered, such as:

- Analysis and design of digital image processing systems
- Picture digitization and coding
- Picture enhancement, noise cleaning and restoration
- Picture preprocessing and data reduction
- Picture segmentation, description, pattern recognition and analysis.

2. Picture processing systems

Back in the early 1960's, many people were surprised by the sharpness and high quality of the television picture sent back from the surface of the moon taken by the NASA unmanned rangers and lunar landers. Surprise was turned over in unbelief by the spectacular comparison with the original (not processed) TV picture from the moon, published by the newspapers.

The original incoming pictures were computerized enhanced and corrected for many TV camera distortion, measured for launching of the spacecraft.

This was essentially the beginning of the applications on a big scale of digital image processing techniques. But it was not the first effort in picture processing. As described by Walton [3] in 1952 a committee of the National Coal Board of Great Britain chaired by Dr. J. Bronowski was started in 1951 to investigate: "the possibility of making a machine to replace the human observer ...". Several patents were generated. Another of the earliest published papers on picture processing is by Kovasznay et al. who introduced in 1954 the use of two-dimensional derivative operators for edge detection and deblurring of a picture [9].

Most of the image processing systems are designed for one or a small set of related applications. Only in some research institutes general processor facilities are available. The systems range from very specialized hardware units which can perform only strictly limited functions to general input systems which can input images to the big scale computers on the computing center.

One of the factors that has greatly affected the design of digital image processing systems has been the enormous amount of data that are present in even a simple image. A full color image with a spatial resolution of 512 x 512 picture elements (e.g. an image from a conventional CRT monitor) contains 768 K bytes (± 6 million bits). For a black and white picture still 256 k bytes are displayed on the monitor every 40 msec. In real time application the entire image is processed and refreshed at 30 Hz (American standards), the required processing speed is in excess of 250 MHz. The extremely high data volume is an inherent requirement in image analysis. This effects drastically the technology and especially in the field of storage devices, data communication, parallel processing, etc...

3. Picture digitization and coding

Webster Dictionary gives a definition of the word image as: "a reproduction or imitation of the form of a person or thing". It is clear that an image requires an object. This object is illuminated by a source of radiant energy or is by itself a source. The radiant energy reflected, transmitted or emitted by the object is propagated through the space, intercepted by the image formation system and transformed in such a way that in the coordinate system (x,y) (referred as image plane) an image is formed.

Assume f(x,y) a two dimensional function which represents a chosen attribute (such as radiant energy, color, gray-scale, etc..). Before a continuous function (of position) can be processed by digital computer it needs to be digitized. This consists of sampling and quantization. The aim of sampling is to represent a continuous picture by a finite string or array of numbers, with the constraint that it should be possible to reconstruct the picture given the samples. Two major techniques, namely, using a sampling lattice and a best approximation using an orthonormal base are described by Rosenfeld [8]. Because of the limited wordlength and finite accuracy of digital computers the picture samples must be quantized. This is a dividing of the total range of the sample values into intervals and all the values within an interval must be represented by a single level.

In order that the picture reconstructed from the quantized samples will be acceptable, the number of quantization levels can be quite high, on the other hand for optimal storage efficiency this number must be low. For an optimal choice of this number, see [8]. A matrix representation of an image is given by:

$$[F] = S\,Q\,f(x,y)$$

where SQ is a non-linear operator, representing sampling and quantization. The dimensions of matrix [F] is the number of sample points N x M.

Given the sample distances in x and y direction as Δx and Δy, the i,j element of the matrix is computed by the relation:

$$f_{ij} = Q\,f(i\Delta x,\, j\Delta y) = Q \sum_{i=1}^{\infty} \sum_{j=1}^{\infty} g(x,y)\, \partial(x-i\Delta x, y-j\Delta y)$$

with $\partial(.)$ the two dimensional Dirac-delta function. Q is defined by the number of quantization levels n, the length of these levels K and the distribution of these levels (uniform, log, etc.).

Example of the Q relation:

$$Q(f_{ij}) = a \quad \text{if} \quad a\,K < f_{ij} \leq (a+1)K$$

The optimal coding question now arises as follows: For a given total number of bits, how should one choose the values for N,M and n in order that the reconstruction error is minimum? (average or in any part of the picture). This problem is handled by Rosenfeld [8].

4. Picture enhancement, noise cleaning and restoration

The aim of analyzing picture is to collect meaningful information about a phenomenon of interest in the picture. Whenever a picture is converted or transformed from one form to another, the quality (useful information context) of the output picture may be degraded. Enhancement is quality improvement of low quality pictures. Rosenfeld states that: "Many enhancement techniques are designed to compensate for the effect of a specific (known or estimated) degradation process. This approach is generally known as image restoration [8].

Image enhancement includes methods of modifying the gray-scale (for contrast enhancement), deblurring, smoothing or removing noise (noise cleaning is a special enhancement technique) and correcting geometrical distortion.

The complicated factor caused by the restoration techniques is the random noise. It is the sole obstacle to perform perfect restoration of the picture. By this noise, the image restoration is ill conditioned, as such it lacks a unique solution. A great variety of methods are available, for a discussion in depth. See [8].

5. Picture processing and data reduction

Rosenfelds' definition of preprocessing: "It involves the transformation of the given picture into modified form designed to facilitate subsequent analysis." [7]. Standard preprocessing techniques include: data reduction (e.g. the background of a picture is only mostly important for edge detection and teaching and can be omitted after these operations), enhancement, filtering, etc. In a more strict sense preprocessing techniques are: The preliminary operation that has to be performed before property measurements can be applied. (e.g. gray-scale normalization, geometrical normalization, etc...)

6. Segmentation and scene analysis

By scene analysis the desired output is a description of the given picture. These descriptions refer mostly to specific parts (regions or objects or parts of objects). To generate a useful description, it is necessary to segment the picture into these parts (segments):

A segment is a part of the total picture and poses two requirements:

(1) Inside the segment (or part) one or more attributes, e.g. gray-scale, gradient, color, texture, etc... are homogeneous over all the segment pixels, in other words the proposition is true, e.g. take the same quantitative value.
(2) All the pixels belonging to one segment are space connected (topological property)

Given the picture with S = the set of all pixels. S_r = segment with index "r", and m = number of segments in the picture.

<u>Definition</u>: A segmentation is the determination of a set s of subsets S_r ; $r \varepsilon \underline{m}$ of S with subjoined listed conditions:

(1) $Us = S$ or $S = S_r | r \varepsilon m$

(2) $(\forall k)(\forall l)[k\varepsilon\underline{m} \wedge l\varepsilon\underline{m}) = S_k \wedge S_l = \emptyset]$

(3) $(\forall k)[k\varepsilon\underline{m} = (l)(l\varepsilon\underline{m} \wedge c(S_k, S_l)]^*$

(4) $(\forall k)[k\varepsilon\underline{m} = p(S_k)]^{**}$

(5) $(\forall k)(\forall l)[k\varepsilon\underline{m} \wedge l\varepsilon\underline{m} \wedge p\neq k = \sim p(S_k U S_l)]^{***}$

* C = proposition, S_k and S_l are connected
** p = proposition is true
*** ~p = proposition p is not true

There are two types of connectedness depending on the manner the neighbourhood is formed by 4 or 8 adjacent pixels (See [7]).

The two main segmentation methods are based on region analysis and edge detection.

<u>Region analysis</u>

Thresholding a picture is dividing it into two segments:

$$S = S_o + U S_k \text{ with } k\varepsilon\underline{m}$$

with S_o = background and US_k all the object (m in total)

$$S_o = \{(i,j) \mid F_{ij} < T_{ij}\}$$

$$US_k = \{(i,j) \mid F_{ij} \geq T_{ij}\}$$

with F_{ij} the chosen attributes in point (i,j) and $p(F_{ij})$ =

$F_{ij} \geq T_{ij}$.

$$S_k = \{(i,j),(r,s) \mid (F_{ij} \geq T_{ij}) \wedge (F_{rs} \geq T_{rs}) \wedge [(i,j) \text{ mutually connected } (r,s)]\}.$$

Every segment S_k is divided into "Border elements" B_k and inner-elements I_k.

$$S_k = B_k \cup I_k$$

This last segmentation can be done by means of a border-follower, detecting all border elements and list them together to a closed contour. This contour represents the total object.

Edge Detection

The total set S (picture) is split in a set of edge points and non-edge points E_o:

$$S = E \cup E_o$$

Example: $E = \{(i,j) \mid D(F_{ij}) \geq D_o\}$ and $E_o = S - E$

D = derivative operator, D_o associated threshold (global or local).

The next step is a tracking of all edge points to form the border between isolated and surrounded background. For more details on algorithms and other segmentation see [8].

PART II. COMPUTERIZED KARYOTYPING OF HUMAN CHROMOSOMES USING A STATISTICAL CLASSIFIER

1. Introduction

In the past five years a system for picture analysis has been built in the Department of Human Genetics of the University of Leuven. With the general interest in biomedical and even in industrial oriented image analysis, the system was designed to handle pictures in a more general way. However, some parts of the system were specifically designed to enable automatic karyotyping.

Because an operational system still requires the presence of the technologist, an accurate, fast and flexible man-computer interaction is a necessity.

A general system structure is shown in Fig. 1, the system configuration for the Leuven system is given in Fig. 2. This can be split up in four main parts:

- the input system
- the metaphase finder
- the picture computer
- the visual output devices + interaction units

2. System organization

2.1 The input system

Several input devices can be used: a drum-scanner, a TV camera, a flying spot scanner.

2.1.1 TV-scanner

When fast access and direct microscope scanning is required, only TV-scanners offer the most realistic possibilities. In the present-day system a high quality commercial TV-camera with selectable line rates has been used. The camera can be equipped with a VIDECON or PLUMBICON tube. Although this scanning device is very cheap and simple, the image quality seems to be much lower compared to that of scanned photo-negatives.

The main reason of this trouble is caused by the large amount of noise in the video signal when scanning chromosomes with very little contrast (specially for Giemsa preparations). For this reason the scanning of photonegatives is more advantageous because of the contrast enhancement technique in the photographic step (long exposure or light integration time.). Several adaptations can be performed to improve the quality of the TV-scanner such as slow scan operations, other sensor tubes, etc. A detailed study on scanners is given in ref. [14]. With the TV-camera connected to the microscope, the sampling frequencies should be chosen according to the resolution capabilities of the microscope. Corresponding to the Shannon sampling theorem for bandlimited functions [4] the maximum sampling distance is 0.125 μm (the resolution of the microscope with an objective 100x, NA = 1.3 is equal to 0.25 μm). In conjunction with the 625-line rate of the TV-camera, a video sampling frequency of 13.6 MHz is taken, which corresponds to 872 pixels on a videoline. Because of the limited throughput of the computer, a video compression is performed to adapt the TV-camera speed to the computer throughput. The grey-value of each pixel is quantized in 128 grey-levels, using a log or linear mode.

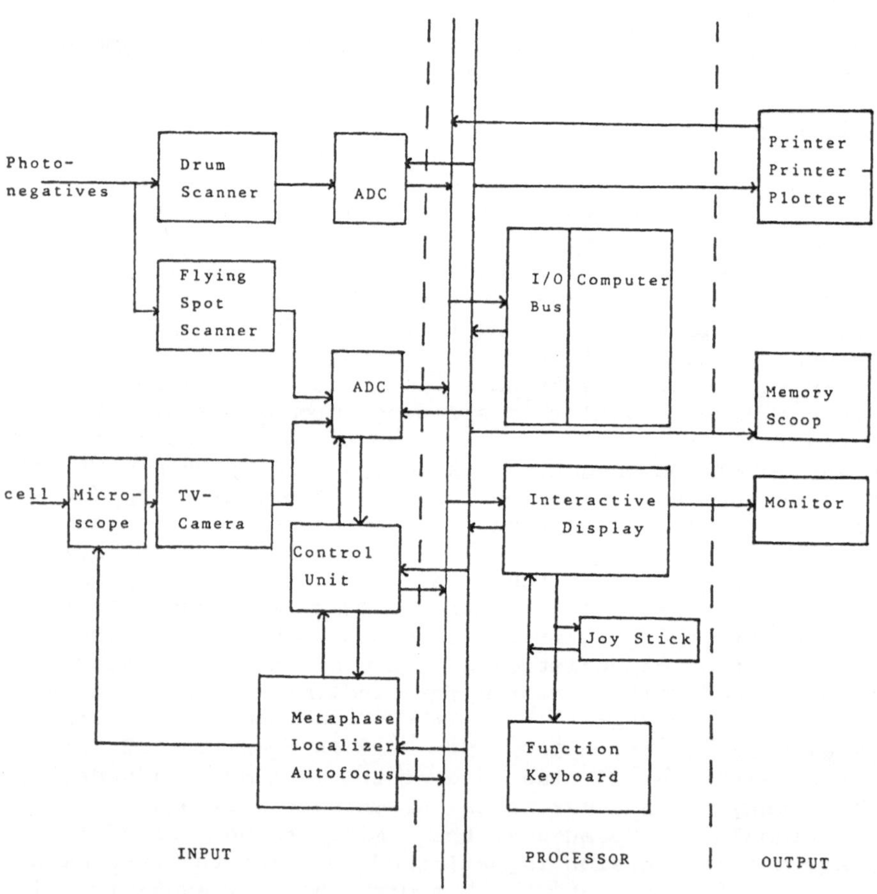

Fig. 1 General picture processing system.

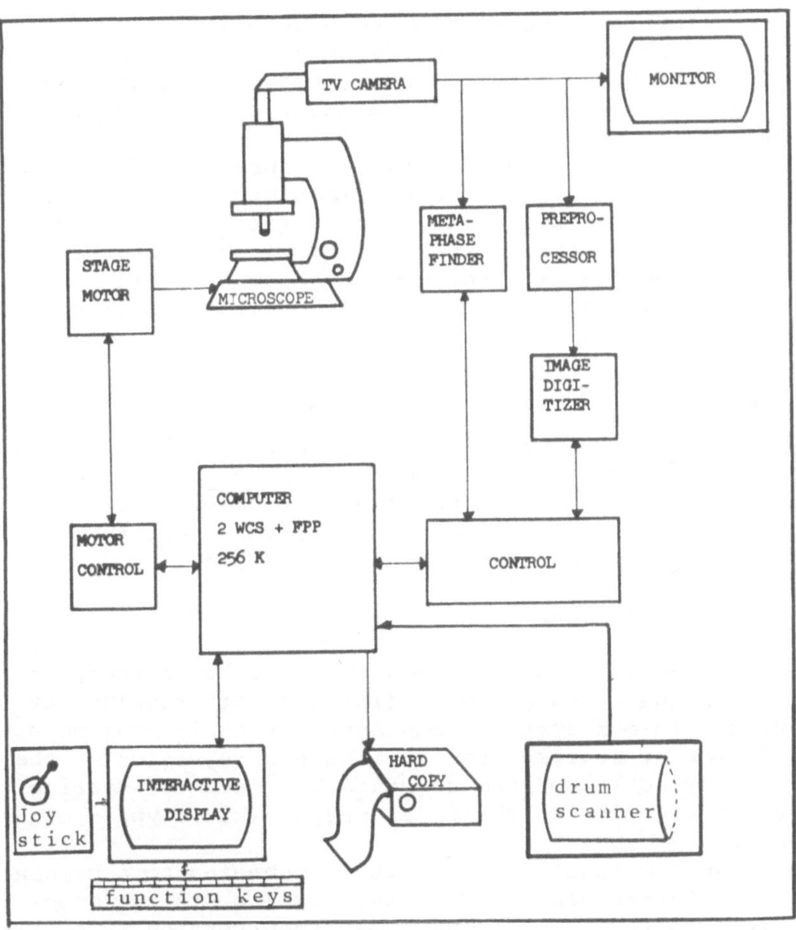

Fig. 2 System configuration.

2.1.2 Drumscanner

This input device can only scan pictures recorded on film. (With a special adaptation also pictures on photo's). The system has a very high signal-to-noise ratio and position accuracy, but the pixels are not randomly accessible and the acquisition is slow.

2.1.3 Flying spot scanner

Used also to scan photonegatives or translucent object. The picture scanning is random, the integration time is adaptable and the space accuracy is high.

2.2 Metaphase finder

The goal of the metaphase finder is to scan automatically a microscope slide, locate and select (classify) a given number of suitable metaphases in the shortest time and take a photo micrograph (if used in conjunction with artisanal method) or file the centerpoint coordinates.

The microscope slide contains metaphase spreads (of different quality) in addition to early prophases, late prophases, small lymphocytes, large transformed cells, debris artifacts,... The system must be capable to change the magnification, center and focus the metaphases and classify these metaphases according to the usability in chromosome classification. A detailed description is given by K. Castleman [13].

2.3 Image computer

A very fast medium size computer is available with a memory capacity of 512 K bytes. Special designed hardware preprocessors are connected to the computer. These parallel processors extract global information in one TV scan time (40 msec) and store this in the computer. For examples see Van Daele et al [15].

2.4 Output systems - interaction modules

In picture processing it is desirable to be able to correct, to interact and to evaluate the pictures (intermediate results) at different levels. In our system human interaction is made possible at three different levels: the scoring and rejection of the metaphases, the correction of the segmentation (contour tracing) and finally the correction of the final output (karyotype - classification result).

For the metaphase finder a TV monitor connected to a camera is used. For the intermediate processing level a storage scope (Tek. 611)* displays the image. The human intervention is possible

* Tektronix, Portland ORE, USA.

through an interactive graphic (pseudo color) display (Ramtek)*. The Ramtek display is a digital storage system which generates a picture with 32 gray-levels on a conventional TV monitor. The resolution is 320 x 256. Also included in the system are: a keyboard, a cursor with controller and a video-look-up table option which permits pseudogray or pseudocolor generation. A hardcopy, customary to the clinician, is produced by photoprinting in a TV-screen.

3. Image processings software

3.1 Minicomputer programming consideration

Because of the 16 bit wordlength and the restricted addressing modes, the total program size is limited to 61 K bytes. The remaining part of the memory is used as a fast bulk memory. For research purpose the different operations (e.g. segmentation, restoration, etc...) are all programmed in divided subroutines and the results of these operations are viewed on the monitor.

AKW	Acquisition monitor
PPM	Preprocessing module
CNTR1	Global localisation
CNTR2	Detailed localisation
IACTC	Interaction upon contours
CNTR3	Feature extraction
CNTR4	Classification
IACTK	Interaction upon karyotype
CNTR5	Manual assignment of classes
CNTR6	Split and merging
CNTR7	Management of statistical file.

<u>AKW</u>: <u>Acquisition monitor</u>

The first part tests the hardware availability of the specified unit (microscope stage, TV interface, drumscanner), initiates the desired shading compensation and assigns the acquisition parameters (stepsize, files, etc...).

The second part transfers the data to magnetic disk, or bulk memory.

<u>PPM</u>: <u>Preprocessor module</u>

The picture is degraded by the limited frequency-response of the input system. This can be partially removed with inverse filtering (enhancing the bands and edge sharpening). Different filter options are available and the selection of the best one can be judged on the visual display.

*Ramtek Corp., Synnyvale, Cal. USA.

CNTR1: Global localisation

The total image is originally scanned with a resolution of 512 x 512 pixels, taking 250 K bytes storage memory. By an averaging of neighbourhoods of equal size a second picture is formed with a resolution of 100 x 100 pixels, needing only 10 K bytes.

A gray-value histogram is formed and a global threshold is calculated or interactively determined. The matrix is scanned for objects, their limits are adapted to the real dimensions and these results are stored upon disk for further processing.

CNTR2: Detailed localisation

The results of the first phase are used to determine the rectangle of interest. This matrix (original resolution) is read into memory while constructing a local histogram which will determine a local threshold (= mean + 2 standard deviations of the gray-levels beneath the global threshold). A detailed localisation is performed with a 4-direction code [1].

Meanwhile, some basic features are calculated (length, perimeter, surface, limits). Inner contours (holes) are linked to their origin and the centers of the object limits are compared to avoid the same chromosome to be found twice. The contours are displayed upon the Ramtek display unit to permit interaction. All intermediate results are stored on disk.

IACTC: Contour interaction

An interaction option is added to the automatic karyotyping system to allow correction for possible errors. Functions used in contour interaction are:

- Delete: to remove an unwished object e.g. a non chromosome
- Zoom: to magnify a specified part on the zoomed picture. All other functions can also be exercised on the zoomed picture.
- Connect: To connect (or merge) two or more fragments
- Search: To locate and find an object in a specified region.
- Review: To redisplay the whole picture without previous interaction markers, etc...

CNTR3: Feature extraction

With aid of the contourcode of the objects, access is possible to all points belonging to the object and the inertia characteristics (moments of inertia, rotation angle, mass, center of gravity) will be calculated. They are used to define length and width and to calculate the symmetry over the object to the axis of the inertia's. For bent chromosomes a parabola-axis is fitted by least-squares. Out of the final axis a projection of the form (width distribution) will in most cases lead to centromere location; a projection of the gray-levels will give a one-dimensional profile description of the bands. For this projection a two-dimensional interpolation method, which has proved to be successful is used [6]. Small objects are analysed with a FFT-procedure to discriminate between a square or triangular form out of the ratio of third and fourth harmonic. For triangular chromosomes a new axis is calculated [2,10].

IACTK: Interaction on the karyotype

The following functions are selectable:

Delete: remove object out of karyotype
Zoom: magnify one pair of or group of chromosomes for detailed consultation.
Turn: rotate the chromosome over a certain angle.
Align: alignment of chromosome to the conventional karyotype form
Move: correction of the chromosome labels.
Mark: mark the specified chromosome (and bring this mark over to the file).

CNTR5: Manual assignment of the class label

In the learning procedure of the classifier, the right class of the chromosomes can be fed into the computer.

CNTR6: Split and merging using semantic information

This program executes an alternative method to locate bands in the chromosomes. See Part 3 and [2].

CNTR7: Management for the statistical file

This program enables easy manipulation of the extracted data of the various metaphases. Following options are available:

AD: add a new metaphase to the existing files
DE: delete last added metaphase
PM: print a metaphase directory
PC: print a chromosome directory
SM: set a metaphase (in)active
SC: set a chromosome (in)active
SV: save files upon a magnetic tape
LD: load file from magnetic tape
SP: activate a special program
ST: stop CNTR7.

4. Feature extraction

A karyotype is a standard format in which the isolated chromosome-pictures are rotated in the labeled output image and placed in pairs (except for the X and Y chromosomes), so as to stand vertical with the longest arm upwards. A total karyotype description is much more than placing the chromosomes in pairs and assign them to one of the 24 different classes. The system (interpreter) must be able to detect all possible structural anomalies of chromosomes, even the smallest part of the smallest chromosome can be missed (deletion) or can be exchanged with part of another chromosome (translocation).

All these defects must be detectable by the classifier. But there is even more, the chromosomes are placed in pairs but they are not identical, they differ in origin and therefore the fine structure (sub-bands) may not be the same. The detection of this fine structure and the relation with the corresponding parental chromosomes can give important genetic information.

Using specific staining methods a band pattern perpendicular to the long arm of the chromosome can be observed (using a conventional light microscope). These banding patterns are different for every chromosome pair (class) and are an aid to identify each pair. Also, the size and the shape of the pairs are different. Because of the immense variation in the staining and preparation methods together with the large cell to cell biological variation, the above statement on the specific banding pattern and size for each class is drastically damaged. The selected features have to reflect the between class variation of the three basic attributes: size, shape and banding pattern. The first two are global, the last one is more local or band oriented feature.

From the contour code the following global features are selected: area, length of perimeter, mass, position of the center of gravity, a set of higher order moments, the axis of inertia and the minimum enclosed rectangle, which with the sides parallel to the axis of inertia gives the length [1].

For the bent chromosomes the symmetry-axis (axis of inertia) divides the chromosome in two unequal parts and is mostly misaligned with the bands. For this reason a medial axis transformation is used provided that this axis is perpendicular to all bands. A parabola is a good approximation of this medial axis, if the chromosomes are not bent too much. Several methods for the computation of chromosome banding profiles can be used, but only the two-dimensional interpolation technique gives a relative maximum error less than 2% and independent from the angle between the scan grid and the projection axis [6]. The effect of misalignment between the chromosome bands and the medial axis can be very high: up to 10% relative error for a misalignment smaller than 7% [6]. The integrated one-dimensional profile is given in 64 points. This dimension of the feature vector is too high to be handled in a classifier, and will be reduced to a limited number, using Gaussian decomposition, Fast Fourier Transform and correlation with standard profiles.

4.1 Decomposition in Gaussian distributions

The total intergrated profile is approximated by the sum of distribution functions and described by a number of triplets each consisting of the peak position, peak height and width of the peak of the distribution function. Other peak distributions can be used if they are mathematically faster. Several methods exist to decompose a density profile in a sum of Gaussian normal functions. The methods used here are maximum likelihood function method and nonlinear least square method [2,5].

The main difficulty is to determine the number of components and to generate a suitable set of initial values [2,11]. Up to now we can either use an interactive method to enter the initial values or we can estimate them from the position of the extrema and the inflection points. This last method results in a variable number of components depending on the quality (number of peaks and shoulders). From the learning set one can extract an average number of components for every class and also the best starting value sets. Usually one has to decompose the profile for 24 different sets of starting values. To reduce the computation time the number of alternatives can be limited using a preliminary classification by computing the correlation coefficient between the profile under computation and the standard profiles [5].

4.2 Fourier descriptors

Although the Fourier coefficients are not invariant, the method has been used frequently because of the low computation time. For physical reasons only the lower order coefficients are under consideration [5,12].

This basic profile vector $X = [X_1, X_2, \ldots X_{64}]^T$ describing the integrated profile in 64 points is - after normalization for offset and magnification - transformed to a vector $\vec{X}^*$ with:

$$\vec{X}^* = [X_1, \ldots X_{64} - X_{64}, \ldots -X_1]^T$$

The periodic extension of $\vec{X}^*$ is continuous in the start and the endpoints; on the contrary the periodic extension of $\vec{X}$ was not. The problem of defining the start and the endpoints of a profile is now unimportant and higher harmonic components that are not related to real information are avoided.

The Fourier coefficients of the vector $\vec{X}^*$ are given as:

$$\vec{Y} = [Y_1, \ldots, Y_{128}]^T \text{ with } Y_p = \frac{1}{128} \sum_{q=1}^{128} X_q \exp[2\pi ipq/128];$$

$$p=1, \ldots, 128, \quad \text{where } i = \sqrt{-1}.$$

5. Classifier

Several versions of classifiers are available but in practice the data from the learning set are insufficient to specify an optimal classifier. Depending on the methods, one uses a parametric or non-parametric classifier [4].

5.1 Parametric classifier

We use a parametric quadratic classifier (Bayesian) where the parameters are substituted by their estimated values defined by the learning set. The loss function is taken symmetrically while the features are assumed to be normally distributed and independent from each other (the learning set is too small to estimate all different covariance parameters).

Although the number of significant features depends on the chromosome class, this number is taken equal for all the classes. The feature set contains the normalized length, the normalized density-area ratio (average density) and an intergrated profile. Several normalizations can be used but only the summed length (or mass/area) of all the chromosomes is taken. This normalization was performed after all overlapping and touching chromosomes were separated (using interaction module) so that all chromosomes are included for the classification experiment.

The accuracy of this Bayesian classifier depends strongly on the number of FFT-harmonics used. This accuracy varies from ± 3% up to 33% and is maximum by the use of 13 harmonics. The average accuracy increases to 46.5% correctly classified chromosomes (24 classes) by the expedient of a combination of 11 harmonics, normalized length and normalized average density. The accuracy is computed on a test set equal to learning set containing 20 normal cells.

5.2 Nonparametric classifier

In the previous section we assumed that the forms for the underlying density functions were known for the 24 classes (23 for female cells). The common parametric forms rarely fit the 24 actually encountered class-condition-densities. In particular the density tends to a bimodal shape (owing to the homologue variations). In this section the unnormalized features length, mass/area ratio and the ordered row of profile-points are used. The serial correlation coefficients used as a similarity measurements between a profile and the reference profile of every class, gain useful information by preserving the positional order of the observations.

Those correlation techniques are mostly called cross-correlation between two (time) series. The cross-correlation-coefficient has the advantage to be independent from the linear transformation of the two series, so normalization for offset and magnification of the profile is unnecessary. By means of the serial correlation with a set of 24 standard profiles, the maximum attainable result is 54% correctly classified chromosomes. Several misclassifications of classes with similar global profiles which differ in length and (or) mass/area ratio can be eliminated using a preclassifier based on one or both previous features. The total classification accuracy reaches 68% for preclassification on length and climbs up to 74% for both attributes.

If the unequal distributed classification power for the different classes is also taken in account, the final accuracy becomes 80%. But to verify this last result the test- and learning sets must be much larger. The above given accuracy is computed on a learning + test set size of 45 cells.

All classifiers given above must handle the problem of the random orientation of the chromosomes (and the resulting profiles). This is solved by the correlation between unknown profiles and all standard profiles in two different orientations or by matching the unknown profile with a sawtooth profile.

5.3 Discussion

The perceived error-rates are quite high compared with the results obtained by some other investigations and much higher than the assumed human error-rates. Granlund [5] reports accuracy rates varying from 64 up to 95% depending on the type of classifier used.

The dramatic low accuracy of our system is due to several reasons:

1. The tests are performed on much larger test-and learning-sets and these two sets are separated.
2. The average quality of the metaphases (and - a fortiori the chromosomes) are the all-day quality of the clinical operation. This leads to a higher variability of the profiles and other attributes.
3. The inclusion of all chromosomes of the metaphases under investigation even those which were interactively corrected such as the touched and overlapped chromosomes. It was observed and also expected that the "within-class variation" of the features for these interactively corrected chromosomes is much higher than for the other ones.

The difference in classification power between parametric and non-parametric is in favour for the last one due to:

1. The assumption of independency of the features for the Bayes classifier.
2. The accepted multi-Gaussian distribution of the features which tends to be bimodal.

The reported error rates are similar to those given by investigators who have also performed the tests on big size metaphase-population. The failure of fully automatic karyotyping is not due to the incorrect use of statistical classifiers but to the impossibility to describe in a statistical way the associated world model for chromosome analysis (world model = description of the chromosomes, number of classes, within- and between class variation and the possible abnormalities). There are two possible ways to improve the accuracy of the classifier and make the use of full automatic systems still possible:

1. Including contextual information and design a new classifier (interpretor); see Part III.
2. Excluding all ill defined chromosomes, but with the consequence that one has to handle incomplete metaphases

This last way can be approached by:

- excluding all bent, touched and overlapped chromosomes (= blobs).
- include for some chromosome blobs automatic splitting programs.
- classify all objects (after segmentation) with a very high accuracy in three classes; namely, isolated chromosomes, blobs and non-chromosomes.
- design a multi-cell-classifier (see [11])

Some additional improvements can be carried out in the segmentation phase. A wrong segmentation affects all features in an unequal manner. The chromosomes are darker regions in a lighter field (background). But unfortunately depending on the staining method and the general picture quality, the contrast ratio between chromosomes and surrounding background is low to very low. Besides this, the boundaries are also very smooth because of the physical structure of the chromosomes.

A more consistent segmentation is approved by:

- edge detector which takes into account the local noise situation and the global quality of the metaphase picture.
- restoration and noise cleaning techniques.
- improvement of picture input quality (hardware, S/N, resolution).
- edge detector guided by semantic information (plan) [1]

5.4 Interactive classifier

Because fully automatic systems are unable to perform usable karyotypes only computer aided systems are trustworthy. For this reason the speed, accuracy and comfort of man-computer interactions are very important. These factors are strongly dependent on the displayed image quality (on the monitor) which is in turn related to monitor resolution in space and gray-scale.

6. Conclusions

In its present form, the system for semi-automatic karyotyping is able to do karyotyping of banded chromosomes, with human assistance with a performance of a skilled technician. The rationale for analyzing cells with a semi-automated system is to collect quantitative data which can be handled in a statistical way; also the speed is an important factor.

The statistical analysis of quantitative data makes it possible to detect consistent events, the system has the potential to detect and describe aberrant chromosomes, even if the detection was not possible for human investigators.

PART III. SYSTEM REALIZATION FOR A TOTAL IDENTIFICATION OF CHROMOSOMES

The design of the system is problem-oriented, the attempt is to solve the karyotyping of human chromosomes. This problem has received recently a new degree of complexity because the bands which already exist now (240 different bands) can be split into relevant (genetic) subbands [16]. With this new technique the number of detectable bands is increased to more than 1500. The development of the new system is limited in the first place to the recognition mode of the interpreter. Because the existing learning procedures (statistical classifier) are so particular and so statistical of nature, they cannot be considered as a structural learning method. This limitation implies a temporary use of a pre-described model structure for every normal chromosome. The design of these models is based on the verbal attributes used in artisanal karyotyping. In a later state, a learning procedure is planned which models normal chromosomes and also takes in account all possible structural variations and abnormalities.

1. Description of the chromosome

The description can be split up in two main parts:

- Two attributes characterizing the chromosomes as a whole;
  - the length of the chromosome
  - the centromere index.

- Two tree structures reflecting the composition of the chromosomes. The method to generate these tree structures are based upon grey-level differences in the chromosome profile between dark and bright intervals. It is the one dimensional analogue of the two-dimensional split and merge procedure described by A. Oosterlinck et al. [2].

Procedure 1: organization of the dark intervals.

a. start from the normalized, intergrated profile. The file is discreted in 64 points (length normalization).
b. threshold upgrading gives new segments of the profile (reject segments whose width is too small). See Fig. 3.
c. stop if the number of new segments is zero, otherwise go to d.
d. link the new fragments to the fragment of origin by a tree structure (split procedure).
e. destroy link if the ratio of the segment width of the parent with respect to descendant is too minute (merge procedure).
f. repeat b.

Procedure 2: organization of the bright intervals.
The differences with regard to the first procedure are:

- the starting point is the maximum threshold.
- threshold is downgrading.

The two previous procedures can be combined so that the two trees are built up simultaneously. The interesting properties of the tree structure are the features associated with each nodes, namely,

- the grey-level (threshold) of the segment
- position parameter (one of two intersections)
- the width of the segment

That implies that the tree can be ordered from the bottom to the top according to the grey-level and from the left to the right according to the position. A tree can also be regarded as a kind of global operator in the sense that it collects elements with a common attribute. Fig. 4 summarizes the total structure of the description.

2. Model

Each chromosome is designed as a subroutine. The main ingredient is the set of relations defined over the elements of the tree structure. The relations are inspired on the features used by the artisanal karyotyping of chromosomes; one can distinguish:

a. global features:
- number of splits
- number of leaves
- position of the leave with the greatest grey-level, etc...

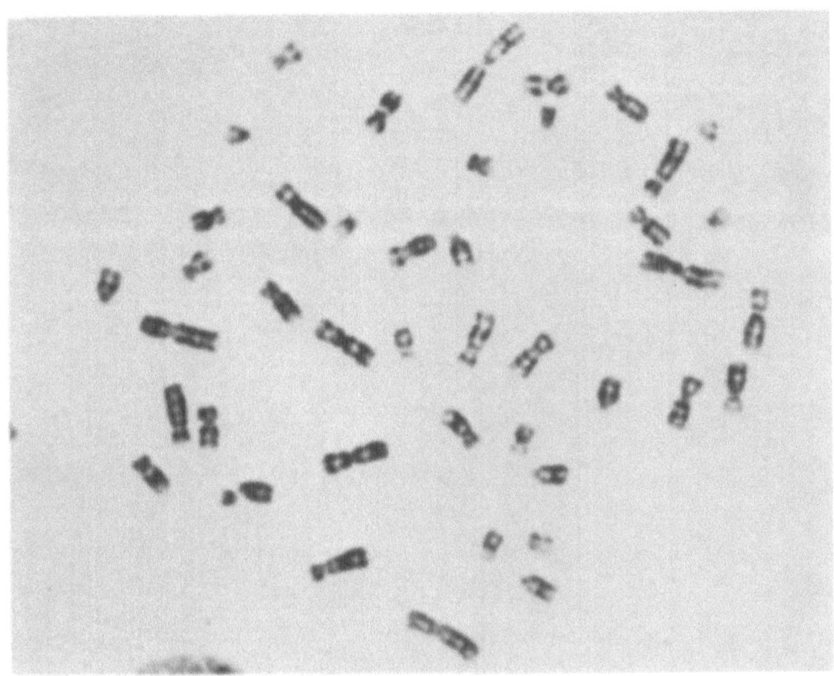

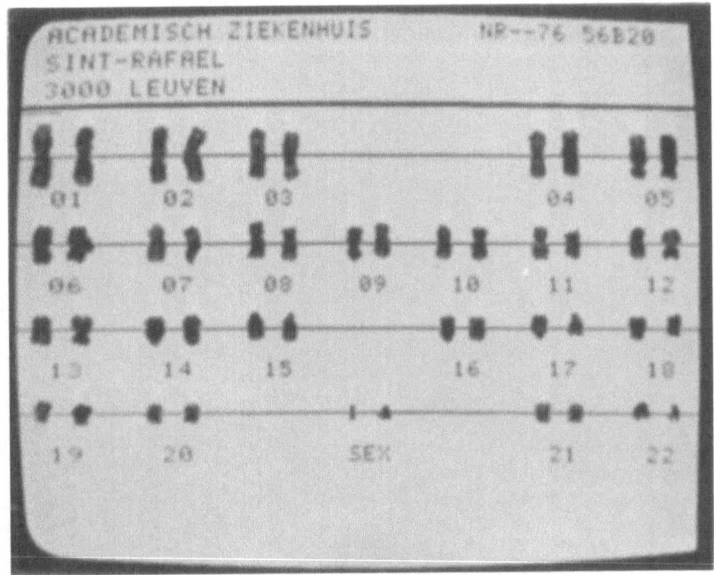

Fig. 3 Input and output picture of the system.

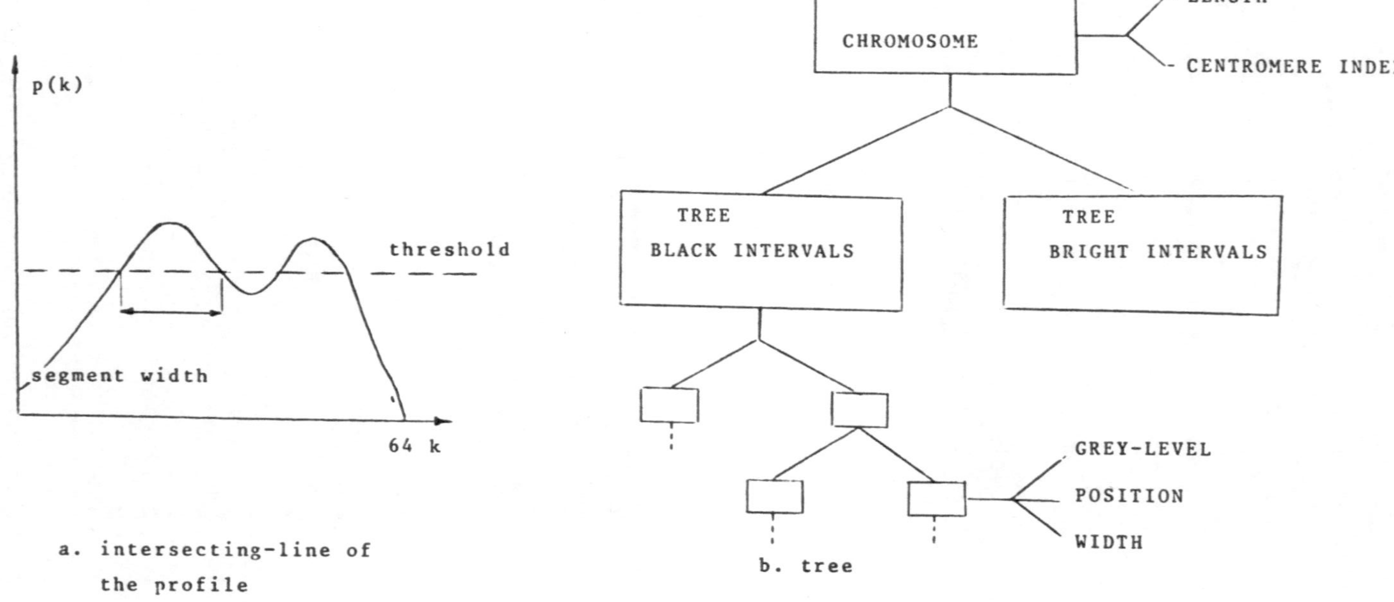

Fig. 4 Parsing tree of a chromosome description.

b. relative features of the form R [$node_i$, $node_j$, f_k, f_l] namely, a relation R defined over the feature f_k of $node_i$ and f_l of $node_j$. It is not necessary that the nodes belong to the same tree structure. Examples of R are:

- ratio of two features
- greater than relation, etc.

3. Interpreter

Three levels of interpretation apply:

- model selection
- individual interpretation
- karyotypic control

The goal of the first step is to select one of the models stored in the world model structures. Therefore the chromosomes of a metaphase are ranked according to their length description. From the inset the ranking number serves as a pointer to a double stack. The first stack contains the addresses of the eligible models, ordered to decreased probability of occurrence. The second stack gives the models which are certainly excluded.

In the second phase the selected model-subroutine evaluates the current description.

Procedure:

a. generation of weights for each calculated relation,
b. total appreciation of the description versus model covering length, centromere index and relation-weights,
c. if the total interpretation is not satisfiable the program is switched to the selection phase to find a new model,
d. if there is not any suitable appreciation, then this model is taken with the highest appreciation level.

The third phase is a control mechanism attempting to avoid conflicts between the current individual classification and the karyotyped chromosomes. Two conflicts arise:

a. there is already attached a chromosome to the class specified by the second phase. No action is undertaken.
b. two chromosomes are already attached to the specified class. Firstly these chromosomes are paired with the highest profile correlation (if the correlation is strong enough) and next we feedback the chromosome or pair of chromosomes to the selection phase.

Note the possibility that at the end of the karyotyping process some chromosomes are not yet associated with a class. That only means that the interpreter was unsure and incompetent to classify them.

4. Concluding remarks

The new approach for the classification of chromosomes is very promising. The use of 24 different world models (one model for each class) helps to overcome the impact of the enormous variation which destroys the effectiveness of the statistical classifier. The most difficult part in the interpretative approach is the construction of the different models. In this paper the compiled artisanal description was used.

REFERENCES

1. Oosterlinck, A., F. Dom, J. De Boer, H. Van Den Berghe, R. Vlietinck, "Automation in Diagnostic Cytogenetics, Proceedings Computer Aided Diagnosis of Medical Images, Nov. 11, 89-98, 1976.
2. Oosterlinck, A., J. Van Daele, F. Dom, A. Reynaerts, H. Van Den Berghe, "Computer Aided Karyotyping of Human Chromosomes", Proceedings IEEE Computer Society Conference on Pattern Recognition and Image Processing, 61-69, Troy, New York, June 6-8, 1977.
3. Walton, W. H., "Automatic Counting of Microscopic Particles", Nature 169: 518-520, 1952.
4. Duda, R. O., P. E. Hart, "Pattern Classification and Scene Analysis", edited by Wiley and Sons, 263-291, 1973.
5. Granlund, G. H., "The Use of Distribution Functions to Describe Integrated Density Profiles of Human Chromosomes", J. T. Biol. Vol. 40, 573-589, 1973.
6. Groen, F. C. A., P. W. Verbeek, G. A. Van Zee, A. Oosterlinck, "Some Aspects Concerning the Computation of Chromosome Banding Profiles," Proc. 3th ICPR, Nov. 1976, Coronado, U.S.A.
7. Rosenfeld, A., J. S. Weszka, "Picture Recognition and Scene Analysis", IEEE, Computer Magazine, May 1976, p. 28-38.
8. Rosenfeld, A., J. S. Weszka, Digital Picture Processing, edited by Academic Press, N. Y. 1976.
9. Kovasznay, L. S. G., H. M. Joseph, "Image Processing", Proc. IRE 43, 1955, pp. 560-570.
10. Groen, F. C. A., "Analysis of DNA Based Measurements Methods Applied to Human Chromosomes Classification", Ph. D. Thesis Delft 1977.
11. Granlund, G. H., G. W. Zack, I. T. Young, M. Eden, "A Technique for Multiple-Cell Chromosome Karyotyping, J. Histo. & Cytochemistry, Vol. 24, N°1, 160-167, 1976.

12. Møller, A. R., H. Nilson, "Computerized Statistical Analysis of Banding Pattern," Nobel Symp. 23: 56, 1973.
13. Castleman, K. R., J. H. Melnyk, "An Automated System for Chromosome Analysis", Final Report JPL, 5040-30, Pasadena, July 4, 1976.
14. Van Daele, J., A. Oosterlinck, H. Van Den Berghe, "Television Scanners", Proc. SPIE (Vd. 130) on the SIRA Int. Sem. on Automation and Inspection Application of Image Processing Techniques, London, 12-13, Sept., 1977.
15. Van Daele, J., A. Oosterlinck, H. Van Den Berghe, "Automatic Visual Inspection of Reed Relais", submitted for _IEEE_, Trans. on Computers. Paper presented at the IEEE Comp. Soc. Workshop on PR and AI, April 12-14, 1978, Princeton, New Jersey.
16. Yunis, J. J., M. Y. Tsai, A. M. Wiley, "Molecular Organisation and Function of the Human Genome", in _Molecular Structure of Human Chromosomes_, (J. J. Yunis ed.), Chapter 1, pp. 1-27, A. P., New York, 1977.

THE USE OF PATTERN RECOGNITION FOR LESION DETECTABILITY IN MEDICAL SCINTIGRAPHIC IMAGES

E. Kahn, R. DiPaola, J. P. Bazin

Institut de Recherches de Radiobiologie
Clinique (INSERM U 66),
94800 Villejuif, France

1. INTRODUCTION

In nuclear medicine, the scintigraphic techniques are based on the administration of a radioactive tracer to a patient. The general purpose is to study the movement of the tracer inside the patient and then pictures of the radioactive distribution are processed for further diagnosis. There must be an answer to the fundamental question: is there a pathological abnormality in the body of the patient or not? and in most cases this answer depends more on the evaluation of the different biological functions emphasized by the tracer (dynamic studies) than on the analysis of the variations of morphology (statics studies).

Improvements in the diagnostic value of procedures used in nuclear medicine depend on the available radiopharmaceuticals and on the instrumentation for the measurement of radiation and scintigraphic data collection, but also on the computer processing of the particular scintigraphic images used as an input to detect the lesions. The resulting computer-aided diagnosis system may be compared to an automatic recognition system of pathological medical patterns the results of which would nevertheless be interpreted by physicians to help them perform their diagnosis.

2. RADIOPHARMACEUTICAL AGENTS AND SCINTIGRAPHIC INSTRUMENTS

Since the first applications of radioisotopes for diagnostic investigations, different radionuclide agents have been used to obtain scintigraphic images of the patient [1,2,3]. They can be distinguished on the basis of the time span of their application (static or dynamic studies).

The choice between radionuclides is dictated by their different properties. Two radionuclides are widely used in clinical routine, ^{131}I and ^{99m}Tc, because they fulfil most of the requirements of nuclear medicine. Their characteristics are outlined below.

1) Iodine-131 has been available for a long time at a low cost and in high specific activity. It has first been used because iodine uptake and organic binding is a normal physiological function of the thyroid gland. In addition, its 8-day half-life makes it amenable to wide distribution, and it can easily be incorporated into a variety of biological and organic agents which extended its usefulness. However, its drawbacks are the higher-than-optimal photon energies (364 KeV) and its relatively long half-life.

2) Technetium-99m ideally satisfies the requirements for a short lived imaging agent and is the most prevalent radionuclide used in present nuclear medical practice. It decays by isomeric transition with 6.0 hour half-life, emitting 140 KeV photons. In addition, Technetium has a multivalent character which gives it versatility for incorporation into various pharmaceutical forms, and its low cost and general availability from a generator system having a convenient half-life overcome the inherent limitations of the short half-life of ^{99m}Tc itself.

The general use of the lower-energy γ rays from ^{99m}Tc(140 KeV), in comparison to the higher-energy γ rays from ^{131}I(364 KeV) and most other radionuclides greatly simplifies the shielding requirements for scintigraphic instruments. It increases the efficiency of both the collimators and the radiation detectors yielding an additional gain in net count rate. This has contributed to advances in scintigraphic instruments: scanners and scintillation cameras.

a. Scanners.

Radioisotope scanning was used first to delineate the thyroid gland and to determine hot or cold nodules in the gland [4]. As ^{131}I was the isotope primarily used in these studies, most scanners were designed for use with that radioisotope. However, radioisotope scanning soon found use in localization of brain tumors, searching for metastases in the liver, delineation of kidneys, lungs and heart, determination of position and size of the placenta, and other clinical applications [5-9].

Two major problems must be solved in the design of a radioisotope scanner:

1) It is necessary to achieve sufficient sensitivity so that the distribution of radioactivity can be determined with adequate statistical precision.

2) It is necessary that the detecting device be capable of sufficient resolution so that the distribution of radioactivity can be reproduced with adequate detail.

As these two requirements are often in conflict, some compromise between the two must be reached.

A radioisotope scanner is made of different components, Fig. 1.

- the focusing collimator: it scans the distribution of radioactivity
- the detector: it converts the γ radiation transmitted by the collimator into pulses, which are then amplified and processed by the electronic components of the scanner.
- the readout system: it displays the processed information for visual interpretation.

In the conventional mechanical scanner, the detection system moves in respect to the patient and the readout is synchronized with the position of the detector. The scanner components are designed according to radioactivity and its distribution. The shape of the source corresponds to the organ to be visualized. The choice of the radioisotope determines the radiation viewed by the scanner, and, most importantly, the flux of radiation emerging from the source which is available to the collimator.

In the first scanners, the collimator was a single cylindrical bore collimator and the radiation detector a calcium tungstate crystal. These scanners had a very low sensitivity, but they could delineate high levels of radioactivity in the thyroid gland. A considerably higher sensitivity was obtained by the use of focusing collimators (consisting of a cluster of tapered apertures focused on one point) instead of the cylindrical bore collimators, and by the use of sodium iodide crystals instead of the calcium tungstate crystals. Most scanners now employ sodium iodide crystals of 3-inch diameter and 2-inch height.

The circuits used for pulse amplification, scaling, and obtaining average count-rates were developed early in scanning history. The incorporation of a pulse-height analyzer provided a substantial step forward in radioisotope scanning (Fig. 1). With a radioisotope that emits γ rays of essentially one energy, pulse-height analysis (that is, the use of the photopeak only) permits the exclusion of radiation which has been scattered in the source. As this scattered radiation will have originated away from the area of focus, the suppression of this radiation results in sharper scans.

The scanning speed and line spacing are variable in all scanners. Scanning speeds between 20 to 400 cm/min are commonly employed depending on the concentration of radioactivity within the organ and the time available for completing the scan. Obviously, the slower the scanning speed, the better will be the statistical

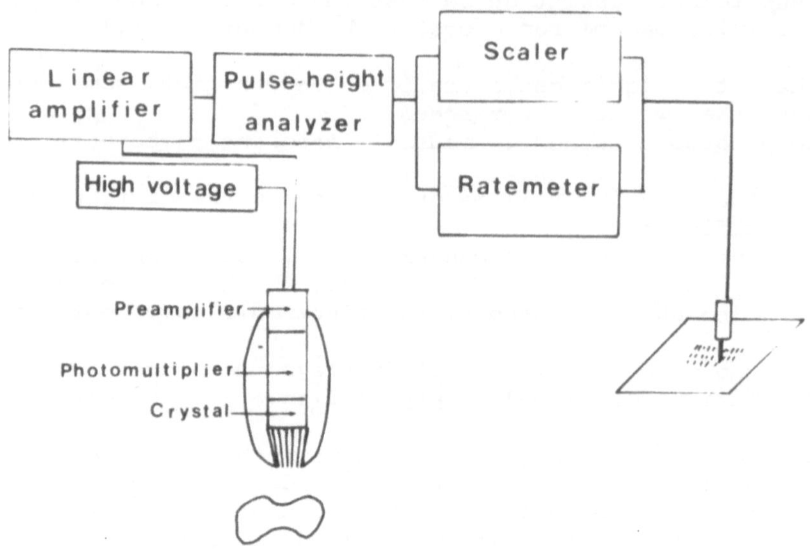

Fig. 1 Block diagram of a radioisotope scanner.

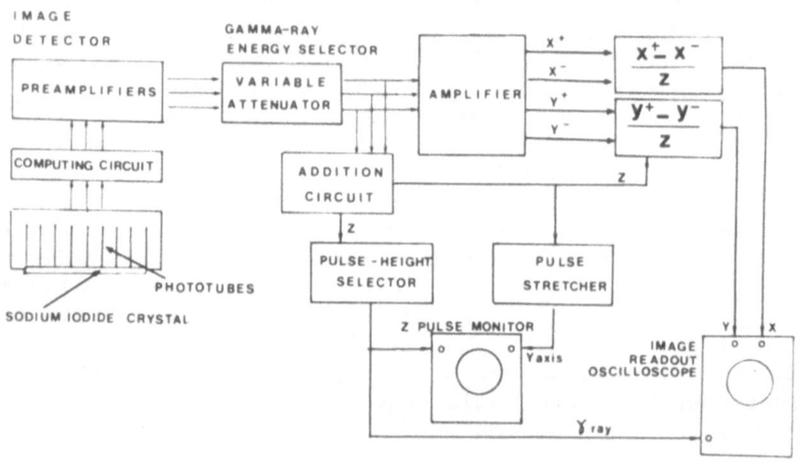

Fig. 2 Block diagram of a γ-ray scintillation camera employing an image detector.

accuracy of the scan. On the other hand, large organs like the liver usually require scanning times of about 10 to 40 min. at relatively fast scanning speeds. The choice of the line spacing is related to that of the scanning speed. If the scanner advances as little as 25 mm for each successive line, a scan readout with no empty space between the lines can be obtained. However, the more lines appear per unit area the faster must be the scanning speed for a given scanning time. The total information contained in a scan is proportional to scanning time independently of the actual number of lines as long as they are not spaced further apart than the resolution distance of the collimator.

A variety of readout systems offers several choices depending on personal preference and on the scanning problem. The earliest and simplest method of scan presentation and one which is still used is the dot record. In this method, a stylus moving over a piece of paper stamps a dot or bar for every fixed number of imput pulses. The resultant scan shows a pictorial density variation proportional to the dot density and consequently to the count-rate at that position.

An electronic method, termed background suppression, eliminates all information from areas in the scan where the count-rate falls below a preselected value. The remainder of the scan is presented usually in a linear fashion. Color scanning has the effect of presenting on one scan contour levels corresponding to different erase levels without loss of any information.

b. Cameras.

Instead of scanning the patient point by point, radioisotope cameras view all parts of the field continuously, and when a subject contains enough radioactivity, they can take pictures at the rate of one per second or faster. The picture-taking speed of scanners is limited by how fast their probes can scan the picture area, and they require at least several minutes to scan most organs.

Shorter exposure times have important clinical advantages. First, rapid sequences of still pictures and motion pictures can be taken to determine the dynamic functions of organs. Second, multiple views of an organ from various angles can be obtained in a short period of time. This improves the visibility of lesions and other abnormalities. Third, more patients can be studied per day with one instrument, an important economic advantage.

The approach to a sensitive camera-type instrument was taken by Anger [10] and resulted in the scintillation camera which employs a solid sodium iodide scintillation crystal viewed by an array of multiplier phototubes, as shown in Fig. 2. A computing circuit, used in combination with the phototube array, senses the position of scintillations in the crystal and sends signals to an image-readout oscilloscope that reproduces the scintillation as point flashes of light. When a scintillation occurs at an intermediate point between phototubes, its position is still determined because of the proportionate division of light among the tubes. The accuracy of determining the position of scintillations in cameras, which employ larger crystals and more multiplier phototubes, allows 1000 picture elements to be resolved at medium γ-ray energies.

The signals from the multiplier phototubes are summed and pulseheight selection is employed to eliminate scintillations not falling within the photopeak of the γ-ray spectrum, thus reducing background by the same method used in conventional radioisotope scanners. Excellent pulseheight resolution results from the efficient collection of light from the sodium iodide crystal.

A time exposure of the oscilloscope screen is taken with a Polaroid camera, and an image of the active areas of the patient results. Pinhole and multichannel collimators are used to image γ-ray emitters.

By the use of the multiplier phototube array, many of the limitations of image amplifiers are avoided. High gain is readily obtained, tube background is negligible, and there is convenient remote reproduction of the images. Coincidence and dot-shifting techniques are easily used in conjunction with positron emitters. Also, there is no practical limit to the size of the camera except for limitations in the manufacture of the sodium iodide crystal.

The useful γ-ray energy range of phototube-array scintillation cameras is about 0.07 to 0.7 MeV. The upper limit is set by the inherent difficulties of collimating and detecting high-energy γ-rays. The lower limit is a result of statistical phenomena.

Because of the use of a solid homogeneous scintillator, rather than a mosaic, the images have no distracting patterns superimposed. Multichannel collimators with the most efficient hole spacing can be used without Moire effect. Furthermore, the 1/2-inch thick scintillator permits, without excessive resolution loss, the use of collimators in which γ rays travel at oblique angles. Also, resolution loss due to γ-ray scattering within the scintillator is minimized.

Recent models employ 16-inch diameter crystals viewed by an array of 61 multiplier phototubes. With the 16-inch camera, pictures of nearly all organs of the body can be taken in a fraction of the time required by conventional radioisotope scanners. Brain tumor pictures have been taken in 10 sec., and stop-motion pictures have been taken of ^{99m}Tc going through the heart and kidneys at the rate of one picture per second.

3. COMPUTER SYSTEMS FOR SCINTIGRAPHIC DATA COLLECTION AND PROCESSING

Digital processing of scintigraphic images in nuclear medicine was first introduced for rectilinear scan image enhancement [11,12,13] and different readout systems of scan data display were employed to obtain maximum useful information. However, these systems were rather limited and efforts are now directed towards the collection and processing of data from scintillation camera. Images of a human thyroid out of a scanner and a camera are given in Fig. 3, where the camera image has also been displayed using multicycle grey levels and contour maps. Correction of image distortion, improvement of image by filtering, quantitative analysis of sequential images in a dynamic function study can now be performed in clinical routine [7-9, 14-21].

An outline of the essential components of our previous scintigraphic data processor in a clinical setting is shown in Fig. 4 [22]. In this system designed for the use of a scintillation camera, data from a scanner can be collected and recorded on a magnetic tape and processed on an off-line computer. Unfortunately, in this case, there is a loss of time between the scintigraphic examination and the output of the processed image. This is why a system connected to the camera has been designed to perform all operations of collection, processing, recording and display. The principal characteristics of this system are:

- choice of the collection-mode: frame or histogram (incremental) mode, and list (sequential) mode, as described in the next section
- choice of the spatial sampling from 32 x 32 to 256 x 256
- digital conversion and recording of energy
- use of a small mini-computer
- easy use by a non-specialist
- connection with a lot of different types of display.

An Informatek SIMIS 3 is now used in clinical routine by the Nuclear Medicine Laboratory of the "Institut Gustave-Roussy" and a DEC-GAMMA 11-34 with an array processor 120 B is used by the Nuclear Medicine Data Processing group of the "Institut de Recherchesde Radiobiologie Clinique (INSERM U66)".

Systems in nuclear medicine have generally been designed on the same model. Their essential components include: 1) a computer, 2) a mass storage system, 3) a data acquisition interface, 4) a display generation module, 5) a control console, 6) a software system.

The hardware components vary widely from one system to the next. Mass storage devices include magnetic disks and tapes. The data acquisition interface serves to translate signals from imaging devices (cameras) into signals acceptable by the computer. A wide variety of data display systems is used, and the control console is usually a keyboard operated unit.

Five common features may be outlined in the software system. They are: 1) collection of static images with optimal computerized image processing, 2) maintenance of patients' files, 3) recording of dynamic function studies, 4) image display for interpretation and region of interest selection, 5) availability of higher level programming languages for further processing.

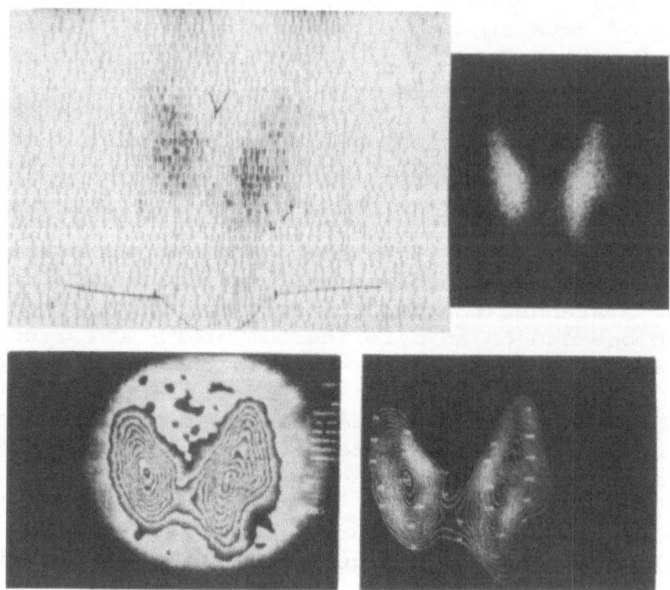

Fig. 3 Scintigraphic images of a human thyroid out of a scanner and a camera. The camera scintigraphic image has also been displayed using multicycle grey levels and contours maps.

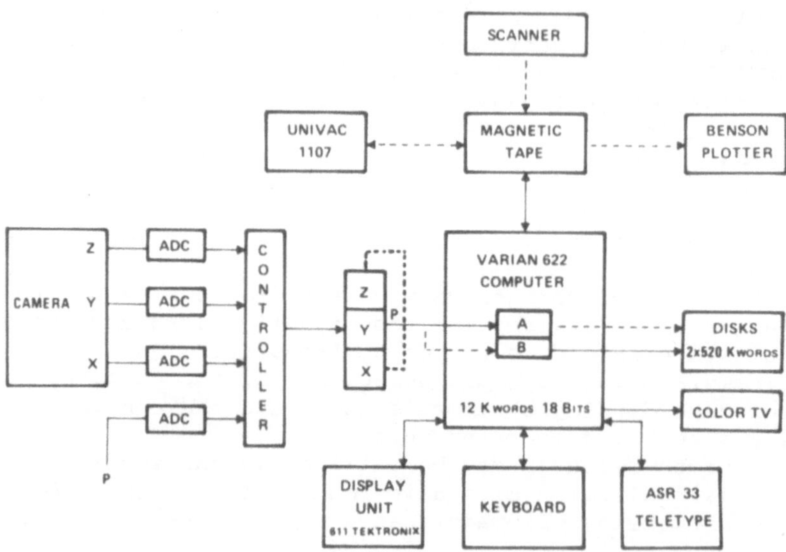

Fig. 4 Outline of the essential components of the system.

4. SCINTIGRAPHIC IMAGE PROCESSING AND PATTERN RECOGNITION

a) Data collection.

The pattern recognition approach is always chosen according to the type of input under consideration. The input is an array of numbers resulting from various measurements performed on the raw data. In nuclear medicine, the raw data are obtained by collecting the photons emitted by the patient, and the result is a scintigraphic image.

The data may be collected 1) in the histogram-mode or incremental-mode or frame-mode form, or 2) in the sequential or list-mode form by the processing system. In the first case, one (more in dynamic cases) n x n image frame is constructed. This n x n image is a 2-D histogram of the counts on every cell of an array. In the second case, the data (X and Y coordinates, time t, energy E, etc...) are stored sequentially and a subset of these data are then reorganized into one (or more) n x n image frames. When the subset of data is reduced to X and Y and there is no constraint on t or E, the image frames correspond to those obtained through the incremental mode form. They are usually called static scintigraphic images. In fact, the existing systems always introduce constraints on E and t: the photons are only collected when their energy falls into some window (E_1, E_2) and during some amount of time (t_1, t_2). The collection of photons ends or restarts when 1) some maximum time is over, or (and) 2) some maximum number of counts has been recorded. The energy is usually taken into account when several radioisotopes are used. The corresponding image frames are called n-uple tracer images. When time is taken into account, the image frames are called dynamic images. A set of dynamic images is in fact a film.

Data collection formats are under the joint control of the central processor unit (CPU) and the data acquisition interface. In the frame-mode, core memory incrementation is generally accomplished through a direct memory access (DMA) facility. This gives the data acquisition interface direct access to a selected memory location without assistance from the CPU. DMA incrementation is done on a cycle-stealing basis. The CPU senses a command from the interface for memory access, and after completing the program step currently being executed it interrupts the normal execution of the program long enough to allow the incrementation. This typically requires two memory cycles or about 2 μs. DMA is used to minimize count rate losses. Software "add one" operations typically involve 10 to 20 program steps requiring 20-30 μs to complete. The basic size of static image matrices used in the frame-mode is therefore usually 64 x 64, (recent studies try to extend it to 128 x 128 or 256 x 256), and for dynamic studies coarser matrices are therefore available to maintain the same statistical validity.

In the list-mode, successive X, Y addresses from the analog-to-digital converters (ADC) are stored in successive memory locations until a predetermined number of memory locations are occupied. Then the memory segment is dumped onto a magnetic disk or tape. Time markers are added within the address listing to permit subsequent reorganization of the data into temporal frames. The advantage of this mode is the ability to collect high resolution images at very fast rates, but it requires large amounts of mass storage space and long playback times for framing or other analysis.

The purpose, in the case of static scintigraphic images, is to obtain points on this scintigram that would be close enough to each other in terms of geometrical distance, and therefore be an obvious medical pattern. The n-uple tracer (radioisotope) images may be compared to multispectral images relatively to energy, and the purpose is still to detect some relevant medical pattern. Dynamic images give the possibility, when medical patterns are detected on each image, to follow their evolution during time. They may also be compared to multispectral images relatively to time.

b) Computer-aided diagnosis and pattern recognition.

Unfortunately, the effort involved in handling these scintigrams without preprocessing is very great. Therefore it is necessary to express the images under consideration in a more appropriate form, without any loss of useful information, and any addition of artefacts. This is also the first step to take in any pattern recognition algorithm. The chosen preprocessing depends very strongly on the type of data.

Some of the most challenging pattern recognition problems occur when data are pictures, since the physical and psychophysiological basis of human visual recognition has not yet been well understood. The existence of optical illusions may lead to wrong interpretations of the content of the picture. This fact, in the case of medical scintigraphic images must be avoided. It is therefore important to find a tool that would help the physician perform his diagnosis and elminiate most errors. This tool can be designed to provide some additional information by superimposing some processed area of the original picture on the displayed image. This superimposition of informations is not usual in nuclear medicine where authors are looking for the most appropriate way of displaying an image. Since the diagnosis is partly based on a visual examination of one or more images, it appeared important to seek a convenient measure for the quality of observer performance. A method providing such a measure, originally introduced in the field of radar detection, is the measurement of a "receiver operating characteristic ROC curve", [23]. This curve plots the probability of a positive response, given that a signal is present (true positive), versus the probability of a positive response given that no signal is present (false positive). Since for some images the observer may receive credit for a correct positive response by missing the actual signal but falsely identify a cluster of noise elsewhere in the image as the signal, true positive decision probability may be expected to drop if the observer is required to correctly determine the position of the original as well as its presence in order to receive credit for a true positive decision. If this more strictly defined "true positive, correct location" decision probability is plotted versus false position decision probability, the resulting "LROC" curve would be expected to lie below the conventional ROC curve. Metz et al., [24], confirm the prediction of this assumed relationship between ROC curves and LROC curves, and this agreement between theoretical prediction and experimental result is also conceptually reassuring by confirming the intuitive notion according to which detection performance with and without localization requirement should be uniquely related. This relationship is of considerable practical importance since in many clinical detection situations the location as well as the presence of a lesion within the image is important, and since experiments to determine the conventional ROC curves are considerably easier to design and carry out than experiments to determine the LROC curves. It has been shown by means of these curves that some specific displays are better in detecting lesions with a minimum number of false positives. These authors have processed the same kind of data to produce a quantitative intercomparison between their visual methods. The pattern recognition approach is different because it provides a method that selects defects without the help of the eye and then gives to the user some different and additional information on the selected defects. The visual methods mentioned above and the enhancement methods proposed by some authors in pattern recognition can be put in the same group: the purpose is to transform an image into some pattern more convenient to the user's interpretation.

Picture segmentation in pattern recognition refers to the operation of analyzing a scene and extracting chosen objects from the background. The picture is thus divided into different parts which have some meaning for the observer. In nuclear medicine, when the physician determines his regions of interest ROI on a scintigraphic image, he roughly performs a segmentation of this image. He usually surrounds an organ with the aid of a joy stick on a first image and this region is not modified on the following image in a dynamic study, by lack of time. The purpose is to find a method of segmentation that is not time-consuming and permits the same computations on the extracted regions.

Algorithms for finding edges or other regions may be classified into local and merging schemes. The local schemes use only the grid, and they label each point according to its number of counts or according to the number of counts of a point at a fixed neighborhood. Thresholding is an example of this methodology. The merging schemes start at the single pixel level (or at an initial segmentation) of the image. Regions are obtained by the addition, or the subtraction, of points to the initial point or region one usually calls a seed, according to some criterion. All these techniques are quite unsatisfactory in nuclear medicine, for the resulting segmentation does not often correspond to human interpretation and physical measures. In many cases, the single operation of thresholding gives the best result, though the value of the threshold has to be chosen carefully. When the regions have some different textures, and the human eye can intuitively insert a line that separates them, merging schemes some authors call region growing methods, may be more suited to solve the problem of segmentation, but unhappily are too expensive to be used in nuclear medicine in routine.

5. CONSTRUCTION OF SCINTIGRAPHIC IMAGES

The scintigraphic images are produced by the detection of a part of the photons emitted by the organs in the direction of the detector (scanner, or scintillation camera). These images represent views of the organs studied along chosen directions. They are obtained, either by the displacement of a detector (scanner) in front of the patient, or by a motionless 2-D detector (gamma camera). However, in the case of a whole body examination, the gamma camera must be moved to obtain a complete image. The coordinates X, Y of each detected photon are obtained from the position of the scanner when it has received a photon, or from the place it has been detected in the field of view of the gamma camera. Conventionally, a scintigraphic image I is formed by the following operation $I(X,Y) = I(X,Y) + 1$, which may be performed by a processor each time a photon is detected. Therefore, a scintigraphic image is created either by a fixed number of photons, or by the number of photons detected in a fixed time.

Other scintigraphic images can be created if the energy of every incident photon is taken into account. This energy must be recorded within a sufficiently large window, such that all relevant information related to all radionuclides of the dose administrated to the patient are collected simultaneously. This basic data can be considered as a 3-D matrix (4-D if time is included) comprising X, Y positional and E energy information, which can be considered as planes of activity distributions at various energies.

In this case, series of two dimensional matrices can be formed using different energy windows selected from the a-priori knowledge of the spectra of the individual radionuclide involved, and the global spectrum obtained from the mixture.

Many practical problems appear in nuclear medicine, consecutive to the characteristics of the radionuclides, to their use in man, and finally to the detector available in radioisotopic studies. The most important problems are:

1) The statistical problem related to the relatively low activity of radionuclide which is possible to give to a patient, with an irradiation dose suitable with its use in medical diagnosis. This activity depends very much on the isotope used, the

organ studied and the irradiation dose of the critical organs, this being also related to the physical and biological half-lifes. Practically, it is not truly possible to use activities higher than 10 millicuries for a clinical study. Generally, this important constraint does not allow us to obtain good statistics in a time short enough to be useful in routine for clinical studies. So it is necessary to process the data to compensate these low statistics.

2) The problem of internal uptake of the radionuclide in the human body. It is a consequence of the presence of the radioisotope in a scattering medium which attenuates the photon activity. This attenuation depends both on the uptake location and on the biological materials which surround it. This so-called self attenuation problem is still under study.

3) The detector problem; the receptor is not perfect as in the theoretical case. It introduces losses in the time and spatial resolutions. Practically, the loss due to time resolution can be neglected. Nevertheless, the loss due to spatial resolution is important: 0.5 to 3 cm for the value of the full width at half maximum (FWHM) of the point spread function (i.e. the detector response to a radioactive point-source). This last resolution depends also on the distance from the point source to the detector and on the mediums interposed between this point and the detector. In all cases, it is necessary to process the data to correct this resolution.

5. PROBLEMS OF SEGMENTATION

Our purpose is the design of an interactive program that will detect medical patterns on a scintigraphic image. Therefore, at different steps, we will have to solve some problems of segmentation: 1) extract the region of interest in the image, and 2) inside the region of interest, detect the defects that might be present. In this paper, we limit our attention to techniques that were applicable to scintigraphic images and could be implemented on a minicomputer with short enough computation time.

1) Edge detection.

In the case of scintigraphic images, we usually do not know the shapes of the organs, but we know that a low number of counts corresponds to some background noise. Therefore, one can say that there is a more or less abrupt change of the number of counts at the boundary between organs and background. Such changes must be detected and evaluated if one wants to extract the limits of the organs in the image.

Changes in the number of counts can be detected by applying some type of derivative operation to the given image. The resulting image should be made of high values where edges are present, and low values elsewhere.

a) Gradient.

A simple derivative operation is the gradient: for the continuous image I(x,y), this is the vector-valued function whose magnitude is: grad I(x,y)

$$\text{grad } I(x,y) = \sqrt{\left(\frac{\partial I}{\partial x}\right)^2 + \left(\frac{\partial I}{\partial y}\right)^2}$$

where

$$\frac{\partial I}{\partial x} \quad \text{and} \quad \frac{\partial I}{\partial y}$$

are the partial derivatives of I(x,y) relatively to x and y.

For the scintigraphic image, these derivatives must be replaced by differences:

$$\text{grad } S(p,q) = \sqrt{d_p^{\,2} + d_q^2}.$$

Another function has first been used in scintigraphy [25]. It is:

$$\text{grad } S(p,q) = \sqrt{d_1^{\,2} + d_2^{\,2} + d_2^{\,2} + d_4^{\,2}} \;,$$

where $d_1 = S(p,q) - S(p+1,q)$; $d_2 = S(p+1,q) - S(p+1,q+1)$ $d_3 = S(p+1,q+1) - S(p,q+1)$; $d_4 = S(p,q+1) - S(p,q)$. Then, in [25], grad S(p,q) is replaced by:

$$\text{Diff } S(p,q) = \frac{\text{grad } S(p,q)}{T}$$

where $T = S(p,q) + S(p+1,q) + S(p+1,q+1) + S(p,q+1)$, and the resulting differential image is smoothed by a 9-point filter.

Other approximations for the gradient may be: $|d_p| + |d_q|$ and $\sqrt{d_p} + \sqrt{d_q}$, or more generally $(d_p)^n + (d_q)^n$, where n is some matched exponent. The differences d_p and d_q can be computed in a large number of ways [26]. One of the simplest ways corresponds to the Roberts operator [27], which is nothing more than: $d_p = S(p,q) - S(p+1,q+1)$ and $d_q = S(p+1,q) - S(p,q+1)$.

The smoothed gradient [28] corresponds to:

$$d_p = I(p,q) + I(p+1,q) + I(p+2,q) - I(p,q+2) - I(p+1,q+2) - I(p+2,q+2).$$

$$d_q = I(p,q) - I(p+2,q) + I(p,q+1) - I(p+2,q+1) + I(p,q+2) - I(p+2,q+2).$$

The Sobel gradient [28] corresponds to:

$$d_p = I(p,q) + 2I(p+1,q) + I(p+2,q) - I(P,q+2) - 2I(p+1,q+2) - I(p+2,q+2).$$

$$d_q = I(p,q) - I(p+2,q) + 2(I(p,q+1) - I(p+2,q+1)) + I(p,q+2) - I(p+2,q+2).$$

A number of authors have given approximations to compute d_p and d_q and have published the resulting gradients on a lot of images. For a recent survey and some improvement of these methods, see [29]. If the purpose is to use a minicomputer, it is interesting to use a simple operator that will be inexpensive and nevertheless have a good performance. Figs. 6a - b are the gradient images of the scintigraphic image of a double isotope phantom [32], shown in Fig. 5. Another approximation of the gradient can be computed by selecting the larger difference [30,31]:

$$\text{grad } S(p,q) = \max (d_p, d_q).$$

We computed d_p and d_q as above. The results are given for the same scintigraphic image, Fig. 6-c.

The double isotope phantom is composed of a container of 18 cm diameter and 2 cm thickness filled with ^{99m}Tc and inside of which a container of 4 cm diameter and 1 cm thickness filled with ^{197}Hg has been placed to represent a tumor. In addition, a third container of 7 cm diameter and 1 cm thickness filled with ^{197}Hg has been placed at the bottom of the first container and tangent to it to represent uptake on the resulting image.

b) Laplacian.

Another simple derivative operation is the laplacian: for the continuous image I(x,y), this is the vector-valued function whose magnitude is:

$$\text{lapl } I(x,y) = \frac{\partial^2 I}{\partial x^2} + \frac{\partial^2 I}{\partial y^2} ,$$

where

$$\frac{\partial^2 I}{\partial x^2} \text{ and } \frac{\partial^2 I}{\partial y^2}$$

are the second derivatives of I(x,y) relatively to x and y.

For the scintigraphic image, these derivatives may be computed according to the approximations most authors [31] use: lapl $S(p,q) = S(p-1,q) + S(p,q-1) + S(p,q+1) + S(p+1,q) - 4S(p,q)$, or lapl $S(p,q) = S(p-1,q-1) + S(p-1,q) + S(p-1,q+1) + S(p,q-1) + S(p,q+1) + S(p+1,q-1) + S(p+1,q) + S(p+1,q+1) - 8S(p,q)$.

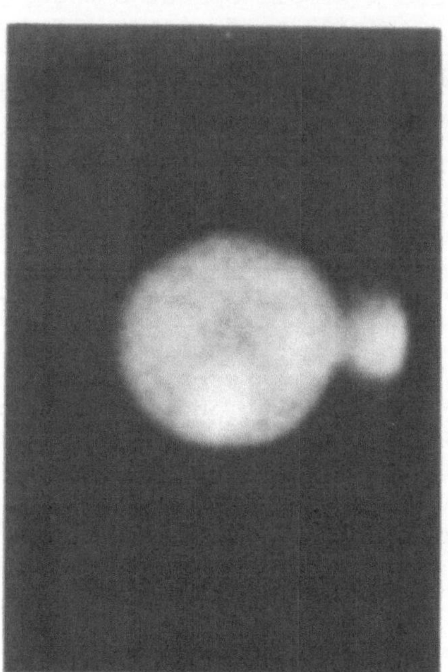

Fig. 5 Scintigraphic image of the double isotope phantom.

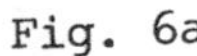

Fig. 6a

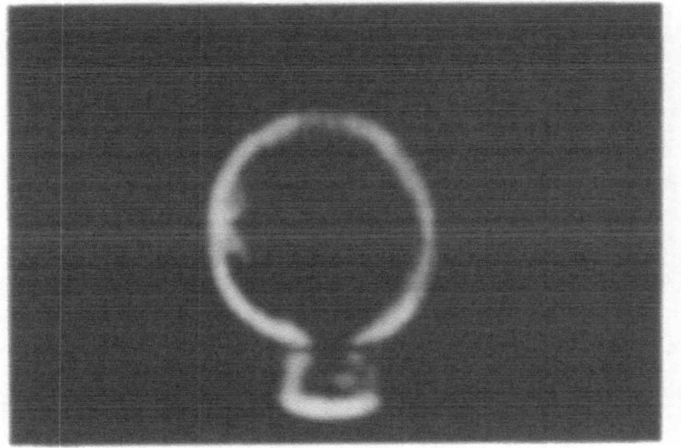

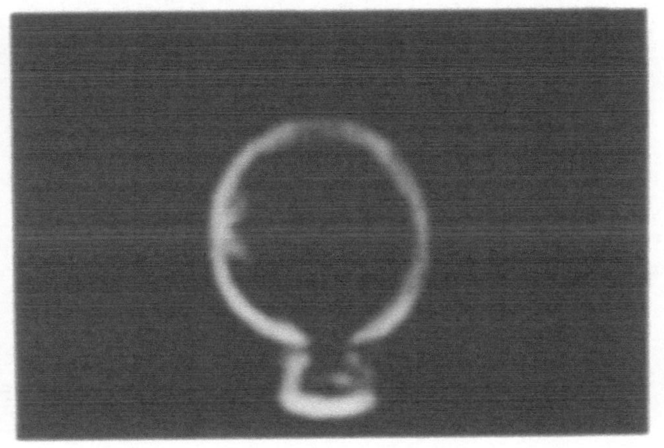

Fig. 6 Gradient images obtained by processing the scintigraphic image in Fig. 5:

a) sum of the absolute values of the Sobel and smoothed differences d_p and d_q.
b) sum of the square roots of the Sobel and smoothed differences d_p and d_q.
c) maximum Sobel and smoothed difference d_p or d_q.

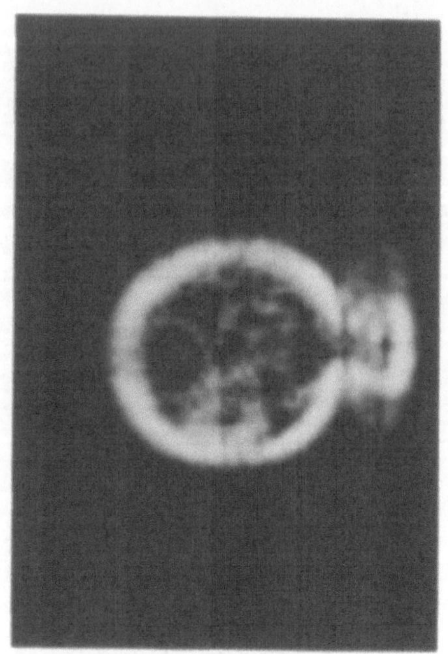

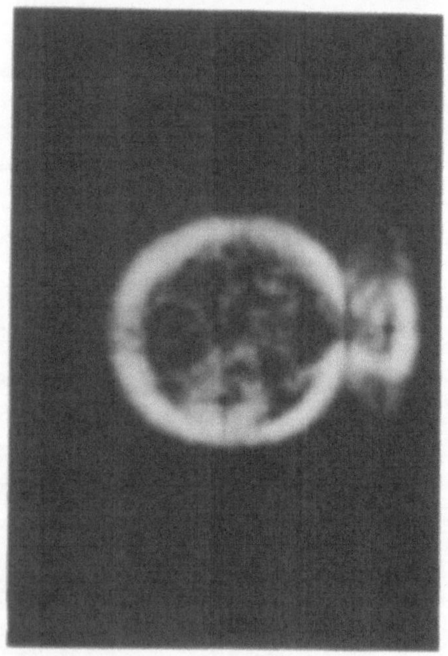

Fig. 6b

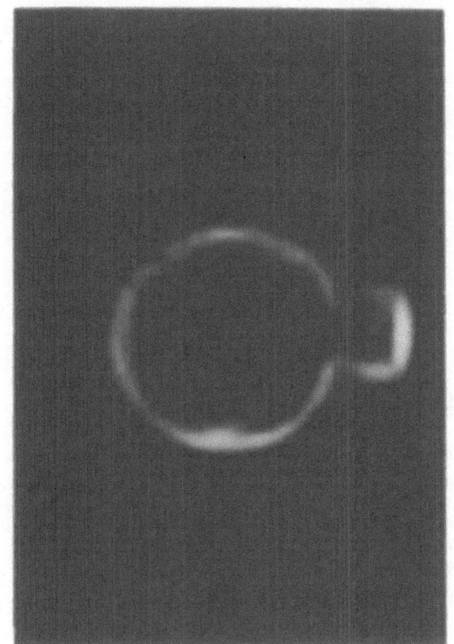

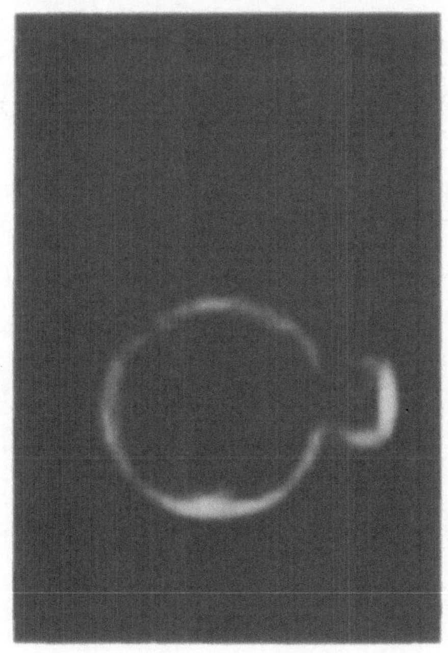

Fig. 6c

A better approximation [26], which is more expensive, we have often used is:

$$\begin{aligned}\text{lapl } S(p,q) = & \frac{1}{12}[S(p-1,q-1) + S(p+1,q+1) + S(p-1,q+1) \\ & + S(p+1,q-1)] \\ & + \frac{10}{12}[S(p-1,q) + S(p,q-1) + S(p,q+1) + S(p+1,q)] \\ & - \frac{44}{12} S(p,q)\end{aligned}$$

The Laplacian image of the scientigraphic image of the double isotope phantom [32], is shown in Fig. 7.

c) Preprocessing.

The gradient image, and much more, the Laplacian image, are largely improved when the scintigraphic image is first low pass filtered.

Low pass filtering is a noise smoothing procedure whose overall effect is to smooth fast gradients and enhance slow gradients. The procedure used here for low pass filtering is either in the image domain, or in the Fourier transform domain.

For a scintigraphic image $I(p,q)$, the low-pass filter may be an averaging process such that: low-pass value at $I(p,q) = \frac{1}{9}[I(p-1,q-1) + I(p-1,q) + I(p-1,q+1) + I(p,q-1) + I(p,q) + I(p,q+1) + I(p+1,q-1) + I(p+1,q) + I(p+1,q+1)]$. The 9-point smoothing (or 9-point binomial filter) corresponds to: low pass value at $I(p,q) = \frac{1}{16}[I(p-1,q-1) + 2I(p-1,q) + I(p-1,q+1) + 2I(p,q-1) + 4I(p,q) + 2I(p,q+1) + I(p+1,q-1) + 2I(p+1,q) + I(p+1,q+1)]$.

We often used the 16-point smoothing (or 16-point binomial filter) that corresponds to:

$$\begin{aligned}\text{Low-pass value at } I(p+\tfrac{1}{2}, q+\tfrac{1}{2}) = & \frac{1}{36}[I(p-\tfrac{3}{2}, q-\tfrac{3}{2}) \\ & + 3I(p-\tfrac{3}{2}, q-\tfrac{1}{2}) + 3I(p-\tfrac{3}{2}, q+\tfrac{1}{2}) + I(p-\tfrac{3}{2}, q+\tfrac{3}{2}) \\ & + 3I(p-\tfrac{1}{2}, q-\tfrac{3}{2}) + 9I(p-\tfrac{1}{2}, q-\tfrac{1}{2}) + 9I(p-\tfrac{1}{2}, q+\tfrac{1}{2}) \\ & + 3I(p-\tfrac{1}{2}, q+\tfrac{3}{2}) + 3I(p+\tfrac{1}{2}, q-\tfrac{3}{2}) + 9I(p+\tfrac{1}{2}, q-\tfrac{1}{2}) \\ & + 9I(p+\tfrac{1}{2}, q+\tfrac{1}{2}) + 3I(p+\tfrac{1}{2}, q+\tfrac{3}{2}) + I(p+\tfrac{3}{2}, q-\tfrac{3}{2}) \\ & + 3I(p+\tfrac{3}{2}, q-\tfrac{1}{2}) + 3I(p+\tfrac{3}{2}, q+\tfrac{1}{2}) + I(p+\tfrac{3}{2}, q+\tfrac{3}{2})]\end{aligned}$$

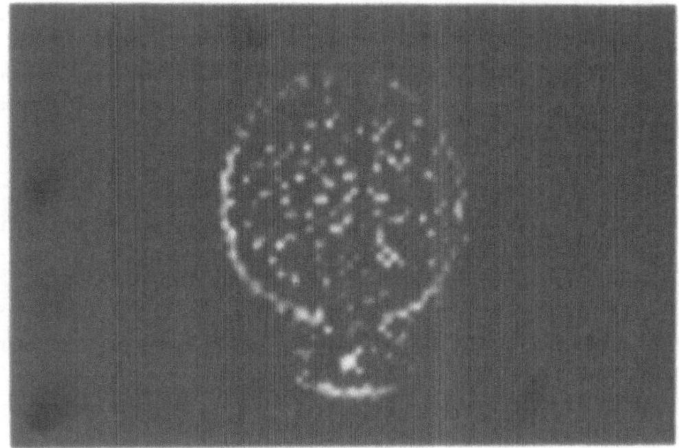

Fig. 7 Laplacian image obtained by processing the scintigraphic image in Fig. 5.

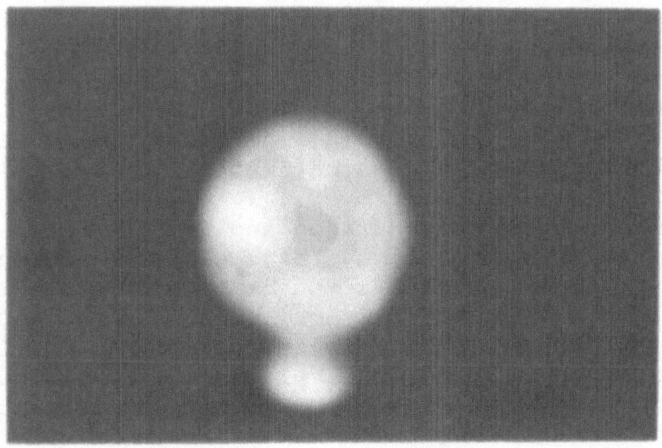

Fig. 8 Pseudo-Wiener filtered image obtained by processing the scintigraphic image in Fig. 5.

The weighting functions of the binomial filters result from the coefficients of the binomial polynomial. Their responses in the frequency domain are always positive on $(0,\nu_N)$, where ν_N is the Nyquist frequency, and they decrease without oscillation from 1 to 0 on $(0,\nu_N)$. The 2-D binomial filters result from the product of 1-D binomial filters [33,34].

A Wiener filter may be easily computed in the Fourier domain, [35,36,37,38] if we put the hypotheses of independence of the spatial resolution with the detector distance and of white noise. The filter used is a 2-D Wiener filter because of the 2-D loss of resolution in the scintigraphic image.

An approximation of a 2-D Wiener filter, taking into account the characteristics of the scintigraphic images so far used has been implemented on our minicomputer Informatek SIMIS 3 in the space domain. This filter we call a pseudo-Wiener filter is fast enough to be used in clinics [38]. The results on the scintigraphic image of the double isotope phantom mentioned above is given in Fig. 8. The laplacian and gradients of the pseudo-Wiener filtered image are given in Fig. 9 a-d.

d) Enhancement.

By making use of local context information for each point of the scintigraphic image, it is possible to enhance edges and lines, and reduce noise. However, our purpose here is not to increase the visually extractable information, or to find the most suited to the eye display of the scintigraphic image, as usually achieved in nuclear medicine. Our aim is to provide a selection of inexpensive programs which can be run prior to segmentation or further processing. An operator called local mean correction and previously used [39] on aerial photographs, gave good results. It adjusts the local average number of counts at each point of the scintigraphic image towards the global mean.

Let μ_o denote the global mean of the image matrix, and suppose that the average number of counts in an m-by-m neighborhood of a pixel is μ. We give the pixel $p(i,j)$ a new number of counts $p(i,j) - \alpha(\mu-\mu_o)$, where α is a constant $(0\leq\alpha\leq1)$. When this new number is negative, we replace it by 0.

We performed an 4 x 4 neighborhood and $\alpha=0.15$ enhancement, Fig. 10-a, of the scintigraphic image of the double isotope phantom, Fig. 5. Fig. 11a is the laplacian of this enhanced image. The result obtained through a 6 x 6 neighborhood and $\alpha = 0.1$ enhancement is given in Fig. 10b. This enhancement is particularly interesting because the thresholded laplacian, Fig. 11b, provides a correct segmentation of the image. This correct segmentation could not be obtained by using the low pass or the Wiener filters. A gradient image of the first enhanced image is also shown in Fig. 11-c.

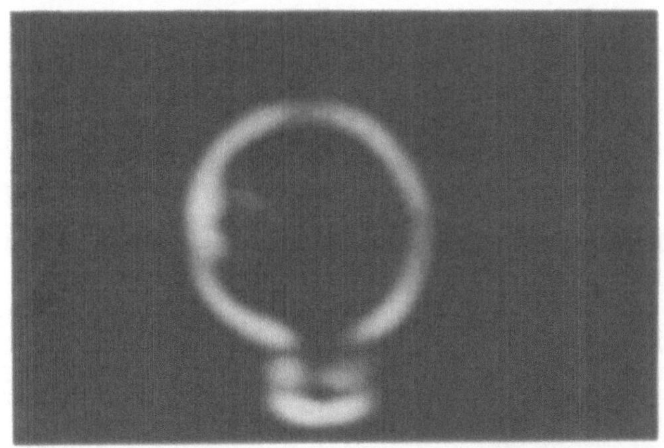

Fig. 9a

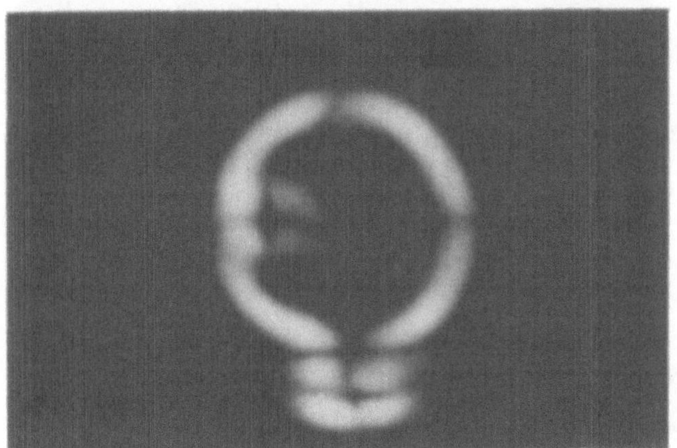

Fig. 9b

Fig. 9 Gradient and Laplacian images obtained by processing the filtered image in Fig. 8.

a) sum of the absolute values of the Sobel differences d_p and d_q.

b) sum of the square roots of the Sobel differences d_p and d_q.

c) maximum Sobel difference d_p or d_q.

d) Laplacian operator: $\begin{bmatrix} 1 & 10 & 1 \\ 10 & -44 & 10 \\ 1 & 10 & 1 \end{bmatrix}$

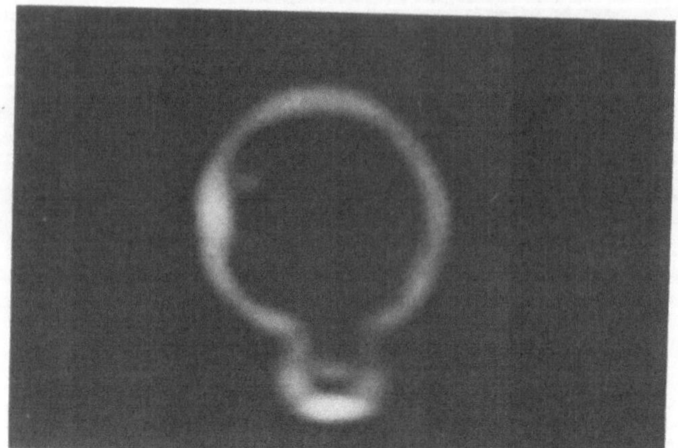

Fig. 9c

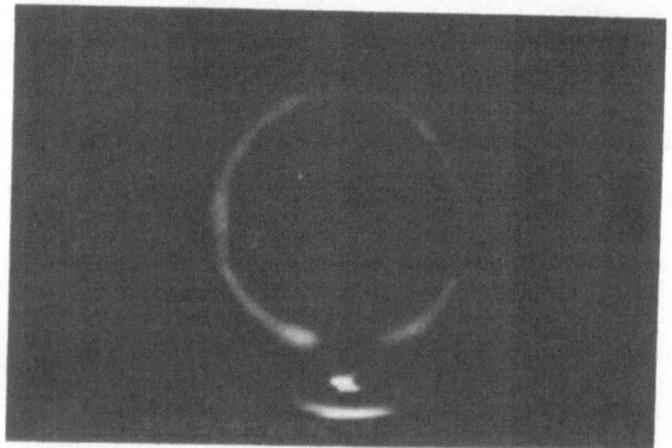

Fig. 9d

Fig. 10a

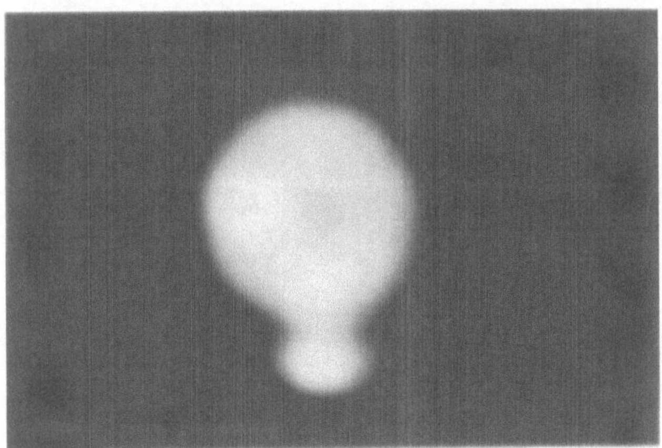

Fig. 10b

Fig. 10 Local mean correction enhanced image obtained by processing the scintigraphic image in Fig. 5.
a) 4 x 4 neighborhood and $\alpha = 0.15$ enhancement
b) 6 x 6 neighborhood and $\alpha = 0.10$ enhancement

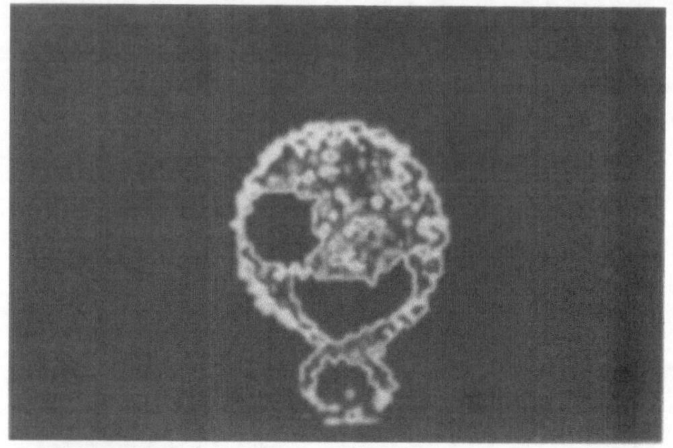

Fig. 11a

Fig. 11 Laplacian and gradient images obtained by processing the enhanced images in Fig. 10.

a) Laplacian image obtianed by processing the enhanced image in Fig. 10a

b) Laplacian image obtained by processing the enhanced image in Fig. 10b

c) Gradient image (sum of the square roots of the Sobel differences d_p and d_q) obtained by processing the enhanced image in Fig. 10a

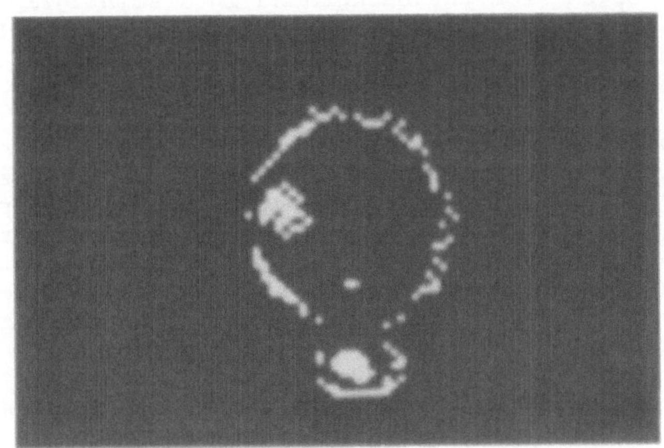

Fig. 11b

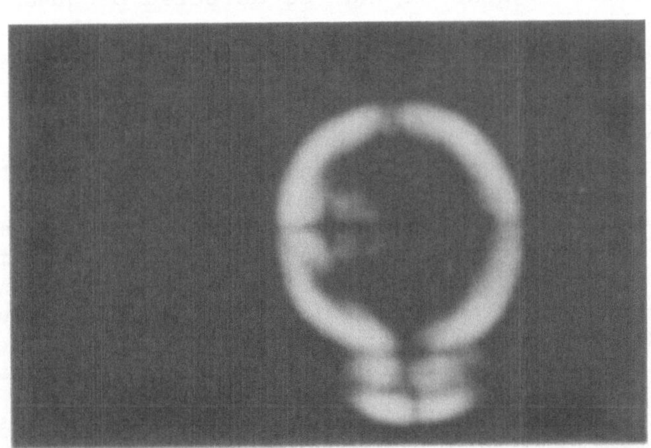

Fig. 11c

e) Thresholding and segmentation through histograms.

A manual thresholding is often convenient and easy to use to separate the organ from the background in a scintigraphic image: the only display of the pixels higher than the threshold gives to the user a straightforward way to find the boundary of the organ. The user also sees by that way which part of the image will not be further processed, as being lower than the threshold. However, such a threshold might take a long time to obtain. A simple way of finding it is by constructing a number of counts histogram of the scintigram (i.e., for each number of counts, the number of pixels that have this number of counts). One can see respectively, Fig. 12, a,b,c, the scintigraphic images of an International Atomic Energy Agency (IAEA) organ phantom [40], of the liver of a patient, and of a geometrical phantom that is similar to the IAEA previous phantoms [41], with their corresponding histograms.

The curve of the number of counts histogram gives information on their distribution in the image. A bimodal histogram, for example, is the result of an image composed of a uniform area and a uniform background. In the case of the liver scintigraphic images, the histograms are almost bimodal, the background noise gives a peak near the origin, and the liver a noisy plateau. The user has therefore to place the threshold between the peak and the plateau to surround the organ. Segmentation by histograms has been used by a number of researchers, [42], including by the author on aerial images [26,43]. The source of the difficulties is that often histogram peaks do not correspond to perceptually distinct areas. This is the case for example when linear gradients are present. Conversely, small regions may not be detected because their contribution to the histogram may be literally overshadowed by the noise from other regions. In the case of scintigraphic images, difficulties come from the facts that, 1) the limits of an object do not correspond to the pixels of maximum gradient, [44] and 2) they are 2-D projections of 3-D emitting objects.

In previous works on aerial photographs [26,43] the author proposed a segmentation scheme based on the extraction of gaussian histograms from the brightness levels histogram: the maximum of each new gaussian histogram is placed at the maximum of the brightness levels histogram from which the preceding gaussian histogram has been subtracted. The standard deviation of the gaussian histogram is a function of the brightness level that was obtained through histograms of calibrated brightness digital images from the device that produced the digital image to segmentate. The mean T_i and the standard deviation t_i of each extracted gaussian histogram define a principal brightness level P_i: all pixels having a brightness level between the two thresholds $(T_i + t_i)$ and $(T_i - t_i)$ take this value P_i. However, some brightness levels might be in the field of several principal levels. The corresponding pixels go through a voting procedure that takes their neighborhood into account: the average of the neighbor pixels is compared

Fig. 12a

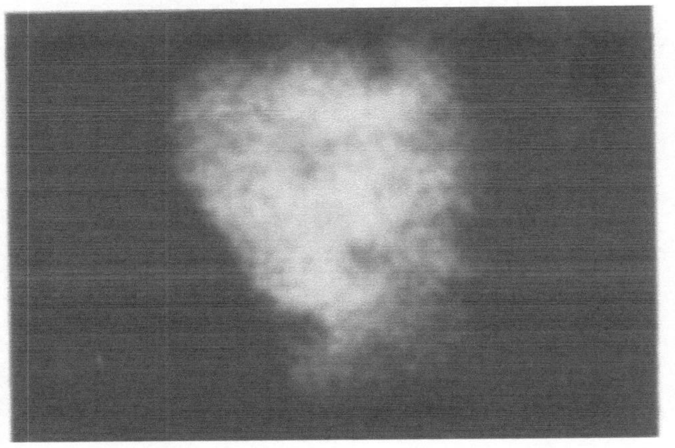

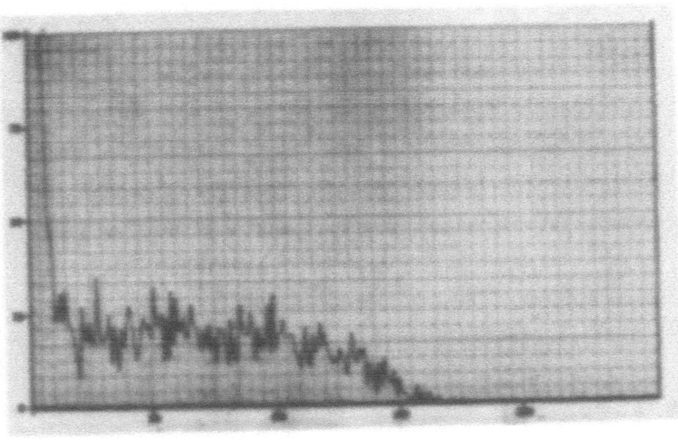

Fig. 12 Scintigraphic images and their corresponding histograms of the number of counts.
a) IAEA organ phantom n°15
b) patient liver
c) geometrical phantom n°7

Fig. 12b

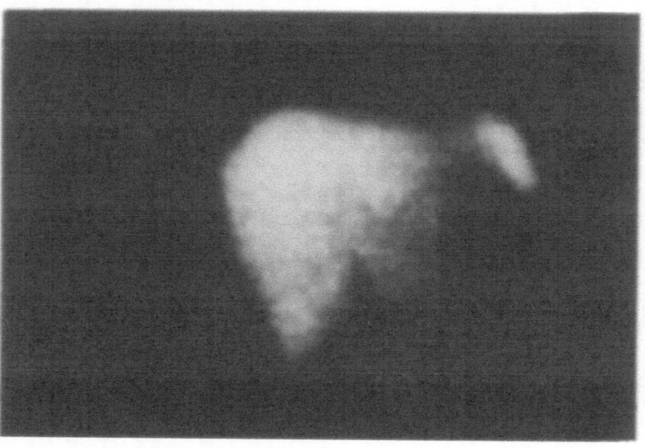

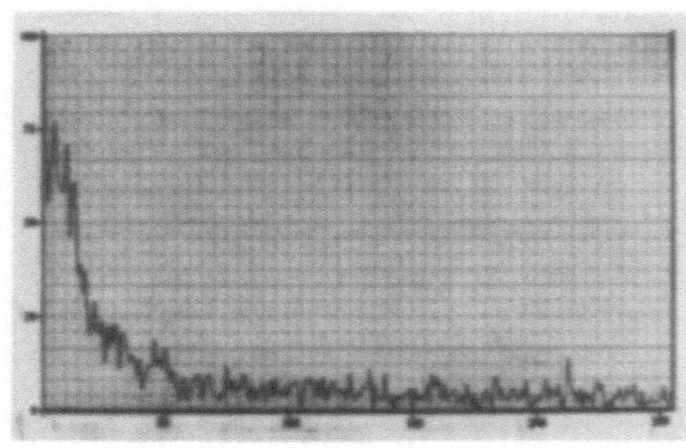

Fig. 12c

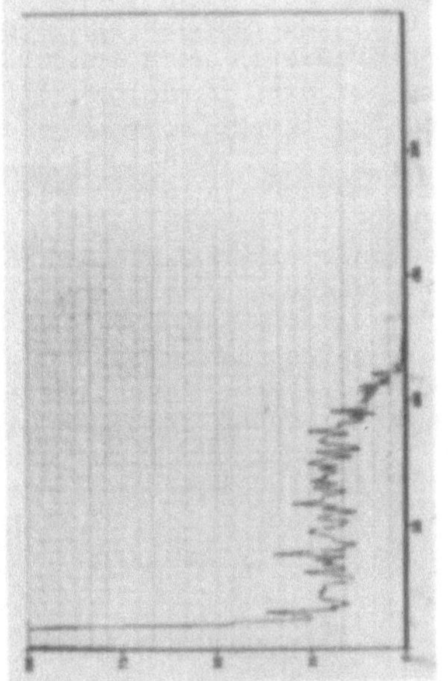

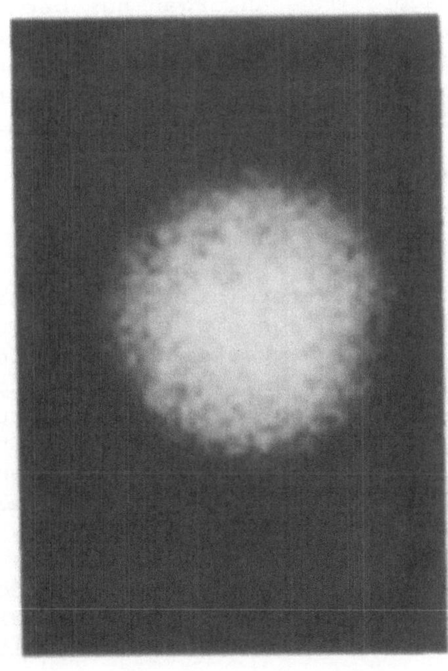

to the challenging principal levels and the pixel receives the value of the nearest principal level. The author proposed a second thresholding method [26,45]: compute and add the gradients of all pixels having the same brightness level, and construct the curve of the sum of gradients as a function of the number of counts. The maximum of this weighted curve gives the threshold. The weighting function may be the square root of the number of pixels at each brightness level, to give an advantage to the pixels having a high gradient. The first method could not be applied to scintigraphic images because there is no way to calibrate the emission of the photons, and the second method failed when we tried to apply it to scintigraphic images because of the low significance of the gradients.

The interactive thresholding scheme gives the physician the possibility of introducing his psychovisual knowledge during the segmentation process. The use of thresholding methods, with sometimes the help of the number of counts histogram of the scintigraphic image, or of some of its corresponding processed images (gradient, laplacian, Wiener filtered, enhanced) to obtain the threshold, happens to be very fruitful in routine.

6. DETECTION OF DEFECTS INSIDE A REGION OF INTEREST

The preceding section was devoted to the techniques of segmentation that are useful to extract the regions of interest on a static scintigraphic image. Once an organ has been enclosed by some contour, a usual processing in the static images is to detect medical patterns that might be further considered as pathological abnormalities. The major constraint in such an approach is that the result of the computer analysis must be obtained in a short time to be useful in clinical routine. Therefore, we designed a program to indicate defects the physician will tell which among them are clinically significant.

In nuclear medicine, the usual approach has been to display the processed images to perform the diagnosis visually [46,47]. Moreover, authors have determined the processed images best suited to the eye for further diagnosis using the ROC and LROC curves [40, 41,48,49]. The aim of the pattern recognition approach is to show the assumed defects as if they were regions of interest, leaving to the physician the responsibility of suppressing the assumed defects he will consider as noise clusters. In such an approach, it is important to obtain all the true positives, even though a lot of false positives are coming out too. An alternative method would be to minimize the number of false positives and to extract the only true positives. In this last case, the process would be completely automatic, but unhappily it is not convenient in clinical routine, because the responsibility of the diagnosis has to be taken by a physician and therefore the result of the computer has to be checked.

The interactive pattern recognition program displays, in two grey levels, the processed areas on the original displayed image in which there are cold lesions. This gives the user a visual means to evaluate the detected defects and reject those he will consider as noise clusters.

To detect and process the areas in which there are lesions and superimpose them on the previously displayed image, the (n x n) image matrix is cut into adjacent submatrices of (p x p) pixels of which the means are computed. The resulting $(\frac{p}{n} \times \frac{p}{n})$ mean matrix is then processed, using different recognition operators.

a) <u>Average operators</u>

Different mean matrices may be obtained according to the average operators used. These may be:

1) The average:

$$m(k,l) = \frac{1}{p^2} \sum_{\varepsilon=0}^{p-1} \sum_{\varepsilon^1=0}^{p-1} s(i+\varepsilon,\ j+\varepsilon^1)$$

where s(i,j) is a pixel of the image matrix, p the size of the submatrix, m(k,l) the resulting mean matrix in which k = [(i-1)/p] + 1 and l = [(j-l)/p] + 1.

The average operator does not attenuate much the medium frequencies and the high frequencies are suppressed by the sampling process when the mean matrix is created. It is a linear filter whose frequency response may easily be plotted [33].

2) The angular moment of order 2:

$$m(k,l) = \sum_{\varepsilon=0}^{p-1} \sum_{\varepsilon^1=0}^{p-1} s^2(i+\varepsilon, j+\varepsilon^1)$$

which has been used by Haralick [50] among a set of operators to classify images of different textures. It is a nonlinear filter where the weighting function is equal to the pixel.

3) The contrast operator:

$$m(k,l) = \sum_{\varepsilon=0}^{p-1} \sum_{\varepsilon^1=0}^{p-1} [(i+\varepsilon) - (j+\varepsilon^1)]^2\ s(i+\varepsilon, j+\varepsilon^1)$$

which has also been used by Haralick [50]. It is a linear filter whose frequency response may easily be computed [33]. It is interesting to use this operator (or its symmetric) to compute the mean, because it does not attenuate the medium frequencies the same way the average filter would do.

4) The binomial operator:

$$m(k,l) = \sum_{\varepsilon=0}^{p-1} \sum_{\varepsilon^1=0}^{p-1} a(\varepsilon+1)a(\varepsilon^1+1)s(i+\varepsilon,j+\varepsilon^1)$$

where a is a p-vector whose values are equal to the coefficients of the $(x+1)^{p-1}$ polynomial. For instance, if $p=3$, $(x+1)^{p-1} = x^2 + 2x + 1$, and $a = (1,2,1)$.

b) Recognition operators

Different recognition operators may be used on the mean matrices, and their efficiency depends on the images that are processed. The following operators were successful on the scintigraphic images.

1) Operator "ICAR": it compares each pixel of the mean matrix to its 8 neighbors. Its value varies from 0 to 4 according to the number of times the considered pixel is lower (or higher) than two neighbor opposite pixels. 8 or 24 additional means may be computed in the neighborhood of the considered submatrix (1 pixel of the mean matrix), and may extend the range of the operator "ICAR" to 8 or 16. The operator "ICAR" computes the number of di rections along which a pixel is a minimum or a maximum.

2) Operator "LW": it computes the sum of the finite differences between the mean of each submatrix and the mean of its 8 neighbors. We have:

$$LW(i,j) = \sum_{1}^{8} (m(i,j) - m(i+\varepsilon,j+\varepsilon^1)) \text{ where } \begin{cases} \varepsilon=[-1,0,1] \\ \varepsilon^1[-1,0,1] \end{cases}$$

and $\varepsilon,\varepsilon^1$ not equal to 0 simultaneously.

In addition, the resulting value LW may be weighted by different functions as

$$\frac{1}{\sqrt{m(i,j)}} \text{ or } \frac{1}{m(i,j)} .$$

This operator is in fact one approximation of the local laplacian. A lot of other approximations have been used in the literature [51,52]. These operators are sometimes called "point templates" or "masks" [29].

3) Operator "CHI2": it computes the CHI2 value over a neighborhood of the considered submatrix versus the linear regression plan of the same neighborhood.

4) Operator "LRP": The linear regression plan is first determined in a neighborhood of (k x k) pixels surrounding the considered submatrix of (p x p) pixels.

Let LRP (i,j) be the value of the linear regression plan in p(i,j).

The means of the considered submatrix of (p x p) pixels and of its neighbors are computed using any average operator. To simplify the presentation, let us use now the average. We have:

$$m(k,l) = \frac{1}{p^2} \sum_{\varepsilon=0}^{p-1} \sum_{\varepsilon^1=0}^{p-1} [s(i+\varepsilon,j+\varepsilon^1) - LRP(i+\varepsilon,j+\varepsilon^1)]$$

$$n_{\eta\eta'}(k,l) = \frac{1}{p^2} \sum_{\varepsilon=-p+\eta}^{-1+\eta} \sum_{\varepsilon'=-p+\eta'}^{-1+\eta'} [s(i+\varepsilon,j+\varepsilon') - LRP(i+\varepsilon,j+\varepsilon')]$$

where m(k,l) is the mean of the considered submatrix, and $n_{\eta\eta'}(k,l)$ the means of its adjacent submatrices.

We have:

$$\eta = \pm p \text{ when } \eta' \neq -p \text{ or } 2p-1$$

and

$$\eta' = \pm p \text{ when } \eta \neq -p \text{ or } 2p-1$$

with

$$\eta = (-p,2p-1) \text{ and } \eta' = (-p,2p-1).$$

The operators ICAR and LW may now be computed with this new set of means m(k,l) and $n_{\eta\eta'}(k,l)$. The value of ICAR (k,l) will be the number of times the mean m(k,l) is lower than the means $n_{\eta\eta'}(k,l)$ and $n_{-\eta-\eta'}(k,l)$ of two opposite adjacent submatrices.

The value of LW will be:

$$LW(k,l) = \sum_{1}^{8} [m(k,l)-n_{\eta\eta'}(k,l)] \text{ where } \begin{cases} \eta = (-p,o,p) \\ \eta' = (-p,o,p) \end{cases}$$

and η, η' not equal to 0 simultaneously.

Once an operator has been computed, a threshold is used on it and all submatrices which result in an operator output higher than the threshold have then to be displayed for visual evaluation.

c) Display operators

Different operators may be used to superimpose a two grey-level display of the selected submatrices on a color display of the analyzed scintigraphic image. This method of superimposition seems to be most convenient to visually distinguish between lesions and clusters of noise. Two operators have been used to display the neighborhood of the selected submatrices.

1) Mean threshold display

The threshold is equal to the mean of all pixels of the neighborhood of the selected submatrices. A neighborhood including all neighbor submatrices was found to be convenient. All pixels whose number of counts is greater than the threshold result in a white point, while all others result in a black point on the output.

2) Linear regression plan display

The threshold is equal to the value the linear regression plan over the neighborhood takes on each pixel of the neighborhood. All pixels whose number of counts are over the linear regression plan result in a white point, while all others result in a black point on the output. A neighborhood including all neighbor submatrices was found to be convenient to better visualize the assumed defects.

The advantage of the second method over the first method (which is faster) is that it gives a view of the defect perpendicular to the tangent plan of the examined organ image. This makes it easier for the physician to evaluate the assumed defects and remove what he will decide to be clusters of noise of morphological defects.

Several results of the program we implemented on our INFORMATEK SIMIS 3 using the described operators on different images of Fig. 12 are given in Fig. 13 to 23 [53].

7. CONCLUSION

The use of pattern recognition techniques for lesion detectability in Nuclear Medicine has become very important, though the use of the psycho-visual evaluation of the displayed scintigrams for diagnosis of lesions is still wide spread in the field. The scintigraphic images are unfortunately very difficult to process because of the lack of statistics, the poor resolution, and the amount of noise. We have presented techniques that happened to be useful for the diagnosis of images: filtering, enhancement, defect detection, etc. To analyze the displayed scintigrams, matching enhancement methods to a clinical context is of interest: these methods must be fast and result in an additional visual tool to the observer. The use of pattern recognition as an alternative method to these psycho-visual methods will extend if the result can be obtained in a very short time. This leads to some interactive program where the user asks the computer for some result he

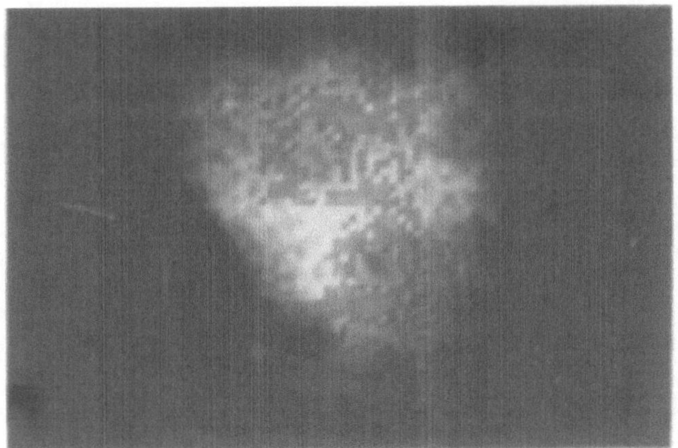

Fig. 13 Detection and 2-level superimposition of defects of the IAEA organ phantom in Fig. 12a, using the linear regression plan display operator.

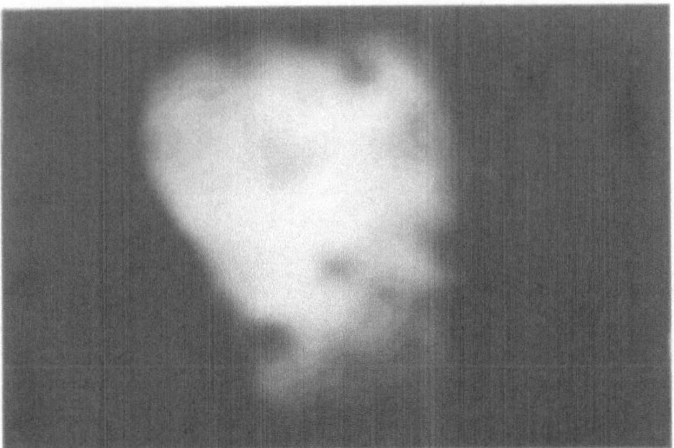

Fig. 14 Pseudo-Wiener filtered image obtained by processing the image in Fig. 12a.

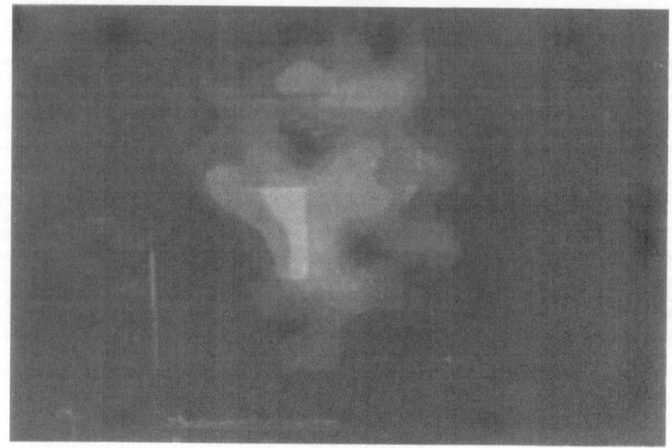

Fig. 15 Detection and 2-level superimposition of defects in Fig. 14 using the mean threshold display operator.

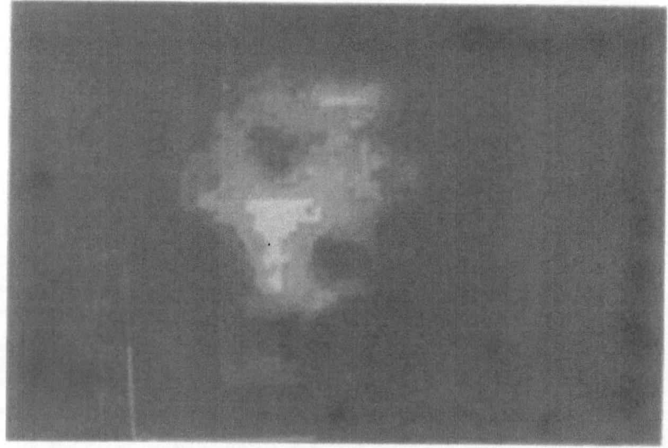

Fig. 16 Detection and 2-level superimposition of defects in Fig. 14 using the linear regression plan display operator.

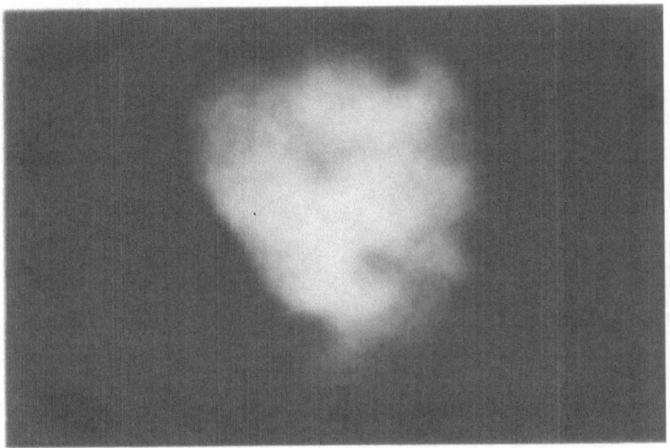

Fig. 17 Local mean correction enhancement image (4 x 4 neighborhood, α = 0.15) obtained by processing the image in Fig. 12a.

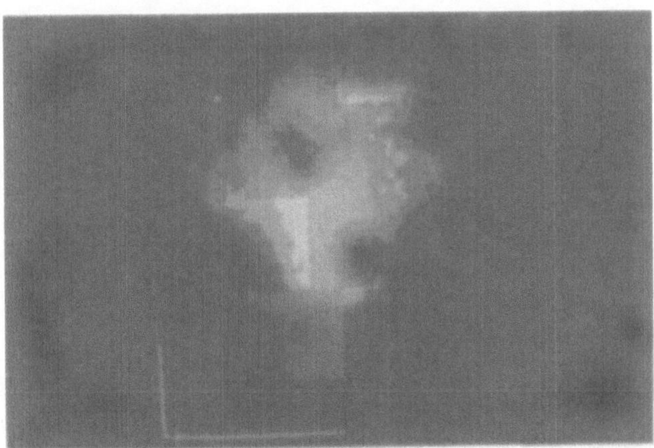

Fig. 18 Detection and 2-level superimposition of defects in Fig. 17 using the linear regression plan display operator.

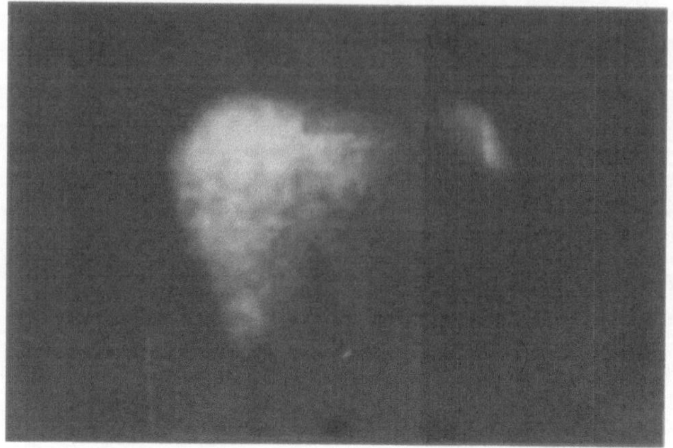

Fig. 19 Detection and 2-level superimposition of defects of the patient liver in Fig. 12b, using the linear regression plan display operator.

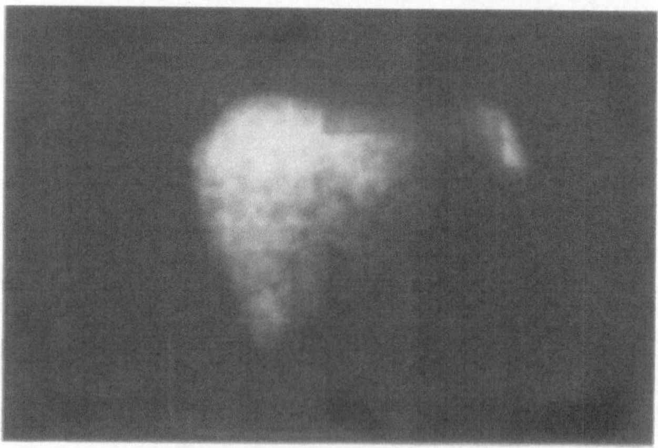

Fig. 20 Detection and 2-level superimposition of defects in Fig. 12b using the mean threshold display operator.

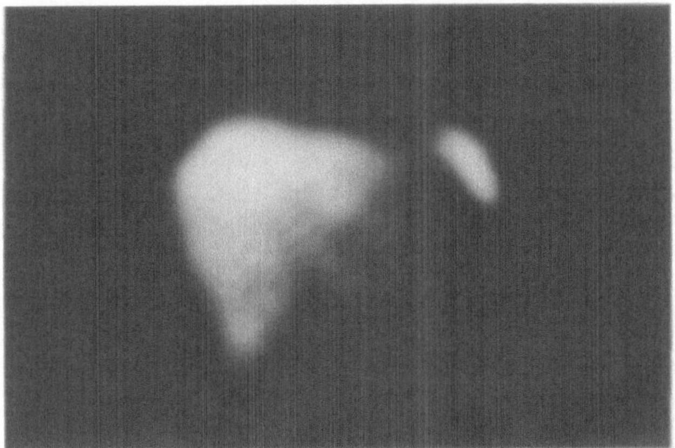

Fig. 21 Pseudo-Wiener filtered image obtained by processing the image in Fig. 12b.

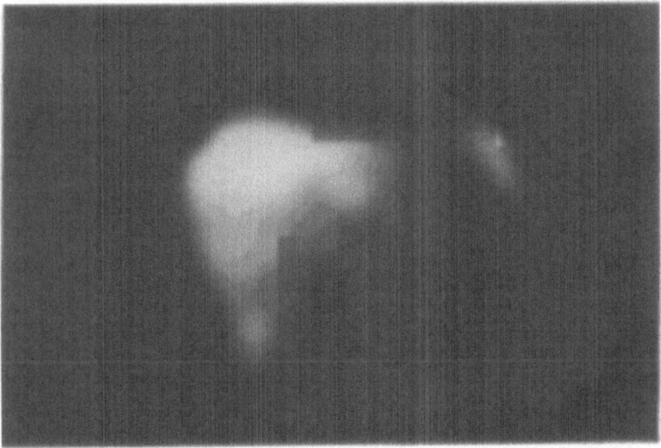

Fig. 22 Detection and 2-level superimposition of defects in Fig. 21 using the linear regression plan display operator.

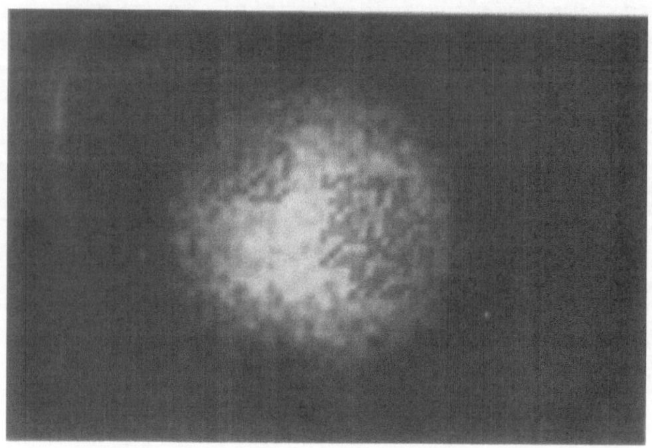

Fig. 23 Detection and 2-level superimposition of defects of the geometrical phantom n°7, using the linear regression plan display operator.

needs. This can be, as in our case, the automatic selection of meaningful areas in some region of interest that may also be selected by the computer; the final selection must nevertheless be performed by a person. Fast methods of segmentation will be very useful, and region growing methods using the texture as a criterion are in prospect. Detection of defects through the use of filters and digital operators was fast enough to be used in clinical routine and happened to be efficient. We are therefore still improving them, using the LROC curves, for an objective comparison of the different operators used.

REFERENCES

1. Hertz S., A. Roberts, R. D. Evans, "Radioactive Iodine as an Indicator in the Study of Thyroid Physiology", Proc. Soc. Exp. Biol. Med. 38 (1938) 510.
2. Hamilton, J. G., M. H. Soley: "Studies in Iodine Metabolism by the Use of a New Radioactive Isotope of Iodine", Am. J. Physiol. 127 (1939) p. 557.
3. Touya, J. J.,"New Applications of Radiopharmaceuticals Labelled with Generator-produced Radionuclides", In Medical Radioisotope Scintigraphy (Proc. Symp. Monte-Carlo, 1972) 2, Vienna, IAEA, pp. 3-25, 1973.
4. Cassen, B., L. Curtis, "The in Vivo Delineation of Thyroid Glands with an Automatically Scanning Recorder", UCLA report n° 130, 1951.
5. Medical Radioisotope Scanning, IAEA, Vienna, 1959.
6. Medical Radioisotope Scanning, vol. 1 and 2, IAEA, Vienna, 1964.
7. Medical Radioisotope Scintigraphy, Vol. 1 and 2, IAEA, Vienna, 1969.
8. Medical Radioisotope Scintigraphy, vol. 1 and 2, IAEA, Vienna, 1973.
9. Medical Radionuclide Imaging, vol. 1 and 2, IAEA, Vienna, 1977.
10. Anger, H. O., "Scintillation Camera", Rev. Sci. Instr. 29, 27 (1958).
11. Kawin, B., F. V. Huston, C. B. Cope, "Digital Processing/Display System for Radioisotope Scanning", Journal of Nuclear Medicine 5, 500-514, 1964.
12. Schepers, H., C. Winkler, "An Automatic Scanning System Using a Tape Perforator and Computer Techniques in Medical Radioisotope Scanning vol. 1, IAEA, Vienna, 321-330, 1964.
13. Brown, D. W., "Digital Computer Analysis and Display of the Radioisotope Scan", Journal of Nuclear Medicine, 5, 802-806, 1964.
14. Kenny, P. J., E. M. Smith eds, "Quantitative Organ Visualization in Nuclear Medicine, University of Miami Press, Coral Gables, Florida, 1971.
15. "Dynamic Studies with Radioisotopes in Medicine, IAEA, Vienna, 1971.

16. "Dynamic Studies with Radioisotopes in Medicine," vol. 1, IAEA, Vienna, 1975.
17. Hine, G. J., J. A. Sorenson Eds, "Instrumentation in Nuclear Medicine, vol. 2, Academic Press, New York, London, 1974.
18. Proceedings of Symposia on Sharing of Computer Programs and Technology in Nuclear Medicine:

CONF - 710425(1971)	Oak Ridge (Tenn.)
CONF - 720430(1972)	Oak Ridge (Tenn.)
CONF - 730627(1973)	Miami (Florida)
CONF - 740531(1974)	Oak Ridge (Tenn.)
CONF - 750124(1975)	Salt Lake City (Utah)
(1976)	Atlanta (Georgia)
CONF - 770101(1977)	Atlanta (Georgia)

19. Metz, C. E., S. M. Pizer, G. L. Brownnell Eds., Information Processing in Scintigraphy, Proc. of the IIIrd Int. Conference, M.I.T., CONF - 730687, USAEC - USERDA, 1973.
20. Raynaud, C., A. Todd-Pokropek Eds., Information Processing in Scintigraphy, Proc. of the IVth Int. Conference, Orsay, C.E.A., 1975.
21. Brill, A. B., R. R. Price, W. J. McClain, M. W. Landay Eds., Information Processing in Medical Imaging, Proc. of the Vth Int. Conference, Nashville, ORNL/BCTIC-2, 1978.
22. Di Paola, R., J. P. Bazin, C. Parmentier, A. Todd-Pokropek, M. Tubiana, C. Zajde, R. Bosshard, "Système en Ligne Adapté au Traitement des Informations Scintigraphiques" in Medical Radioisotope Scintigraphy 1, Vienna, IAEA, 571-582, 1973.
23. Green, D. M., J. A. Swets, Signal Detection Theory and Psychophysics, New-York, Wiley, 1966; reprint with corrections: R.E. Krieger Pub. Co., Huntington, N. Y., 1974.
24. Metz, C. E., S. J. Starr, L. B. Lusted, K. Rossman, "Progress in Evaluation of Human Observer Visual Detection Performance Using the ROC Curve Approach", in Information Processing in Scintigraphy (Proc. IVth Int. Conf.) Orsay, edited by C. Raynaud and A. Todd-Pokropek, pp. 420-439, 1975.
25. Nagai, T., T. Iinuma, "A Comparison of Differential and Integral Scans", Journal of Nuclear Medicine 9, pp. 202-204, 1968.
26. Kahn, E., "Prétraitement en Reconnaissance des Formes: Application à l'extraction de Contours, à la Simplification et à la Corrélation Numérique", Thèse de Doctorat d'Etat, Université Pierre et Marie Curie, Paris (France), 1976.
27. Roberts, L. G., "Machine Perception of Three-Dimensional Solids", in"Optical and Electro-Optical Information Processing", J. T. Tippett et al Eds., Cambridge, MA, MIT Press, 1965.
28. Duda, R. O., P. E. Hart, "Pattern Classification and Scene Analysis", New York, Wiley, 1971.
29. Frei, W., C. C. Chen, "Fast Boundary Detection, a Generalization and a New Algorithm", IEEE Transactions on Computers C-26, pp. 988-998, 1977.

30. Schachter, B. J., A. Rosenfeld, "Some New Methods of Detecting Step Edges in Digital Pictures", Communications of the ACM 21, 172-176, 1978.
31. Rosenfeld, A., "Picture Processing by Computer", New York, Academic Press, 1969.
32. Soussaline, F., A. E. Todd-Pokropek, R. Di Paola, J. P. Bazin, "Techniques for Combining Isotopic Images Obtained at Different Energies", in Information Processing in Scintigraphy (Proc. 4th Int. Conf.) Orsay, Edited by C. Raynaud and A. Todd-Pokropek, pp. 17-42, 1975.
33. Kahn, E., "Le Filtrage Digital", Thèse de Doctorat de 3è cycle, Université de Paris (France), 1970.
34. Berche, C., R. Di Paola, "Analyse Fréquentielle et Filtrage des Images Scintigraphiques", Journal de Biologie et de Médecine Nucléaires, 2e partie, ATEN, no 37, 1974.
35. Hunt, W. A., W. J. Lorenz, H. G. Meder, H. Luig, P. Pistor, P. Schmidlin, G. Walch, M. G. Schmitt, "Digital Processing of Scintigraphic Images," Wissenschaftliches Zentrum Heidelberg, Germany, IBM Technical Report 70.03.001, 1970.
36. Pistor, P., "Digital Processing of Scintigraphic Images by Two Dimensional Recursive Wiener Filters", Wissenschaftliches Zentrum Heidelberg, Germany, IBM Technical Report 70.12.006,1970.
37. Pistor, P., P. Georgi, G. Walch, "The Heidelberg Scintigraphic Image Processing System", Proceedings of Second Symposium on Sharing of Computer Programs and Technology in Nuclear Medicine, Oak Ridge, Tennessee, 1972, pp. 411-427.
38. Berche, C., R. Di Paola, "Analyse Fréquentielle et Filtrage des Images Scintigraphiques," Journal de Biologie et de Médecine Nucléaires, ATEN, no38, 1974.
39. Hummel, R., "Image Enhancement by Histogram Transformation", Computer Graphics and Image Processing 6, pp. 184-195, 1977.
40. INTERNATIONAL ATOMIC ENERGY AGENCY, "IAEA Co-ordinated Research Programme on the Intercomparison of Computer-assisted Scintigraphic Techniques, Third Progress Report "In Medical Radionuclide Imaging (Proc. Symp. Los Angeles, 1976) 1, Vienna, IAEA, pp. 585-615, 1977.
41. INTERNATIONAL ATOMIC ENERGY AGENCY, "IAEA Co-ordinated Research Programme on the Intercomparison of Computer-assisted Scintigraphic Techniques: Progress Report", In Medical Radioisotope Scintigraphy (Proc. Symp. Monte-Carlo, 1972) 1, Vienna, IAEA, p. 727, 1973.
42. Pavlidis, T., "Structural Pattern Recognition", Springer-Verlag, New-York, 1977.
43. Kahn, E., "Un Système d'analyse Optique: le CALIFE", Nouvelle Revue d'optique Appliquée 3,pp. 209-215 et pp. 273-279, 1972.
44. Aubert, B., "Etude Quantitative de la Fixation d'isotopes Radioactifs dans l'organisme", Thèse de Doctorat 3è cycle, Université Paul Sabatier, TOULOUSE (FRANCE), 1974.

45. Kahn, E., D. Piquet-Pellorce, "Etude de Corrélation Numérique Automatique en Vue de l'obtention d'un Fichier Altimétrique", Actes du Symposium de la Société Internationale de Photogrammétrie (commission IV), Paris, 1974, Bulletin de la Société Francaise de Photogrammétrie 57, 1975, pp. 30-33.
46. Pizer, S. M., G. L. Brownell, D. A. Chesler, "Scintigraphic Data Processing", in Instrumentation in Nuclear Medicine 2, G. J. Hine, J. A. Sorenson eds, Academic Press, New York, 1974, pp. 229-262.
47. Todd-Pokropek, A. E., S. M. Pizer, "Displays in Scintigraphy" In Medical Radionuclide Imaging 1 (Proc. Symp. Los Angeles, 1976), Vienna, IAEA, 1977, pp. 505-537.
48. INTERNATIONAL ATOMIC ENERGY AGENCY, "IAEA Co-ordinated Research Programme on the Intercomparison of Computer-assisted Scintigraphic Techniques: Second progress Report", In Medical Radionuclide Imaging (Proc. Symp. Los Angeles, 1976) 1, Vienna, IAEA, pp. 571-584, 1977.
49. Houston, A. S., M. A. Macleod, "An Intercomparison of Computer Assisted Data Processing and Display Methods in Radioisotope Scintigraphy Using Mathematical Tumours", Phys. Med. Biol. 22, pp. 1097-1114, 1977.
50. Haralick, R. M., "Texture-tone Study with Application to Digitized Imagery", Lawrence, Kansas, the University of Kansas Center for Research Inc., RSL-TR-182-6, 1974.
51. Rosenfeld, A., A. C. Kak, "Digital Picture Processing", Academic Press New-York, 1976.
52. Amman, W. W., R. Vaknine, "Structure Analysis - a New Method for Evaluating Scintigrams" In Information Processing in Scintigraphy (Proc. IVth Int. Conf.) C. Raynaud, A. Todd-Pokropek Eds. Orsay, 1975, pp. 66-79.
53. Kahn, E., R. Di Paola, J. P. Bazin, "An Interactive Program for Lesion Detectability in Scintigraphic Images", submitted to the Journal of the Nuclear Medicine.

A CONTRIBUTION TO THE AUTOMATIC PROCESSING OF ELECTROCARDIOGRAMS USING SYNTACTIC METHODS

Gustavo Belforte
Istituto di Elettrotecnica Generale,
Politecnico di Torino

Renato De Mori
Istituto di Scienza dell'Informazione
Università di Torino

Franco Ferraris
Centro per l'Elaborazione Numerale dei Segnali,
CNR, Torino

ABSTRACT. A syntactic method for analysing ECG waveforms considered as a multichannel process is presented. A procedure is designed in order to extract the time evolution of the rhythm using the energy of the ECG derivatives.

This method performs a coding by table look-up methods. The parsing is performed with simple finite state automata inferred by experiments and suitable to be updated during experiment execution. The possible ambiguities can be solved introducing fuzzy or stochastic linguistic variables processed by fuzzy or stochastic automata. Some experimental results are presented.

1. INTRODUCTION

Investigations on the automatic analysis of ECG waveforms have recently received a remarkable attention particularly for the following applications: rapid analysis of long records, on-line patient monitoring, extraction of informations from exercise electrocardiograms [1-5] and, perhaps, automatic diagnosis [6].

Research performed in this area has been primarily devoted to the design of special circuits or systems generally built up around a digital computer. Examples of works oriented towards practical applications can be found in [7-9].

Other research has been developed for conceiving ECG methodologies involving theories of signal processing [10-12], functional approximation of waveforms [13-15], data compression via transforms [16-19], feature extraction [20-22], and pattern recognition [23-28].

This paper proposes a methodology for automatic processing of ECG waveforms based on syntactic algorithms. Motivation for using a syntactic approach resides in the fact that human inspection of ECG waveforms is firstly an extraction of structural and qualitative information. Once this information has been obtained and some typical forms (like a QRS complex) have been recognized, the numerical values of the durations and amplitudes useful for diagnosis are measured. Similar approaches have been recently followed for the analysis of other biological waveforms (see [27] for example) and speech patterns [29-39]. Other interesting descriptions of syntactic pattern recognition methods can be found in [34-36].

Some of the basic ideas underlying the project described in this paper follow. The presentation of such ideas will also contribute to properly allocate this paper in the current literature on automatic processing of ECGs.

The first important point of this design is that a functional approximation of the waveform is performed after an automatic extraction of the fundamental period. This allows to make a "beat-synchronous" analysis with advantages, well-known to experts in signal processing [30], when performing, for example, a Fourier transformation. Furthermore, even if a piece-wise polynomial approximation of the waveform has to be performed (for discussion on this technique for ECG analysis, see Horowitz [31]) "beat-synchronous" analysis gives remarkable advantages. In fact, the approximation can be done, period-by-period, using iterative algorithms of discrete optimization. The advantages of using such algorithms for a precise evaluation of the position of some important points of the ECG waveform has been demonstrated by Pavlidis [14]. On the other hand, the application of an iterative method is practically acceptable only if a short interval of the entire ECG signal is approximated.

A second important point of this design is that it makes use of the redundancy implicit in the fact that the ECG signals are taken from different leads. Therefore these signals appear as the outputs of a multichannel information system. The same physical event must produce, even with different evidence, synchrounous features in all the channels. This allows to extract very carefully the time evolution of the rhythm before approximating the ECG waveform. Although the details of the algorithm will be described later in this paper, it is worth noticing that the first derivative of the ECG signal is a too crude feature to be used alone for extracting the QRS complexes. Nevertheless it becomes a very reliable feature when more ECG waveforms coming from different leads are analyzed simultaneously.

The following paragraphs will describe in detail a syntactic algorithm for the extraction of the rhythm in ECG waveforms considered as outputs of a multichannel system. A similar methodology can be used for extracting other types of information after waveform approximation.

2. PREPROCESSING, FEATURE EXTRACTION AND CODING

ECG waveforms are recorded from triplets of leads simultaneously filtered with Butterworth four poles filters, having 100 Hz as cutoff frequency, sampled and quantized at a sampling rate of 1000 Hz and stored into files on a 21MX Hewlett-Packard computer. Then, each signal is passed through a notch filter (see [10] for the design of such filter). Such filter does not require too much computation time because the computer is equipped with a Fast-Fortran processor. These operations, as well as others described in the following, may not be strictly necessary in an on-line processor; they are performed in the actual project because its task is that of verifying the methodology and investigating the real usefulness of each component.

The filtered waveforms are then normalized in order to make the samples of each waveform range between the minimum and maximum values allowed in a fixed-point representation of the samples.

Let T be the sampling period and $x_i(nT)$ the n-th sample of the normalized waveform taken at the i-th lead. Then, for each sample of each waveform, the digital derivative $y_i(nT)$, defined as follows:

$$y_i(nT) = k \frac{x_i(nT) - x_i(nT - T)}{T} \text{ ; for all } n, i \qquad (1)$$

is taken.

The calculation of the digital derivative as performed in Eq. (1) is very crude; it is easy to see that such digital differentiator has a frequency response approaching that of the ideal one only for small values of $2\pi fT$; thus an acceptable differntiator, for a signal having a given bandwidth, can be obtained by making T small enough as it happens approximately in our case.

Fig. 1 shows an example of ECG waveforms, its derivative and the pulses corresponding to the squares $y_i^2(nT)$ of the derivative.

Using such preprocessing, tracking the time evolution of the rhythm is performed by recognizing sequences of peaks, suitably time-aligned, at the output of the multichannel system.

Some elemental considerations can be done just by inspection of Fig. 1: the rhythm cannot be tracked by a simple peak-picking technique on any waveform. The peaks of the squares of the derivative appear as various types of sequences for QRS events.

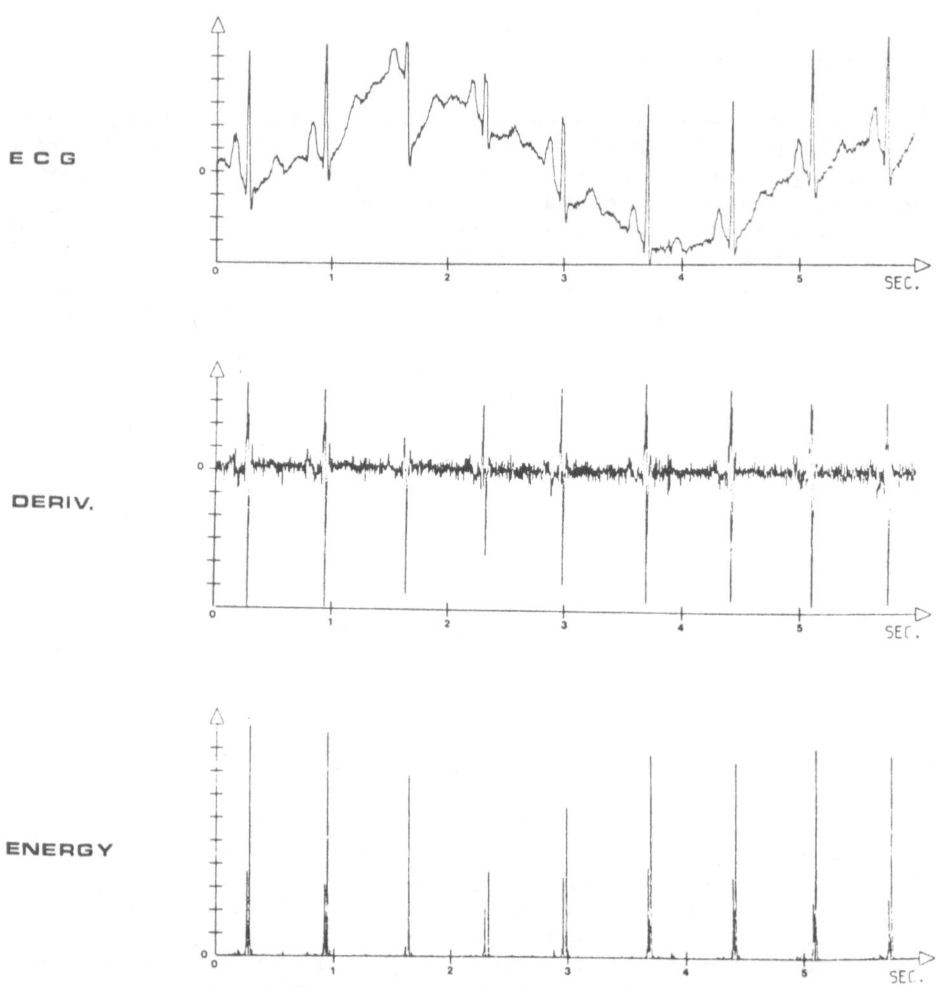

Figure 1. Preprocessing of an ECG Einthoven leads:
ECG: ECG waveform;
DERIV.: first derivative;
ENERGY: energy of the derivative.
The amplitudes are normalized.

After processing, all the peaks of $y_i^2(nT)$ are represented by points on a plane whose coordinates are the duration (d) and maximum amplitude of each peak (a). This representation allows one to consider, after having performed an extensive set of experiments, some dense areas corresponding to the peaks assigned during learning to QRS events and areas of peaks corresponding to other events or noise. Each area that appears in the a-d plane can be assigned a linguistic variable that represents the coding for all the peaks whose representative points fall into that area.

As in most of the pattern recognition problems of this type, the areas corresponding to positive events (QRS) and to negative events often overlap. This would imply that the linguistic variables assigned to each point should be fuzzy linguistic variables and semantic rules should be applied together with syntactic rules for recognizing the events that one wishes to characterize in the recognition process. As the syntactic rules that will be introduced in the following can be formulated in terms of regular expressions, the semantic rules that eventually should accompany the syntactic rules are based on the usual min-max operations [32].

For the particular application considered in this paper it has been decided to perform a purely deterministic coding, by introducing more linguistic variables and a very simple tabular method for their assignment; this assignment is defined by Table I. Such simplification implies that more complex grammars, with more terminal symbols, have to be introduced in the following steps. An a-posteriori analysis of the inferred grammars confirms that they are still very simple and this makes the initial choice acceptable.

3. SYNTACTIC RECOGNITION OF QRS COMPLEXES

The sequences of energy peaks of the derivatives of the ECG waveforms corresponding to different leads are coded into string of messages according with the coding system defined by Table I. Peaks are considered belonging to different events every time the distance between them is higher than 80 msec. Thus, strings are separated by end of string symbols issued every time no significant peaks are detected in a signal for a time interval higher than 80 msec.

Fig. 2 shows three ECG signals taken at three Einthoven leads. Each ECG signal is represented with the corresponding derivative and the energy of the derivative.

All the peaks above a fixed threshold of the energy of the derivatives are coded according with Table I and the strings so obtained are written below the sequences of peaks. The "end of string" symbol is represented by W. Not all the strings obtained in this way correspond to QRS complexes; some of such strings never appear to describe a QRS complex. A string that *may* correspond to a QRS complex is called QRS *hypothesis*. A QRS hypothesis on the i-th lead signal will be indicated by Q_i. A Q_i is recognized by a grammar that has been inferred by experiments using the approach described in Section 4.

TABLE I

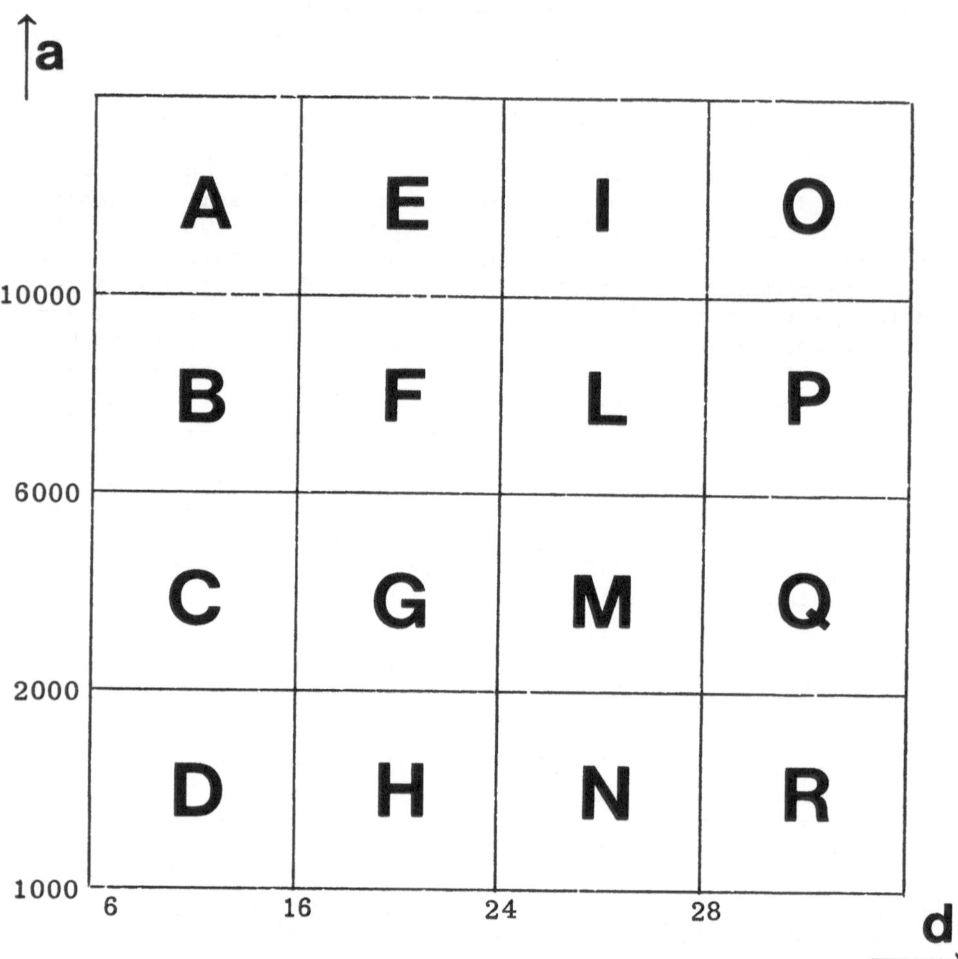

Coding system for the syntactic recognition of QRS complexes

a - maximum amplitude of each peak (normalized units)

d - duration of peak; the unit is the number of samples (1000 Hz of sampling rate)

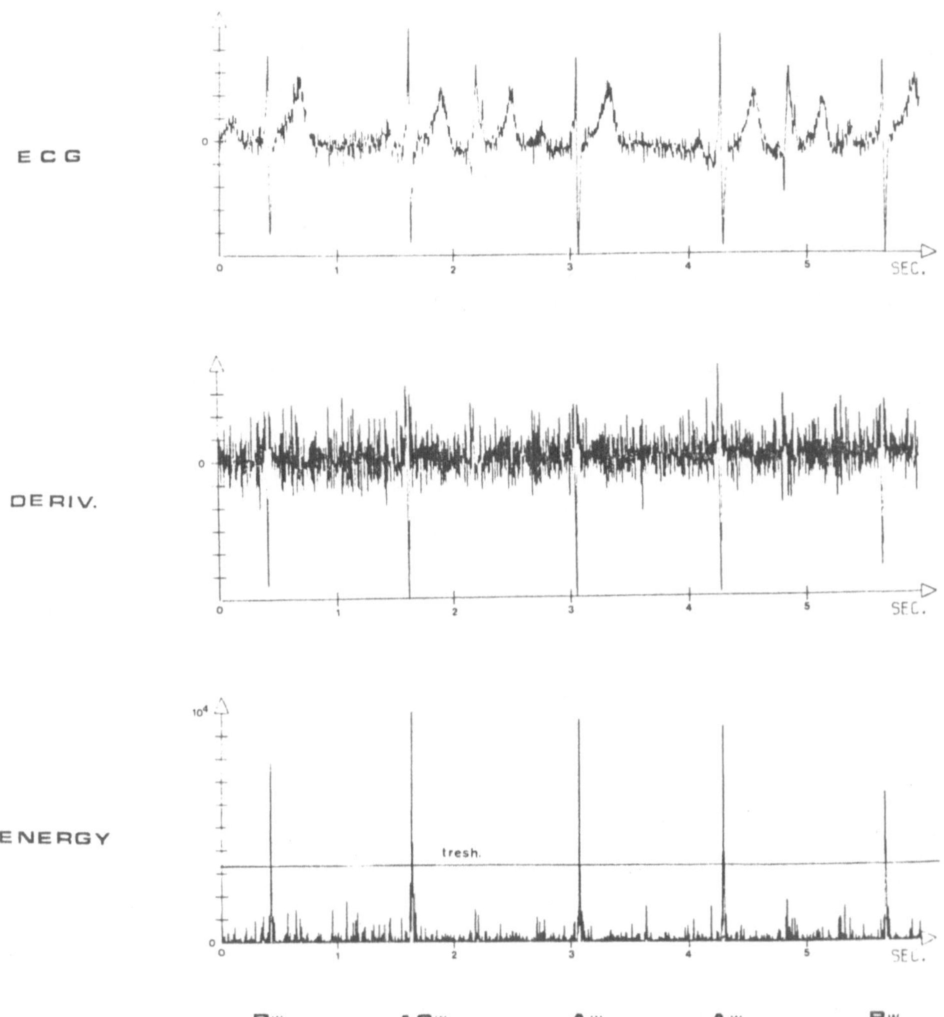

Figure 2. Coding system of ECG Einthoven leads:

Fig. 2A - First lead

Fig. 2B - Second lead

Fig. 2C - Third lead

The amplitude of the ECG signal and its first derivative are normalized. The amplitudes of the energy of the derivative are consistent with values in Table I.

tresh. is the threshold (see sec. 3).

Below are reported the strings according with Table I.

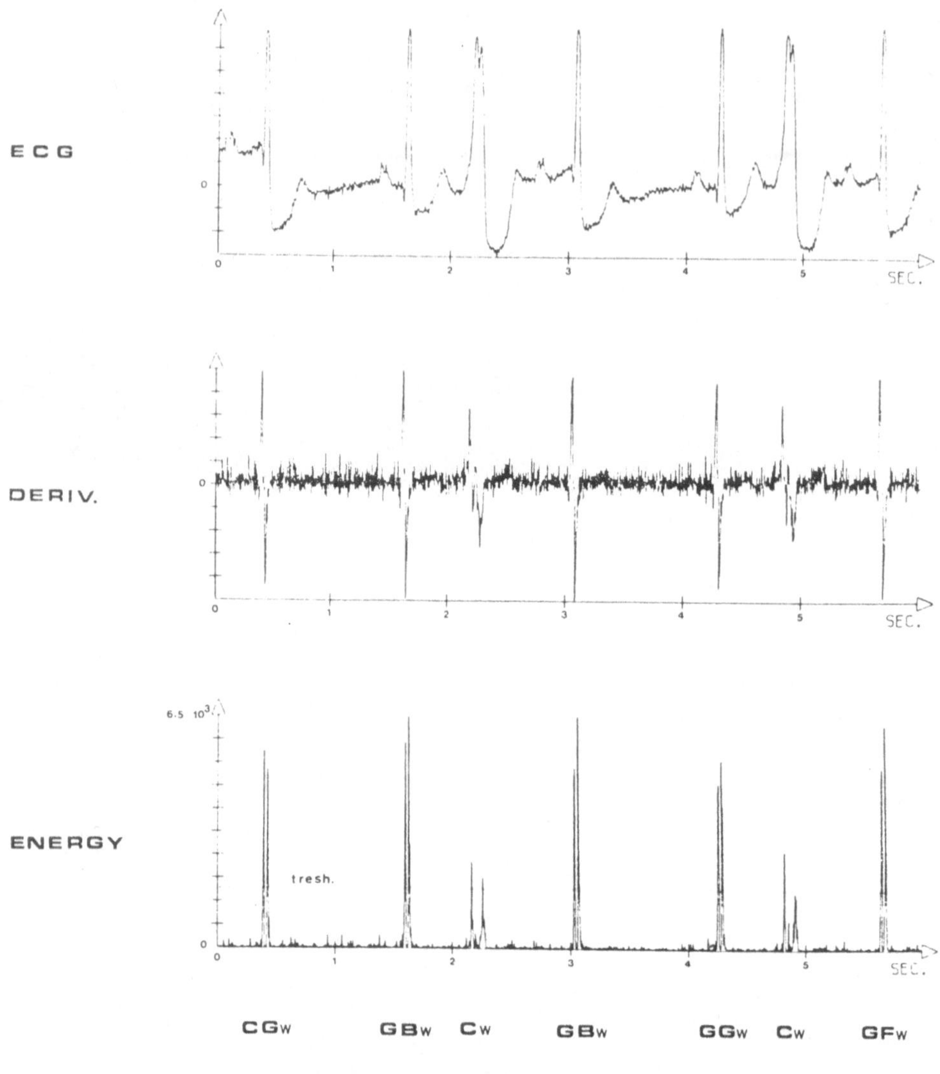
ECG
DERIV.
ENERGY
6.5 10^3
tresh.
SEC.
CGw GBw Cw GBw GGw Cw GFw

Fig. 2C

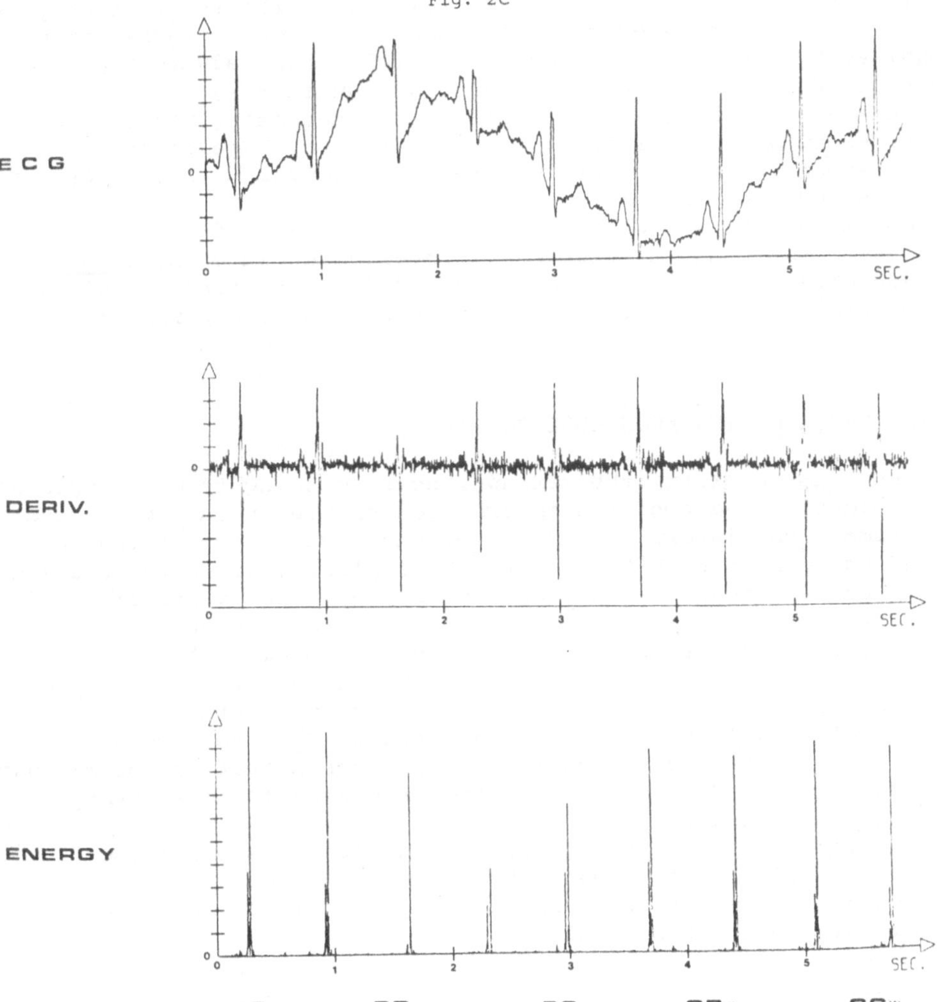

In order to allow a larger flexibility in the recognition process, the syntactic category Q_i has been reserved to those strings recognized by a grammar derived from samples that always appeared with true QRS complexes during learning. Another syntactic category, named Z_i, has been introduced for representing the strings recognized by another grammar, also inferred by experiments whose samples appeared in the learning process, sometimes as true QRS complexes and sometimes as false-alarms.

Combining the symbols Q_i and Z_i $(i = 1,2,3)$ into syntactic rules, a grammar that transforms QRS hypotheses into QRS events has been obtained. Such a grammar has QRS as starting symbol and will be presented in Section 5 together with some experimental results.

4. INFERENCE OF THE KNOWLEDGE SOURCE

For each lead waveform, the strings, corresponding to QRS complexes, are considered as a positive information sample and are processed by a grammatical inference algorithm; analogously the strings corresponding to events that are not QRS complexes, are saved and are considered as a negative information sample to be eventually used in a further generalization step of the inference procedure (see Fu and Booth [33] for the details of this problem). In this way it is possible to build two grammars; the first one generates only positive sequences (corresponding to QRS events), starting from a nonterminal symbol that will be indicated as Q_i; the second one generates sequences corresponding to hypotheses that may or may not correspond to a QRS complex, starting from a nonterminal symbol that will be indicated as Z_i.

The learning algorithm infers linear grammars based upon formal derivatives and k tails (see Fu and Booth [33] sect. III-G); it is implemented with about 3000 assembler instructions and is based on the following definitions and steps.

Definitions

Let $G^j(n)$ be the j-th grammar after the presentation of n samples of the j-th language. Let $I^j(n+1)$ be the string describing the (n+1)-th sample of the j-th language.

If $I^j(n+1)$ is recognized by the automaton that accepts $L(G^j(n))$, the productions of the grammar are not altered.

If $I^j(n+1)$ is not in the language $L(G^j(n))$, then a new grammar is generated by the following algorithm AL1.

Algorithm AL1

Step 1. The regular expression $R^j(n)$ is obtained from the automaton that recognizes $L(G^j(n))$.

Step 2. The regular expression:

$$R^j(n+1) = R^j(n) + I^j(n+1) \tag{2}$$

is considered.

Step 3. The set:

$F[R^j(n+1)] = \{x_i/x_i$ is a symbol of $R_j(n+1)$ that is not preceded by other symbols$\}$ is derived.

Step 4. Eq. (2) is rewritten as follows:

$$R^j(n+1)=x_1 D_{x_1}[R^j(n+1)]...+x_i D_{x_i}[R^j(n+1)]...+x_{F1} D_{x_{FN}}[R^j(n+1)] \tag{3}$$

where $D_{x_i}[R^j(n+1)]$ is the derivative of the regular expression $R^j(n+1)$ with respect to x_i and FN is the number of elements of the set $F[R^j(n+1)]$.

Step 5. Steps 3 and 4 are repeated for all the derivatives in Eq. (3) and for the derivatives of the derivatives and so on, until no more sets are found. At this point a deterministic finite state automaton can be found in a straightforward way from $R^j(n+1)$. Notice that steps 3, 4 and 5 of AL1 realize all the possible left factorizations on $R^j(n+1)$.

The automaton does not generally have the minimum number of states. State minimization is carried out with a simplified version of the general theory because the automaton recognizes only strings of finite length. Thus each state can be assigned a distance from the final state equal to the maximum length of the string that causes a move from this state to the last final state.

Step 6. Detection of equivalent states is carried out easily considering states having the same distance from the final state, starting from those having distance one and continuing with sets of states having increasing distance.

Step 7. A right linear grammar is obtained from the reduced automaton using the straightforward procedure that can be found in [33].

In order to infer the grammar of the possible QRS hypotheses, segments of ECG signals of healthy and ill patients have been analyzed.

A set of three patients have been considered for learning. For each patient segments of 30 sec. duration of the three Einthoven leads have been described.

Each peak, in the energy of the derivative of the signals, has been coded accordingly with Table I and intervals between two successive peaks having duration higher than 80 msec. have been coded by an "end-of-string" symbol W.

A set L_1 of different strings has been obtained from the set of ECG signals used for learning. The strings of L_1 are represented, without the "end-of-string" symbol, in Table II. Such strings have the property that they do not belong to the negative information sample, i.e. they never appear to describe false alarm peaks.

Using the algorithm AL1, the grammar $G_1 = (V_{T_1}, V_{N_1}, P_1, Q)$ has been inferred.

The components of G_1 are defined as follows:

$$V_{T_1} = \{A,B,C,D,E,F,G\} \ ;$$

$$V_{N_1} = \{S_1, S_2, \ldots, S_i, \ldots, S_{19}, Q\} \ ;$$

Q is the symbol that represents the recognition of a QRS hypothesis from the description of the energy peaks of the derivative of any whatever lead signals; P_1 is the set of the inferred rewriting rules listed in Tab. III.

The grammar G_1 has been generalized with heuristic considerations to accept an infinite number of strings and a new grammar, named $G_2 = (V_{T_2}, V_{N_2}, P_2, Q)$ has been obtained.

The components of $G_2=(V_{T_2}, V_{N_2}, P_2, Q)$ are defined in the following:

$V_{T_2} = \{A,B,C,D,E,F,G,H\}$

$V_{N_2} = \{U_1, U_2, U_3, U_4, Q\}$

P_2 : $Q \to CU_1$; $Q \to GU_1$; $Q \to DU_2$; $Q \to HU_2$;

$Q \to AU_4$; $Q \to BU_4$; $Q \to EU_4$; $Q \to FU_4$;

$U_1 \to DU_2$; $U_1 \to HU_2$; $U_1 \to CU_3$; $U_1 \to GU_3$;

$U_1 \to AU_4$; $U_1 \to BU_4$; $U_1 \to EU_4$; $U_1 \to FU_4$;

$U_2 \to CU_1$; $U_2 \to GU_1$; $U_2 \to DU_2$; $U_2 \to HU_2$;

$U_2 \to AU_4$; $U_2 \to BU_4$; $U_2 \to EU_4$; $U_2 \to FU_4$;

$U_3 \to DU_2$; $U_3 \to HU_2$; $U_3 \to AU_4$; $U_3 \to BU_4$;

$U_3 \to EU_4$; $U_3 \to FU_4$; $U_3 \to CU_4$; $U_3 \to GU_4$;

$U_4 \to AU_4$; $U_4 \to BU_4$; $U_4 \to CU_4$; $U_4 \to DU_4$;

$U_4 \to EU_4$; $U_4 \to FU_4$; $U_4 \to GU_4$; $U_4 \to HU_4$;

$U_4 \to A$; $U_4 \to B$; $U_4 \to C$; $U_4 \to D$

$U_4 \to E$; $U_4 \to F$; $U_4 \to G$; $U_4 \to H$

TABLE II

Strings belonging to the language L_1		
CCCC	CCB	BAC
CCCB	GB	AB
CCBC	GCCC	BCC
CCG	BF	CA
CCF	FF	CCA
CBCA	FG	CE
CGGG	FBB	F
CF	GF	BE
CCBB	BB	BCA
BCG	BC	CBA
CB	AC	AAA
B	AA	ABA
A	BA	ABCB

TABLE III

List of rewriting rules P_1 inferred from L_1

$Q \to CS_1$;	$Q \to B$;	$Q \to AS_3$;	$Q \to GS_4$;
$Q \to FS_5$;	$Q \to BS_2$;	$Q \to A$;	$Q \to F$;
$S_1 \to CS_6$;	$S_1 \to BS_7$;	$S_1 \to GS_8$;	$S_1 \to F$;
$S_1 \to A$;	$S_1 \to E$;	$S_1 \to B$;	
$S_2 \to CS_9$;	$S_2 \to C$;	$S_2 \to F$;	$S_2 \to B$;
$S_2 \to AS_{10}$;	$S_2 \to A$;	$S_2 \to E$;	
$S_3 \to C$;	$S_3 \to AS_{11}$;	$S_3 \to A$;	$S_3 \to BS_{12}$;
$S_3 \to B$			
$S_4 \to B$;	$S_4 \to CS_{13}$;	$S_4 \to F$;	
$S_5 \to F$	$S_5 \to G$;	$S_5 \to B$;	
$S_6 \to CS_{15}$;	$S_6 \to BS_{16}$;	$S_6 \to B$;	$S_6 \to G$;
$S_6 \to F$;	$S_6 \to A$;		
$S_7 \to CS_{17}$;	$S_7 \to A$;		
$S_8 \to GS_{18}$;			
$S_9 \to G$;	$S_9 \to C$;	$S_9 \to A$;	
$S_{10} \to C$;			
$S_{11} \to A$;			
$S_{12} \to A$;	$S_{12} \to CS_{14}$;		
$S_{13} \to CS_{19}$;			
$S_{14} \to B$;			
$S_{15} \to C$;	$S_{15} \to B$;		
$S_{16} \to C$;	$S_{16} \to B$;		
$S_{17} \to A$;			
$S_{18} \to G$;			
$S_{19} \to C$.			

The grammar G_2 is supposed to recognize a language that is a generalization of L_1. Another grammar $G_3=(V_{T_3},V_{N_3},P_3,Z)$ has been inferred and heuristically generalized for representing sequences of peaks that may be or may not be generated by QRS waveforms. Attention has been paid in order to ensure that:

$$L(G_2) \cap L(G_3) = \emptyset$$

where $\emptyset$ is the empty set.

The details of G_3 are listed below.

$V_{T_3} = \{C,G,D,H\}$

$V_{N_3} = \{Y_1,Y_2,Z\}$

P_3 : $Z\to DY_1$; $Z\to HY_1$; $Z\to CY_2$; $Z\to GY_2$;

$Y_1\to DY_1$; $Y_1\to HY_1$; $Y_1\to CY_2$; $Y_1\to GY_2$;

$Y_2\to DY_2$; $Y_2\to HY_2$; $Y_2\to C$; $Y_2\to G$

After inspection of G_2 and G_3 it results that, grouping the input variables, more simple grammars may be obtained without affecting the recognition power of the system.

The following renaming of the variables may be performed:

$$a = A \vee B \vee E \vee F$$

$$b = C \vee G \qquad (4)$$

$$c = D \vee H$$

where v is the "inclusive or" operation.

With the substitutions indicated in (4) the grammar G_2 is transformed into another grammar that will be indicated as $G_Q = (V_{T_Q},V_{N_2},P_Q,Q)$, where:

$V_{T_Q} = \{a,b,c\}$

P_Q : $Q\to bU_1$; $Q\to cU_2$; $Q\to aU_4$;

$U_1\to cU_2$; $U_1\to bU_3$; $U_1\to aU_4$;

$U_2\to bU_1$; $U_2\to cU_2$; $U_2\to aU_4$;

$U_3\to cU_2$; $U_3\to aU_4$; $U_3\to bU_4$;

$U_4\to aU_4$; $U_4\to bU_4$; $U_4\to cU_4$;

$U_4\to a$; $U_4\to b$; $U_4\to c$.

With the same simplifications (4), the grammar G_3 reduces to a grammar $G_Z = (V_{T_Z}, V_{N_3}, P_Z, Z)$, where:

$V_{T_Z} = \{b,c\}$

P_Z : $Z \to cY_1$; $Z \to bY_2$;

$Y_1 \to cY_1$: $Y_1 \to bY_2$;

$Y_2 \to cY_2$; $Y_2 \to b$.

The simplification introduced to obtain the grammars G_Q and G_Z show that Table I can be reduced to Table IV for which less accurate measurements on the peak amplitudes are required. Furthermore the informations about peak duration become unnecessary.

5. RECOGNITION OF QRS EVENTS FROM QRS HYPOTHESES AND EXPERIMENTAL RESULTS.

Let $Q_i^{(t_{i1},t_{i2})}$ (i=1,2,3) be a QRS hypothesis emitted on the i-th signal lead for the time interval (t_{i1},t_{i2}). Such hypotheses are emitted under the control of a grammar G_Q applied to the input strings obtained after coding the energy peaks according with Table IV. Let $Z_i^{(t_{i1},t_{i2})}$ (i = 1,2,3) be analogous QRS hypotheses emitted under the control of the grammar G_Z.

Table V reports the number of false-alarms and of issued QRS events for 16 healthy and ill patients depending on the expressions of Q_i and Z_i used for recognizing a QRS event. The following expressions have been considered:

$$x = Q_i \wedge Q_j$$

$$y = Q_i \wedge [Q_j \vee (Z_j \wedge Z_k)] \qquad i,j,k = 1,2,3, \quad i \neq j \neq k$$

$$h = Q_i \wedge [Q_j \vee Z_j]$$

where $\wedge$ is the logical "and".

From the data shown in Table V it appears that the expression $y = Q_i \wedge [Q_j \vee (Z_j \wedge Z_k)]$ gives no false alarms for all the patients and 1,5% of missed events.

Fig. 3 shows the time evolution of the same ECG signals as in Fig. 2. Each sequence of energy peaks is coded according with

TABLE IV

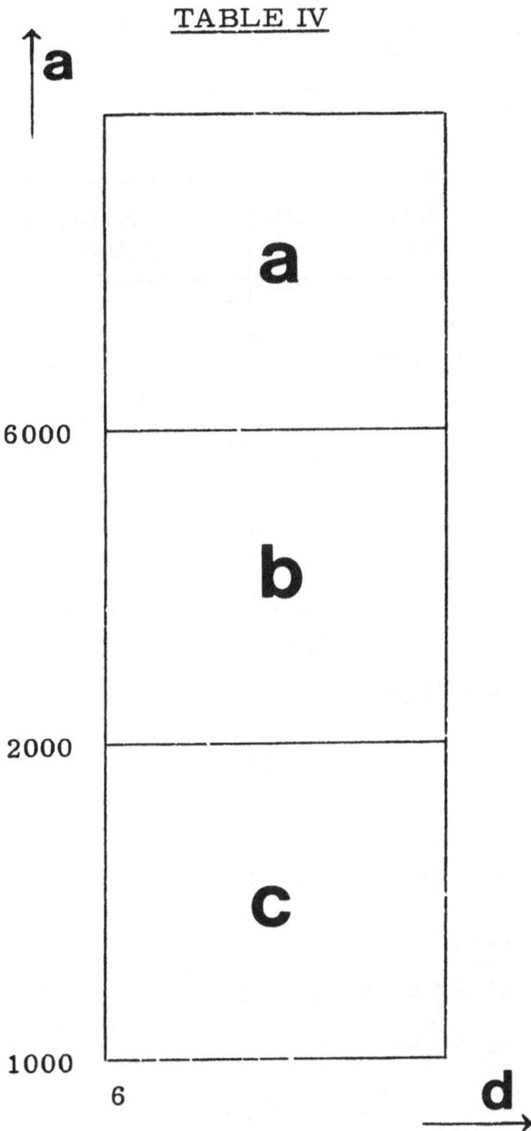

Coding system for the syntactic recognition of QRS complexes same as Table I.

TABLE V

	Number of QRS events recognized	Number of false alarms
x	593/620	-
y	611/620	-
h	620/620	2

The total number of QRS is 620 for 16 patients.

$$\mathbf{x} = Q_i \wedge Q_j$$
$$\mathbf{y} = Q_i \wedge \left[Q_j \vee (Z_j \wedge Z_k) \right]$$
$$\mathbf{h} = Q_i \wedge \left[Q_j \vee Z_j \right]$$

$i, j, k = 1, 2, 3 \quad i \neq j \neq k$

$\wedge$ = logical and

$\vee$ = logical inclusive or

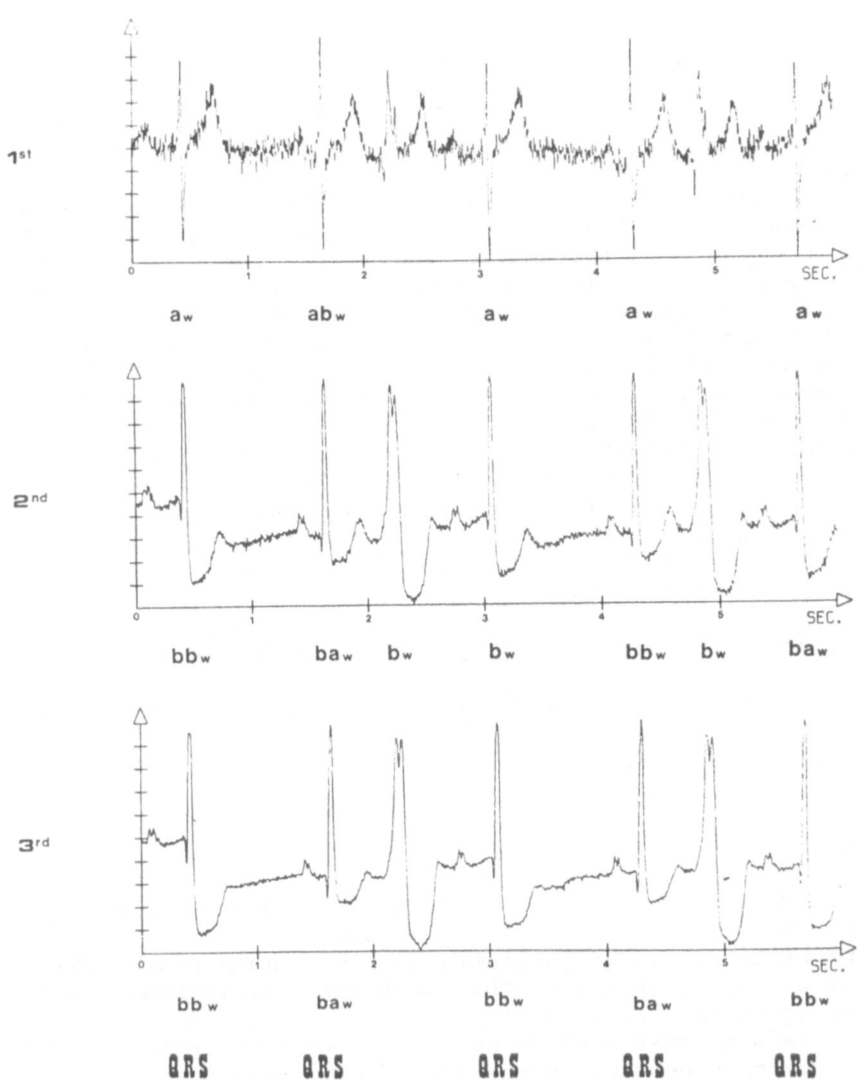

Figure 3. Recognition of QRS events.

The three Eithoven leads are reported. Amplitudes are normalized. Under each signal the simplified strings according with Table IV are listed. Below the recognized QRS events are reported.

Table IV and the hypotheses emitted by the system are shown below the descriptions. Using the rule y for the recognition of QRS events, the recognized QRS are marked by QRS in the plot below the entire set of diagrams. Fig. 4 shows the time evolution of the heart rhythm obtained connecting all the recognized QRS events.

6. EXTENSION OF THE METHOD FOR THE CASES OF POSSIBLE AMBIGUITIES

A crucial problem, related to the inferred grammar, is whether it is ambiguous or not. It is ambiguous only if it may recognize strings that sometime correspond to positive events and sometime not. Those ambiguities may be avoided considering more lead waveforms simultaneously. Although this approach represents theoretically a good solution to the ambiguity problem, it is unsuitable in practice because it limits the area of applications of the method itself. A more concrete approach would consist in considering the grammar as a stochastic or a fuzzy grammar and to use a-priori knowledge on the time of the evolution of the rhythm for making the final decision on the ambiguities (for more details on stochastic grammars, see [34-38]).

Although such ambiguities never appeared in the testing of the method, an algorithm and the corresponding programs must be available because the learning process can continue indefinitely and ambiguities may appear. The algorithm is very simple because if a point $A(\tau)$ is candidate to be a sample of the rhythm curve λ and the following definitions hold:

α: $A(\tau)$ is consistent with the previous samples of λ,

β: $A(\tau)$ has been recognized as corresponding to a QRS event,

the following relation defines the acceptance of $A(\tau)$ as sample of λ:

$$R = \alpha \wedge \beta \ .$$

In a framework of stochastic grammars, the following approach can be taken into account for time constraints.

Let p^+ be the class of configurations of energy peaks corresponding to a real event and p^- the configurations of peaks that do not correspond to a real event.

Let G^+ be a grammar generating the set of descriptions of p^+ and G^- be a grammar generating the set of descriptions of p^-.

Let d be a description of energy peaks to be classified as corresponding to a real event or not. The stochastic grammar G^+ gives the a-priori probability:

$$p(d/p^+)$$

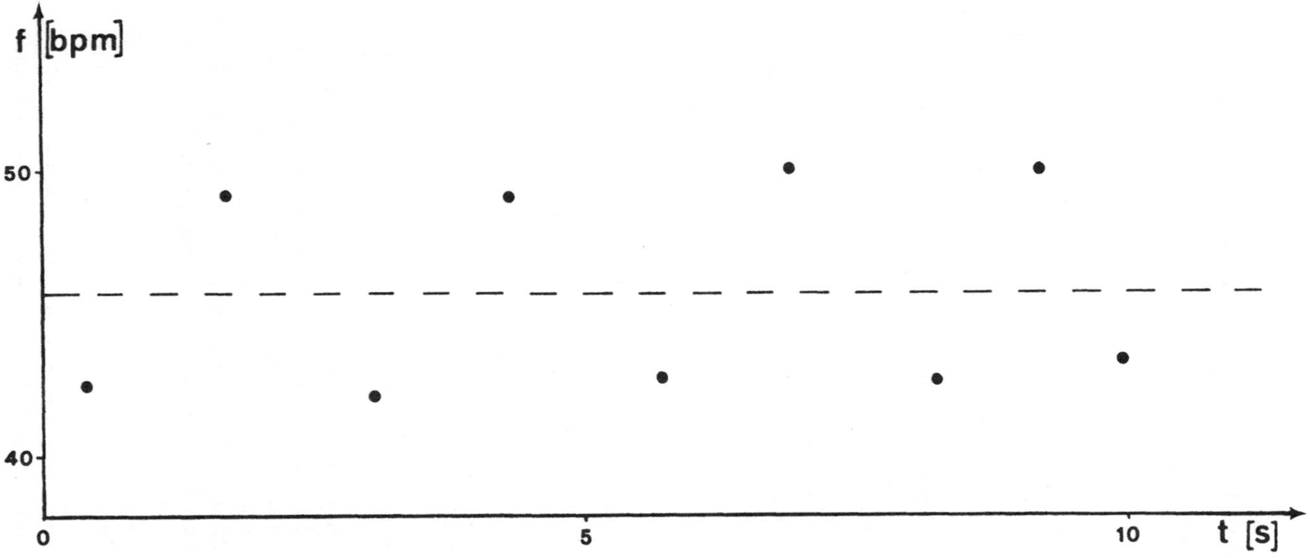

Figure 4. Time evolution of cardiac rhythm. V_m = mean value. f = frequency in beats per minute.

while, for recognition, it is important to know:

$$p(p^+/d)$$

that is the probability that, given the description d, a real event generated the description p^+.

Using well-known probability relations one gets:

$$p(p^+/d) = \frac{p(d/p^+)p(p^+)}{p(d)} \quad (5)$$

The probability of the description p(d) can be easily computed considering that:

$$p(d) = p(d/p^+)p(p^+) + p(d/p^-) \cdot p(p^-) \ .$$

Notice that if the description d is not recognized as belonging to p^-, i.e. as a false alarm, the probability $p(p^+/d)$ is 1. Otherwise, in order to compute (5) it is necessary to know $p(p^+)$ and $p(p^-) = 1 - p(p^+)$.

The (5) can be rewritten as a probability further conditioned by the previous history of the event evolution; thus the probability $p(p^+/\text{event history})$ has now to be computed.

This probability can be computed considering the constraints the time evolutions of the event have to respect.

7. CONCLUSIONS

A syntactic method for processing ECG waveforms considered as a multichannel process has been presented. The method performs a coding with a very simple table look-up. Furthermore, parsing is performed with simple finite state automata inferred by experiments and suitable for updating as the system is running. Possible ambiguities can be solved introducing fuzzy or stochastic linguistic variables processed by fuzzy or stochastic automata.

The entire system could operate in real-time on a mini or a network of microcomputers with a memory size of 64 Kwords.

The results reported in Table V show that a syntactic approach to the recognition of QRS events is very promising. Moreover, further investigations are required in order to make the system capable of handling different types of illnesses and to work with signals affected by various sources of noise.

ACKNOWLEDGEMENTS

This work was carried out at the "Centro per l'Eleborazione Numerale dei Segnali" (CNR), Torino, Italy and was supported by the

"Consiglio Nazionale delle Ricerche", the National Research Council of Italy. The authors are also indebted with Dr. Bonamini for providing the experimental data and the assistance of medical experience.

REFERENCES

1. Clark, K. W., F. M. Nolle, J. R. Cox, Jr., and G. C. Oliver, "High Performance Computer Programs for Rapid Analysis of Long ECG Records", in Proceedings of the San Diego Biomedical Symposium, S. Diego, Calif., vol. 13, 1974.
2. Swenne, C. A., J. H. van Bemmel, S. J. Hengeveld, and M. Hermans, "Pattern Recognition for ECG-Monitoring: An Interactive Method for the Classification of Ventricular Complexes", Computers and Biomedical Research, vol. 5, pp. 150-160, 1973.
3. Yanowitz, F., P. Kinias, D. Rawling, and H. A. Fozzard, "Accuracy of a Continuous Real-Time ECG Dysrhythmia Monitoring System", Circulation, vol. 50, pp. 65-72, 1974.
4. Oliver, G. C., F. M. Nolle, G. A. Wolff, J. R. Cox, Jr., and H. D. Ambox, "Detection of Premature Ventricular Contractions With a Clinical System for Monitoring Electrocardiographic Rhythms", Computer and Biomedical Research, vol. 4, pp. 523-541, 1971.
5. Wolf, H. K., P. J. MacInnis, S. Stock, R. K. Helppi, and P. M. Rantaharju, "Computer Analysis of Rest and Exercise Electrocardiogram", Computers and Biomedical Research, No. 5, pp. 329-346, 1972.
6. __________, "Clinical Electrocardiography and Computers", C. A. Caares and L. S. Dreifus editors, New York, Academic Press, 1970.
7. Fitzgerald, J. W., R. J. Clappier, and D. C. Harrison, "Small Computer Processing of Ambulatory Electrocardiograms", Computer, vol. 8, n. 7, pp. 48-54, July 1975.
8. Simoons, M. L., E. Smallenburg, C. Zeelmberg, "On-Line Analysis of Exercise Electrocardiograms", Computer, vol. 8, n. 7, pp. 42-45, July 1975.
9. Cox, J. R., F. M. Nolle, H. A. Fozzard and G. C. Oliver Jr., "Aztec, a Preprocessing Program for Real-Time ECG Rhythm Analysis", IEEE Trans. Biomedical Engineering, BME-15, pp. 128-129, April 1968.
10. Weaver, C. S., J. Von Der Groben, P. E. Mantey, J. G. Toole, C. A. Coole, Jr., J. W. Fitzgerald, R. W. Lawrence, "Digital Filtering With Applications to Electrocardiogram Processing", IEEE Trans. of Acoustics, AU-16, n. 3, pp. 350-389, September 1968.
11. McFee, R., G. M. Baule, "Research in Electrocardiography and Magnetocardiography", Proceedings IEEE, vol. 60, n. 3, pp. 290-321, March 1972.

12. Cox, J. R. Jr., F. M. Nolle, R. M. Arthur, "Digital Analysis of the Electroencephalogram, the Blood Pressure Wave, and the Electrocardiogram", Proceedings IEEE, vol. 60, n. 10, pp. 1137-1164, October 1972.
13. Pavlidis, T., "Linguistic Analysis of Waveforms", in Software Engineering, J. T. Tou Ed., vol. 2, pp. 203-225, New York, Academic Press, 1971.
14. Pavlidis, T., "Waveform Segmentation Through Functional Approximation", IEEE Trans. Comput., vol. C-22, n. 7, pp. 689-697, July 1973.
15. Pavlidis, T., S. L. Horowitz, "Segmentation of Plane Curves", IEEE Trans. of Computers, vol. C-23, n. 8, pp. 860-870, August 1974.
16. Young, T. Y., W. H. Huggins, "On the Representation of Electrocardiograms", IEEE Trans. on Bio-Medical Electronics, BME-10, pp. 86-95, July 1963.
17. Young, T. Y., W. H. Huggins, "The Intrinsic Component Theory of Electrocardiography", IRE Trans. On Bio-Medical Electronics, BME-9, pp. 214-221, October 1962.
18. Hambley, A. R., R. L. Moruzzi, C. L. Feldman, "The Use of Intrinsic Components in an ECG Filter", IEEE Trans. Bio-Med. Eng., BME-21, n. 6, pp. 469-473, November 1974.
19. Ahmed, N., P. J. Milne, S. G. Harris, "Electrocardiographic Data Compression Via Orthogonal Transforms", IEEE Trans. on Biomed. Eng., BME-22, n. 6, pp. 484-492, November 1975.
20. Amazeen, P. G., R. L. Moruzzi, D. L. Feldman, "Phase Detection of R Waves in Noisy Electrocardiograms", IEEE Trans. on Biomed. Eng., BME-19, pp. 63-66, January 1972.
21. Goovaerts, H. G., H. H. Ros, T. J. Van Den Akker, H. Schneider, "A Digital QRS Detector Based on the Principle of Contour Limiting", IEEE Trans. on Biomed. Eng., BME-23, pp. 154-160, March 1976.
22. Maitra, S., S. Zucker, "A Concise Parametric Representation of Electrocardiograms", IEEE Trans. on Biomed. Eng., BME-22, pp. 350-355, July 1975.
23. Young, T. Y., W. H. Huggins, "Computer Analysis of Electrocardiograms Using a Linear Regression Technique", IEEE Trans. on Biomed. Eng., BME-11, pp. 60-67, July 1964.
24. Steimberg, C. A., S. Abraham, C. A. Caceres, "Pattern Recognition in the Clinical Electrocardiogram", IRE Trans. on Biomed. Electronics, BME-9, pp. 23-30, January 1962.
25. Okajima, M., L. Stark, G. Whipple, S. Yasui, "Computer Pattern Recognition Techniques: Some Results With Real Electrocardiographic Data", IEEE Trans. on Biomed. Electronics, BME-10, pp. 106-114, July 1963.
26. Horowitz, S. L., T. Pavlidis, "Picture Segmentation by a Tree Trasversal Algorithm", Journal of the Association for Computing Machinery, vol. 23, n. 2, pp. 368-388, April 1976.

27. Stockman, G., L. Kanal, M. C. Kyle, "Structural Pattern Recognition of Carotid Pulse Waves Using a General Waveform Parsing System", Communications of ACM, vol. 19, n. 12, pp. 688-695, December 1976.
28. Albus, J. E., "Electrocardiogram Interpretation using Stochastic Finite-State Model", in Syntactic Pattern Recognition Application, Editor K. S. Fu, Springer-Verlag, Berlin, 1976.
29. DeMori, R., P. Laface, E. Piccolo, "Automatic detection and description of syllabic features in continuous speech", IEEE Trans. Acoust. Speech and Sign. Proc., ASSP-24, n. 5, pp. 365-380, Oct. 1976.
30. Schafer, R. W., L. R. Rabiner, "Digital Representations of Speech Signals", Proceedings IEEE, vol. 63, n. 4, pp. 662-677, April 1975.
31. Horowitz, S. L., "A Syntactic Algorithm for Peak Detection in Waveforms with Applications to Cardiography", Comm. of ACM, vol. 18, n. 5, pp. 281-285, May 1975.
32. Zadeh, L. A., "A Fuzzy-algorithmic Approach to the Definition of Complex or Imprecise Concepts", Memorandum No. ERL-M474, Electronics Research Laboratory, University of California, Berkeley, October 1974.
33. Fu, K. S., T. L. Booth, "Grammatical Inference: Introduction and Survey - Part I", IEEE Trans. Systems, Man and Cybernetics, SMC-5, pp. 95-111, January 1975.
34. Fu, K. S., "Syntactic Methods in Pattern Recognition", Academic Press, 1974.
35. Fu, K. S., ed., "Digital Pattern Recognition", Springer Verlag, Berlin, New York 1976.
36. Fu, K. S. ed., "Syntactic Pattern Recognition Applications", Springer Verlag, Berlin, New York, 1977.
37. Fu, K. S., T. L. Booth, "Grammatical Inference. Introduction and Survey - Part II", IEEE Transactions on Syst. Man and Cybernetics, Vol. SMC-5, No. 4, July 1975, pp. 409-423.
38. Maryanski, F. J., T. L. Booth, "Inference of Finite-State Probabilistic Grammars", IEEE Transactions on Computers, vol. C-26, No. 6, June 1977, pp. 521-536.
39. De Mori, R., P. Laface, V. A. Makhonine, M. Mezzalama, "A Syntactic Approach for the Recognition of Glottal Pulses in Continuous Speech", Pattern Recognition, Dec. 1977 (in press).

OBJECT RECONSTRUCTION FROM PROJECTIONS AND SOME NON-LINEAR EXTENSIONS

M. Tasto, H. Schomberg

Philips GmbH Forschungslaboratorium Hamburg,
D-2000 Hamburg 54
W. Germany

ABSTRACT. The reconstruction of objects from projections has had a vast impact on medical X-ray diagnostics during the last years. Other applications are in electron microscopy, nuclear medicine, spin mapping, astronomy, materials testing. The inherent mathematical problem is the solution of integral equations using line integrals over object density along straight paths through the object. When X-ray beams are replaced by ultrasoundwaves, light, microwaves, or even DC-currents, line integrals along curved paths are obtained. These paths are object dependent, and hence their coordinates are not known in advance. As a result, the reconstruction process involves the solution of a non-linear system of equations. New methods and results are presented for the example of the reconstruction of the spatial distribution of electric resistivity of objects from resistance "projections".

1. INTRODUCTION

The reconstruction of objects from projections has had a vast impact on medical X-ray diagnostics during the last years. Other applications of the basic mathematical principles used are in the field of electron microscopy, nuclear medicine, spin mapping, astronomy, and materials testing. A survey of the mathematical methods used for reconstruction can be found in [1]. Let ρ (x,y) be an unknown density distribution of a plane two-dimensional slice of a three-dimensional object. Let $b(r,\emptyset)$ be a line integral of the density along a straight line passing through that slice at angle $\emptyset$ and at distance r from the center of the object. In the X-ray case, $\rho(x,y)$ would be the spatial distribution of the X-ray absorption coefficient, and $b(r,\emptyset)$ is obtained from the total absorption of an X-ray beam being transmitted at $r,\emptyset$ through the object. Radon [2] first showed that $\rho(x,y)$ can be reconstructed from the known $b(r,\emptyset)$, if the latter are available for all r and $\emptyset$. In order to evaluate this and other analytic solutions, they must be discretized and then calculated numerically.

An alternative approach is to discretize the problem first, without using an analytic solution, and then solve the problem numerically. This results in a linear system of equations

$$A \cdot \rho = b \tag{1}$$

where A is a coefficient matrix, ρ is a vector of the object densities at the points of a square two-dimensional grid (arranged into a one-dimensional vector), and b is a vector of all available line integrals at discrete values of r, $\emptyset$, again arranged into a one-dimensional vector. ρ is essentially unknown, b is known from measurements, and A can be determined from the geometry of the rays passing through or near the grid points of the object; details can be found in [1]. Hence, solving for the unknown ρ requires the solution of a linear system of equations. Often this is done iteratively by a method called algebraic reconstruction technique (ART), a survey of which is given in [3].

The linearity of the problem results from the fact that A is not a function of ρ. We wish to generalize the problem to cases where this is no longer true. If instead of X-rays DC fields, microwaves, ultrasound waves or the like are applied to an object, the currents or waves will pass through the object along straight lines only if the object is homogeneous. Otherwise, there will be some interaction between object and field resulting in some sort of bending of the current or wave paths. Three important facts are observed in this case:

a) The paths along which line integrals are measured usually are not straight.
b) In addition, the geometry of these paths depends on the object and, hence, is not known in advance.
c) There is a unique relationship between object and path, i.e. if the object is known, then the paths of the line integrals can be determined.

The resulting system of equations is of the form

$$A(\rho) \cdot \rho = b \tag{2}$$

i.e., the coefficient matrix A depends on the solution ρ, and hence the problem is nonlinear. Nevertheless, there is a unique relationship between solution ρ and coefficient matrix $A(\rho)$.

In the following two sections we will describe the reconstruction of the spatial distribution of resistivity of objects from external resistance measurements as an example of a non-linear reconstruction problem.

2. RESISTANCE PROJECTIONS

Our goal is to determine the spatial distribution of resistivity, and for simplicity we assume a two-dimensional problem, i.e. an object of constant height h with resistivity homogeneous in z-direction and variable in x- and y-directions. In order to obtain information about the internal resistance of objects, a measurement device according to Fig. 1 is used. A square tank is filled with salt water, and the object is placed in the tank. Two opposite sides of the tank as well as the bottom consist of insulating material, whereas the rear is completely covered with a metal electrode. The front consists of many (20 to 40) metallic electrodes insulated from each other. As shown in Fig. 2, these electrodes are connected via current meters of negligible internal resistance with one pole of a DC or low frequency AC source. The large electrode is connected to the other pole of the source. Due to the negligible resistance of the current meters, all small electrodes have equal potential. For each small electrode, the current is then measured and used to determine the resistance of the current path passing from the corresponding small electrode to the large electrode, as shown in Fig. 3. A somewhat similar measurement device was described in [4], yet no reconstruction was attempted. The set of resistance values obtained from all small electrodes and for one orientation of the object is called a "resistance profile". The object is then rotated by an angle $\emptyset$ relative to the tank as shown in Fig. 4, and a new resistance profile is obtained, which usually differs from the other ones, unless the object is rotationally symmetric. Typically, the number of resistance profiles obtained at various angles is in the same order as the number of small electrodes, odd, and the angles are equally spaced over the range of 360°.

The electric potential $u(\xi,\eta)$ in the tank with object at angle $\emptyset$ is determined by the solution of the partial differential equation

$$\operatorname{div}\left[\delta_{\emptyset}(\xi,\eta) \cdot \operatorname{grad} u_{\emptyset}(\xi,\eta)\right] = 0 \tag{3}$$

where $\delta_{\emptyset}(\xi,\eta)$ equals the spatial distribution of the electric conductivity in the tank ($\delta_{\emptyset}(\xi,\eta) = 1/\rho_{\emptyset}(\xi,\eta)$ with $\rho_{\emptyset}(\xi,\eta)$ the spatial distribution of resistivity). The following boundary conditions must be satisfied:

$$u(\xi, -r) = 0$$
$$u(\xi, +r) = u_0 \quad \text{with } u_0 \text{ the voltage of the source.} \tag{4}$$

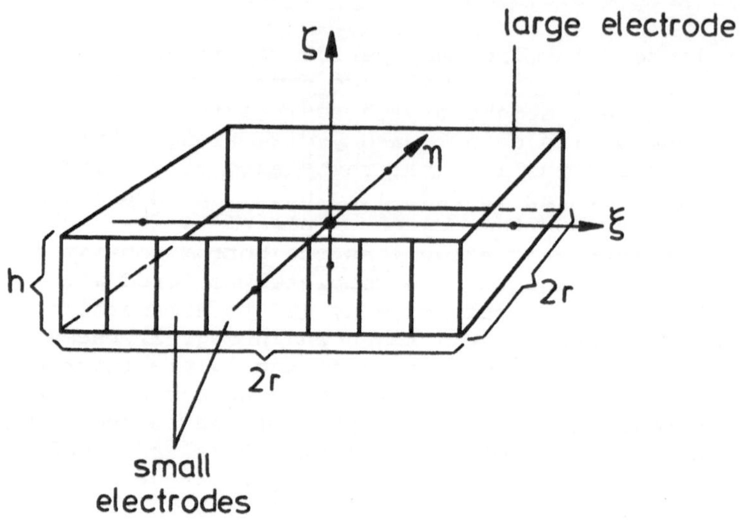

Fig. 1. Water tank with electrodes for resistance measurements.

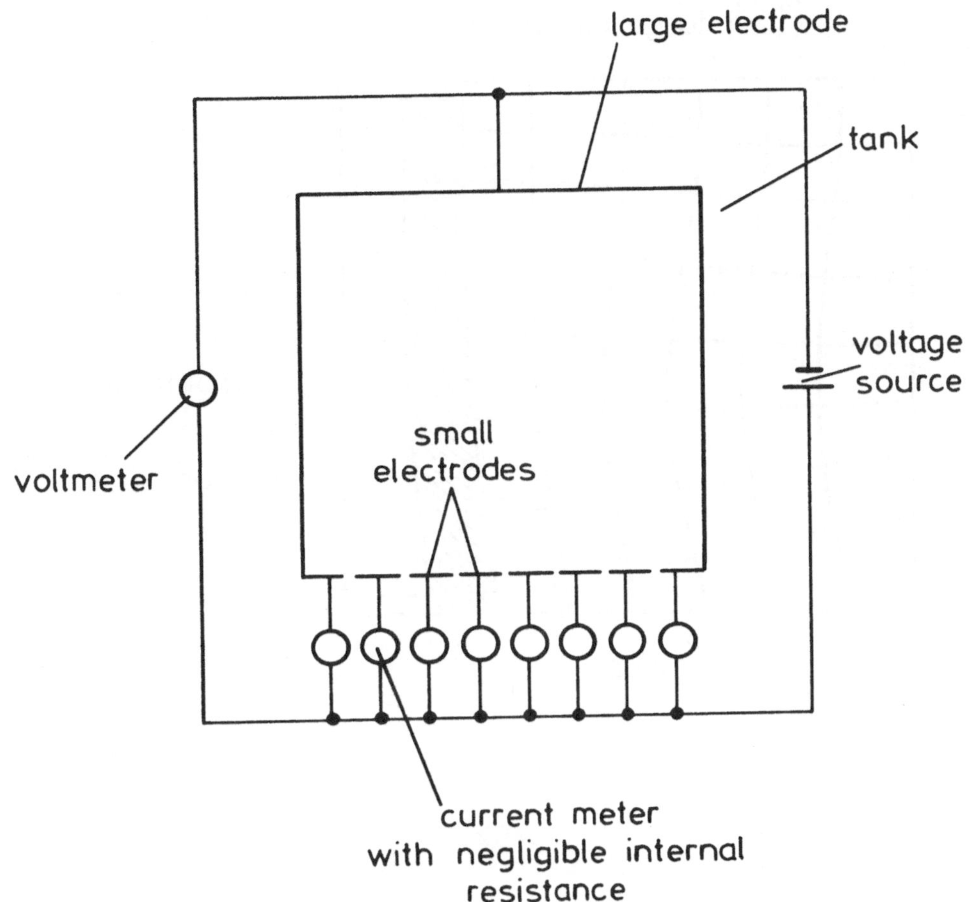

Fig. 2. Electrical set-up to take resistance measurements from the tank.

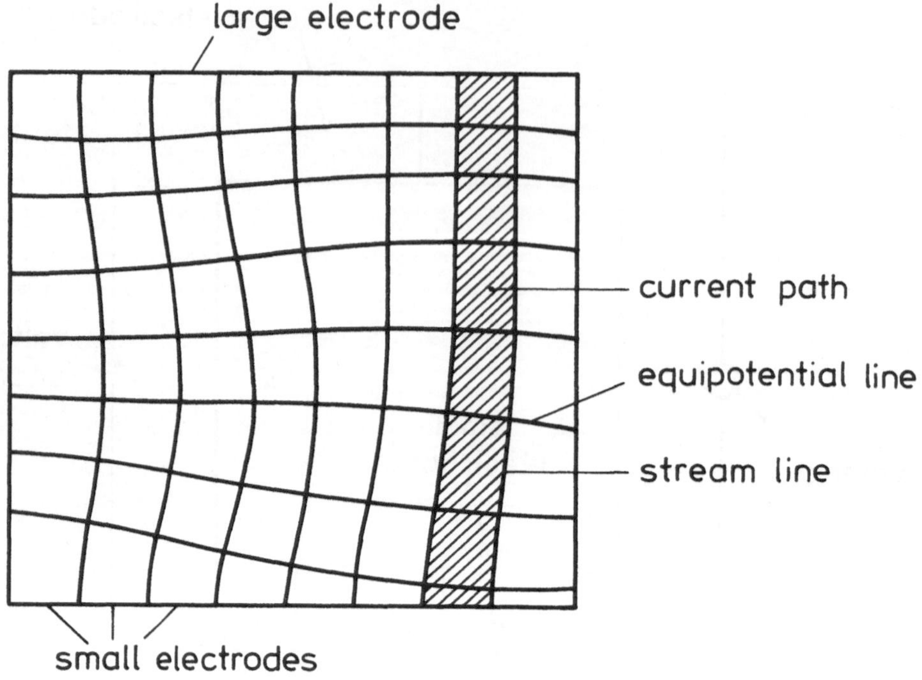

Fig. 3. Equipotential lines and stream lines in the tank when some inhomogeneous object is held into it.

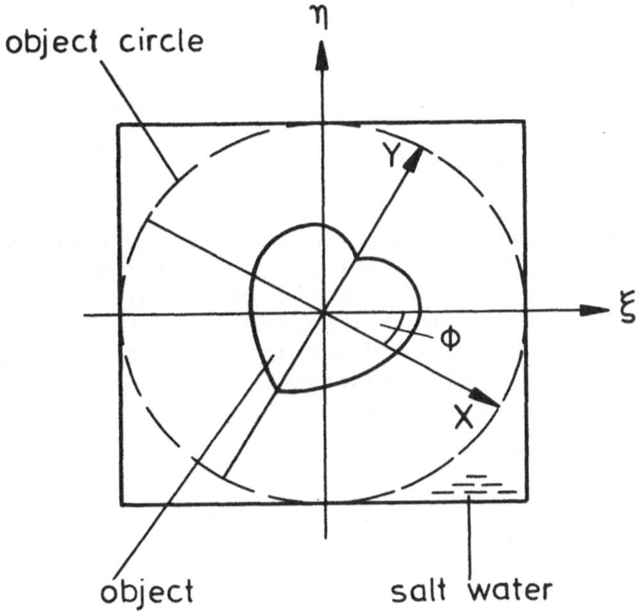

Fig. 4. Rotating an object relative to the tank.

$$\left.\frac{\partial u_{\phi}}{\partial \xi}\right|_{y=+r} = 0$$

$$\left.\frac{\partial u_{\phi}}{\partial \xi}\right|_{y=-r} = 0 \qquad (5)$$

The subscript ϕ has been attached to u to indicate its dependence on the angular orientation of the object in the tank. The streamlines in the tank follow the gradient of the potential field. Hence, we can determine the path of all streamlines starting from the edges of the small electrodes by determining the solution of the system of ordinary differential equations

$$\frac{\partial \xi(s)}{\partial s} = -\frac{\partial u[\xi(s),\eta(s)]}{\partial \xi}$$

$$\frac{\partial \eta(s)}{\partial s} = -\frac{\partial u[\xi(s),\eta(s)]}{\partial \eta} \qquad (6)$$

with the initial condition chosen such that the streamline starts at one edge of the small electrode considered and perpendicular to the wall.

We can now approximately relate the resistance measurement R for one electrode (i.e. the voltage across the tank divided by the current flowing through the electrode) to the inner structure of the object by a line integral, somewhat similar to the line integrals known from X-ray reconstruction:

$$R \approx \frac{1}{h} \int_{\text{Path } S} \frac{\rho(s)}{w(s)}\, ds \qquad (7)$$

where the path S is the streamline starting at the center of the small electrode considered, w(s) is the width of the current path as determined by the two streamlines starting from the edges of the small electrode and $\rho(s)$ is the specific resistance of the object along the path of integration with the assumption that the current path is so narrow that $\rho(s)$ does not vary within the cross-section of the path.

The measurements required for the reconstruction of one object cross-section, for a two-dimensional case, are $R_{r,\phi j}$ with $r = 1, 2, \ldots M$ the index of the small electrodes and ϕ_j $j = 1,2,\ldots K$ the different orientations of the object.

The spatial resistance distribution is assumed unknown within the "object circle" shown in Fig. 4, and known outside the object circle.

3. NON-LINEAR RECONSTRUCTION PROCEDURE

The reconstruction procedure proposed is an extension of the well-known algebraic reconstruction technique (ART), an iterative procedure [1], [3]. Traditionally, the ART method is realized by making an initial guess of the object to be reconstructed, e.g. a homogeneous distribution, computing projections from this guess, comparing them with the measured projections, making a correction by backprojecting the difference, and proceeding to the next orientation angle. This procedure is repeated until convergence is obtained.

In our situation, this method cannot be applied, for we do not know the paths along which the measured line integrals have been obtained. Therefore, the following modification is proposed:

An initial guess of the object is made, and for this guess the streamlines and resistances along the current paths are calculated for the first angle of orientation. The resistance values are compared to the measured resistance projections, and the guessed object is corrected by backprojecting the difference along the calculated current paths, i.e. generally along curved paths. The same procedure is repeated for all other orientations of the object, each time using the corrected object of the previous step. Usually, sufficient convergence is not obtained with one revolution, hence, several or many have to be made. Figure 5 shows a flow chart of the basic method.

We now describe the method in some detail. In order to do calculations by computer, the object as well as the parameters characterizing field and streamlines must be discretized. Three different coordinate systems and corresponding index pairs have been introduced. The system x,y and the grid with indices p,q is used to represent the object (Fig. 6). The second coordinate system ξ, η is fixed relative to the tank rather than to the object, hence, it is rotated by the orientation angle $\emptyset$ relative to the coordinate system x,y, as shown in Fig. 4. This second system is required for the solution of the potential equation for each angle of orientation. The third coordinate system is curvilinear with grid points ξ_{rs}, η_{rs} relative to the second coordinate system, as shown in Fig. 7. The grid points along a curve with r = constant correspond to a streamline starting at the left edge of the r'th small electrode. The grid points with s = constant correspond to the s'th equipotential line. Hence, this coordinate system can only be determined after the potential lines and the current paths have been computed. This is then used to determine the calculated resistance profiles and to do the backprojection of the correction profiles. It is apparent that between different steps of the procedure parameters must be transformed from one coordinate system to the other. Since the grid points do not usually coincide, the well-known bilinear interpolation or related methods are used.

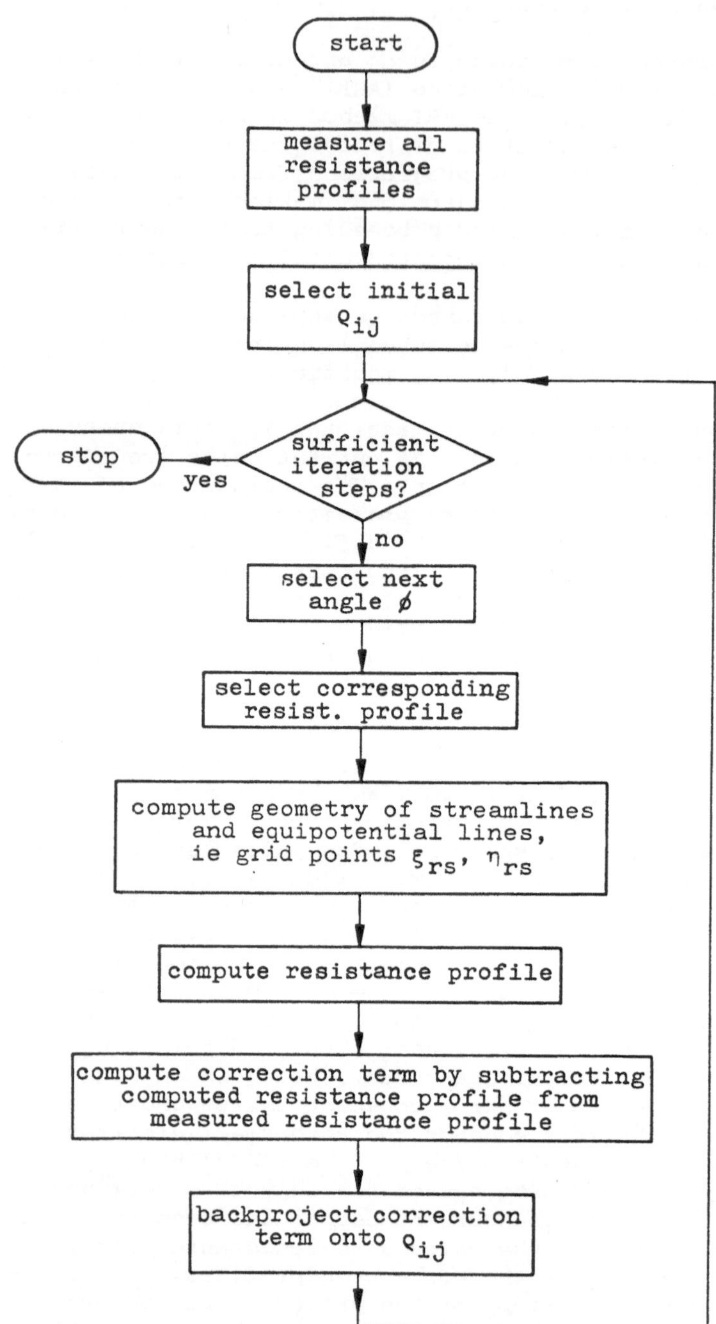

Fig. 5. Flow-chart of the method.

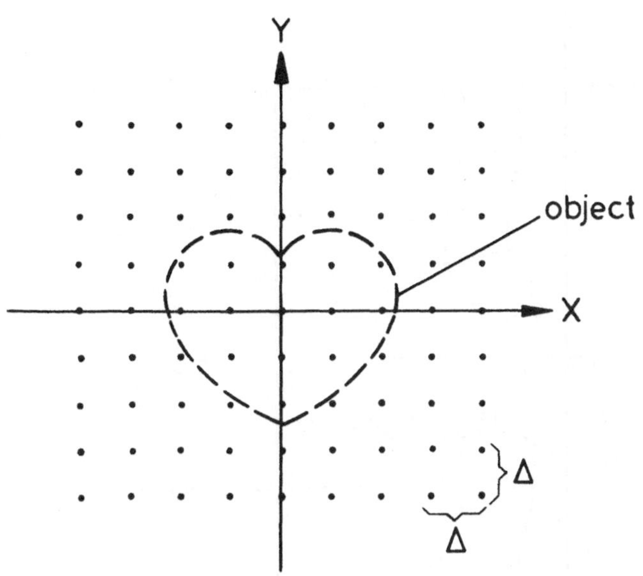

Fig. 6. Discrete representation of the object.

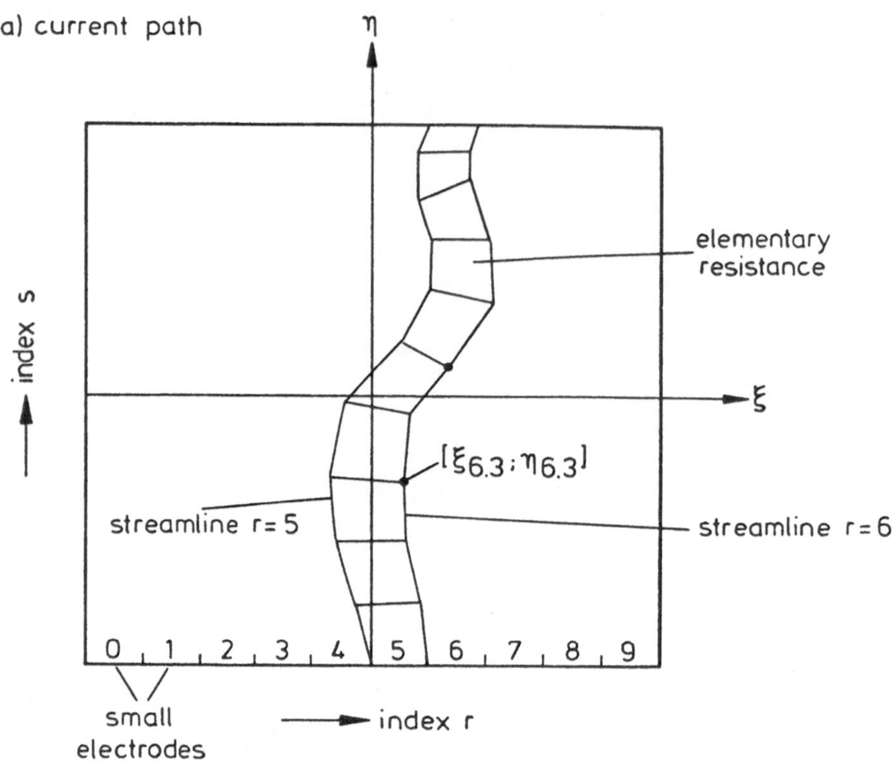

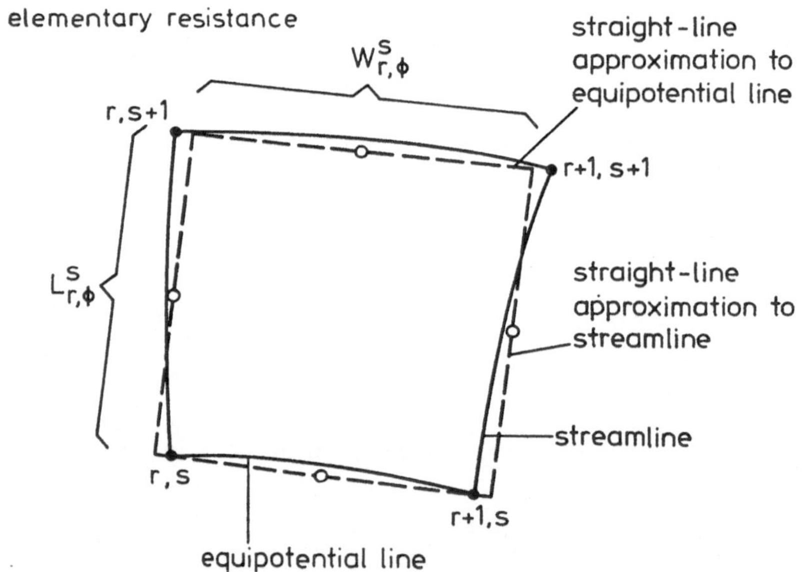

Fig. 7. Description of the current path and elementary resistance.

The electric potential is determined by numerical solution of a two-dimensional elliptic boundary value problem using finite difference techniques and successive overrelaxation.

The determination of the current paths is done by assuming that the distortion is not "too strong", reducing (6) to one single ordinary differential equation and solving it by a fourth order Runge Kutta method. The equipotential lines are determined by tracing the streamlines and checking, where certain specified potential values are attained.

Once we know the grid points ξ_{rs}, η_{rs}, we are able to compute the resistance profile for the corresponding distribution of specific resistance. Consider the r'th current path corresponding to the r'th electrode. The resistance $R_{r,\emptyset}$ for this path, and for the angle of rotation $\emptyset$ is determined as follows: First, the current path is determined by selecting the proper ξ_{rs}, η_{rs} grid points of the curvilinear coordinate system describing its location, as shown in Fig. 7. Then the values of all elementary resistances $R_{r,\emptyset}^{s}$ are determined by assuming the streamlines and equipotential lines to be straight between grid points and by setting

$$R_{r,\emptyset}^{s} = \frac{L_{r,\emptyset}^{s} \cdot \rho_{r,\emptyset}^{s}}{w_{r,\emptyset}^{s} \cdot h} \tag{8}$$

Here, $\rho_{r,\emptyset}^{s}$ is the resistivity at the center point of the trapezoid considered, $L_{r,\emptyset}^{s}$ and $W_{r,\emptyset}^{s}$ are average length and width, and h is the height of the current path. These elementary resistances are then summed to form the total resistance $R_{r,\emptyset}$ of one current path:

$$R_{r,\emptyset} = \sum_{s} R_{r,\emptyset}^{s} \tag{9}$$

Once the comparison of computed and measured resistances has been done, the differences must be "backprojected" along the current paths. This is done as follows: First we determine, which elementary resistances of the path considered fall into the "object circle". Only these elements are considered for correction, since those outside the circle are known and fixed. The resistance difference is equally distributed over the elementary resistances of the path within the circle, and from this the correction term for the specific resistance of each trapezoid can be determined.

The number of different "projection angles" used is in the order of the number of small electrodes used in the measurement tank. In the iterative procedure angular increments of between 45° to 90° were used rather than using consecutive angles. No definite stopping criterion has been selected so far; instead, a preselected number of iteration steps was made in order to study the convergence behaviour.

To improve convergence of the method, the principle of "under-relaxation" techniques and intermediate smoothing of the reconstructed resistivity in each iteration step was used. The latter was especially useful when noisy measurements were taken for reconstruction.

4. RESULTS

First experiments with the reconstruction technique described were carried out with simulated data. Objects were "invented" by assigning numbers to the grid points of $\rho(\xi,\eta)$; from these objects resistance projections were calculated, and then these projections were used to reconstruct the object, thus allowing easy comparison of original and reconstructed objects. In Fig. 8 a sequence of iteration steps together with the original object is shown for a simple circular object. Fig. 9 shows an object with corresponding potential field lines. Fig. 10 and 11 show the original and reconstruction of a "hat" and a more complex object. It is apparent that some deviation of the reconstructed object from the original one occurs, namely some blurring and local overshoots. The mean square reconstruction error for the examples considered lies in the area of 10 to 30% of the variance of the original object. This appears somewhat high, and improvements of the numerical technique will hopefully improve these results. We found that with our technique only objects represented by up to about 25 x 25 grid points could be reconstructed. Improvements up to 40 x 40 matrices could be obtained by reconstructing a coarse grid of say 20 x 20 elements, and using the result as the initial "guess" for a reconstruction of a 40 x 40 grid, applying proper interpolation techniques.

In addition to the computer simulation described, a first experiment was made where actually measured data rather than simulated ones were used for reconstruction. Instead of using a water tank, however, we used a square sheet of carbonated conductive paper, with a second piece pressed onto it serving as the object. The electrodes were painted onto the paper using conductive silver paint. The results were satisfactory for simple objects such as the "hat", but not for complex objects. This was attributed to the inhomogeneity of the carbonated paper. More elaborate experiments using a salt water tank with precision electrodes are in preparation.

5. CONCLUSIONS

It has been demonstrated that, in principle, the reconstruction of spatial distribution of resistivity from resistance measurements taken without penetrating the object can be done. Much work remains to be done, however, to make these principles applicable to the medical or material testing area. The extension to three

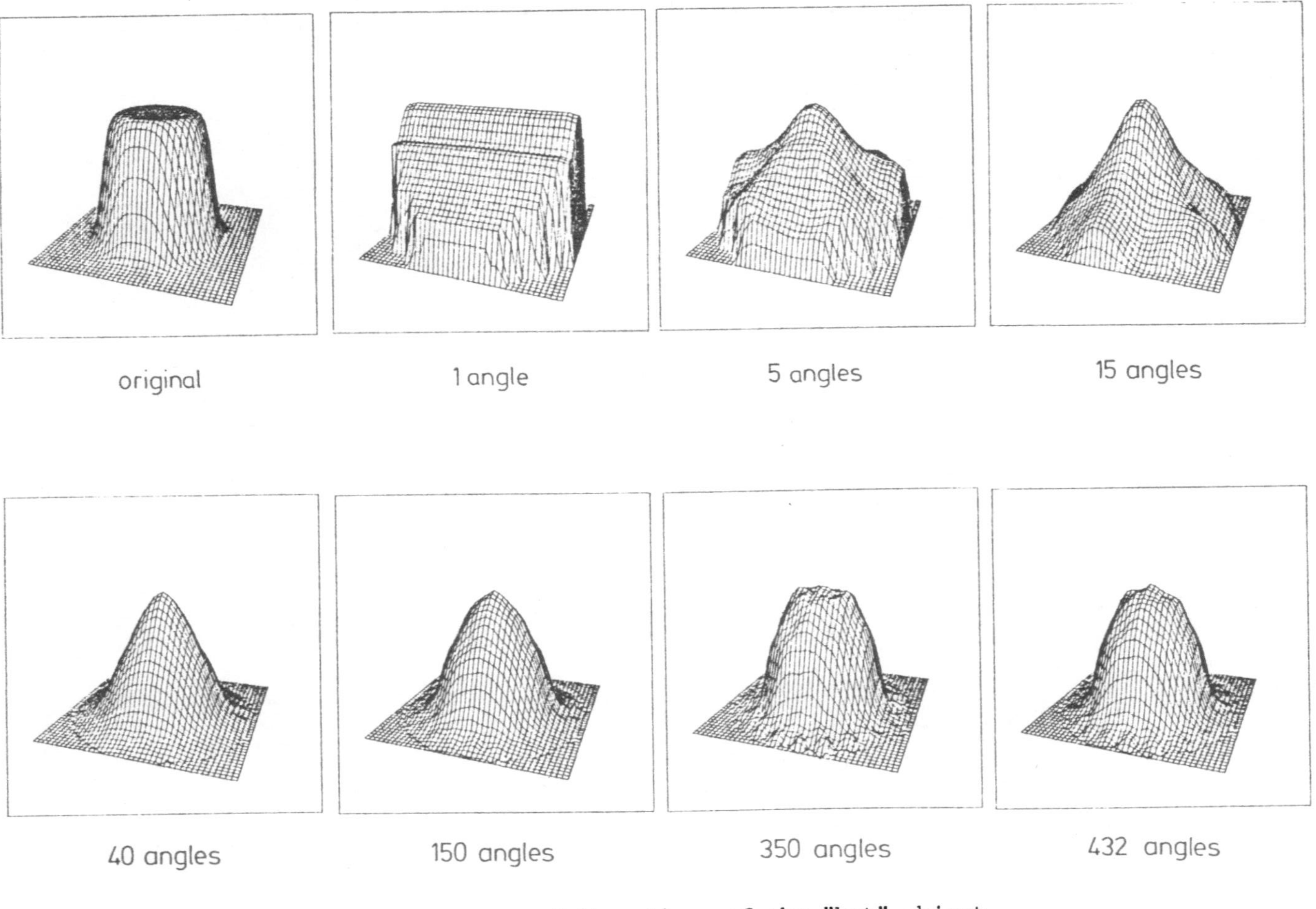

Fig. 8. Sequence of iterations of the "hat" object.

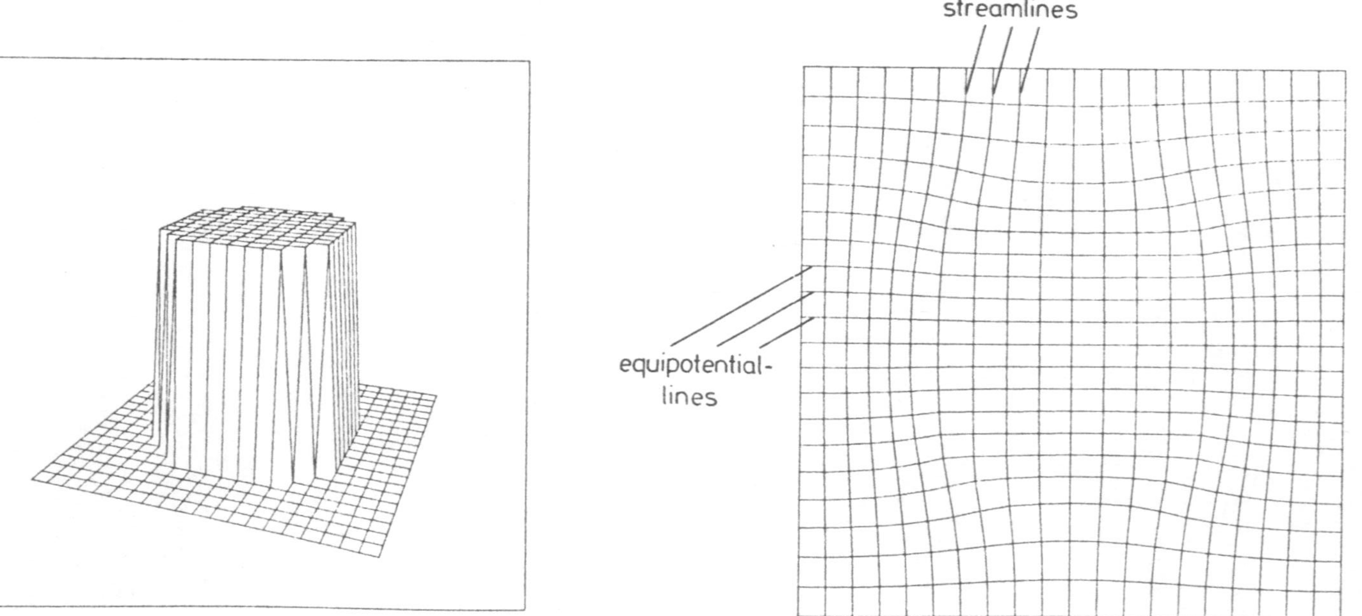

Fig. 9. Example of an object and corresponding equipotential lines and streamlines.

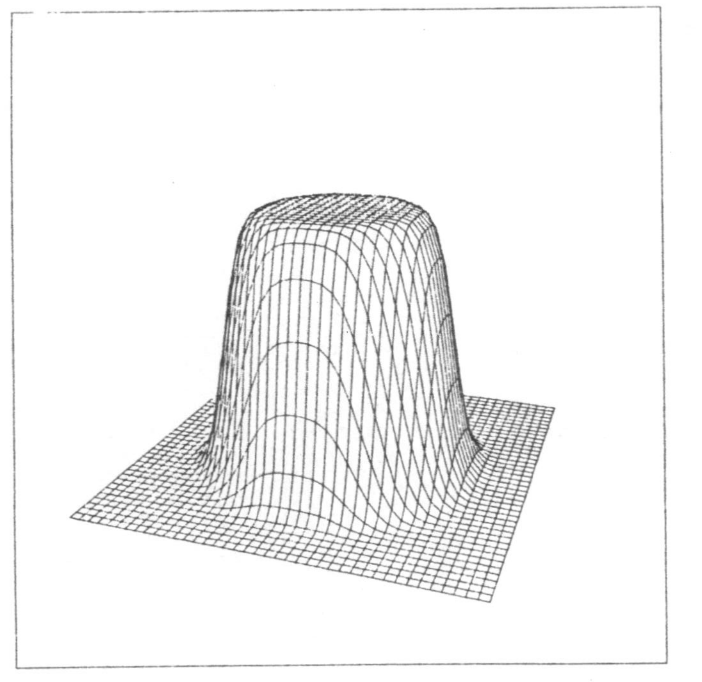

original object

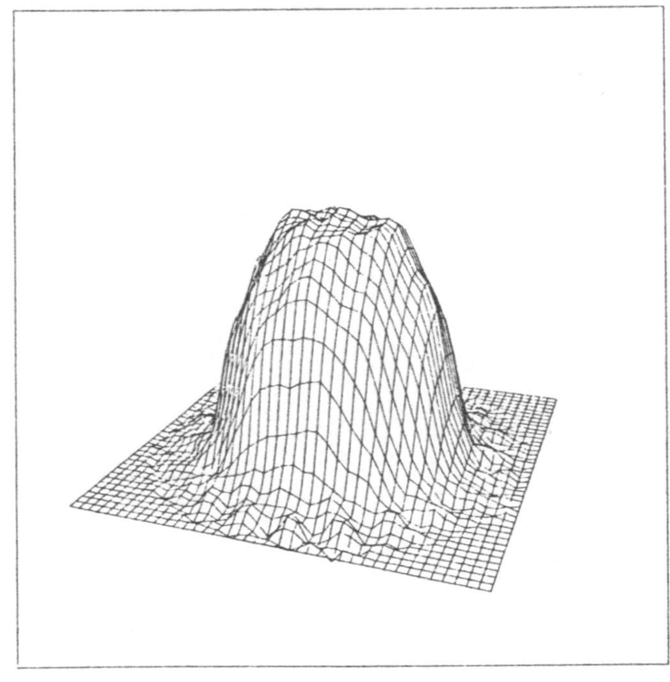

reconstructed object

Fig. 10. Example of the original and reconstructed rotationally symmetric "hat" object.

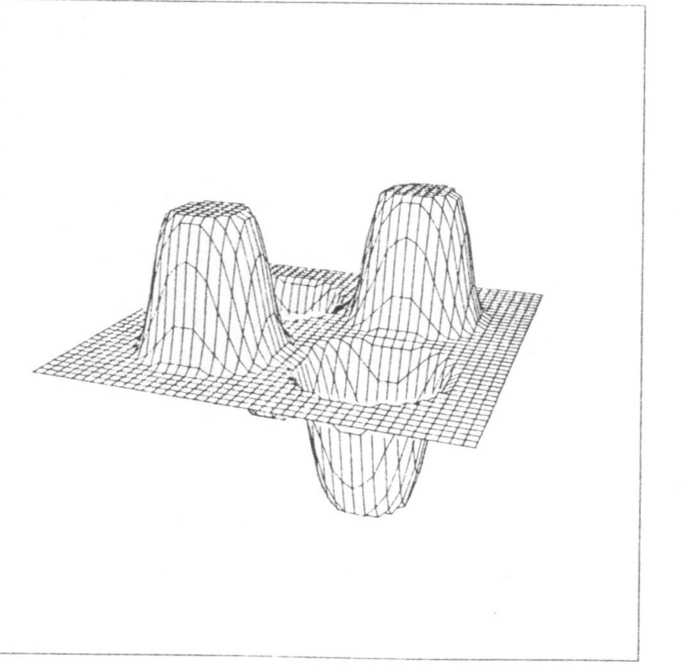

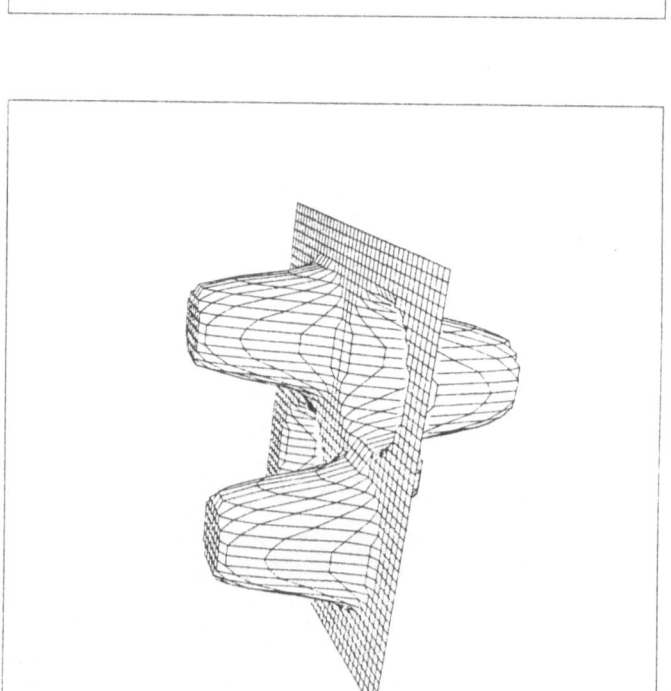

original object

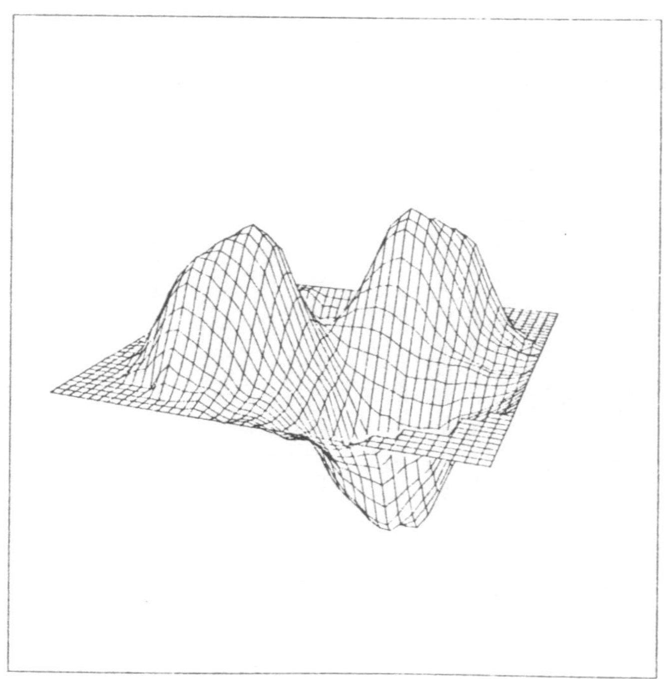

reconstructed object

Fig. 11. Example of original and reconstructed object with no rotational symmetry.

dimensions must be done, and problems of nonlinearity and anisotropy may occur in medical applications [5]. Also, a problem of interest that remains to be solved is the application of higher frequency AC where, for example, the effects of dielectricity become noticeable in addition to the object resistivity.

Finally, we would like to remark that rather similar methods as those described can be used to reconstruct objects from time-of-flight or phase measurements of waves passing through an object, e.g. microwaves, light waves or ultrasound. Our experience in those areas shows that the reconstruction algorithms behave better, and that reconstruction quality is better than in the resistance reconstruction case.

6. ACKNOWLEDGEMENTS

We would like to thank Mr. Havixbeck and Mr. Klinger for writing the programs and for carrying out all the experiments with them.

REFERENCES

1. Brooks, R. A., G. DiChiro, "Principles of Computer Assisted Tomography (CAT) in Radiographic and Radioisotopic Imaging", Phys. Med. Biol., 1976, Vol. 21, No. 5, pp. 689-732.
2. Radon, J., "Uber die Bestimmung von Funktionen durch ihre Integralwerte Längs Gewisser Mannigfaltigkeiten", Ber. Verhandl. d. Sächs. Ak. D. Wiss., Leipzig, Bd. 69, 1917, pp. 262-277.
3. Gordon, R., "A Tutorial on ART (Algebraic Reconstruction Techniques)", IEEE Trans. on Nucl. Sci., Vol. NS-21, June 1974, pp. 78-93.
4. Henderson, R. P., J. G. Webster, "An Impedance Camera for Spatially Specific Measurements of the Thorax", IEEE Trans. on Biomed. Eng., Vol. BME-25, No. 3, May 1978, pp. 250-254.
5. Geddes, L. A., L. E. Baker, Principles of Applied Biomedical Instrumentation, Sec. Ed., Wiley, 1975, Chapter 10.

NEW ALGORITHMS FOR IMAGE RECONSTRUCTION FROM FAN BEAM PROJECTIONS

C. E. Goutis and T. S. Durrani
Department of Electronic Science and Telecommunications
University of Strathclyde
Glasgow G1 1XW, U.K.

ABSTRACT. The reconstruction of an object from its parallel ray projections frequently suffers from artifacts due to the object motion during the data acquisition interval. The recently introduced fan beam reconstruction system overcomes this difficulty by reducing mechanically moving parts and thus collecting the data speedily. In this paper we propose a set of algorithms for the reconstruction of slices (images) from their fan beam projections. The problem is formulated in a constrained optimization scheme and some well established performance criteria are optimized. Next, employing the 'minimum variance' cost function, a non-recursive algorithm is presented which exploits the many useful properties of a block circulant matrix structure. Finally, some new results are quoted which include a new projection slice theorem for fan beam configurations and a convolution algorithm based on this theorem.

1. INTRODUCTION

Reconstruction of images from their projections forms one of the more important and growing areas of application for signal processing techniques. There are a large number of diverse fields where a requirement exists for the estimation of the internal structure of multidemensional objects from their one-dimensional projections. Thus, application of image reconstruction techniques arise in x-ray Computerised Tomography (CT) [1]; radio astronomy [2]; n.d.t analysis [3]; electron micrography [4] etc.

Fig. 1 is an illustration of the conventional system which employs parallel beams of radiation to generate projections of irradiated objects. The sensor outputs yield discretised values of the

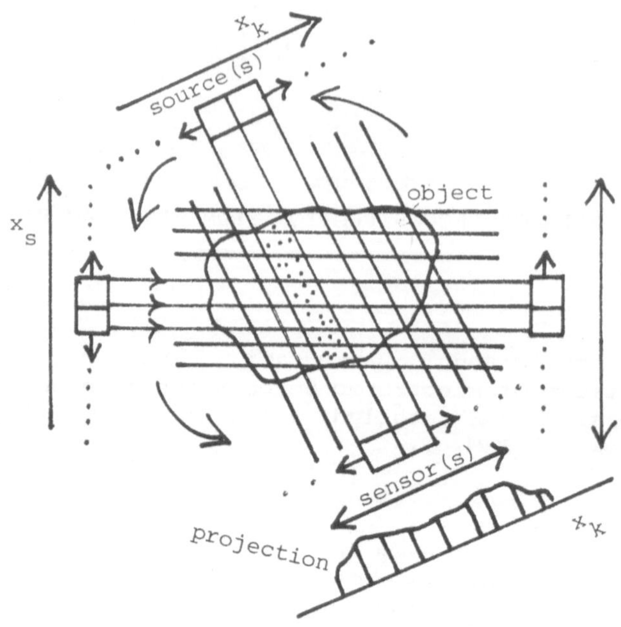

Fig. 1. Source(s) and sensor(s) travels along x_k to obtain the measurements of one projection. They rotate and then travel along x_s to obtain another projection.

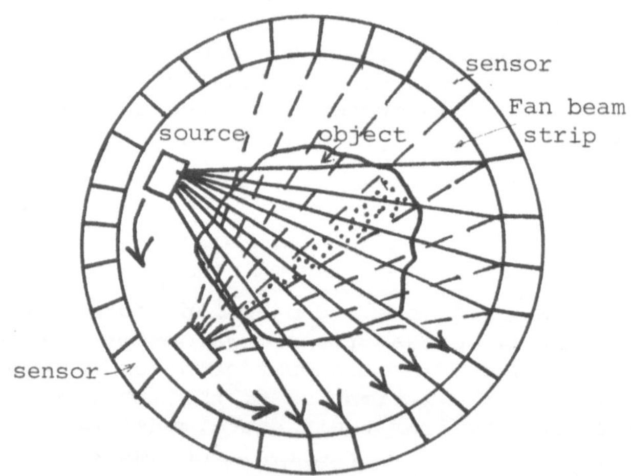

Fig. 2. Fan beam projection system. The only mechanical motion involved is the rotation of the source.

projections. By shifting the assembly of sources and sensors through small angles a large number of projections can be obtained. Several computing techniques have been developed for reconstructing images from such a set of axial projections of which the Algebraic Reconstruction Technique (ART) and its variants; and the convolution algorithm are much in favour. In this context we have proposed a set of algorithms which allow optimal image reconstruction from transaxial projections by satisfying specified performance criteria [5, 6].

Due to the finite amount of time required for the acquisition of all projection data, parallel beam systems have been seen to be effected by any object (or body, as in CT) movement. This has normally led to a distorted reconstruction of the image. To minimise this effect fan beam systems (Fig. 2) have been recently introduced [8] which consist of far fewer rotating mechanical parts, and they employ only a single source producing a fan beam of incident radiation. The latter is collected, after transmission through the object, by a set of fixed sensors placed on a circle. From the sensor measurements we obtain the polar or fan beam projections of the object as the single source moves through small angles on the periphery of the circle.

The existing image reconstruction algorithms for the parallel ray geometry can be used to obtain fan beam reconstructions only when the projection measurements are taken at a restricted set of angles, and if the divergent rays (Fig. 2) are approximated by parallel rays. Herman et. al. [7] have presented a new convolution algorithm for the divergent ray geometry which exploits the projection slice theorem of the parallel ray geometry and it includes several approximations. Pang et. al. [8] have also attempted to solve the divergent ray reconstruction problem by proposing a numerical solution to the associated integral equations.

In this paper we develop a constrained optimisation approach to the solution of the divergent ray reconstruction problem. This is an extension of our earlier work on the parallel-ray geometry [5, 6]. As there is no unique solution to the reconstruction of the image from discrete fan beam projections, we employ the Euler-Lagrange method of undetermined multipliers to minimise a general performance criterion, (Section 2). Applying this to a specific cost function we obtain a relationship (model) between the image and the Lagrange multipliers.

For a 'minimum variance' cost function a linear relationship is developed in Section 3 between the fan beam projections and the Lagrange multipliers. This relationship is embedded in a block circulant matrix which can be readily inverted by exploiting its block eigenvalue decomposition. It leads to a non-recursive algorithm for image reconstruction which requires a computing effort comparable to that for the convolution algorithm [9].

We have recently derived a new projection slice theorem for the fan beam geometry and have been further able to construct a new convolution algorithm which arises naturally from this theorem.

This is outlined in Section 4. Detailed derivations will be presented elsewhere.

2. CONSTRAINED OPTIMISATION OF A GENERAL COST FUNCTION

Fig. 3 represents the fan beam geometry of the projection system of Fig. 2. Mathematically, the fan beam projection $g_\psi(\theta')$, with the source at an angle ψ is given by

$$g_\psi(\Theta') = \int_{\ell_1(\Theta')}^{\ell_2(\Theta')} f(x,y)\, dr' \tag{1}$$

Here f(x,y) denote the intensity at any point (x,y) of the object (or slice) to be reconstructed, where $x,y \in D$, the domain of the object is considered here as a circle of radius A, and

$$\Theta' = \tan^{-1}\left(\frac{-x\cos\psi - y\sin\psi + R}{-x\sin\psi + y\cos\psi}\right)$$

$$r' = (R^2 + r^2 - 2Rr\cos(\psi-\theta))^{\frac{1}{2}}$$

with $\ell_1(\Theta')$, $\ell_2(\Theta')$ and R, as shown in Fig. 3. (r,Θ) are the polar coordinates of any point $(x,y) \in D$.

If the projections are taken at discrete angles $\psi = \psi_k$, $k = 1,2 \ldots N$, then $\{g_k(\Theta')\}$ represent the set of projections from which the object $\{f(x,y)\}$ has to be reconstructed. Employing the Euler Lagrange method the problem of constrained optimisation takes the form:

Estimate $\{f(x,y)\}$ by optimising the cost function

$$J = \iint_D F(f(x,y)r\,dr\,d\theta + \sum_{k=1}^{N} \int_0^{2\pi} \lambda_k(\theta')[g_k(\theta') - \int_{\ell_1(\theta')}^{\ell_2(\theta')} f(x,y)\,dr']\,d\theta' \tag{2}$$

where F is a function of f(x,y); (r,θ) and (r',θ') are the polar coordinates of any point within the object with respect to the origin at O and O' respectively. N is the number of available fan beam projections $\{g_k\}$ which form the N constraints for the optimisation, as Eq. (1) has to be satisfied by all reconstructions. $\{\lambda_k(\theta')\}$ denote the N Lagrange multipliers associated with the given constraints. Note that each Lagrange multiplier $\lambda_k(\theta')$ is a one dimensional function corresponding to each 1-D projection $g_k(\theta')$.

As the Jacobian of the transformation between the polar coordinates (r,θ) and (r',θ') is unity, Eq. (2) may be recast as

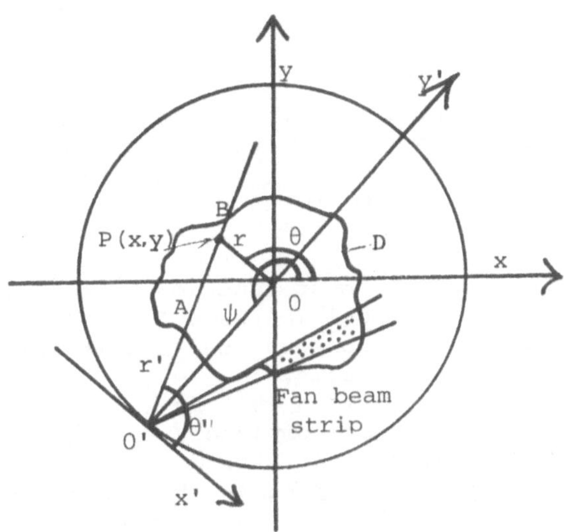

Fig. 3. Fan beam geometry; $\ell_1(\theta')=0'A, \ell_2(\theta')=0'B$; $R=0'0$. Radiating source positioned at 0'.

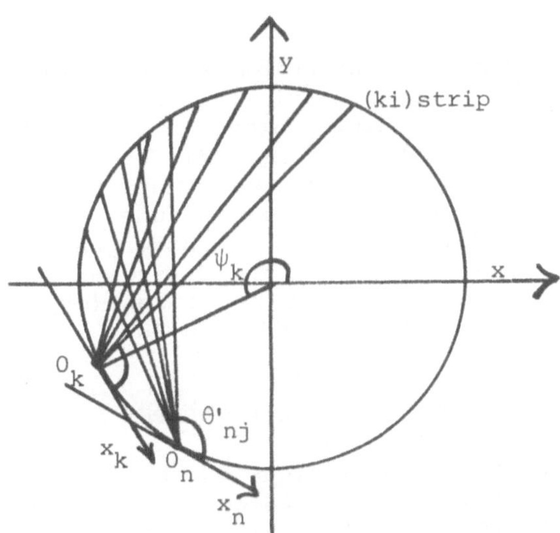

Fig. 4. The contribution of $\{\lambda_{nj}\}$ to $\{g_{ki}\}$ depends only on $\{\psi_k-\psi_n=2\pi(k-n)/N\}$ since the projections are taken at equal angles, and the strips have equal angular width.

$$J = \iint_D [F(f) + \sum_{k=0}^{N} \lambda_k(\theta')[g_k(\theta')\delta(r' - \frac{\ell_1 + \ell_2}{2}) - \frac{f(x,y)}{r'}]]rdrd\theta \quad (3)$$

where $\delta(\cdot)$ is a delta function, and (r',θ') are as defined earlier. Setting the first derivative of (3) w.r.t. $f(x,y)$ to zero gives

$$\frac{\partial F}{\partial f} - \sum_{k=1}^{N} \frac{\lambda_k(\theta')}{r'} = 0 \ , \ (x,y) \in D \quad (4)$$

Though the choice of the cost function is arbitrary, it must be such that the resulting model (between f and λ) is valid for the class of objects to be reconstructed, and it should also lead to computationally fast algorithms.

It is worth stating that if $g_\psi(\theta')$ is known for all continuous values of ψ and θ', then it completely specifies a unique object for reconstruction within a finite domain. In practice as the projections $\{g_\psi(\theta')\}$ are only known at a finite number of angles ($\psi = \psi_k$; $k = 1,2, \ldots N$), the above optimisation procedure aims at specifying the missing projections in accordance with the prescribed performance criterion. Moreover, each projection $g_k(\theta')$ is available as an averaged value over a narrow divergent ray (see Fig. 3), i.e. as

$$g_{ki} = \int_{\theta_i}^{\theta_{i+1}} g_k(\theta')d\theta' \qquad \begin{array}{l} k = 1,2,\ldots N \\ i = 1,2,\ldots P \end{array} \quad (5)$$

where $\theta_i = \pi(i-1)/P$ and P is the number of measurements per projection, consequently an interpolation scheme (w.r.t. θ') must be incorporated in any reconstruction algorithm.

Table I lists the solution of Eq. (4) for cost functions corresponding to 'minimum variance' and 'maximum entropy' (two definitions) of the reconstruction, and it includes models (between f and λ) for both the parallel ray and the divergent ray geometries. It has been shown earlier [10] that the first and third models for parallel ray geometry in Table I are the image (object) models adopted in the multiplicative and additive ART.

3. THE NON-RECURSIVE RECONSTRUCTION ALGORITHM

The (discretised) Lagrange multipliers of the variance model for the divergent ray system (Table I) is related to the projection data by a system of linear equations which can be incorporated in a block circular matrix. By exploiting the properties of this circulant matrix we propose a non-resursive algorithm for image reconstruction from fan beam data.

The Lagrange multipliers $\{\lambda_k(\theta')\}$ have to be estimated from the projection data $\{g_{ki}\}$ using Eqs. (1) and (5); i.e.

$$g_{ki} = \int_{\theta_{i-1}}^{\theta_i} \int_{\ell_1(\theta')}^{\ell_2(\theta')} \left(\sum_{n=1}^{N} \frac{\lambda_n(\theta'_n)}{r'_n} \right) dr'_k d\theta' \tag{6}$$

with k = 1,2 ... N; i = 1,2 ... P; where $\{(r'_k, \theta'_k),\ k = 1,2 \ldots N\}$ are the polar coordinates of any point on the (object) image intensity with respect to the origin of the axes at $\{O'_k\}$ (see Fig. 4).

By considering each $\{\lambda_n(\theta')\}$ as constant within each of the divergent ray strip associated with the source resident at $\{O'_n\}$, the above equation may be recast in discretised terms as

$$g_{ki} = \sum_{n=1}^{N} \sum_{j=1}^{P} C_{ki,nj} \lambda_{nj}, \quad \begin{matrix} k = 1,2 \ldots N \\ i = 1,2 \ldots P \end{matrix} \tag{7}$$

where

$$C_{ki,nj} = \iint_S \frac{1}{r'_n r'_k} r'_k dr'_k d\theta'_k \tag{8}$$

S is the common area of the (ki) and (nj) divergent strips in Fig. 4 and λ_{nj} represents discretised values of $\{\lambda_n(\theta');\ \theta'=2\pi j/P\}$. The double integral of Eq. (8) depends only on the domain of image definition (D), and the strip geometry. As these are usually known for specific applications, the coefficients $\{C_{ki,nj}\}$ need be calculated once only.

If the divergent ray projections are equispaced i.e. $\{\psi = 2\pi(k-1)/N;\ k = 1,2 \ldots N\}$ and all the strips have equal angular widths i.e. $\theta_i - \theta_{i-1} = \pi/p$; for all i, the equations in (7) may be written in matrix form as

$$\underline{g} = C\underline{\lambda} \tag{9}$$

or as it can be justified from the fan beam geometry

TABLE I

cost function	model	
	parallel ray geometry	divergent ray geometry (fan beam)
$f(x,y)\log f(x,y)$ (Entropy I)	$f(x,y)=e^{-1}\exp(\sum_{k=1}^{N} \lambda_k(x_k))$	$f(x,y)=e^{-1}\exp(\frac{\sum_{k=1}^{N} \lambda_k(\theta')}{r'})$
$\log f(x,y)$ (Entropy II)	$f(x,y)=\frac{1}{\sum_{k=1}^{N} \lambda_k(x_k)}$	$f(x,y)=\frac{1}{\sum_{k=1}^{N} \lambda_k(\theta')/r'}$
$f^2(x,y)$ (Variance)	$f(x,y)=\frac{1}{2}\sum_{k-1}^{N} \lambda_k(x_k)$	$f(x,y)=\frac{1}{2}\sum_{k-1}^{N} \frac{\lambda_k(\theta')}{r'}$

$x_k = x\cos\psi + y\sin\psi$

ψ = projection angle

r', θ' as in (1)

$$\begin{bmatrix} \underline{g}_1 \\ \underline{g}_2 \\ \cdot \\ \cdot \\ \cdot \\ \cdot \\ \underline{g}_k \\ \cdot \\ \cdot \\ \underline{g}_N \end{bmatrix} = \begin{bmatrix} C_0 & C_1 C_2 & \cdots\cdots & C_{N-1} \\ C_{N-1} & C_0 C_1 & \cdots\cdots & C_{N-2} \\ \cdot & \cdot\ \cdot & & \cdot \\ \cdot & \cdot\ \cdot & & \cdot \\ \cdot & \cdot\ \cdot & & \cdot \\ \cdot & \cdot\ \cdot & & \cdot \\ C_{N-k+1} & & \cdots\cdots & C_{N-k} \\ \cdot & \cdot\ \cdot & & \cdot \\ \cdot & \cdot\ \cdot & & \cdot \\ C_1 & C_2 C_3 & \cdots\cdots & C_0 \end{bmatrix} \begin{bmatrix} \underline{\lambda}_1 \\ \underline{\lambda}_2 \\ \cdot \\ \cdot \\ \cdot \\ \cdot \\ \underline{\lambda}_k \\ \cdot \\ \cdot \\ \underline{\lambda}_N \end{bmatrix}$$

where

$$\underline{g}_k^T = [\ g_{k1},\ g_{k2}\ \cdots\ g_{ki}\ \cdot\cdot\ g_{kP}\ ]$$

$$\underline{\lambda}_k^T = [\ \lambda_{k1},\ \lambda_{k2}\ \cdots\ \lambda_{ki}\ \cdot\cdot\ \lambda_{kP}\ ]$$

C_k = P x P submatrix whose elements are given by (8)

$C_i = C_{N-i}^T$, i = 1,2 ... (N-1); T denote transpose, and each

C_i is symmetric w.r.t. to the second diagonal.

As the matrix (of Eq. (9)) is block circulant, using its block Jordan Canonical form [6], we obtain

$$\underline{\lambda} = [U \otimes I]\ \Lambda^+\ [U^* \otimes I]\underline{g} \tag{11}$$

where U is the discrete Fourier transform (DFT) matrix, $\otimes$ denotes Kronecker product, * denotes transposition, Λ^+ is a block diagonal matrix, which is a generalised inverse of Λ, and Λ is also block diagonal with submatrices

$\{y_i;\ i = 0,1,2\ \ldots\ (N-1)\}$, i.e., $\Lambda = \text{diag.}\ |y_o, y_1, \cdots\ y_{N-1}|$

such that

$$y_i = \sum_{k=0}^{N-1} C_k \exp\left(-j\frac{2\pi ki}{N}\right). \tag{12}$$

Eq. (11) forms the direct or non-recursive algorithm for image reconstruction from fan beam data. It exploits the symmetry properties and the eigenvalue decomposition of the system matrix C to devise a fast inversion procedure for computing $\underline{\lambda}$. The final reconstruction is achieved by employing a weighted (by 1/r') back projection of the Lagrange multipliers $\{\underline{\lambda}_k\}$ (see Table I) in an angular, and not parallel manner.

The algorithm based on Eq. (11) may be implemented as follows:

STEP 1: Given the fan beam projection $\underline{g}$, compute the inverse block DFT of $\underline{g}$. i.e. compute $\underline{Z} = \{U^* \otimes I\}\underline{g}$. This can be achieved by computing P N-length DFTs of the series $g_{1i}, \ldots g_{Ni}$; i = 1,2 .... P, and assigning the mth term of the nth series to Z_{mn}.

STEP 2: Compute Λ by calculating the block eigenvalue $\{\underline{Y}_k\}$ from Eq. (12).

STEP 3: Invert Λ to obtain Λ^+. Specifically pseudo-invert each block eigenvalue $\{Y_k\}$. If the values N and P are maintained constant for any fan beam system i.e. for a fixed geometry, STEPS 2 and 3 need be performed only once, and $\{\Lambda_i^+\}$ stored for all future reconstruction.

STEP 4: Compute $\underline{Z} = \Lambda^+\underline{Z}$, equivalently calculate $Z_k' = Y_{k-1}Z_k$, k = 1,2 ... N, where $\{Z_k'\}$ is available from STEP 1.

STEP 5: Perform the angular weighted backprojection of $\{\lambda_{ki}\}$ according to the variance model for the fan beam geometry in Table I to produce the final reconstruction.

A block diagram for this algorithm is given in Fig. 5. If the eigenvalues $\{Y_i\}$ are precomputed and stored the number of arithmetic operations (multiplications) needed is approximately $2PN\log_2 N + 7NP^2$, as STEPS 1 and 5 require $PN\log_2 N$ operations each, STEP 4 requires NP^2 multiplications, and $6NP^2$ multiplications are allowed for the weighted angular back projections. The storage requirements for the algorithm are $NP^2/4 + 2NP$ words, as $NP^2/4$ are needed for storing the block eigenvalues, NP for the Lagrange multipliers, and NP for the projection data.

4. ANGULAR PROJECTION TRANSFORM AND THE FAN BEAM CONVOLUTION ALGORITHM.

According to the conventional Projection Slice Theorem [9] the Fourier transform of a parallel ray projection at any angle is exactly equal to the 'slice' of the 2-D Fourier transform of the

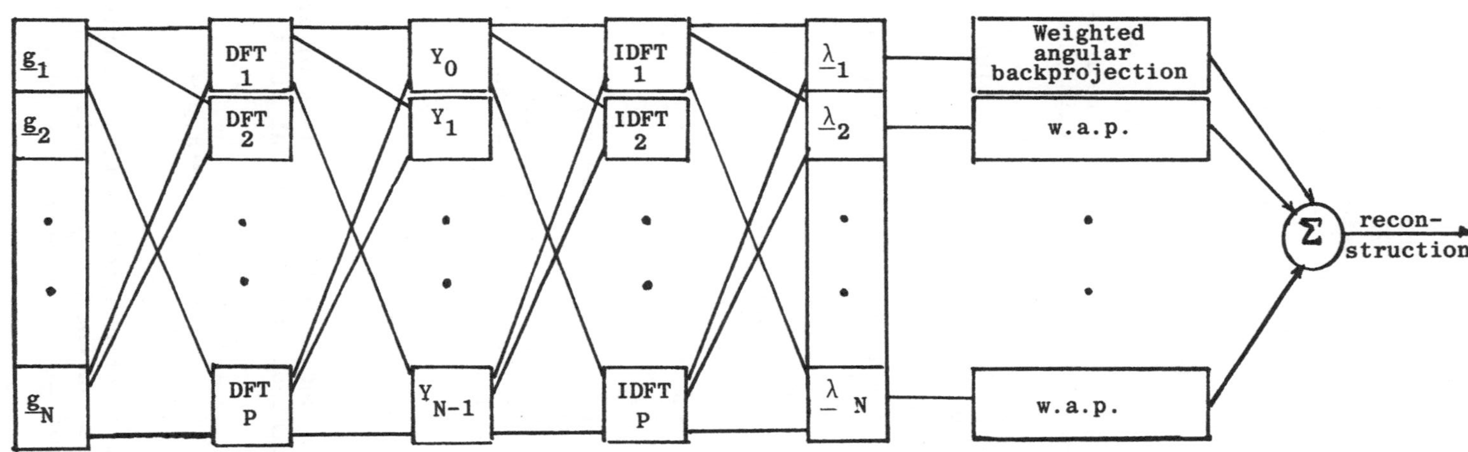

Fig. 5. Block diagram for the non-recursive algorithm.

object taken at the same angle. Here we show that by employing the Angular Projection Transform, defined below, an identical relationship holds between the Fourier transform of the fan beam projections and 'slices' of the 2-D Angular Projection Transform of the object.

Defining the Angular Projection transform of the object $f(x,y)$ as

$$A(\psi,n) = \iint_D f(x,y)e^{-jn\theta'}dr'd\theta', \ \forall \ \psi,\theta' \tag{13}$$

with ψ,θ',r' as in Eq. (1), and where D is the domain of the object; and using Eqs. (1) and (13) we obtain

$$\begin{aligned} A(\psi,n) &= \int_0^{2\pi} d\theta' e^{-jn\theta'} \int_{\ell_1(\theta')}^{\ell_2(\theta')} f(x,y)dr' \\ &= \int_0^{2\pi} g_\psi(\theta')e^{-jn\theta'}d\theta' = G_n(\psi) \ \text{(by definition)} \end{aligned} \tag{14}$$

where $G_n(\psi)$ is the nth Fourier series coefficient of $g_\psi(\theta')$ since $g_\psi(\theta')$ is periodic over both ψ and θ', i.e.

$$g_\psi(\theta') = \frac{1}{2\pi} \sum_k G_k(\psi)e^{jk\theta'} \tag{15}$$

From Eq. (14) it may be observed that the slice of the 2-D Angular Projection Transform, $A(\psi,n)$ (for a specific value of n), is exactly equal to the nth Fourier series coefficient of $g_\psi(\theta')$.

A Fourier series decomposition also exists for $\{\lambda_\psi(\theta')\}$, as it is periodic in θ' with $(0<\theta'<2\pi)$, i.e.

$$\lambda_\psi(\theta') = \frac{1}{2\pi} \sum_m L_m(\psi)e^{jm\theta'} \tag{16}$$

It may be shown that for a class of images which can be expressed in terms of one dimensional functions $\{\lambda_\psi(\theta')\}$ as

$$f(x,y) = \int_0^{2\pi} \frac{\lambda_\psi(\theta')}{r'} d\theta', \ -\infty<x,y<\infty \tag{17}$$

where r',θ' are defined in Eq. (1), a strict mathematical relationship holds between the Fourier coefficients of $\{\lambda_k\}$ and the Fourier coefficients of the fan beam projections $\{g_k\}$, namely,

$$L_m(\psi) = |m| \; G_m(\psi) \tag{18}$$

Note that any image (object) can be expressed by the model of Eq. (17) within a finite domain (such as D in Fig. 3). There exists a general class of images for which this model holds for the complete x-y plane. The weighting in Eq. (18) is identical to the frequency weighting included in the convolution algorithm for the parallel ray system.

Eqs. (16)-(18) form a convolution algorithm for fan beam systems. Taking the Fourier transform of the series $\{g_{ki}, i=1,2 \,..\, P\}$ gives $G_n(k)$. $\{\lambda_k(m)\}$ may be determined from Eq. (16) by inserting the value of $\{L_m(k)\}$ calculated from Eq. (18). By employing the weighted angular back projection of $\{\lambda_k(m)\}$ according to Eq. (17) the reconstruction is generated.

It is interesting to note that if the radius of the circle (O,R) in Fig. 3 increases, i.e. the source moves away from the object, the fan beam strips tend to become parallel strips and the above algorithm tends to the convolution algorithm for the parallel ray geometry. For $R=\infty$, the two algorithms become identical.

5. COMPUTED RESULTS

Fig. 6 illustrates a plot from a Tetronix 4010 of an 80x80 (image) matrix which was produced by digitizing the original on a P-1000 PHOTOSCAN microdensitometer. Different grey levels were obtained by varying the number of dots plotted within each pixel of the Tetronix CRT screen.

Fig. 7 is the reconstruction of the image shown in Fig. 6 obtained from the convolution algorithm of Eq. (18), using 40 projections with 80 strips per projection. Similarly, Fig. 8 is the reconstruction where 60 projections were employed. The projections used in these reconstructions were computer generated from Fig. 6, using Eqs. (1) and (5) with $\{y_k = 2\pi(k-1)/N\}$. In both cases fan beam strips of equal angular width were taken.

The 'streaks' observed in Fig. 7 are due to the 'star' type point spread function which appears in all methods that employ the backprojection operation to obtain a reconstruction. Note that this 'streaking' is much less severe in Fig. 8. Obviously, by increasing the number of projections this effect can be completely suppressed.

6. ACKNOWLEDGEMENT

The research reported here was carried out under a U.K. S.R.C. grant.

Fig. 6. Original (80x80 pixels)

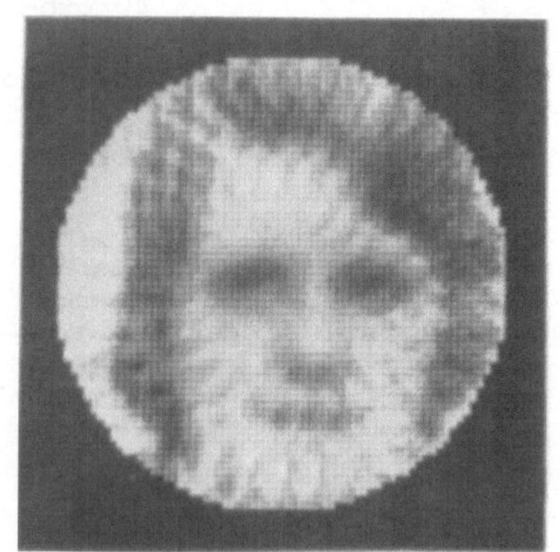

Fig. 7. Reconstruction from 40 projections

Fig. 8. Reconstruction from 60 projections

REFERENCES

1. Cho, Z. H., "General Views on 3-D Image Reconstruction and Computerized Transverse Axial Tomography", Trans. IEEE, NS-21, pp. 44, 1974.
2. Bracewell, R. N., and A. C. Riddle, "Inversion of Fan-Beam Scans in Radio Astronomy", The Astrophys. J., 150 (2), pp. 427, 1967.
3. Sweeney, D. W., "Reconstruction of Three Dimensional Refractive Index Fields from Multi Directional Inter-Ferometric Data", Applied Optics, 12, pp. 2649, 1973.
4. Gilbert, P. F. C., "The Reconstruction of a Three-Dimensional Structure from Projections and its Applications to Electron Microscopy: II Direct Methods", Proc. Roy. Soc., London, ser. B. 182, pp. 89, 1972.
5. Durrani, T. S. and C. E. Goutis, "Frequency Domain Techniques for Image Reconstruction from their Projections", in Digital Image Processing and Analysis, Ed. J. C. Simon and A. Rosenfeld, pp. 93, Noordhoff Pub. Leyden, 1977.
6. Goutis, C. E. and T. S. Durrani, "Digital Image Reconstruction by Circulant-Like Matrix Algorithms", Proc. IERE/IEEE Conf. on Digital Processing of Signals in Communications, Loughborough Univ., U.K., 37, pp. 285, 1977.
7. Herman, G. T., A. V. Lakshminarayanan, and A. Naparstek, "Reconstruction Using Divergent-Ray Shadowgraphs" in Reconstruction Tomography in Diagnostic Radiology and Nuclear Medicine, Ed. M. M. Ter-Pogossian et. al., University Park Press Baltimore, pp. 105, 1977.
8. Pang, S. C. and S. Genna, "A Fourier Convolution Fan-Geometry Reconstruction Algorithm: Simulation Studies, Noise Propagation, and Polychoromatic Degradation", ibid No. 7, pp. 120.
9. Shepp, L. A., B. F. Logan, "The Fourier Reconstruction of a Head Section", IEEE Trans. Nucl. Sci. NS-21, pp. 21, 1974.
10. Goutis, C. E., "Constrained Optimisation Approach to the Reconstruction from Projections Problem," Southampton University Report, U. K., Aug. 1976.

DIFFERENT ASPECTS OF LINEAR MODELLING FOR SIGNAL PROCESSING

G. Carayannis

Institut de Phonétique, Université Libre de Bruxelles,
50, Avenue F. D. Rossevelt - 1050 Bruxelles - Belgium.

ABSTRACT. The purpose of this paper is to give an evaluation of the techniques employed for discrete signal processing using a linear model. Many physical phenomena behave in "a linear manner", so this model will describe them sufficiently well. The state of the art will be discussed here with special emphasis given to the algorithms used to obtain the model parameters. These algorithms are very often much more than simple mathematical tools. For example, access to some of the intrinsic variables of the algorithm enable us to measure the regularity or the stability of the process structure or to make some important decisions. Decision making relative to system order determination and event detection will be viewed in more detail. Linear prediction (LP) techniques will be discussed extensively because of their important properties (efficient solution producing algorithms, spectral matching properties, etc.) Forward and backward LP are discussed with emphasis on their possible connection. On the other hand a general mathematical framework will be given which is common to all the methods (covariance, autocorrelation, and the sequential covariance method). A special form of the linear model (factorial model) corresponding to a non-predictive normalisation of the parameter vector is also examined. Some important properties of this model are discussed and a physical meaning is found for the eigen-values of the signal autocorrelation matrix. The same model is also used in data reduction techniques in a very similar manner. Finally, ARMA modelling is examined from a critical point of view and a new algorithm is given in order to obtain a separate solution, for the AR-part only, of the connected forward-backward problem.

1. INTRODUCTION

Time series analysis using a linear model is being developed into a subject of intensive study. Important literature is available, with particular reference to autoregressive modelling and linear prediction [24],[29],[31]. This paper is a specialized survey of some useful properties and features of the algorithms which are used to obtain the model parameters. Thus in the case of linear prediction, for example, we will not discuss optimality criteria and possible relations among least squares, minimum variance, maximum likelihood, maximum entropy, spectral matching or autocorrelation matching formulations. These topics are fully discussed in the literature. For the optimization step, only the least squares approach will be used here because of its simplicity. The problem of the normalization of the parameter vector will be studied and some simple "immediate" normalizations will be examined. We qualify as "immediate" those normalizations not requiring any supplementary a priori information. An important decision for processing is related to the selection of the signal observation matrix. Several possibilities exist which are related to different windowings of the signal. A windowing assuring zero initial and final conditions results in a symmetric Toeplitz autocorrelation matrix. One of the several important consequences of this choice is the fact that the same model can be used for forward and backward prediction, thus an alternative structure - the lattice structure - is possible for the prediction filter. The properties of these structures will be discussed together with the generalization of their use.

It is important in signal processing to be able to make a number of decisions explaining some aspects of the variable structure of the signal. The most important decisions are in relation to parsimonious model order (PMO) determination and to event detection. The PMO is related to the number of peaks (and/or valleys) of the spectrum but this relation is not very simple. Generally, a statistical criterion is used and the PMO is defined at the minimum of this criterion. The use of a recursive algorithm (order recursion) for the solution enables us to define stopping rules for the PMO evaluation.

We define as events some structural discontinuities of the signal. For example, voiced sounds are produced after excitation of the vocal tract by discontinuous impulse character input. It is important to be able to decide about discontinuity location in order to avoid integrating it into the model. Some criteria are defined here which make it possible to take this kind of decision. It is possible to carry out this time domain selection at the same time as the solution by using sequential methods (time recursion). We will also discuss how to set up algorithms for combined time and order recursion which are useful for automatic time domain and order selection.

Signal analysis by linear modelling is generally obtained by a convenient factorization of the signal covariance matrix. This concept has much in common with factor analysis in data reduction. Linear prediction techniques correspond to only one possible factorization of this matrix, triangularization. Other possibilities exist as for example the diagonalization of the covariance matrix. This is possible by a different choice of coefficient vector norm, a process which leads to an equivalent of the Karhunen-Loeve transformation applied to time series. Some important properties of the method are given for Toeplitz approximation of the covariance matrix. This model has been labelled the factorial model.

Some original aspects of ARMA modeling are examined here with special emphasis on the connections between forward and backward ARMA models and the repercussion of these connections on the formulation of the solution. A new algorithm which constitutes a generalisation of Levinson's recursion is given for the AR-part estimation.

It is important to note that many of the results discussed here relative to one channel can be extrapolated to cover multichannel signals. In particular,the different algorithms of order and time recursion have analogous forms for multichannel signal processing.

Section 2 concerns general assumptions for modelling and the problem of normalisation of the coefficients of the model. Section 3 is given over to the predictive model. Factorial modelling is discussed in Section 4 and ARMA models in Section 5.

2. GENERAL LINEAR MODEL ASSUMPTIONS AND NORMALISATIONS.

Some basic ideas and hypotheses are reviewed in this chapter. The selection of an adequate model is usually a very important and difficult step. Most of the time the evaluation of the quality of its representativity is carried out after very tedious experimentation. For the speech signal, the evaluation of the representativity of the linear model was easy and direct, as this model was used to generate synthetic speech.

It may be assumed that our signal is generally made up of two kinds of components, one random and one organized. Furthermore, using the system theory approach, we consider that if an organized part exists, it is because the signal carries some information characteristic of the system through which it has passed. No particular assumptions will be made about this system except that of linearity and stationarity over the analysis frame. More specific characteristics of the system as, for example, overall resonant behavior or important zeros contribution can be verified after analysis. Later in this section we will discuss assumptions concerning system excitation. Consider now the sufficiently general approach for signal linear modeling which is the following:

$$\sum_{i=0}^{p} a_i s(n-i) = \sum_{i=0}^{q} b_i e(n-i) \qquad (1)$$

where s(n) is the signal sample and e(n) an unknown input. This equation is the discrete equivalent of the following differential equation used to describe continuous linear systems.

$$a_p \frac{d^p s}{dt^p} + a_{p-1} \frac{d^{p-1} s}{dt^{p-1}} + \ldots + a_o s(t) = b_q \frac{d^q e}{dt^q} + \ldots + b_o e(t) \quad (2)$$

Using a z-transform with zero-initial conditions, equation (1) can be written:

$$S(z) = E(z)\, H(z) \quad (3)$$

with

$$H(z) = \frac{b_o + b_1 z^{-1} + \ldots + b_q z^{-q}}{a_o + a_1 z^{-1} + \ldots + a_p z^{-p}} \quad (4)$$

We thus have a transfer function equivalent to the linear model. This is a typical pole-zero model or, using terminology from the statistical literature an ARMA model (Mixed autoregressive-moving average). For $z = \exp(j2\pi fT)$, where T is the sampling period, a frequency response can be computed from (4).

Some particular cases of this model are interesting to examine. For $b_i = 0$, $\forall i \geq 1$ and $b_o \neq 0$, the numerator of the transfert function becomes a constant quantity. In this case we have an all pole or autoregressive (AR) model. This model is adequate for the processing of signals having an almost resonant behavior (for example: non-nasalized sounds). The algorithmic solution facility in this case motivates the use of this model, even if zeros exist. Valleys corresponding to large bandwidths are approximated by some number of poles [30]. Consequently we have a substantial increase of the AR-model order p.

Another particular case occurs when $a_i = 0, \forall i \geq 1$ and $a_o \neq 0$. In this case we have an all-zero or moving average (MA) model. It is not easy to use this model when the measurements of the input signal are not available. If, however, these measurements are available, finding the model is equivalent to the problem of discrete impulse response identification. One method of obtaining a solution is by the discrete equivalent of Wiener-Hopf equations.

For the general ARMA model, it is also far from easy to arrive at a solution directly. Optimization gives a set of non-linear equations which can only be solved iteratively without any guarantee of convergence to the absolute minimum. Separate computation of the AR and the MA-part is easier because the AR-part can be obtained from a linear system.

We will examine the AR-model here in more detail:

$$a_o\, s(n) + a_1\, s(n-1) + \ldots + a_p\, s(n-p) = e(n) \tag{5}$$

p is the order or memory of the model and e(n) an unknown input or an error, because the relation of the first member may not have been verified exactly. Let us examine the input hypothesis first. For an input equivalent to an uncorrelated sequence (white noise), model evaluation can be presented as a particular identification of a system. Another case when the model identification concept can be used is for impulse-like input. In this case, it is possible to conceive identification in the free oscillation time. Both cases (white noise - impulse-like input) are present in speech signals (fricatives - voiced sounds). The identification idea is interesting for speech because the model can be viewed as a means of realizing a deconvolution. This is possible if it represents the system well. (p is equal to the exact order, where the system has an overall resonant behavior, etc.). For real signals it is very difficult to realize all these conditions exactly; the error computed after evaluation of the model is not exactly the input assumed.

For all these reasons most authors prefer the linear regression theory to identification as a theoretical support. For this theory e(n) is simply an error "everything which cannot be modeled".

In the following we will be looking at two important and fundamental problems: model vector normalization and assumptions needed for finite length frame modeling. Let us write equation (5) for all known samples, say from n = M until n = N (with M < N).

$$\begin{bmatrix} s(M+p) \\ s(M+p+1) \\ \cdot \\ \cdot \\ \cdot \\ s(N) \end{bmatrix} a_o + \begin{bmatrix} s(M+p-1) & \ldots & s(M) \\ s(M+p) & \ldots & s(M+1) \\ \cdot & \ldots & \cdot \\ \cdot & \ldots & \cdot \\ \cdot & \ldots & \cdot \\ s(N-1) & \ldots & s(N-p) \end{bmatrix} \begin{bmatrix} a_1 \\ a_2 \\ \cdot \\ \cdot \\ \cdot \\ a_p \end{bmatrix} = \begin{bmatrix} e(M+p) \\ e(M+p+1) \\ \cdot \\ \cdot \\ \cdot \\ e(N-p) \end{bmatrix} \tag{6}$$

or in condensed form:

$$\underline{s}\, a_o + S\, \underline{a} = \underline{e} \tag{7}$$

The definition of quantities appearing in this last equation is obvious after comparison with (6). The name observation matrix will be used in what follows to designate matrix S.

The number of signal samples involved in (6) is equal to N-M + 1. We must generally take N-M+1 >> p. For a deterministic signal generated by an AR-model and associated with a system in free oscillation, we can obtain an exact solution for $a_o = 1$, from (7). In this case we need N - M + 1 = 2p samples for the solution. This is the lowest limit for the number of samples. If the previous conditions are not satisfied (which is the case for real signals) we must look for a least squares solutions of (7) by minimization of an error criterion.

We define:

$$\underline{\tilde{a}}^T = [a_o \; \underline{a}^T] \tag{8}$$

$$\tilde{S} = [\underline{s} \quad S] \tag{9}$$

$$\underline{\tilde{s}}_p^T(n) = \tilde{s}^T(n) = [s(n) \quad s(n-1) \; . \; . \; . \; s(n-p)] \; . \tag{10}$$

Thus the total squared error can be written:

$$\begin{aligned} \varepsilon &= \sum_n e^2(n) = \sum_n (\underline{s}^T(n)\underline{\tilde{a}})^2 = \underline{\tilde{a}}^T \; (\sum_n \underline{s}(n) s^T(n))\underline{\tilde{a}} \\ &= \underline{a}^T \; \tilde{S}^T\tilde{S} \; \underline{a} = \underline{a}^T \; \tilde{R} \; \underline{a} \end{aligned} \tag{11}$$

where,

$$\tilde{R} = \tilde{S}^T\tilde{S} \tag{12}$$

is the covariance matrix of the signal.

The annulation of derivatives gives a necessary minimum condition:

$$\frac{\partial \varepsilon}{\partial \underline{a}} = \tilde{R} \; \underline{\tilde{a}} = 0 \; . \tag{13}$$

Because this equation only has trivial solutions, it seems necessary to introduce constraints in form of the the normalization of the coefficient vector. Different possible normalizations exist. We are only interested here in "immediate" normalizations (i.e. those which do not require any a priori knowledge relative to the process).

Thus, norms like:

$$\underline{a}^T \; H \; \underline{a} = 1$$

where H is known a priori or computed from the signal matrix, will not be studied here. These norms are studied elsewhere [10] and used for signal segmentation purposes. The simplest normalizations one can imagine are perhaps $a_o = 1$ or $a_p = 1$. The first is the one which is the basis of forward linear prediction techniques (FLP). A sample at instant n can be predicted from p previous samples; an error term is used for correction.

$$s(n) = - \sum_{i=0}^{p} a_i^+ \, s(n-i) + e^+(n) \tag{14}$$

The second normalization is relative to backward linear prediction (BLP). A sample at instant n is "predicted" from p future samples:

$$s(n) = - \sum_{i=0}^{p} a_i^- \, s(n+i) + e^-(n) \tag{15}$$

The exponent (+) is employed to designate the FLP model and the exponent (-) for the BLP model.

To obtain the FLP solution, let us write the total squared error in the following way: (the exponent (+) is omitted here for convenience).

$$\varepsilon = \underline{\tilde{a}}^T \tilde{S}^T \tilde{S} \, \underline{\tilde{a}} = \begin{bmatrix} 1 & \underline{a}^T \end{bmatrix} \begin{bmatrix} \underline{s}^T \\ S^T \end{bmatrix} \begin{bmatrix} \underline{s} & S \end{bmatrix} \begin{bmatrix} 1 \\ \underline{a} \end{bmatrix} \tag{16}$$

$$\varepsilon = \underline{a}^T S^T S \, \underline{a} + 2\underline{a}^T \, S^T \underline{s} + \underline{s}^T \underline{s} \tag{17}$$

The derivation of this relation gives:

$$\frac{\partial \varepsilon}{\partial \underline{a}} = S^T \underline{s} + S^T S \, \underline{a} = 0 \, . \tag{18}$$

Thus by way of a solution we have:

$$S^T S \, \underline{a} = - S^T \, \underline{s} \tag{19}$$

or

$$R \, \underline{a} = - \underline{r} \tag{19b}$$

where R is a covariance matrix and $\underline{r}$ is a covariance vector.

On the other hand, for the minimum error we obtain:

$$\varepsilon = \underline{s}^T \underline{s} + \underline{a}^T S^T \underline{s} = \underline{s}^T \underline{s} - \underline{a}^T S^T S \, \underline{a} \tag{20}$$

These relations will be used in the next sub-sections as a point of departure to obtain some interesting solution algorithms.

Another reasonable normalization of the model is the following:

$$\underline{\tilde{a}}^T \underline{\tilde{a}} = 1 \tag{21}$$

The resulting constrained minimization problem can be solved using the Lagrangian:

$$\mathcal{L} = \underline{\tilde{a}}^T\tilde{S}^T\tilde{S}\underline{\tilde{a}} - \lambda(\underline{\tilde{a}}^T\underline{\tilde{a}} - 1) \tag{22}$$

The derivation of this relation results in the following optimality conditions:

$$\tilde{S}^T\tilde{S}\ \underline{a} = \lambda\underline{\tilde{a}} \tag{23}$$

$$\min\ (\varepsilon) = \lambda^{\min} \tag{24}$$

Thus we have an eigen-value problem. The properties of this solution are discussed in section 4.

Here we want to discuss another important problem, that of signal windowing and its influence on the structure of the matrix S^TS. Windowing reduces some terms of the observation matrix S to zero. If the p terms at the beginning and the p terms at the end of the signal are reduced to zero, matrix S^TS takes on the Toeplitz structure. Let S^o be the S observation matrix with only zero initial conditions, S_o the same matrix with zero final condition and S_o^o with zero-initial and zero-final conditions. The choice of one of the matrices S, S^o, S_o or S_o^o is important for the analysis.

The above considerations are valid for finite frame length. But if N-M + 1 becomes more and more important in comparison to p, the model memory, (N-M + 1 >>> p), matrix S^TS is reasonably well-approximated by the Toeplitz structure. Thus this structure also appears as an upper theoretical limit when more and more samples are taken into account. In a recent paper [35] matrix S^TS is viewed as being close to Toeplitz, whilst $S_o{}^TS_o$ is considered to be closer, and $(S_o^o)^T\ (S_o^o)$ is Toeplitz. When $a_o = 1$ the method resulting from the S_o^o choice of the observation matrix is known in the literature as the autocorrelation method. The general method resulting from the S choice of the observation matrix is known as the covariance method.

Writing the solution for the autocorrelation method we have:

$$R_A\ \underline{a} = -\ \underline{r}_A \tag{25}$$

with R_A a symmetric Toeplitz matrix (autocorrelation matrix) and $\underline{r}_A$ an autocorrelation vector:

$$R_A = (S_o^o)^T(S_o^o) = \begin{bmatrix} r_o & r_1 & r_2 & \dots & r_{p-1} \\ r_1 & r_o & r_1 & \dots & r_{p-2} \\ \cdot & \cdot & \cdot & \dots & \cdot \\ \cdot & \cdot & \cdot & \dots & \cdot \\ \cdot & \cdot & \cdot & \dots & \cdot \\ r_{p-1} & \cdot & & \dots & r_o \end{bmatrix} \tag{26}$$

$$\underline{r}_A^T = [r_1 \; r_2 \; \ldots \; r_p] = (S_o^o)^T \; \underline{s}_o^o \tag{27}$$

where

$$r_j = \sum_{n=M}^{N-1-j} s(n) \; s(n+j) \qquad (j > 0), \tag{28}$$

Being Toeplitz the R_A matrix has a number of properties. Below we only give some of the properties of the R_A matrix when operated by the self-reciprocal operator J defined here:

$$J = \begin{bmatrix} 0 & 0 & . & . & . & 0 & 1 \\ 0 & 0 & . & . & . & 1 & 0 \\ . & . & & & & . & . \\ . & . & & & & . & . \\ 1 & 0 & . & . & . & 0 & 0 \end{bmatrix} \tag{29}$$

The J operator follows the relations:

$$J^T = J \; , \quad J^2 = I \tag{30}$$

where I is the unitary matrix.

Matrix R_A properties are the following

$$R_A J = JR_A \; , \; R_A^{-1} J = JR_A^{-1} \text{ and } JR_A J = R_A \tag{31}$$

$$R_A J = \text{a Hankel matrix} \tag{32}$$

These properties will be useful in the rest of the development of this paper.

For the covariance method we have:

$$R_c \underline{a} = -\underline{r}_c \tag{33}$$

where R_c is a general symmetric positive definite matrix (covariance matrix) and $\underline{r}_c$ is a covariance vector:

$$R_c = \begin{bmatrix} r_{11} & r_{12} & \ldots & r_{1p} \\ r_{21} & r_{22} & \ldots & r_{2p} \\ . & . & \ldots & . \\ . & & & \\ . & & & \\ r_{p1} & r_{p2} & \ldots & r_{pp} \end{bmatrix} = S^T S \tag{34}$$

and

$$\underline{r}_c^T = [r_{o1} \;\; r_{o2} \;\; \ldots \;\; r_{op}] = S^T \underline{s} \tag{35}$$

with

$$r_{ij} = \sum_{n=p+M}^{N-1} s(n-i)\, s(n-j) \tag{36}$$

The next section studies the predictive AR-model in more detail.

3. THE PREDICTIVE MODEL.

An analysis will be given here relative to the possible connections between backward and forward prediction which lie behind the lattice structure implementation of the model. On the other hand, the different solution algorithms will be classified into two categories, order recursion and time recursion. Both of these classes are important for the applications of signal analysis. Connected time and order recursion will be examined in detail.

3.1 Backward and Forward Prediction

In this section we will try to establish some connections between the backward and the forward prediction models. In some cases discovering these connections can be useful as is shown in the subsequent discussion.

Let us once again consider the fundamental equations for backward and forward prediction from the previous section (equations (14) and (15)). Assuming that we know the signal between samples M and N; thus s(M) is the first and s(N) the final samples known. Assuming that we have no knowledge before M and after N, the two previous equations can be used for prediction in two different ways. The first way is to realize FLP and BLP between samples (M+p) and (N-p). In this case, exactly the same samples are predicted by both methods. The second way is to realize FLP between (M+p+1) and N and BLP between samples M and (N-p-1). In this latter case, error minimization is not carried out for exactly the same samples, but observation matrices are very close together.

Figure 1 shows the two different choices between finite FLP and BLP intervals.

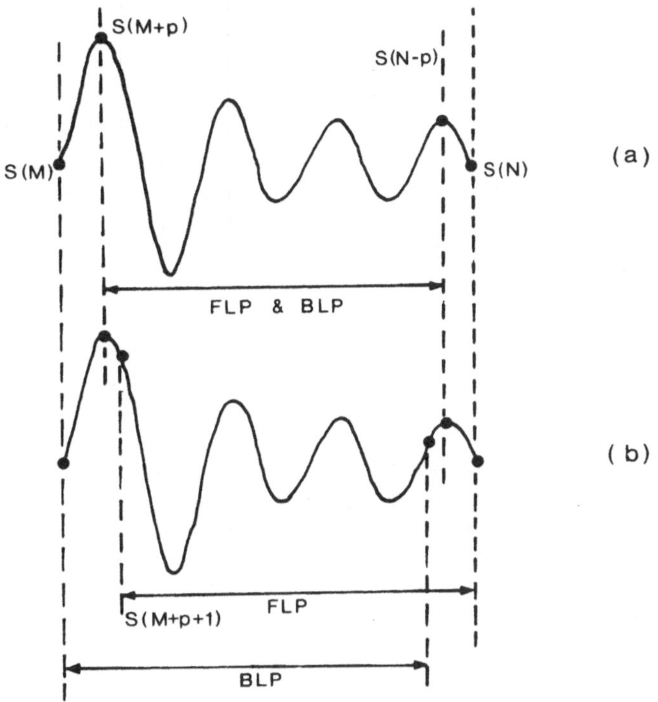

Figure 1. Two different choices between finite FLP and BLP intervals.

Writing equations (14) and (15) for all samples of the first choice we have:

$$\begin{bmatrix} s(M+p) \\ s(M+p+1) \\ \cdot \\ \vdots \\ s(N-p) \end{bmatrix} + \begin{bmatrix} s(M+p-1) & \dots & s(M) \\ s(M+p) & \dots & s(M+1) \\ \cdot & \dots & \cdot \\ \vdots & \vdots & \vdots \\ s(N-p-1) & \dots & s(N-2p) \end{bmatrix} \begin{bmatrix} a_1^+ \\ a_2^+ \\ \cdot \\ \vdots \\ a_p^+ \end{bmatrix} = \begin{bmatrix} e^+(M+p) \\ \cdot \\ \cdot \\ \vdots \\ e^+(N-p) \end{bmatrix} \tag{37}$$

$$\begin{bmatrix} s(M+p) \\ s(M+p+1) \\ \cdot \\ \vdots \\ s(N-p) \end{bmatrix} + \begin{bmatrix} s(M+p+1) & \dots & s(M+2p) \\ s(M+p+2) & \dots & s(M+2p+1) \\ \cdot & \dots & \cdot \\ \vdots & \vdots & \vdots \\ s(N-p+1) & \dots & s(N) \end{bmatrix} \begin{bmatrix} a_1^- \\ a_2^- \\ \cdot \\ \vdots \\ a_p^- \end{bmatrix} = \begin{bmatrix} e^-(M+p) \\ \cdot \\ \cdot \\ \vdots \\ e^-(N-p) \end{bmatrix} \tag{38}$$

or equivalently:

$$\underline{s}' + S_f' \, \underline{a}^+ = e'^+ \tag{37a}$$

$$\underline{s}' + S_b' \, \underline{a}^- = e'^- \tag{38a}$$

As we observe, matrices S_f' and S_b' are fundamentally different. The (p+1) first lines and the (p+1) last lines of the two matrices are different. The use of zero-initial and zero-final conditions for the signal does not make up for the differences. The resulting backward and forward models are different ($\underline{a}^+ \neq \underline{a}^-$).

A different situation appears if we agree to shift our BLP and FLP intervals slightly in a number of samples equal to the prediction memory (see Figure 1b). Writing equations (14) and (15) for all samples of each interval, we obtain:

$$\begin{bmatrix} s(M+p+1) \\ s(M+p+2) \\ \cdot \\ \vdots \\ s(N) \end{bmatrix} + \begin{bmatrix} s(M+p) & \dots & s(M+1) \\ s(M+p+1) & \dots & s(M+2) \\ \cdot & \dots & \cdot \\ \vdots & \vdots & \vdots \\ s(N-1) & \dots & s(N-p) \end{bmatrix} \begin{bmatrix} a_1^+ \\ a_2^+ \\ \cdot \\ \vdots \\ a_p^+ \end{bmatrix} = \begin{bmatrix} e^+(M+p+1) \\ \cdot \\ \cdot \\ \vdots \\ e^+(N) \end{bmatrix} \tag{39}$$

$$\begin{bmatrix} s(M) \\ s(M+1) \\ \cdot \\ \vdots \\ s(N-p-1) \end{bmatrix} + \begin{bmatrix} s(M+1) & \dots & s(M+p) \\ s(M+2) & \dots & s(M+p+1) \\ \cdot & \dots & \cdot \\ \vdots & \vdots & \vdots \\ s(N-p) & \dots & s(N-1) \end{bmatrix} \begin{bmatrix} a_1^- \\ a_2^- \\ \cdot \\ \vdots \\ a_p^- \end{bmatrix} = \begin{bmatrix} e^-(M) \\ \cdot \\ \cdot \\ \vdots \\ e^-(N-p-1) \end{bmatrix} \tag{40}$$

or in condensed notation:

$$\underline{s}^{+} + S_f\underline{a}^{+} = \underline{e}^{+} \tag{41}$$

$$\underline{s}^{+} + S_b\underline{a}^{-} = \underline{e}^{-} \tag{42}$$

We can now easily see that there is a very simple relation between matrices S_f and S_b

$$S_f = S_b\ J \tag{43}$$

where J is the self-reciprocal operator defined by (29). It is interesting to note that the sample s(M) is not used in equation (39) in the same way as sample s(N) is not used in equation (40).

Writing a LS solution for the FLP case, we obtain:

$$S_f^T S_f\underline{a}^{+} = -\ S_f^T\ \underline{s}^{+} \quad \text{or} \quad R_f\underline{a}^{+} = -\ \underline{r}^{+} \tag{44}$$

In an equivalent manner for the BLP case, we obtain:

$$S_b^T\ S_b\ \underline{a}^{-} = -S_b^T\ \underline{s}^{-} = \underline{r}^{-} \tag{45}$$

or using (43) and (30):

$$JS_f^T\ S_fJ\ \underline{a}^{-} = -JS_f^T\ \underline{s}^{-} \quad \text{or} \quad JR_fJ\underline{a}^{-} = -\underline{r}^{-} \tag{46}$$

There is an interesting case when zero initial and final conditions are used for the signal (windowing). In this case the covariance matrix R_f takes on a symmetrical Toeplitz structure and obeys relations (31). On the other hand, it is possible to verify the following relation:

$$\underline{r}^{-} = \underline{r}^{+} \tag{47}$$

Thus equation (46) can be written:

$$JR_fJ\underline{a}^{-} = -\underline{r}^{+} \quad \text{or} \quad R_f\underline{a}^{-} = -\underline{r}^{+}$$

(using relations (31) for Toeplitz matrices).

Consequently:

$$R_f\ \underline{a}^{-} = -\underline{r}^{+} \tag{48}$$

and:

$$\underline{a}^{-} = \underline{a}^{+} \tag{49}$$

In the autocorrelation case we thus have the same model for FLP and BLP.

Some interesting properties of the autocorrelation and the covariance methods for combined FLP-BLP will be discussed with a recursive solution (order recursion) in the next subsection.

3.2 Order Recursion

A simple matrix formulation is given here for the recursive estimation of the AR-Model. Firstly we will examine the covariance method; the autocorrelation method is viewed as a special case only. Let us write system (44) (omitting index f for convenience) and for an order m of the AR-Model.

$$R_m \underline{a}_m^+ = - \underline{r}_m^+ \tag{50}$$

where:

$$(\underline{a}_m^+)^T = \begin{bmatrix} a_{m1}^+ & a_{m2}^+ & \dots & a_{mm}^+ \end{bmatrix} \tag{51}$$

$$(\underline{r}_m^+)^T = \begin{bmatrix} r_1^+ & r_2^+ & \dots & r_m^+ \end{bmatrix} \tag{52}$$

$$(\underline{r}_m^-)^T = \begin{bmatrix} r_1^- & r_2^- & \dots & r_m^- \end{bmatrix} \tag{52a}$$

writing a matrix R recursion, we now have:

$$R_m = \begin{bmatrix} R_{m-1} & \underline{\rho}_m \\ \underline{\rho}_m^T & \rho_m \end{bmatrix} \tag{53}$$

An inverse matrix recursion can be obtained using a matrix inversion by partitioning Lemma.

$$R_m^{-1} = \begin{bmatrix} R_{m-1}^{-1} + \underline{w}_m \underline{w}_m^T \alpha_m^{-1} & \underline{w}_m \alpha_m^{-1} \\ \underline{w}_m^T \alpha_m^{-1} & \alpha_m^{-1} \end{bmatrix} \tag{54}$$

where:

$$\underline{w}_m = - R_{m-1}^{-1} \underline{\rho}_m \tag{55}$$

$$\alpha_m = \underline{\rho}_m^T \underline{w}_m + \rho_m = \frac{\det R_m}{\det R_{m-1}} \tag{56}$$

By defining $\underline{a}'^- = J\,\underline{a}^-$, (57)

we obtain for the recursive solution [8]:

$$L\,\underline{a}_m^+ = \underline{a}_{m-1}^+ + k_m^+ \underline{w}_m \tag{58}$$

$$a_{mm}^+ = k_m^+ \tag{59}$$

$$L\,a_m'^- = a_{m-1}'^- + k_m^- \underline{w}_m \tag{60}$$

$$a_{mm}'^- = k_m^- \tag{61}$$

where L is a matrix operator obtained from the unit matrix when the last row is omitted. When a vector is pre-multiplied by this operator it is deprived of its last component. k-coefficients are defined as follows:

$$k_m^+ = -\frac{\beta_m^+}{\alpha_m} \qquad k_m^- = -\frac{\beta_m^-}{\alpha_m} \tag{62}$$

with:

$$\beta_m^+ = \underline{w}_m^T \underline{r}_{m-1}^+ + r_m^+ \tag{63}$$

$$\beta_m^- = \underline{w}_m^T J \underline{r}_{m-1}^- + r_m^- \tag{64}$$

Another important recursion is related to error computation. Using the definition of equation (20) we can show that :

$$\varepsilon_m^+ = \varepsilon_{m-1}^+ - (k_m^+)^2\,\alpha_{m-1} \tag{65}$$

$$\varepsilon_m^- = \varepsilon_{m-1}^- - (k_m^-)^2\,\alpha_{m-1} \tag{66}$$

From (65) and (66), we can find that:

$$\frac{\Delta\,\varepsilon_m^+}{\Delta\,\varepsilon_m^-} = \frac{k_m^+}{k_m^-} \qquad \text{where } \Delta\varepsilon_m = \varepsilon_m - \varepsilon_{m-1}$$

For the Toeplitz case, the covariance matrix recursion is written:

$$R_m = \begin{bmatrix} R_{m-1} & J\underline{r}_{m-1} \\ \underline{r}_{m-1}^T J & r_o \end{bmatrix} \tag{67}$$

Thus we have:

$$\underline{\rho}_m = J\underline{r}_{m-1} \quad \text{and} \quad \rho_m = r_o \tag{68}$$

Consequently:

$$\underline{w}_m = - R_{m-1}^{-1} J\underline{r}_{m-1} = J\underline{a}_{m-1} \tag{69}$$

On the other hand,

$$\underline{a}^+ = \underline{a}^- = \underline{a} \tag{70}$$

$$\alpha_m = \varepsilon_m \tag{71}$$

and, in this case, the recursive procedure for the solution can be written:

$$L\ \underline{a}_m = \underline{a}_{m-1} + k_m J\underline{a}_{m-1} \tag{72}$$

$$a_{mm} = k_m \tag{73}$$

$$\varepsilon_m = \alpha_m = \alpha_{m-1}\ (1-k_m^2) \tag{74}$$

This procedure is known as Levinson's recursion.

The relation $|k_m| \leq 1$ implies the positive definiteness of matrix R (all principal minors must be positive; thus $\alpha_m > 0$ and $|k_m| \leq 1$). It can be demonstrated that the same relation is the necessary and sufficient condition for the stability of the predictor filter. On the other hand, relations (72) and (73) establish an equivalence between a_j coefficients and k_j coefficients. The k_j coefficients define a lattice structure equivalent to the direct prediction filter (see Fig. 2).

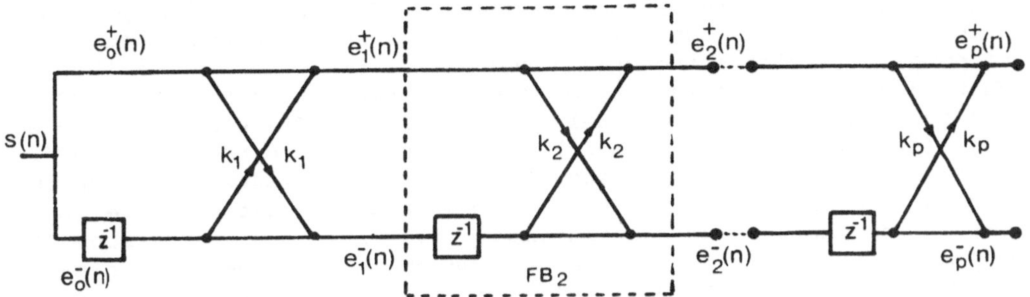

Figure 2. A lattice equivalent of the predictive model.

Let us define:

$$\underline{s}_m^T (n) = [\, s(n-1) \;\; s(n-2) \;\; \ldots \;\; s(n-m)\,] \tag{75}$$

and the following FLP and BLP errors

$$e_m^+(n) = s(n) + \underline{a}_m^T \underline{s}_m(n) \tag{76}$$

$$e_m^-(n-1) = s(n-m-1) + \underline{a}_m^T J \, \underline{s}_m(n) \tag{77}$$

Using Levinson's recursion (72), with (75), (76) and (77) it is easy to show that:

$$e_{m+1}^+(n) = e_m^+(n) + k_{m+1} e_m^-(n-1) \tag{78}$$

$$e_{m+1}^-(n) = k_{m+1} \, e_m^+(n) + e_m^-(n-1) \tag{79}$$

with

$$e_o^+(n) = e_o^-(n) = s(n) \tag{80}$$

Equations (78) and (79) can be used to define the lattice structure in Figure 2. It is obvious that each part FB_i of the lattice filter corresponds to a new step in the recursive procedure traducing an increase in order. On the other hand, BLP and FLP concepts are combined in this structure.

Unfortunately for the covariance method such a structure cannot result from the recursive procedures (58)-(61). The interesting properties of the lattice structure and especially the possibilities for stability control ($|k| \leq 1$) motivate researchers to try to compute lattice structures directly for the covariance method. What is usually done is to look for the solution of the backward and forward problems separately; k_i coefficients are computed as a mean (e.g. geometric, harmonic) of the forward and backward coefficients. It is possible to show that these mean values have the important property $|k| \leq 1$. Linear prediction coefficients are computed from the k_j mean-coefficients using Levinson's recursion. It has been demonstrated [30] that the harmonic-mean computation can be obtained by the minimization at each step m of an error criterion:

$$\varepsilon_m = \varepsilon_m^+ + \varepsilon_m^-$$

All these possible lattice methods are in some way "bounded" by the autocorrelation method and spectra obtained are found between those of autocorrelation and covariance methods [30]. For the computation of k parameters, covariances are used instead of autocorrelations but backward and forward residuals appearing in these lattice filters are intrinsic variables to the filters and not the backward and forward residuals of equations (39) and (40).

One important property of order recursion is related to the definition of stopping rules for selecting the parsimonious model memory. This memory is related to the number of dominant frequencies in the spectrum, but this relation cannot be simple. Most of the criteria for order determination are related to the minimum total squared error ε or to the determinant of the covariance matrix [1], [12], [28]). As has been shown previously, this error can be computed recursively. The recursive computation is also possible for the determinant of the covariance matrix using the ratio α defined by (56). In statistical literature an accepted criterion for PMO evaluation is as follows [29]:

$$1 - \frac{\varepsilon_{m+1}}{\varepsilon_m} \leq \delta \tag{81}$$

Some more sophisticated criteria give a minimum for the PMO. Figure 3 shows Akaike's criterion as it varies with the filter memory. This criterion can be written:

$$I_m = \text{Log}\ \varepsilon_m + \frac{2m}{N_e} \tag{82}$$

where N_e is an effective window length.

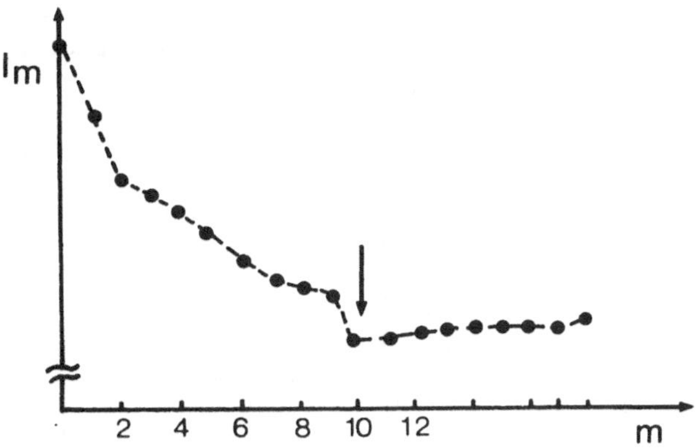

Figure 3. Akaike's criterion computed on a French vowel.

It is interesting to note that for the autocorrelation method the squared error ε is equivalent to the determinant ratio α. The quantity:

$$d_{\varepsilon}^{\alpha} = \sum_{i=1}^{p} (\varepsilon_i - \alpha_i)^2 \tag{83}$$

can be used as a distance measure between the autocorrelation and covariance methods. Another possible distance measure is [8]:

$$d_{w}^{a} = \sum_{i=1}^{p} (\underline{w}_i - J\underline{\underline{a}}_i)^T (\underline{w}_i - J\underline{\underline{a}}_i) \tag{84}$$

Quantities d_w^a and d_ε^α can also be used to define a distance between a covariance and a Toeplitz autocorrelation matrix.

The next subsection deals with time-recursion for model determination. This recursion is examined in connection with the problem of event detection and time selection.

3.3. Time Recursion (sequential methods)

Only the forward prediction problem will be examined here. Some signals have a general discontinuous behavior (because of discontinuous excitation, for example). In this case, the general stationnarity assumptions used in the previous section are no longer valid. Time recursion can be very useful for these signals, because it enables us to define some criteria for discontinuity detection.

The following definition of this recursive solution will be used:

$$\underline{a}(n+1) = \underline{a}(n) + k(n+1)\,\underline{w}(n+1) \tag{85}$$

Vector $\underline{w}(n+1)$ and scalar $k(n+1)$ will be computed subsequently for a sequential least squares (SLS) minimisation.

Let us define:

$$S(n) = \begin{bmatrix} s(M+p) & \cdots & s(M+1) \\ s(M+p+1) & \cdots & s(M+2) \\ \cdot & \cdots & \cdot \\ \vdots & \cdots & \vdots \\ s(n-1) & \cdots & s(n-p) \end{bmatrix} \tag{86}$$

and

$$\underline{s}^T(n) = [\, s(M+p+1) \quad s(M+p+2) \quad \ldots \quad s(n)\,] \tag{87}$$

$$\underline{s}^T(n+1) = [\, s(n) \quad s(n-1) \quad \ldots \quad s(n-p+1)\,] \tag{88}$$

using equation (19) the solution at instant n can be written:

$$\underline{a}(n) = [\,S^T(n)\; S(n)\,]^{-1}\; S^T(n)\; \underline{s}(n) = R^{-1}(n)\; \underline{r}(n) = P(n)\; \underline{r}(n) \tag{89}$$

with

$$P(n) = R^{-1}(n) \tag{90}$$

Let us now write a time recursion for matrix S and vector $\underline{s}$.

$$S(n+1) = \begin{bmatrix} S(n) \\ \\ \underline{s}^T(n+1) \end{bmatrix} \tag{91}$$

and

$$\underline{s}(n+1) = \begin{bmatrix} \underline{s}(n) \\ \\ s(n+1) \end{bmatrix} \tag{92}$$

the solution at instant (n+1) can be written:

$$\underline{a}(n+1) = [R(n) + \underline{s}(n+1)\underline{s}^T(n+1)]^{-1}[r(n) + \underline{s}(n+1)s(n+1)] \tag{93}$$

Using a matrix inversion Lemma (the rank annulation method), we have:

$$P(n+1) = [R(n) + \underline{s}(n+1)\ \underline{s}^T(n+1)]^{-1}$$

$$= P(n) - \frac{P(n)\underline{s}(n+1)\underline{s}^T(n+1)P(n)}{1 + \underline{s}^T(n+1)P(n)\underline{s}(n+1)} \tag{94}$$

or:

$$P(n+1) = P(n) - \underline{w}(n+1)\underline{w}^T(n+1)\ \alpha^{-1}(n+1) \tag{95}$$

where:

$$\underline{w}(n+1) = P(n)\ \underline{s}(n+1) \tag{96}$$

$$\alpha(n+1) = \frac{\det R(n+1)}{\det R(n)} = \underline{s}^T(n+1)\ \underline{w}(n+1) + 1$$

$$= 1 + \underline{s}^T(n+1)\ P(n)\ \underline{s}(n+1) \tag{97}$$

On the other hand, k(n+1) is given by the following relation:

$$k(n+1) = \frac{s(n+1) - \hat{s}(n+1)}{\alpha(n+1)} = \frac{s(n+1) - \underline{a}^T(n)\ \underline{s}(n+1)}{\alpha(n+1)} = \frac{e(n+1)}{\alpha(n+1)} \tag{98}$$

Some generalisations of this algorithm are possible, for example, we can take:

$$\alpha(n+1) = \sigma_e^2 + \underline{s}^T(n+1)\ P(n)\ \underline{s}(n+1) \tag{99}$$

where σ_e^2 is defined by the relation:

$$E(\underline{e}\ \underline{e}^T) = \sigma_e^2\ I \tag{100}$$

where E is the expected value, σ_e^2 is thus the variance of the prediction error supposed white.

Another generalisation is possible for formula (95) which can be written:

$$P(n+1) = P(n) - \underline{w}(n+1)\underline{w}^T(n+1)\ \alpha^{-1}(n+1) + \sigma_q^2\ I \tag{101}$$

σ_q^2 is the variance of a state noise from the formula:

$$\underline{a}(n+1) = I\ \underline{a}(n) + \underline{q}(n) \tag{102}$$

with

$$E(\underline{q}\ \underline{q}^T) = \sigma_q^2\ I \tag{103}$$

These generalisations can be better understood, if we consider this recursive algorithm as a simple formulation of a Kalman filter. The use of the formula (101) makes the algorithm more adaptive. Both the adaptive properties and the problem of algorithm initialization are discussed elsewhere [4], [34]. A completely adaptive scheme is possible after recursive subtraction of the information relative to the past. In this case we must define an analysis frame length of Δ samples. Let us define the following quantities:

$$\underline{w}(n-\Delta) = P(n)\ \underline{s}(n-\Delta) \tag{104}$$

$$\alpha(n-\Delta) = \frac{\det\ R^{\Delta}(n)}{\det\ R(n)} = -\ \underline{s}^T(n-\Delta)\underline{w}(n-\Delta) + 1 \tag{105}$$

The exponent Δ is used to denote the covariance matrix (or inverse covariance matrix) after subtraction of the information concerning the past samples. We then obtain:

$$P^{\Delta}(n) = P(n) + \underline{w}(n-\Delta)\ \underline{w}^T(n-\Delta)\ \alpha^{-1}(n-\Delta) \tag{106}$$

On the other hand,

$$\underline{a}(n+1) = \underline{a}(n) + k(n+1)\ \underline{w}(n+1) \tag{107}$$

where:

$$\underline{w}(n+1) = P^{\Delta}(n)\underline{s}(n+1) \quad , \ \alpha(n+1) = 1 + \underline{s}^T(n+1)\underline{w}(n+1) \tag{108}$$

$$P(n+1) = P^{\Delta}(n) - \underline{w}(n+1)\ \underline{w}^T(n+1)\ \alpha^{-1}(n+1) \tag{109}$$

$$k(n+1) = \frac{e(n+1)}{\alpha(n+1)} \tag{110}$$

We thus have a double recursion. This algorithm is equivalent to the covariance method used with a moving window of fixed length.

Unfortunately the sequential methods given above need a matrix inversion at each time instant and are consequently not very fast. Faster sequential methods exist. These methods can be grouped in the general category of stochastic approximation. The algorithms examined before are the most general possible. The choice of vector $\underline{w}$ was determined by the formula:

$$\underline{w}(n+1) = P(n)\ \underline{s}(n+1) \qquad (111)$$

where P(n) is the inverse of the covariance matrix of the signal. If a less complicated quantity is chosen for matrix P, less operations are required. For example, a possible choice for P(n) could be "a diagonal matrix" or even a "scalar" ([44],[11]). These algorithms may be useful for event detection [33], or for simple signal processing [20], but, unfortunately, for more complicated signals such as speech, they are very difficult to implement to reach an optimum. Sometimes, they approximate the optimum and give quite good dominant frequency estimation.

An important feature of the sequential solution is that it is possible to define some "structural" criteria for event detection. By the word "structural" we mean "related to the model or to some intrinsic variables of the algorithm explaining the structural properties of the process." Some external criteria such as prediction error will not be considered here.

We give below two structural criteria, the first one is related to the model evolution;

$$\delta(n+1) = [\underline{a}(n+1) - \underline{a}(n)]^T\ [\underline{a}(n+1) - \underline{a}(n)] \qquad (112)$$

and the second $\alpha(n)$ to the covariance matrix determinant ratio:

$$\alpha(n+1) = 1 + \underline{s}^T(n+1)\ P(n)\ \underline{s}(n+1) \qquad (113)$$

Both criteria have been used with a high degree of efficiency to determine pitch. Their important property in comparison with external criteria such as prediction error:

$$e(n+1) = s(n+1) - \hat{s}(n+1) \qquad (114)$$

is "selectivity". "Selectivity" is taken to mean that these criteria amplify differences due to real discontinuity in the signal structure (presence of input, artefacts, etc...) and minimize differences due to noises. Figure 4 gives $\delta(n)$'s criterion variations for a stationary formulation of the SLS-algorithm. Periodical reinitialization of the algorithm is needed in this case.

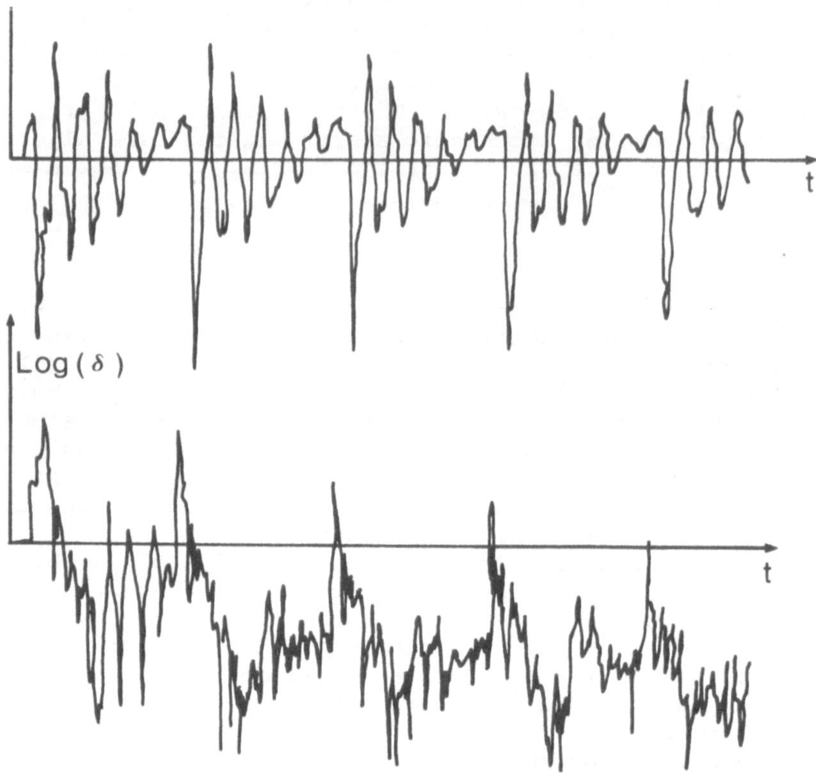

Figure 4. Pitch detection using the $\delta(n)$ criterion and a stationary formulation of a sequential algorithm.

Figure 5 gives the same criterion with the adaptive formulation of relations (104)-(110). These criteria can be very useful for speech processing not only for pitch detection, but also for time domain selection ([42],[11]).

Time domain selection may be very useful for the accurate resonance representation of speech spectra. It is known that for voiced sounds, input has an impulse character. Analysis of a frame taken in free oscillation time has the following properties:

a) Input influences are minimized (thus the best possible deconvolution obtained).
b) Zeros influences are minimized.
c) An increase is obtained in the percentage of stable models.
d) Bandwidths computed are closer to values predicted by the results of the speech production theory.

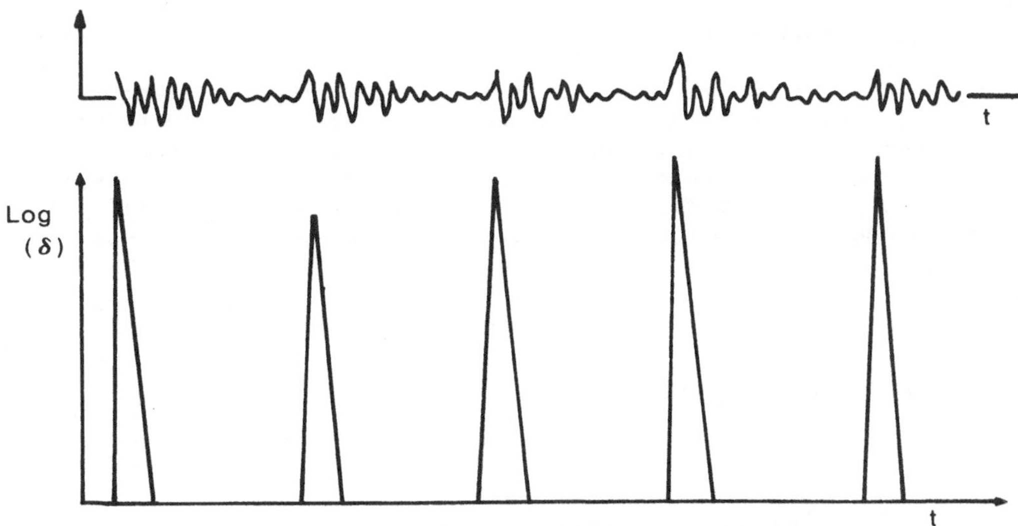

Figure 5. Pitch detection using the $\delta(n)$ criterion and an adaptive formulation of a sequential algorithm.

A more sophisticated criterion for time selection is Sobakin's criterion [43] which is based on the determinant of the covariance matrix. Let us define:

$$\underline{s}_j^T = [s(n-j)\ s(n-j+1)\ \dots\ s(n-j+\Delta)]$$

For good signal predictability (absence of input) the following relation holds:

$$\underline{s}_o = - \sum_{j=1}^{p} a_j \underline{s}_j$$

In this case the determinant of the covariance matrix R:

$$R = \{\underline{s}_i^T \underline{s}_j\} \qquad 0 \le i \le p \quad , \quad 0 \le j \le p$$

must be zero, thus traducing the linear dependance of vectors $\underline{s}_j$. For real signals, the minima of the determinant are characteristic of the small influence of the input.

Using a sequential algorithm, the determinant is computed at every instant. If a stationary formulation is used, we have:

$$\det R(n+1) = \alpha(n+1) \det R(n) \tag{115}$$

If an adaptive formulation is used, we have:

$$\det R(n+1) = \alpha(n+1)\ \alpha(n-\Delta)\ \det R(n) \tag{116}$$

Figure 6 shows the variation of this determinant on a French vowel. Strube's observations [43] were used. The selection of the model corresponding to points B of the figure has the four advantages stated above. The minima of the determinant correspond to regions of the signal where the prediction has been thoroughly verified and the maxima to regions where the input influence is more important.

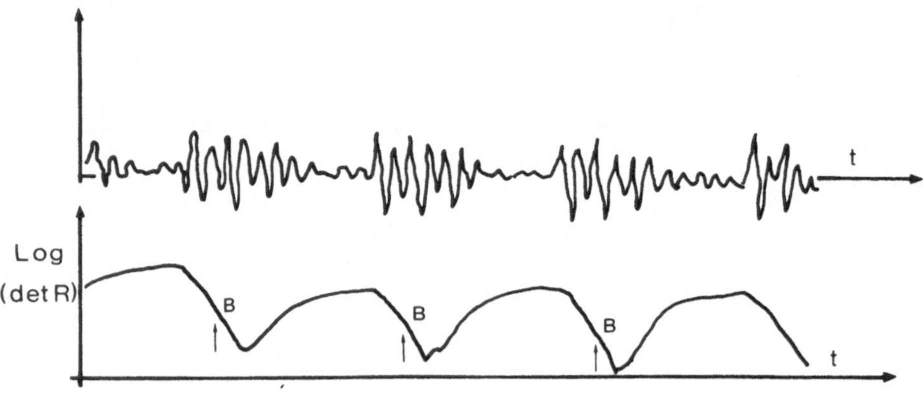

Figure 6. Time domain selection using the determinant of the covariance matrix.

3.4 Combined Time and Order Recursion.

It is perhaps clear from the two previous sub-sections and from the simple observation of formulae, that important algorithmic affinities exist between order and time recursion. The following table illustrates the procedures of computation where these affinities appear clearly.

It is important to try to realize simultaneous time and order recursion. One advantage of this combined recursion is a possible combined time and order selection, using for example Sobakin's criterion for time selection and, amongst others, Akaike's criterion for order selection.

This is possible without any major modification of time recursion. We have only added some supplementary operations corresponding to a step-down order recursion.

ORDER RECURSION	TIME RECURSION
L $\underline{a}_m = \underline{a}_{m-1} + k_m \underline{w}_m$ $a_{mm} = k_m$	$\underline{a}(n) = \underline{a}(n-1) + k(n)\,\underline{w}(n)$
$\underline{w}_m = -R^{-1}_{m-1}\underline{\rho}_m = -P_{m-1}\underline{\rho}_m$	$\underline{w}(n) = P(n-1)\,\underline{s}(n)$
$k_m = -\frac{\beta_m}{\alpha_m} = -\frac{\underline{w}_m^T\,\underline{r}_{m-1} + r_m}{\alpha_m}$	$k(n) = \frac{s(n) - \hat{s}(n)}{\alpha(n)}$ $= \frac{s(n) - \underline{a}^T(n)\underline{s}(n)}{\alpha(n)} = \frac{e(n)}{\alpha(n)}$
$\alpha_m = \frac{\det R_m}{\det R_{m-1}} = \underline{w}_m^T\underline{\rho}_m + \rho_m =$ $= -\underline{\rho}_m^T P_{m-1}\underline{\rho}_m + \rho_m$	$\alpha(n) = \frac{\det R(n)}{\det R(n-1)} = \underline{s}^T(n)\underline{w}(n) + 1 =$ $= 1 + \underline{s}^T(n)\,P(n-1)\,\underline{s}(n)$
$P_m = \left\{\begin{matrix} P_{m-1} + \underline{w}_m\,\underline{w}_m^T\,\alpha_m^{-1} & x \\ x & x \end{matrix}\right\}$	$P(n) = P(n-1) - \underline{w}(n)\underline{w}^T(n)\alpha^{-1}(n)$
EQUIVALENCES	
$P_m = R_m^{-1}$, $\underline{w}_m$, α_m	$P(n) = R^{-1}(n)$, $\underline{w}(n)$, $\alpha(n)$
$\underline{\rho}_m$, $\rho(m)$	$\underline{s}(n)$, -1 or $-\sigma_e^2$
β_m	$e(n)$

Table 1. Equivalences between time and order recursive procedures.

This can be done after observing in relation [54] that the last element of matrix P(n) is equal to α_p^{-1}. Vector $\underline{w}_p$ can be obtained by simply multiplying the last row of matrix P(n) by α_p and matrix $P_{p-1}(n)$ after subtracting a matrix of rank one by the formula:

$$P_{p-1}(n) = P_p(n) - \underline{w}_p \underline{w}_p^T \alpha_p^{-1} \tag{117}$$

On the other hand, we have:

$$a_{pp} = k_p \tag{118}$$

and

$$\underline{a}_{p-1} = L\underline{a}_p - k_p \underline{w}_p \tag{119}$$

Error computation can be carried out by the formula:

$$\varepsilon_p = \varepsilon_{p+1} + k_p^2 \alpha_p \tag{120}$$

4. THE FACTORIAL MODEL

This model makes it possible to find a simple physical interpretation for the eigen-values of the autocorrelation matrix. These values are identified here to "a representation error".

The approaches examined above have already produced many useful results for speech and other signals. But it has been noted [30], [31] that if the resonance frequencies of the spectrum are correctly retrieved, this is not always the case for the corresponding dampings, especially if the autocorrelation method is used. Since these dampings are useless in many applications it would be interesting to work out, on the same basis, a special method for computing resonant frequencies only.

We will use the following reasonable normalization of the parameter vector:

$$\underline{\tilde{a}}^T \underline{\tilde{a}} = 1 \tag{121}$$

and no longer the predictive norm ($a_o = 1$).

If ε is the total squared representation error:

$$\varepsilon = \underline{\tilde{a}}^T \tilde{R}\underline{\tilde{a}} \tag{122}$$

we have a constrained minimization problem which can then be solved using the Lagrangian (see equation (22)). This problem leads to the following optimality condition:

$$\tilde{R}\tilde{\underline{a}} = \lambda\tilde{\underline{a}} \quad \text{and} \quad \min(\varepsilon) = \lambda^{\min} \tag{123}$$

Consequently, the a_i coefficients are given by the eigen-vectors $\tilde{\underline{a}}$ associated with the smallest eigen-value of $\tilde{R}$. This particular choice will now be referred to as factorial linear modeling. This approach is clearly related to the well-known Karhunen-Loeve expansion.

This method takes all its major features from the properties of correlation Toeplitz matrices. The invariance property of $\tilde{R}$ when operated by J:

$$J\tilde{R}J = \tilde{R} \tag{124}$$

where J is defined by (29), results in a special structure for any eigen-vector $\tilde{\underline{a}}$:

$$\tilde{a} = \pm J \tilde{a} \tag{125}$$

This property implies that the roots of the polynomial

$$A(z) = \sum_{i=o}^{p} a_i z^{-i} \tag{126}$$

which are the poles of the model lie on the unit circle [5].

Applications of this model to speech analysis are given elsewhere [5]. Comparison of this method with linear prediction and cepstral techniques is given in [7], where the name "factorial model" is better justified. In fact, this method follows the fundamental factor theorem [21]. This is not the case with LP methods, if the factorization is assumed to be the triangularization of the covariance matrix.

In factor analysis approximation, an observation matrix S is written as a function of a "factor" matrix multiplied by a matrix of some weighting coefficients (usually referred to as the pattern matrix).

$$S = EA^{-1} \tag{127}$$

Let us examine two different cases of pattern matrices. Firstly:

$$A = A_T = \begin{bmatrix} 1 & & & & \\ 1 & a_{11} & & & \\ 1 & a_{21} & a_{22} & & \\ \cdot & \cdot & \cdot & & \\ 1 & a_{p1} & \cdot & \cdot\ \cdot & a_{pp} \end{bmatrix} \tag{128}$$

The pattern matrix is made up from the prediction coefficients which are the solutions of increasing order models.

Secondly:

$$A = A_D = \begin{bmatrix} a_{oo} & \cdots & a_{po} \\ a_{o1} & \cdots & \cdot \\ \cdot & \cdots & \cdot \\ \vdots & \vdots & \vdots \\ a_{op} & \cdots & a_{pp} \end{bmatrix} \tag{129}$$

This matrix is made up from all the eigen-vectors corresponding to all the eigen-values of the autocorrelation matrix from λ^{min} to λ^{max}.

The equation

$$\underline{s}_n^T A = \underline{e}_n^T \tag{130}$$

is valid for $A = A_D$ as well as $A = A_T$.

Defining:

$S^T = [\underline{s}_1 \; \underline{s}_2 \; \cdots \; \underline{s}_N]$ we now have:

$$SA = E \tag{132}$$

and $$S = EA^{-1} \tag{133}$$

For the factorial model, we have:

$$S^T S = A_D^{-T} E^T E A_D^{-1} = A_D E^T E A_D^T = A_D \Lambda A_D^T \tag{134}$$

thus

$$E^T E = \Lambda = \text{diag}(\lambda_o \; \lambda_1 \; \cdots \; \lambda_p) \tag{135}$$

For the linear prediction model, we obtain:

$$E^T E = A_T^T W^{-1} DW^{-T} A_T \tag{136}$$

where:

$$W = \begin{bmatrix} 1 & & & \\ a_{11} & 1 & & \\ a_{22} & a_{12} & 1 & \\ \cdot & \cdot & \cdot & \\ a_{pp} & \cdot & \cdot & 1 \end{bmatrix} \tag{137}$$

$$D = \text{diag}\ (\alpha_o \quad \alpha_1 \quad \ldots \quad \alpha_p) \tag{138}$$

It is evident that the error covariance matrix is not in this case diagonal. In this sense LP does not follow the fundamental factor theorem [21]. To obtain a diagonal error covariance matrix, we must choose:

$$A_T = A_{TF} = W^T \tag{139}$$

Some useful comparisons between LP and factorial models are summarized in Table 2.

PREDICTIVE MODEL	FACTORIAL MODEL
$R = W^{-1}\ D\ W^{-T}$	$R = A_D^T \Lambda A_D = \sum_{i=0}^{p} \lambda_i \underline{a}_i \underline{a}_i^T$
$R^{-1} = W^T D^{-1} W = \sum_{i=0}^{p} \frac{1}{\alpha_i} (J\underline{a}_i)(J\underline{a}_i)^T$	$R^{-1} = A_D^T \Lambda^{-1} A_D = \sum_{i=0}^{p} \frac{1}{\lambda_i} \underline{a}_i \underline{a}_i^T$
$\det(R) = \prod_i \alpha_i$	$\det(R) = \prod_i \lambda_i$
$\text{trace}(R) = p\alpha_o$	$\text{trace}(R) = \sum_i \lambda_i$

Table 2. Some comparisons between predictive and factorial models.

Other interesting properties of this model are related to the behavior of the eigen-values of the covariance matrix. Criteria like

$$1 - \frac{\lambda_{m+1}^{min}}{\lambda_m^{min}} < \delta' \tag{140}$$

can be used for the PMO evaluation, as was the case for criterion (81). Unfortunately we have not yet found any recursive procedure (order recursion) for computing these eigen-values. The lack of such an algorithm results in long and drawnout computation for criterion (140).

Eigen-vectors corresponding to the maximum eigen-values are also interesting because the spectra corresponding to these vectors have a frequency constitution which is as far from the signal as it is possible to get. This property may be useful for peak enhancement.

5. THE ARMA MODEL.

A more general model is the mixed autoregressive-moving average (ARMA) model of equation (1). In this equation, p is the order of the AR-part and q the order of the MA-part, e(n) being an uncorrelated sequence. It can be demonstrated that in this case the equations giving the solution are non-linear [30]. The solution algorithm can only be iterative without any guarantee of convergence to a global minimum. This procedure is usually very sensitive to initial parameter values [30].

In practice what is generally done is to separate the computation of the AR-part and the MA-part and solve two distinct problems. Even if they are not the most satisfactory, the estimates obtained in this way can at least be used as initializations of the iterative procedure. A number of methods exist for MA-parameter estimation after computation of AR-parameters. ([3], [37], [30]). Here we will only examine the problem of AR-part estimation when $a_o = 1$. It is known [12], [14], that an asymptotically unbiased estimator of AR parameters can be obtained for the FLP-model as the solution of a modified set of Yule-Walker equations when the system is excited by an unobserved uncorrelated sequence. This estimation can be obtained by once again solving a linear system.

$$R\, \underline{a}^{+} = -\, \underline{r}^{+} \quad \text{or} \quad R_{qp}\, \underline{a}^{+}_{qp} = -\, \underline{r}_{qp} \tag{141}$$

The structure of the matrix R and vector $\underline{r}^{+}$ are somewhat different when compared to the AR-model solution.

$$R_{qp} = \begin{bmatrix} r_q & r_{q-1} & \cdots & r_{q-p+1} \\ r_{q+1} & r_q & \cdots & r_{q-p+2} \\ \vdots & \vdots & & \vdots \\ r_{q+p-1} & & & r_q \end{bmatrix} \tag{142}$$

$$\underline{r}^T_{qp} = [r_{q+1} \quad r_{q+2} \quad \quad r_{q+p}] \tag{143}$$

It is easy to justify this choice. By multiplying both sides of equation (1) by s (n - j) and taking the expectation, we obtain:

$$\sum_{i=0}^{p} a_i \, r_{n-i} = 0 \text{ for } n>q \tag{144}$$

$$\sum_{i=0}^{p} a_i \, r_{n-1} \neq 0 \text{ for } n\leq q \tag{145}$$

r_n is the autocorrelation function of the signal, as defined in (28). We can easily see that matrix R_{qp} is not symmetric. Matrix R_{qp} is Toeplitz and is symmetric only about the main cross-diagonal. For q = 0, R_{qp} becomes an autocorrelation symmetric Toeplitz matrix.

Non-symmetric Toeplitz matrices have the following properties when operated by J:

$$RJ = JR^T, \; JRJ = R^T, \; JR^TJ = R$$

$$R^{-1}J = JR^{-T}, \; JR^{-1} = R^{-T}J, \; JR^{-1}J = R^{-T}, \; JR^{-T}J = R^{-1} \tag{146}$$

The choice of equations (141) with R_{qp} and $\underline{r}_{qp}$ given by (142) and (143) is only one possibility for model estimation, with the advantage of having a minimum error covariance matrix [14]. Other alternatives are any set of p equations taken from (144). In this case we again have a non-symmetric Toeplitz matrix R_{sp} similar to R_{qp} with s > q.

Let us introduce what may be called a "backward" ARMA model:

$$\sum_{i=0}^{p} \bar{a}_i \, s(n+i) = \sum_{i=0}^{q} \bar{b}_i \, e(n+i) \quad (\text{with } a_o = 1) \tag{147}$$

It is easy to see that for this model, s(n) is expressed as the sum of a linear combination of future samples s(n + 1), s(n + 2), ... s(n + p) and a linear combination of residual samples. This justifies the use of the term "backward". The normal equations for the "backward" model can be written

$$R^T_{qp} \, \underline{\bar{a}}_{qp} = - \underline{r}_{q,-p} \tag{148}$$

where

$$\underline{r}^T_{q',-p} = [r_{q-1} \; r_{q-2} \; . \; . \; . \; r_{q-p}] \tag{149}$$

It is possible to show ([9], [10]) that important connections exist between the two predictive schemes forward and backward and that solutions can be obtained in parallel by the formulae:

$$L\,\underline{a}_m^+ = \underline{a}_{m-1}^+ + k_m^+\, J\, \underline{a}_{m-1}^- \tag{150}$$

$$a_{mm}^+ = k_m^+ \tag{151}$$

$$L\,\underline{a}_m^- = \underline{a}_{m-1}^- + k_m^-\, J\, \underline{a}_{m-1}^+ \tag{152}$$

$$a_{mm}^- = k_m^- \tag{153}$$

where:

$$k_m^+ = -\frac{\beta_m^+}{\alpha_m} \qquad k_m^- = -\frac{\beta_m^-}{\alpha_m} \tag{154}$$

and:

$$\beta_m^+ = \underline{r}_{q,m-1}^T\, J\, \underline{a}_{m-1}^+ + r_{q+m} \tag{155}$$

$$\beta_m^- = \underline{r}_{q,1-m}^T\, J\, \underline{a}_{m-1}^- + r_{q-m} \tag{156}$$

$$\alpha_m = \alpha_{m-1}\,(1 - k_m^+\, k_m^-) = \frac{\det R_{q,m}}{\det R_{q,m-1}} \tag{157}$$

This method of arriving at a solution is a generalization of Levinson's recursion. A two multiplier lattice form (see Fig. 2) may also be used in this case instead of the direct predictor filter, with the difference that multiplication coefficients are no more the same for FLP and BLP.

This recursive procedure may be very useful for the PMO evaluation using Chow's [12] or Lindberger's [28] method, based on the autocorrelation matrix determinant. The determinant is computed recursively using equation (157). The recursion examined here is a recursion in the AR-part order only. A less simple recursion exists in both MA and AR orders [9].

The model obtained from the solution of equation (141) does not always correspond to a stable filter. A partial solution consists of taking into account more samples of the autocorrelation function and solving a least squares problem [6]. From equations (144) we can see that for white noise excitation and n>q, the same model is used both for autocorrelation and signal prediction. The least squares solution gives models more often stable. We believe that the stabilization of the AR-part of a mixed ARMA model is an interesting subject for future research.

6. CONCLUSION

Some selected topics in the domain of linear signal modeling have been studied here. This study covers autoregressive modeling with emphasis on predictive and factorial models. More emphasis has been given to the forward and backward prediction connections which lie behind model lattice structures. Decision making relative to parsimonious order determination and time domain selection has been studied with application to speech signal. We believe that the determinant criterion may also be interesting for signal segmentation purposes. Some original aspects of ARMA modeling have been examined and a new recursive algorithm is given for the AR-part determination. This algorithm constitutes a generalization of Levinson's recursion. ARMA-models have not lost their attraction for future research.

REFERENCES

1. Akaike, H., "A New Look at Statistical Model Identification", IEEE-Trans. on Automat. Contr., Vol. AC-19, pp. 716-723, Dec. 1974.
2. Atal, B. S., S. L. Hanauer, "Speech Analysis and Synthesis by Linear Prediction of the Speech Wave", J.A.S.A. 50, 1971, pp. 637-655.
3. Atashroo, M. A., Autocorrelation prediction. Proceedings of 1977 - ICASSP, Hartford U.S.A., May 1977, pp. 5-9.
4. Carayannis, G., "Analyse de la Parole par Identification Récurrente d'un Modèle du Système de Phonation", Thèse de Docteur Ingénieur., Univ. de Paris, 1973.
5. Carayannis, G., C. Gueguen, "The Factorial Linear Modelling. A Karhunen-Loeve Approach to Speech Analysis", Proceedings of the IEEE International, Conference on Acoustics, Speech and Signal Processing, Philadelphia 1976, pp. 489-492.
6. Carayannis, G., P. Jospa, "On the Analysis of Autocorrelation Function for Speech Spectra Estimation. Application for Nasality Detection." 1977 - IEEE, International Conference on Acoustics, Speech and Signal Processing. Hartford U.S.A., May 1977, pp. 754-757.
7. Carayannis, G., "Analyse du Signal par Décomposition Triangulaire et Diagonalisation des Matrices de Covariance". L'Onde Electrique. 1977, vol. 57, n° 8-9, pp. 525-531.
8. Carayannis, G., "An Alternative Formulation for the Recursive Solution of the Covariance and Autocorrelation Equations". IEEE, Transactions on ASSP. Vol. 25, n° 6 - December 1977, pp. 574-577.
9. Carayannis, G., "Nouvelles Conceptions Pour la Résolution Récurrente des Équations Yule-Walker dans le Cas des Modèles ARMA", C. R. de l'Académie des Sciences de Belgique, pp. 363-378, mars 1977.

10. Carayannis, G., "Analyse Comparative du Signal de la Parole". Proceedings of the 7th GALF Seminar on Speech. Nancy - FRANCE, may 1976, Vol. I, pp. 275-283.
11. Carayannis, G., "Contributions à la Modelisation Linéaire des Signaux". Thèse Doctorat d'Etat. Univ. de Paris Sud, 1978.
12. Chow, J. C., "On Estimating the Orders of an Autoregressive Moving-Average Process with Uncertain Observations", IEEE, Transactions on Automatic Control, pp. 707-709, October 1972.
13. Gantmacher, F. R., "Theorie des Matrices", tome 1, Dunod, Paris 1966.
14. Gersch, W., "Estimation of the Autoregressive Parameters of a Mixed Autoregressive Moving-Average Times Series", IEEE, Transactions on Automatic Control, pp. 583-588, October 1970.
15. Gibson, J., J. Melsa, S. Jones, "Digital Speech Analysis Using Sequential Estimation Techniques, IEEE, Trans. on ASSP, vol. ASSP-23, n° 4, 1975.
16. Gray, A. H., J. D. Markel, "Digital Lattice and Ladder Filter Synthesis", IEEE-Transactions on Audio and Electroacoustics, Vol. AU-21 n° 6, pp. 491-500, December 1973.
17. Gueguen, C., G. Carayannis, "Analyse de la Parole par Filtrage Optimal de Kalman", Automatisme, Tome XVIII, n° 3, pp. 99-105, 1973.
18. Gueguen, C., "Sur une Approche Système de la Reconnaissance des Formes", Actes du Congrès AFCET-IRIA, Chatenay-Malabry, France, Fevrier 1978, pp. 491-501.
19. Griffiths, J. W. R., P. L. Stocklin, C. Van Schooneveld, "Signal Processing" a NATO Advanced Study Institute, Academic Press - 1973.
20. Griffiths, L. J., "Rapid Measurement of Digital Instantaneous Frequency", IEEE, Trans. on ASSP, vol. 23, n° 2. April 1975, pp. 207-222.
21. Harman, H. H., Modern Factor Analysis, Second Edition - Revised. The University of Chicago Press, Chicago and London, 1968.
22. Huggins, A. W. F., R. Viswanathan, J. Makhoul, "Quality Ratings of LPC Vocoders: Effects of Number of Poles, Quantization and Frame Rate." Proceedings of 1977 - ICASSP; Hartford, U.S.A. May 1977, pp. 413-416.
23. Itakura, F., S. Saito, "Digital Filtering Techniques for Speech Analysis and Synthesis", 7th International Congress of Acoustics, Budapest, 1971, 2 SC 1, pp. 261-264.
24. Jenkins, G. M., D. G. Watts, Spectral Analysis and Its Applications, Holden-Day, San Francisco, Cambridge, London, Amsterdam, 1968.
25. Kopec, G. E., A. V. Oppenheim, J. M. Tribolet, "Speech Analysis by Homomorphic Prediction, IEEE-Trans. on ASSP. Vol. 25, N° 1, February 1977, pp. 40-49.
26. Leroux, J., C. J. Gueguen, "A Fixed Point Computation of Partial Correlation Coefficients", IEEE - Trans. on ASSP, Vol. 25, N° 3, June 1977, pp. 257-259.

27. Lienard, J. S., Les Processus de la Communication Parlée, Masson, Paris, Barcelone, Milan. 1977.
28. Lindberger, N. A., Comments on "On Estimating the Orders of an Autoregressive Moving-Average Process with Uncertain Observations", IEEE, Trans. on AC., Vol. 18, N° 6, Dec. 1973, pp. 689-691.
29. Makhoul, J., "Linear Prediction: A Tutorial Review", Proceedings of the IEEE, Vol. 63 N° 4, 1975.
30. Makhoul, J., "Stable and Efficient Lattice Methods for Linear Prediction", IEEE - Trans. on ASSP, Vol. 25, N° 5, Oct. 1977, pp. 423-428.
31. Markel, J. D., A. H. Gray, Jr., Linear Prediction of Speech, Springer-Verlag Berlin, Heidelberg, New York, 1976.
32. Markel, J. D., A. H. Gray, "On Autocorrelation Equations as Applied to Speech Analysis", IEEE, Transactions on Audio and Electroacoustics, Vol. AU-21, N° 2, April 1973, pp. 69-79.
33. Maksym, J. N.,"Real time Pitch Period Extraction by Adaptive Prediction of the Speech Waveform", Proceedings of IEEE - 1972. Conference on Speech Communication and Processing, Boston, U.S.A. April 1972, pp. 70-73.
34. Mathieu, M., "Analyse de l'Electro-encéphalogramme par Prédiction Linéaire", Thèse de Docteur Ingénieur, ENST, Paris, Septembre 1976.
35. Morf, M., B. Dickinson, T. Kailath, A. Vieira, "Efficient Solution of Covariance Equations for Linear Prediction", IEEE - Trans. on ASSP, Vol. 25, N° 5, Oct. 1977, pp. 423-433.
36. Oppenheim, A. V., R. W. Schafer, Digital Signal Processing, Prentice-Hall, Englewood Cliffs, N. J. - 1975.
37. Oppenheim, A. V., G. E. Kopec, J. M. Tribolet, "Signal Analysis by Homomorphic Prediction, IEEE, Trans. on ASSP, Vol. 24, N° 4, August 1976, pp. 327-332.
38. Picinbono, B., "Reconnaissance des formes et Traitement du signal", Actes du Congrès AFCET-IRIA, Châtenay-Malabry, France, Fevrier 1978, pp. 881-895 (conférencier invité).
39. Protonotarios, E., G. Carayannis, "Efficient Speech Storage for Speech Synthesis", NATO-Seminar, Athens, Grece, April 1974.
40. Richalet, J., A. Rault, R. Pouliquen, Identification des Processus par la Méthode du Modèle, Gordon and Breach, Paris, London, N.Y.
41. Schmid, C., "A Direct Method for Sequentially Updating Linear Predictor Coefficients for the Covariance Method", IEEE - International Conf. on Acoustics Speech and Signal Processing, Philadelphia - April 1976.
42. Steiglitz, K., B. Dickinson, The use of Time-Domain Selection for Improved Linear Prediction, IEEE - Trans. on ASSP., Vol. 25, N° 1, February 1977, pp. 34-39.
43. Strube, A. W., "Determination of the Instant of Glottal Closure from the Speech Wave", JASA, Vol. 56, N° 5, Nov. 1974.
44. Tsypkin, Y. Z., "Adaptation and Learning in Automatic Systems", Academic Press, New York and London, 1970.

45. Viswanathan, R., J. Makhoul, "Quantization Properties of Transmission Parameters in Linear Predictive Systems", IEEE - Trans. on ASSP., Vol. 23, N° 3 - June 1975.

ADAPTIVE LATTICE METHODS FOR LINEAR PREDICTIVE SIGNAL PROCESSING*

R. Viswanathan and J. Makhoul

Bolt Beranek and Newman Inc.
Cambridge, Mass. 02138, U.S.A.

ABSTRACT. The lattice structure has been recently employed in the development of all-pole or autoregressive modelling and spectral estimation methods. This paper presents a general method for adaptive estimation of the coefficients of the lattice model, for use in the analysis of nonstationary signals. The method is given as one of two sequential estimation methods, the other being a block sequential estimation method. The fast convergence of adaptive lattice algorithms relative to the conventional tapped-delay-line implementation is seen to be due to the orthogonalization and decoupling properties of the lattice. The paper suggests a number of applications for adaptive lattice methods in areas such as speech signal processing, channel equalization and noise cancelling.

1. INTRODUCTION

The lattice method of linear prediction was first introduced by Itakura [1,2] for speech analysis. A similar algorithm was proposed independently by Burg [3] in geophysics. Recently, Makhoul [4] showed the existence of a class of lattice methods of which the methods of Itakura and Burg are special cases. All these methods guarantee the stability of the corresponding all-pole filter, with or without windowing of the signal, independently of the stationarity properties and duration of the signal, and with finite

* This work was sponsored by the Advanced Research Projects Agency and monitored by Rome Air Development Center (ETC) under Contract No. F19628-78-C-0136.

wordlength computations. Also, Makhoul [4] developed the so-called covariance-lattice methods, which compute the lattice model parameters from the covariance of the signal, with a 3-4 fold saving in computation over the methods of Itakura and Burg.

This paper explores the sequential estimation of the parameters of the lattice in a nonstationary environment. A sequential estimation method, by our definition, provides a new estimate for the lattice parameters upon receiving each signal sample. The paper is organized as follows. In Section 2, we review the lattice formulation for linear prediction. Section 3 introduces the two sequential estimation methods: 1) Block method, and 2) Adaptive method. The two methods are described in detail, respectively, in Sections 4 and 5. We compare the convergence of the two methods through a spectral estimation example in Section 6. Finally, we present a number of applications for adaptive lattice methods in Section 7.

2. LATTICE FORMULATION

In the sequel we assume that the signal spectrum is to be modelled by an all-pole transfer function

$$H(z) = G/A(z), \tag{1}$$

where

$$A(z) = 1 + \sum_{k=1}^{p} a(k) z^{-k} \tag{2}$$

is the "inverse filter", G is a gain factor, a(k) are the predictor coefficients, and p is the model order.

The inverse filter A(z) can be implemented as a lattice filter, as shown by the signal flow-graph in Fig. 1 [5,6]. From Fig. 1, the following relations hold:

$$f_0(n) = g_0(n) = x(n), \tag{3a}$$

$$f_m(n) = f_{m-1}(n) + K_m g_{m-1}(n-1), \tag{3b}$$

$$g_m(n) = K_m f_{m-1}(n) + g_{m-1}(n-1), \tag{3c}$$

where x(n) is the input signal, $f_m(n)$ is the "forward" residual at stage m, $g_m(n)$ is the "backward" residual, and K_m is the reflection coefficient, with $1 \le m \le p$. There are p stages in the lattice in Fig. 1, and each stage involves two multiplies. This two-multiplier lattice model is originally due to Itakura and Saito [1]. (For one-multiplier and other types of lattice structures, refer to [6].)

Forward and backward transfer functions at stage m are defined by:

$$A_m(z) = F_m(z)/X(z), \tag{4a}$$

$$B_m(z) = G_m(z)/X(z), \tag{4b}$$

where $F_m(z)$, $G_m(z)$ and $X(z)$ are z-transforms, respectively, of $f_m(n)$, $g_m(n)$ and $x(n)$. Let

$$A_m(z) = \sum_{k=0}^{m} a_m(k) z^{-k}, \quad a_m(0) = 1, \tag{5}$$

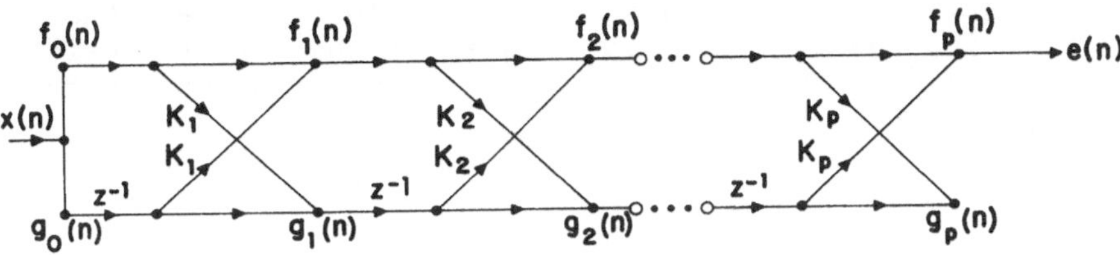

Fig. 1. Lattice inverse filter.

Then, from (4a), we can write

$$f_m(n) = x(n) + \sum_{k=1}^{m} a_m(k) x(n-k), \tag{6}$$

where the second term on the right is the (negative of the) forward prediction of $x(n)$. Hence, we call $f_m(n)$ the forward residual and $a_m(k)$ the predictor coefficients of the m-th order forward predictor. From (3) and (5), it can be shown [6] that

$$B_m(z) = \sum_{k=0}^{m} a_m(m-k)\, z^{-k}. \tag{7}$$

That is, $B_m(z)$ is the reverse polynomial corresponding to $A_m(z)$. From (4b) and (7), we have:

$$g_m(n) = x(n-m) + \sum_{k=1}^{m} a_m(k) x(n-m+k), \tag{8}$$

which clearly shows that $g_m(n)$ is the backward residual resulting from the backward prediction of $x(n-m)$. The predictor coefficients $a_m(k)$ are computed from the reflection coefficients using the recursion

$$a_m(m) = K_m$$

$$a_m(k) = a_{m-1}(k) + K_m a_{m-1}(m-k), \quad 1 \le k \le m-1. \tag{9}$$

The all-pole filter $1/A_p(z)$ is stable i.e., the zeros of $A_p(z)$ lie inside the unit circle in the z-plane, if and only if

$$|K_m| < 1, \quad 1 \le m \le p. \tag{10}$$

There are a number of methods for estimating the reflection coefficients which satisfy the stability condition (10). These methods involve minimizing the energy (or variance) of the forward residual $f_m(n)$ or the backward residual $g_m(n)$, or a combination of the two [4].

2.1 Stationary or Windowed Case

When the input signal $x(n)$ is stationary, or in the case of a finite data interval -- when the signal is suitably windowed such that the values outside this data interval are zero -- the forward and backward residual energies become equal. Consequently, the various estimation methods mentioned above give identical values for the reflection coefficients. Furthermore, the all-pole modelling problem of globally minimizing the energy of the lattice output residual $e = f_p$ with respect to all the reflection coefficients reduces to a sequence of local minimization problems, one at each lattice stage [6]. Thus, the optimal value of K_m is obtained by minimizing the residual (forward or backward) energy at the output of stage m with respect to K_m, but assuming that all previous stage residual signals are independent of K_m. The optimal value of K_m can be shown to be equal to the negative of the correlation coefficient between $f_{m-1}(n)$ and $g_{m-1}(n-1)$, which are the two input signals of stage m:

$$K_m = - \frac{\overline{f_{m-1}(n)\, g_{m-1}(n-1)}}{\sqrt{\overline{f^2_{m-1}(n)}\ \overline{g^2_{m-1}(n-1)}}} \tag{11}$$

where the overbars denote expected values or time-averages. Since the above correlation coefficient is less than 1 in magnitude, the stability condition (10) is satisfied.

An important observation in the stationary case is the following orthogonalization property of the backward residuals [6]:

$$\overline{g_i(n) g_j(n)} = E_i \delta_{ij} \tag{12}$$

where E_i is the minimum residual energy at the i-th stage. It is this orthogonalization process that results in the decoupling of the global minimization problem into p simpler, stage-by-stage local minimization problems. Further, the orthogonalization property (12) is quite useful in adaptive estimation.

2.2 Nonstationary or Unwindowed Case

For a nonstationary or unwindowed case, minimizing either the forward residual energy or the backward residual energy may not result in a stable filter. However, as shown in [4], the reflection coefficient estimates resulting from these two minimizations can be combined in an infinite number of ways to guarantee stability. We shall consider here only the method due to Burg, which forms the harmonic mean of the forward and backward estimates. The Burg estimate has the unique distinction that it can be derived by explicitly minimizing the sum of the energies of the forward and backward residuals:

$$E_m = \overline{f_m^2(n)} + \overline{g_m^2(n)}. \tag{13}$$

The reflection coefficient estimate in Burg's method is given by

$$K_m = -2\, \overline{f_{m-1}(n) g_{m-1}(n-1)} / [\overline{f_{m-1}^2(n)} + \overline{g_{m-1}^2(n-1)}] \tag{14}$$

In the nonstationary or unwindowed case, any stage-by-stage minimization method, such as Burg's method, yields in general only a suboptimal solution in the sense that the computed reflection coefficients may not minimize the energy of the output residual [7]. For many applications, this is of no consequence.

2.3 Further Advantages of the Lattice Model

Besides the advantages of guaranteed stability and orthogonalization property discussed above, there are other factors that favor employing the lattice model in general. The lattice form implementation of (1) produces a lower sensitivity to roundoff noise than, for example, the direct form implementation [8]. The reflection coefficients, which are the parameters of the lattice model, were found to be the best for use in speech transmission systems [9]. Also, for data compression purposes, quantization of the reflection coefficients may be accomplished within the lattice recursion. Finally, as a result of the orthogonalization process, an (m+1)-stage lattice has its first m reflection coefficients identical to those of the m-stage lattice. Using this property and a suitable criterion, an estimate of the "true" order of the model for a given signal sequence may be readily obtained [10,11]. In fact, such an estimate was employed in the design of variable order linear prediction as a data compression technique [10].

3. SEQUENTIAL ESTIMATION METHODS

A sequential estimation method, as mentioned above, provides a new estimate for the reflection coefficients at each time instant n. We differentiate two types of sequential estimation methods [12]:

(1) Block estimation,
(2) Adaptive estimation.

Block estimation is the usual method of linear prediction analysis, where one value of each reflection coefficient is estimated for a whole block of data. The analysis is repeated over again as each signal sample is added to the block of data. In contrast to block estimation, adaptive estimation determines a new estimate at time n as a function of the last estimate at time n-1 and a "measurement" at time n.

For sequential estimation, we distinguish three types of estimator memory. By memory, we mean the dependence of the current estimate on past signal samples. The three memory types are:

(a) Fixed memory,
(b) Growing memory,
(c) Fading memory.

The extent of the estimator memory is constrained to be constant for (a); as new signal samples arrive, the estimator memory is updated such that the signal samples furthest in the past are discarded to make room for the most recent signal samples. For case (b), the size of the estimator memory increases as new data is processed. Fading memory methods, which form case (c), can have either a fixed or growing memory span, but the most recent data is given greater emphasis than the data further back in time. One can combine the different memory types. It is common to use either a growing-fading memory or a fixed-nonfading memory.

Below, we present both block and adaptive sequential estimation methods. The resulting estimators have weighting sequences which can be appropriately chosen to yield one of the above memory conditions.

4. BLOCK SEQUENTIAL ESTIMATION

We assume that x(n) is a time-varying signal, and that we wish to estimate the reflection coefficients K_m at each instant of time n. We shall take advantage of the decoupling property of the lattice (even though it is only approximately true in the nonstationary case), and determine each $K_m(n)$ by minimizing some function of the forward and backward residual energies at that stage. Furthermore, since in a time-varying situation we are mainly interested in the most recent history of the signal, it is reasonable to weight the residuals such that the more recent values are given greater importance. We are thus led to minimizing a mean-square type of error of the form:

$$E_m(n) = \sum_{k=-\infty}^{n} w(n-k)\; e_m^2(k) \tag{15}$$

where w(n) is the weighting sequence, or window, and $e_m^2(k)$ is the instantaneous residual energy at time k. We shall have more to say about the window later on. As for the residual energy, we shall consider here only one case, the sum of the forward and backward residual energies:

$$e_m^2(k) = f_m^2(k) + g_m^2(k). \tag{16}$$

Substituting (16) and (3) in (15) and minimizing $E_m(n)$ with respect to $K_m(n)$ results in:

$$K_m(n) = -\frac{2\sum_{k=-\infty}^{n} w(n-k)\; f_{m-1}(k)\; g_{m-1}(k-1)}{\sum_{k=-\infty}^{n} w(n-k)\;[\,f_{m-1}^2(k) + g_{m-1}^2(k-1)\,]} \tag{17a}$$

$$= -\,C_m(n)/D_m(n). \tag{17b}$$

The value of K as given by (17) is always guaranteed to obey (10). Other possibilities exist for defining K such that (10) is guaranteed [4], but they will not be discussed here.

In the block method, $K_1(n)$ is computed first, using (17) and the input signal (see (3a)). Then, the residuals $f_1(k)$ and $g_1(k)$ are computed using (3) for <u>all</u> time up to n. Then, $K_2(n)$ is computed from (17), followed by the computation of $f_2(k)$ and $g_2(k)$, and so on for all stages. The whole process is then repeated at time n+1, with the residuals having to be completely reevaluated for all time up to n+1. The amount of computation is clearly large, and so is the apparent amount of storage. However, one can effect substantial savings in both by using the covariance lattice method instead [4], and by recursively updating the signal covariance [12]. It is important to note that, in the block method, the value of $K_m(n+1)$ does not depend in any simple way on $K_m(n)$. This is to be contrasted with the adaptive method described in Section 5.

Weighting Windows. We point out at the outset that windowing of the error in (15) is very different from windowing of the signal. Windowing the signal results in a stationary signal, while windowing the error does not effect the stationarity of the signal; it merely weights the different error values. Signal or data windows may be quite arbitrary, and may take on positive and negative values. In contrast, the error window in (15) must be always nonnegative. In particular, we must have

$$w(n) \geq 0, \quad n\geq 0,$$

$$w(n) = 0, \quad n<0. \tag{18}$$

Negative values are not allowed since they will result in cancellation of errors, which is generally undesirable.

As examples, we shall give one finite impulse response (FIR) window and one recursive window. The FIR window is the usual rectangular window of width M:

$$w_1(n) = 1, \quad 0\leq n\leq M-1,$$

$$\quad = 0, \quad \text{otherwise}. \tag{19}$$

This window has some bad effects as a signal window but has good properties as an error window. Notice that use of this window yields a fixed memory estimator. The recursive window is the impulse response of a single real pole:

$$w_2(n) = \beta^n, \quad n\geq 0, \quad 0<\beta\leq 1$$

$$\quad = 0, \quad n<0. \tag{20}$$

From (20) and (17), one can compute $C_m(n)$ and $D_m(n)$ recursively using

$$C_m(k) = \beta\, C_m(k-1) + 2f_{m-1}(k)g_{m-1}(k-1) \tag{21a}$$

$$D_m(k) = \beta\, D_m(k-1) + f^2_{m-1}(k) + g^2_{m-1}(k-1) \tag{21b}$$

for all k up to n and $1\leq m\leq p$. Other recursive windows may be defined, but because of condition (18), all such windows must be the impulse responses of lowpass filters with positive real poles.

Updating relations for the correlations C and D similar to those in (21) can also be derived for the case where the FIR window (19) is used [12]. These relations are given as follows:

$$C_m(k) = C_m(k-1)+2f_{m-1}(k)g_{m-1}(k-1)-2f_{m-1}(k-M)g_{m-1}(k-M-1), \tag{22a}$$

$$D_m(k) = D_m(k-1)+f^2_{m-1}(k)+g^2_{m-1}(k-1)-f^2_{m-1}(k-M)-g^2_{m-1}(k-M-1). \tag{22b}$$

5. ADAPTIVE ESTIMATION

In adaptive estimation, we assume given $K_m(n)$, $1\leq m\leq p$, at time n, and the forward and backward residuals up to time n. The problem is then to estimate $K_m(n+1)$, $1\leq m\leq p$, at time n+1 using the given quantities. We shall employ the estimate in (17) but in a different manner:

$$K_m(n+1) = -C_m(n)/D_m(n). \tag{23}$$

Given $K_m(n)$ and $g_m(n-1)$, $1\leq m\leq p$, one computes $f_m(n)$ and $g_m(n)$ for all stages using (3). Then $K_m(n+1)$, $1\leq m\leq p$, are computed from (23), and so on. In contrast with the block method, the residuals are computed only once for each point in time. Stated differently, the adaptive approach "ripples" each new signal sample, and only that sample, through the entire lattice to compute the estimate of all the reflection coefficients. Thus, the previous sample-by-sample (or instantaneous) estimates of the reflection coefficients determine the residuals which in turn are used for computing the current estimate.

The weighting windows $w_1(n)$ and $w_2(n)$ may also be used in adaptive estimation. For example, with $w_2(n)$ we have a growing-fading memory adaptive estimator described completely by the equations (3), (21) with k=n, and (17b). A simple inspection of these equations reveals that with nonfading memory, only 5 multiplications (neglecting multiplication by 2) and 1 division are needed to compute each of the reflection coefficients at each point in time, and with fading memory, the number of multiplications goes up by 2 due to the factor in (21a) and 21b). In addition, the necessary storage is minimal. Similarly, the adaptive scheme with the rectangular window in (19)(fixed-nonfading memory) is described by (3), (22) with k=n, and (17b). If we store the quantities $f_m^2 + b_m^2$ and $f_m b_m$, then the number of computations per coefficient per signal sample becomes 5 multiplies and 1 divide. The increase in necessary storage is proportional to M, the window width. In contrast, the growing memory block estimation requires, when implemented efficiently using the covariance lattice method, about $p^2/2 + 2p$ multiplies per coefficient per signal sample [12]. For p=10, this number becomes 70. Therefore, the main advantage of adaptive estimation over the block method is the reduced computation, and the reduced storage when using the recursive window. The price to be paid is that adaptive estimation is noisier; we view the adaptive estimation method as an approximation to the block method.

A spectral estimation example illustrating the difference between the two methods is given in Section 6.

An adaptive algorithm using $w_2(n)$ was used by Itakura in his original hardware realization of the lattice in a speech vocoder system [13], and by Srinath and Viswanathan [14]. A vocoder similar to Itakura's has been designed by Kang [15].

5.1 Least Mean Squares (LMS) Interpretation

Using $w_2(n)$ and therefore (21), one can show that $K_m(n+1)$ for this special window may be written as an update on $K_m(n)$:

$$K_m(n+1) = K_m(n) - [f_{m-1}(n)g_m(n)+g_{m-1}(n-1)f_m(n)]/D_m(n) \tag{24}$$

where $D_m(n)$ is given recursively by (21b). For the special case $\beta=1$, $D_m(n)$ increases continuously and the correction term in (24) tends to zero as n goes to infinity. In this case K_m tends to its optimal value with probability 1, assuming a stationary signal. For $\beta<1$, one can show that (24) becomes similar to the LMS lattice estimate recently given by Griffiths [16,23] with a step size $\alpha = 1-\beta$.

5.2 Kalman Filter Interpretation

One can show [12] that (24) can be rewritten in the form of a Kalman filter:

$$K_m(n+1) = K_m(n) + L_m(n+1)[K_m^s(n+1)-K_m(n)] \tag{25}$$

where

$$K_m^s(n+1) = -2f_{m-1}(n)g_{m-1}(n-1)/[f_{m-1}^2(n)+g_{m-1}^2(n-1)] \tag{26}$$

can be viewed as a single "measurement" at time n+1, and $L_m(n+1)$ is a gain term at n+1 given by:

$$L_m(n+1) = d_m(n)/D_m(n), \tag{27}$$

where

$$d_m(n) = f_{m-1}^2(n) + g_{m-1}^2(n-1), \tag{28a}$$

and

$$D_m(n) = \beta\, D_m(n-1) + d_m(n). \tag{28b}$$

$d_m(n)$ may be interpreted as the instantaneous residual variance, while $D_m(n)$ is the total variance.

5.3 Convergence Properties

For stationary signals, as mentioned above in Section 5.1, the adaptive lattice algorithm converges to the optimal values of the reflection coefficients with probability one. Considering the rate of convergence, the conventional tapped-delay-line implementation of the adaptive LMS algorithm has a convergence rate that decreases with the eigenvalue spread ($\lambda max/\lambda min$) of the signal correlation matrix [18]. For the adaptive lattice schemes, due to the orthogonalization and decoupling properties of the lattice given in Section 2, the convergence rate is independent of the eigenvalue spread of the input signal, as noted by Griffiths [16]. For time-varying or nonstationary signals, the orthogonalization and decoupling properties hold only approximately, leading to a slightly slower convergence rate in general.

6. A SPECTRAL ESTIMATION EXAMPLE

The data x(n) was generated by passing white Gaussian noise through an 11-pole filter whose power spectrum is shown in Fig. 2. The true reflection coefficients of this filter and the power spectrum were in fact obtained from a linear prediction analysis of a real speech signal corresponding to the unvoiced sound [s]. The minimum possible residual energy E_p^* for this synthetic data is given by

$$E_p^* = R(0) \prod_{i=1}^{p} (1 - K_i^2) \tag{29}$$

where p=11 for this example, K_i are the true reflection coefficients and R(0) is the zeroth autocorrelation coefficient of the signal x(n). We employed both the growing-memory (nonfading) block estimation scheme (Section 4) and the growing-memory (nonfading) adaptive estimation scheme (Section 5) for this spectral estimation example. The goodness of fit after each iteration j was measured using the residual energy ratio $E_p(j)/E_p^*$, where $E_p(j)$ is the energy of the residual obtained by inverse-filtering the data x(n) with the filter $A_p^j(z)$ -- the predictor coefficients of this filter were computed from the reflection coefficients at iteration j. The residual energy ratio is always greater than or equal to 1; it is equal to 1 if the computed filter is identical to the true filter. In Fig. 3, we have plotted the average of the values of this residual measure corresponding to 15 different data records as a function of iteration or sample number. Fig. 3b corresponds to the block method, while Fig. 3a corresponds to the adaptive lattice method. For this example, the two methods lead to nearly the same performance. From Fig. 3, we also see that the speed of convergence is generally proportional to p, but is not affected by the spectral dynamic range or equivalently the eigenvalue spread of the signal, which is about 1000 in this example.

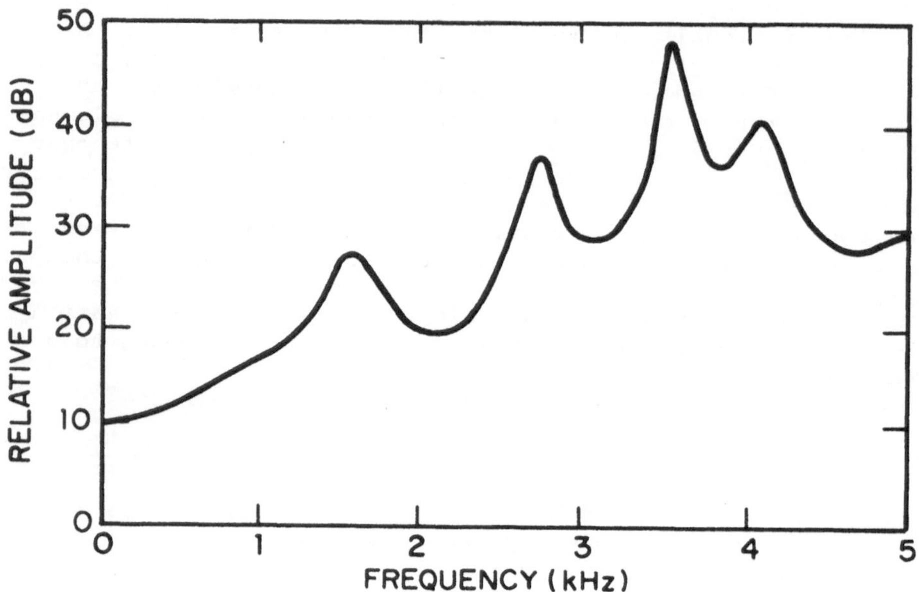

Fig. 2. 11-pole spectrum corresponding to an [s] sound.

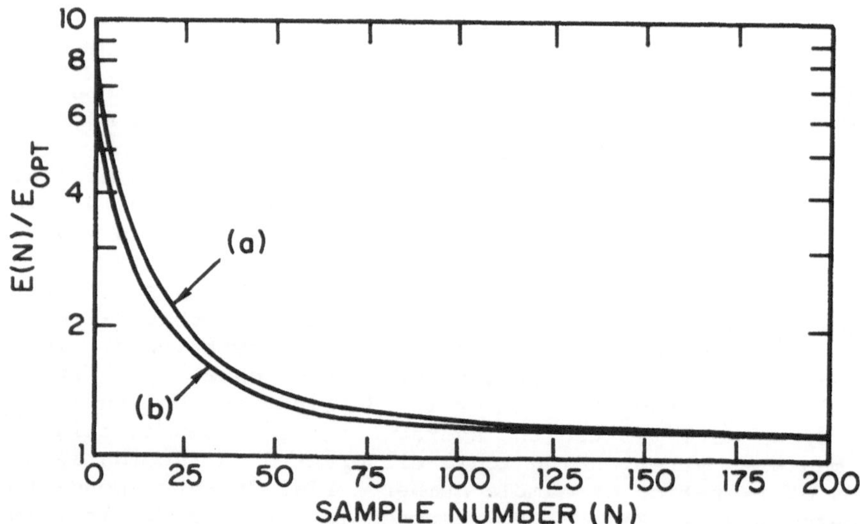

Fig. 3. Performance of sequential estimation methods as a function of sample number. (a) Adaptive lattice method. (b) Block method.

7. APPLICATIONS

Below, we suggest a number of applications for adaptive lattice methods.

7.1 Speech Processing

Determination of the instants at which certain speech events occur, such as glottal closure, may be accomplished through fixed memory adaptive estimation [19]. If the ratio of mean-squared prediction error e(n) to mean-squared signal x(n) is used as a measure, and if the estimator memory is short compared with the pitch period, then the measure will often show sharp peaks whenever the time segment representing the estimator memory contains a glottal closure.

Adaptive estimation has been used in pitch extraction schemes by Maksym [20]. Next, we consider the application to efficient speech transmission systems. In linear predictive speech transmission systems, one set of p reflection coefficients ($p \sim 12$ for 10 kHz signal sampling rate) is computed for every data frame (typically 20 ms long), and transmitted to the receiver. Employing an adaptive estimator that is initialized at the start of a data frame and terminated at the end of the data frame, one has as many sets of reflection coefficient estimates as there are speech samples in the data frame. Transmission of all of those estimates would tremendously increase the transmission bit rate. One may select a best estimate in some sense and transmit that estimate only. Kang [15] has suggested transmitting the estimate that produces the minimum mean square value for the residual e(n). Use of this selection procedure with the growing memory adaptive estimation method was found to reduce the "wobble" quality usually present in steady state regions of voiced sounds that are synthesized using regular linear prediction methods [15]. Alternately, one may select the median of the estimates or mode of the probability histogram formed from the sample-by-sample estimates.

In high-quality music synthesis, lattice parameters are updated every sample instant [21]. Presently, this is done by using interpolated parameters obtained from two successive data frames. Adaptive lattice methods can be very naturally employed in this case.

7.2 A Fast Start-Up Equalizer

As an application of adaptive lattice methods to adaptive Wiener filtering, we recently proposed a new fast start-up equalizer [6, 17].

During a start-up period prior to data transmission, the tap coefficients of an adaptive equalizer are adjusted automatically to optimal values that minimize the distortion or mean-square error of the received pulses. It is desirable that the start-up time used for this initial adaption process be as small as possible. Chang [22] proposed an equalizer structure that reduces the start-up time drastically. His method employs an eigenvector-based orthonormalization process, and requires N^2 equalizer tap coefficients and an equal number of multiplies, where N is the dimension of the channel correlation matrix. The new equalizer uses the adaptive lattice method, with a suitably normalized set of the backward residual signals providing the required orthonormal signals [6,17]. The number of equalizer coefficients and the number of multiplies vary linearly with N for the lattice structure.

7.3 Other Applications

Adaptive lattice methods promise to be useful in areas where transversal, predictive, or finite impulse response filters are used in an adaptive manner. Applications include adaptive noise cancelling [23], adaptive line tracking and adaptive beamformers.

REFERENCES

1. Itakura, F., and S. Saito, "Digital Filtering Techniques for Speech Analysis and Synthesis", Proc. 7th Int. Cong. Acoust., Budapest, Paper 25-C-1, pp. 261-264, 1971.
2. Itakura, F., and S. Saito, "On the Optimum Quantization of Feature Parameters in the PARCOR Speech Synthesizer", IEEE Conf. on Speech Comm. and Processing, Newton, MA, pp. 434-437, April 1972.
3. Burg, J. P., "Maximum Entropy Spectral Analysis", Ph.D. dissertation, Geophysics Dept., Stanford Univ., Stanford, CA, May, 1975.
4. Makhoul, J., "Stable and Efficient Lattice Methods for Linear Prediction", _IEEE Trans. Acoustics, Speech, and Signal Processing_, Vol. ASSP-25, pp. 423-428, Oct. 1977.
5. Gray, A. H., and J. D. Markel, "Digital Lattice and Ladder Filter Synthesis", _IEEE Trans. Audio and Electroacoust._, Vol. AU-21, pp. 491-500, December 1973.
6. Makhoul, J., "A Class of All-Zero Lattice Digital Filters: Properties and Applications". To appear in _IEEE Trans. Acoustics, Speech and Signal Processing_, Aug. 1978.
7. Makhoul, "Lattice Methods in Spectral Estimation", Proceedings of the Spectral Estimation Workshop, Rome Air Development Center, Rome, New York, May 24-26, 1978.
8. Markel, J. D., and A. H. Gray, Jr., "Roundoff Noise Characteristics of a Class of Orthogonal Polynomial Structures", _IEEE Trans. Acoustics, Speech, and Signal Processing_, Vol. ASSP-23,

9. Viswanathan, R., and J. Makhoul, "Quantization Properties of Transmission Parameters in Linear Predictive Systems", IEEE Trans. Acoustics, Speech, and Signal Processing, Vol. ASSP-23, pp. 309-321, June 1975.
10. Makhoul, J., R. Viswanathan, L. Cosell and W. Russell, Natural Communication with Computers, Final Report, Vol. II, Speech Compression Research at BBN, Report No. 2976, Dec. 1974. (NTIS No. AD/A003478/5GA, 104 pp.)
11. Makhoul, J., "Linear Prediction: A Tutorial Review", Proc. IEEE, Vol. 63, pp. 561-580, April 1975.
12. Viswanathan, R., and J. Makhoul, "Sequential Lattice Methods for Stable Linear Prediction", Proc. EASCON '76, pp. 155A-155H, Sept. 1976.
13. Itakura, F., personal communication.
14. Srinath, M. D., and M. M. Viswanathan, "Sequential Algorithm for Identification of Parameters of an Autoregressive Process", IEEE Trans. Automatic Control, pp. 542-546, Aug. 1975.
15. Kang, G. S., personal communication. (See G. S. Kang, "Linear Predictive Narrowband Voice Digitizer", Proc. 1974 EASCON Conf., Washington, D. C., pp. 51-58, Oct. 1974, for a description of the hardware system.)
16. Griffiths, L. J., "A Continuously-Adaptive Filter Implemented as a Lattice Structure", Proc. IEEE Int. Conf. Acoustics, Speech, and Signal Processing, Hartford, Conn., pp. 683-686, May 1977.
17. Makhoul, J., and R. Viswanathan, "Adaptive Lattice Methods for Linear Prediction", Proc. IEEE Int. Conf. Acoustics, Speech, Signal Processing, Tulsa, OK, pp. 83-86, April 1978.
18. Gitlin, R. D., J. E. Mazo and M. G. Taylor, "On the Design of Gradient Algorithms for Digitally Implemented Adaptive Filters", IEEE Trans. Circuit Theory, Vol. CT-20, pp. 125-136, March, 1973.
19. Strube, H. W., "Determiniation of the Instant of Glottal Closure from the Speech Wave", J. Acoust. Soc. Amer., Vol. 56, pp. 1625-1629, Nov. 1974.
20. Maksym, J. N., "Real-Time Pitch Extraction by Adaptive Prediction of the Speech Waveform", IEEE Trans. Audio and Electroacoust., Vol. AU-21, pp. 149-154, June 1973.
21. Moorer, J. A., personal communication.
22. Chang, R. W., "A New Equalizer Structure for Fast Start-up Digital Communication", Bell Syst. Tech. J., pp. 1969-2014, July-Aug. 1971.
23. Griffiths, L. J., "An Adaptive Lattice Structure for Noise-Cancelling Applications", Proc. IEEE Int. Conf. Acoustics, Speech, and Signal Processing, Tulsa, OK, pp. 87-90, April 1978.

ON A THRESHOLD MODEL*

H. Tong

Department of Mathematics, The University of Manchester Institute of Science and Technology
P. O. Box 88, Manchester M60 1QD England

SUMMARY. After a brief summary of some pertinent properties of the threshold models introduced by Tong (1978), we describe a statistical method for their identifications.

Both simulated data and real data are used as illustrations.

1. THE THRESHOLD AUTOREGRESSIVE MODEL

We have introduced a class of non-linear time series models, which we called the *Threshold Autoregressive Models (TAR)*. (See Tong 1977, 1978.)

Definition 1: A time series $\{X_t\}$ is called a TAR1$(1;k_1,\ldots k_\ell)$ if it is given by the following piece-wise linear submodels.

$$X_t = \begin{cases} a_{01} + \sum_{j=1}^{k_1} a_{j1} X_{t-j} + \varepsilon_t, \text{ if } -\infty < Y_{t-1} < r_1, \\ a_{02} + \sum_{j=1}^{k_2} a_{j2} X_{t-j} + \varepsilon_t, \text{ if } r_1 \le Y_{t-1} < r_2, \\ \cdots\cdots\cdots\cdots \\ a_{0\ell} + \sum_{j=1}^{k_\ell} a_{j\ell} X_{t-j} + \varepsilon_t, \text{ if } r_{\ell-1} \le Y_{t-1} < r_\ell = \infty, \end{cases} \tag{1.1}$$

* This research was supported in part by a grant from the Science Research Council, U.K.

where $\{Y_t\}$ is an instrumental time series, $\{\varepsilon_t\}$ is a stationary white noise sequence with zero mean and finite variance, σ^2, say.

Some conditions on the coefficients a_{ji} are usually needed to ensure stability. (c.f. Jones, 1977)

Definition 2: A bivariate time series $\{X_{1t}, X_{2t}\}$ is called a TAR 2 $(\ell_1; k_{11}, k_{12}, \ldots, k_{1\ell_1}; \ell_2; k_{21}, \ldots, k_{2\ell_2})$ if $\{X_{1t}\}$ is a TAR 1 $(\ell_1; k_{11}, k_{12}, \ldots, k_{1\ell_1})$ with $\{X_{2t}\}$ being the instrumental time series and if $\{X_{2t}\}$ is a TAR 1 $(\ell_2; k_{21}, k_{22}, \ldots, k_{2\ell_2})$ with $\{X_{1t}\}$ being the instrumental time series. The white noise sequences $\{\varepsilon_{1t}\}$ and $\{\varepsilon_{2t}\}$ are assumed uncorrelated with each other.

We, (Tong, 1978) have shown that if X_t denotes the riverflow at time t and Y_t the rainfall at time t around the catchment area of the river, then one way of implementing Sugawara's (1961) idea, which expresses the non-linear relationship between the input Y_t and the output X_t via a tank model, is to treat $\{X_t\}$ as a TAR 1 with $\{Y_t\}$ being the instrumental time series. In the same paper, we have shown how a TAR 2 may be used to model a predator-prey system in ecology.

Some of the important properties of the TAR models are now summarized. For full details, see Tong (1978).

(P1) The systematic part (i.e. the part in the absence of the white noise) of a TAR can give rise to limit cycles, the existence and stability check of which may be studied by the standard technique of "broken line approximation" and Lameré diagram in the stability theory of piecewise linear differential equations.

(P2) Non-linear least-squares prediction may be easily obtained in the case of one step ahead. It reduces to that of a linear least-squares prediction of the appropriate piece-wise linear AR submodel. For more than one step ahead, recursive prediction may be obtained on successively taking conditional expectations.

(P3) A time dependent spectrum may be defined as the collection of the AR spectra of the piece-wise linear AR submodels.

(P4) A TAR is usually expected to be time irreversible.

2. MODEL IDENTIFICATION

Throughout this and subsequent sections, we assume that the white noise sequences are all Gaussian. Suppose that $\ell, k_1, \ldots, k_\ell$

are pre-specified, then it is clear that after ignoring the 'transient' term, the log likelihood function of the parameters of a TAR 1 is

$$-(2\sigma^2)^{-1} \sum_1^N \varepsilon_t^2 ,$$

where N is the sample size. The maximum likelihood estimation (m.l.e.) method requires the determination of a vector $\hat{\underset{\sim}{\theta}}$ such that $\| \underset{\sim}{X} - \underset{\sim}{A}\hat{\underset{\sim}{\theta}} \|$ is a minimum, where $\underset{\sim}{X}$ is the data vector, $\underset{\sim}{A}$ is the design matrix whose entries are 1's, 0's and 'past' X_t's with their positioning being determined by the instrumental time series $\{Y_t: t=0,1,\ldots,N-1\}$ and $\| \cdot \|$ is the Euclidean norm. $\hat{\underset{\sim}{\theta}}$ has to be determined for each set of trial thresholds $(r_1,r_2,\ldots,r_{\ell-1})$, or in short $\underset{\sim}{r}$, from which $\hat{\sigma}^2(\underset{\sim}{r})$ is obtained on dividing $\| \underset{\sim}{X} - \underset{\sim}{A}\hat{\underset{\sim}{\theta}} \|^2$ by N. Minimizing $\hat{\sigma}^2(\underset{\sim}{r})$ over all possible $(r_1,r_2,\ldots,r_{\ell-1})$ yields the m.l.e. of $\underset{\sim}{r}$ denoted by $\hat{\underset{\sim}{r}}$. The associated $\hat{\underset{\sim}{\theta}}(\hat{\underset{\sim}{r}})$ and $\hat{\sigma}^2(\hat{\underset{\sim}{r}})$ are then the m.l.e. of the a_{ji}'s and σ^2 respectively. For the case of $\ell=2$, $k_1=k_2=1$, Tong (1978) has given sufficient conditions for $\hat{\underset{\sim}{\theta}}$ to be 'best asymptotically normal'. For the general case, we conjecture that relatively straight-forward extension of the above conditions will suffice.

For the estimation of ℓ, $k_1,\ldots,k_\ell$, we use the newly developed approach of Entropy Maximization Principle due to Akaike (1977a). One simple implementation of this principle is obtained by minimizing the following Akaike's Information Criterion with respect to $\ell,k_1,\ldots,k_\ell$.

$$\mathrm{AIC}(\ell,k_1,\ldots,k_\ell) = N\ln\hat{\sigma}^2(\hat{\underset{\sim}{r}}) + 2 \sum_{i=1}^{\ell} (k_i + 1). \tag{2.1}$$

However, we find it more convenient to carry out the search in three stages as follows:

(S1) Fix $(\ell,r_1,\ldots,r_{\ell-1})$ and minimize AIC $(\ell,k_1,\ldots,k_\ell)$ with respect to $k_1,\ldots,k_\ell$, obtaining their estimates $\hat{k}_1(\ell,\underset{\sim}{r}),\ldots,\hat{k}_\ell(\ell,r)$.

(S2) Scan through all possible $(\ell-1)$-tuples of $(r_1,\ldots,r_{\ell-1})$ and obtain its estimate, $\hat{\underset{\sim}{r}}(\ell)$, which corresponds to that $(\ell-1)$-tuple whose associated AIC is a minimum among the AIC's associated with all $(\ell-1)$-tuples.

(S3) Repeat (S1) and (S2) for $\ell=1,2,\ldots,L$, where L is the highest conceivable number of thresholds likely to be entertained.

3. COMPUTATIONAL ASPECTS

First, in order to save computation time, we have found it necessary to restrict $r_1,\ldots,r_{\ell-1}$ in (S1) to differ by at most 1 such that $r_i \geq r_j$ if $i \leq j$. Next, in (S2), we have to limit ourselves to a finite set of possible $(\ell-1)$-tuples.

Computationally speaking, the most crucial problem is the determination of $\hat{\ell}$ because the matrix $\underset{\sim}{A}$ is commonly ill-conditioned by its very nature of being a least squares problem and, more seriously, rather sparse. A numerically stable method which we have found particularly useful for our purpose is the one described by Golub (1965). It is based on an upper triangularization of $\underset{\sim}{A}$ by successive orthogonal Householder transformations.

We now give a simulated example with N = 1000.

Example 1:

$$X_t = \begin{cases} 1 + 0.3X_{t-1} + 0.2X_{t-2} + \varepsilon_t \text{ if } -\infty<Y_{t-1}< 10 \\ 2 + 1.79X_{t-1} - 0.81X_{t-2} + \varepsilon_t \text{ if } 10 \leq Y_{t-1}< \infty \end{cases} \quad (3.1a)$$

$$Y_t = \begin{cases} 5 + 1.89Y_{t-1} - 0.9025Y_{t-2} + \eta_t \text{ if } -\infty<X_{t-1}< 7 \\ 1 + 0.5Y_{t-1} + 0.1Y_{t-2} + \eta_t \text{ if } 7 \leq X_{t-1} < \infty \end{cases} \quad (3.1.b)$$

where $\{\varepsilon_t\}$ and $\{\eta_t\}$ are white noise sequences, uncorrelated with each other, and each with a N (0,1) distribution. Figure 1 gives a typical part of a realization of $\{X_t,Y_t\}$.

Table 1 gives $(\hat{k}_1,\hat{k}_2)$ and the minimum AIC values for r_1 = 4(3)22, for the Y_t series.

r_1	$(\hat{k}_1,\hat{k}_2)$	$\hat{\sigma}^2$	AIC
4	(6,6)	67.3795	4200.45
7	(2,2)	1.0288	40.15
10	(3,2)	63.5252	4128.07
13	(6,5)	77.4674	4336.71
16	(7,6)	80.6429	4380.52
19	(8,7)	82.9510	4412.49
22	(8,7)	84.3573	4429.15

TABLE 1

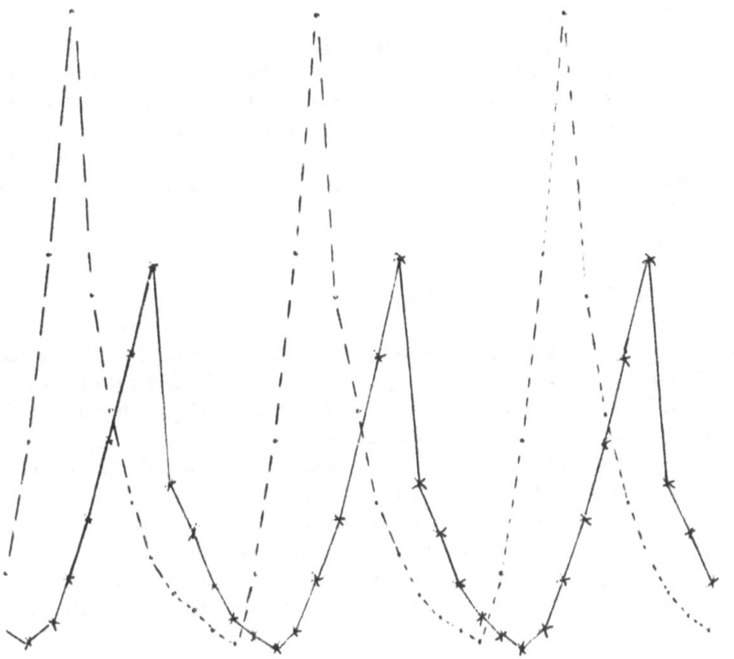

Fig. 1

Minimizing AIC with respect to r_1, clearly gives $\hat{r}_1 = 7$. We observe that the estimation of r_1, based on Minimum AIC method, is usually very reliable and simple. The estimated coefficients are 5.1003 (5), 1.885 (1.89), -0.8995 (-0.9025), 0.9730 (1), 0.5078 (0.5), 0.0963 (0.1). (True values are bracketed). Results for the X_t series are similarly very encouraging. However, we do find that when the true values of k_i's are rather low, such as in the present case, minimum AIC method may lead to some slight overestimation of them. Of course, in reality k_i's are never very low. A Baysian extension of minimum AIC is recently proposed by Akaike (1977b).

4. ANALYSES OF SOME REAL DATA

Example 1: (A further analysis of the Kanna River data). In Tong (1978) we have given a preliminary TAR model for the 366 data points of the riverflow (X_t)-rainfall (Y_t) record of the Kanna River (Japan) in 1956, in which only one threshold is used, giving rise to a 'wet' model and a 'dry' model. Introducing one more threshold, we find that both the $\hat{\sigma}^2$ and the AIC values are substantially reduced, the former being about half of its previously estimated value; the newly fitted model is comparable with Ozaki's fitted model (1978) based on a different approach. The two threshold estimates are 8.5867 mm/day and 25.7600 mm/day and the AR parameter estimates are

	$\hat{a}_0$	$\hat{a}_1$	$\hat{a}_2$	$\hat{a}_3$	$\hat{a}_4$
Below 1st threshold ('dry' model)	0.0271	0.8160	0.0089	0.0148	0.0898
Between thresholds ('medium' model)	0.2236	1.1568	-0.4455	0.0641	0.1045
Above 2nd Threshold ('wet' model)	0.3278	0.4155	1.9196	-1.1208	

$\hat{\sigma}^2 = 0.003532$ AIC = -2015.85

The results are shown in Figs. 1a, 1b, 1c. It is interesting to note that if the information of the rainfall (Y_t) is not used in the modelling, the fit is substantially inferior, whether a linear AR or a TAR with X_t as the instrumental series is used. ($\hat{\sigma}^2 > 0.01$ in each case). On the other hand, by using the rainfall information to a fuller extent, Ozaki's model (1978) gives $\hat{\sigma}^2 = 0.003176$ involving roughly the same number of parameters. This shows the almost imperative use of the rainfall information in the modelling of the riverflow. Contrast this attitude with the models discussed in Lawrance and Kottegoda (1977).

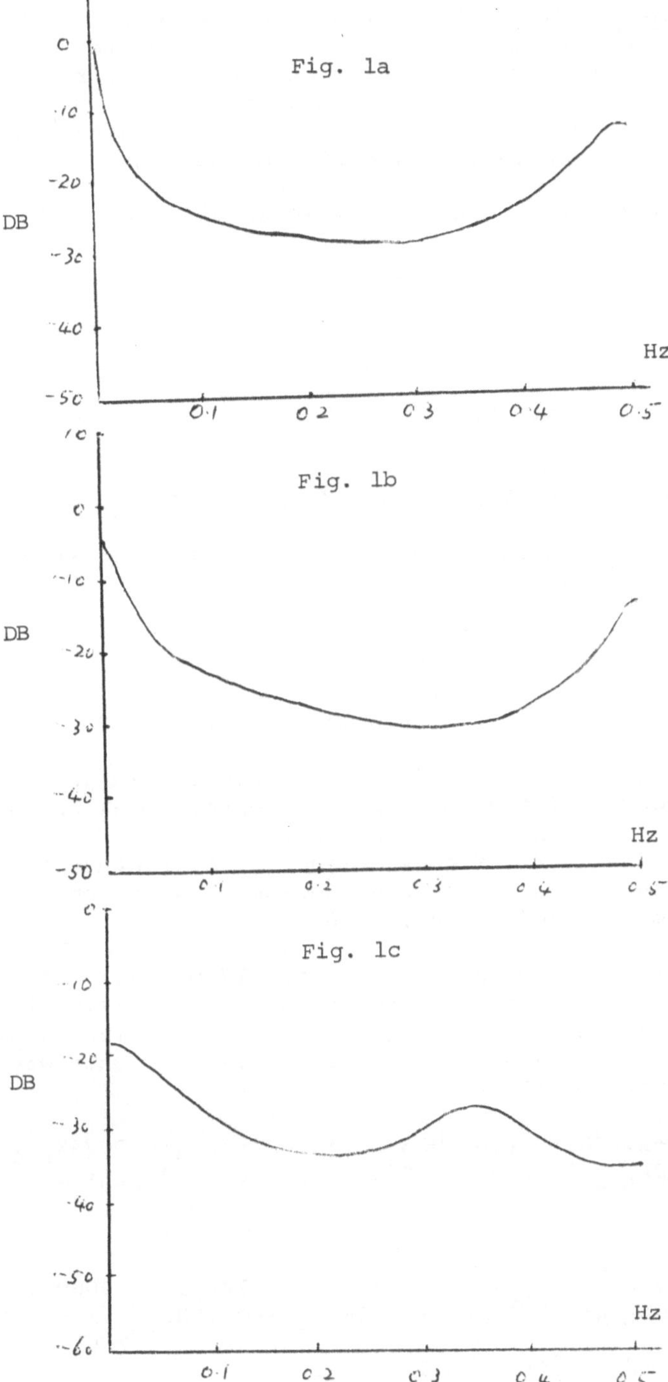
Fig. 1a
DB
Hz
Fig. 1b
DB
Hz
Fig. 1c
DB
Hz

Example 2: (Mink - Muskrat data of Canada). We have fitted a TAR 2 to the mink-muskrat data of Canada for the period of 1767-1849 recorded by the Hudson's Bay Co. After a first-differencing of the logarithmically transformed data, we have 82 data points. Let $\{X_t\}$ and $\{Y_t\}$ denote the transformed mink data and transformed muskrat data respectively. The fitted model is

$$X_t = \begin{cases} 0.0256-0.5886X_{t-1}-0.2134X_{t-2}+0.6017X_{t-3}+\varepsilon_t & \text{if } Y_{t-1}<-0.2017 \\ 0.2273-0.7077X_{t-1}-0.1998X_{t-2}-0.3485X_{t-3}+\varepsilon_t & \text{if } Y_{t-1}\geq-0.2017 \end{cases}$$

$$\hat{\sigma}^2_\varepsilon = 0.175052.$$

The fitted AR spectra are given in Figs. 2a and 2b.

$$Y_t = \begin{cases} 0.1213 - 0.8290\, Y_{t-1}-0.6433\, Y_{t-2}+\eta_t & \text{if } X_{t-1} <-0.0337 \\ -0.0955 - 0.0420\, Y_{t-1} + \eta_t & \text{if } X_{t-1} \geq-0.0337 \end{cases}$$

$$\hat{\sigma}^2_\eta = 0.326594.$$

The fitted AR spectra are given in Figs. 2c and 2d.

It is interesting to note that the deterministic part (i.e. the systematic part) of the above predator-prey model has a (stationary) limit point, rather than a limit cycle, at $X = 0.1008$ and $Y = -0.0916$ on the X - Y phase plane. This feature is shared by the lynx - hare data of roughly the same region, and over roughly the same time period. [Tong, 1978].

Example 3: (Wolf's Sunspot Data, 1700-1945). To these 246 data points, we have fitted the following 'self-limiting' TAR, in which the sunspot series itself is used as the instrumental series, because no other obvious series is available. Data were obtained from Waldmeier (1961).

$$X_t = \begin{cases} 7.2050+1.4280X_{t-1}-0.5923X_{t-2}-0.0278X_{t-3}+0.0074X_{t-4} \\ \quad +0.0028X_{t-5}-0.1118X_{t-6}+0.0637X_{t-7}+0.1123X_{t-8}+\varepsilon_t & \text{if } X_{t-1}<56.613 \\ \\ 37.7035+0.5968X_{t-1}-0.2521X_{t-2}-0.1323X_{t-3}+0.2060X_{t-4} \\ \quad -0.4239X_{t-5}+0.2765X_{t-6}+0.0585X_{t-7}+0.4032X_{t-8}+\varepsilon_t & \text{if } X_{t-1}\geq 56.6133 \end{cases}$$

$$\hat{\sigma}^2_\varepsilon = 145.77$$

The estimated AR spectra are given in Figs 3a and 3b.

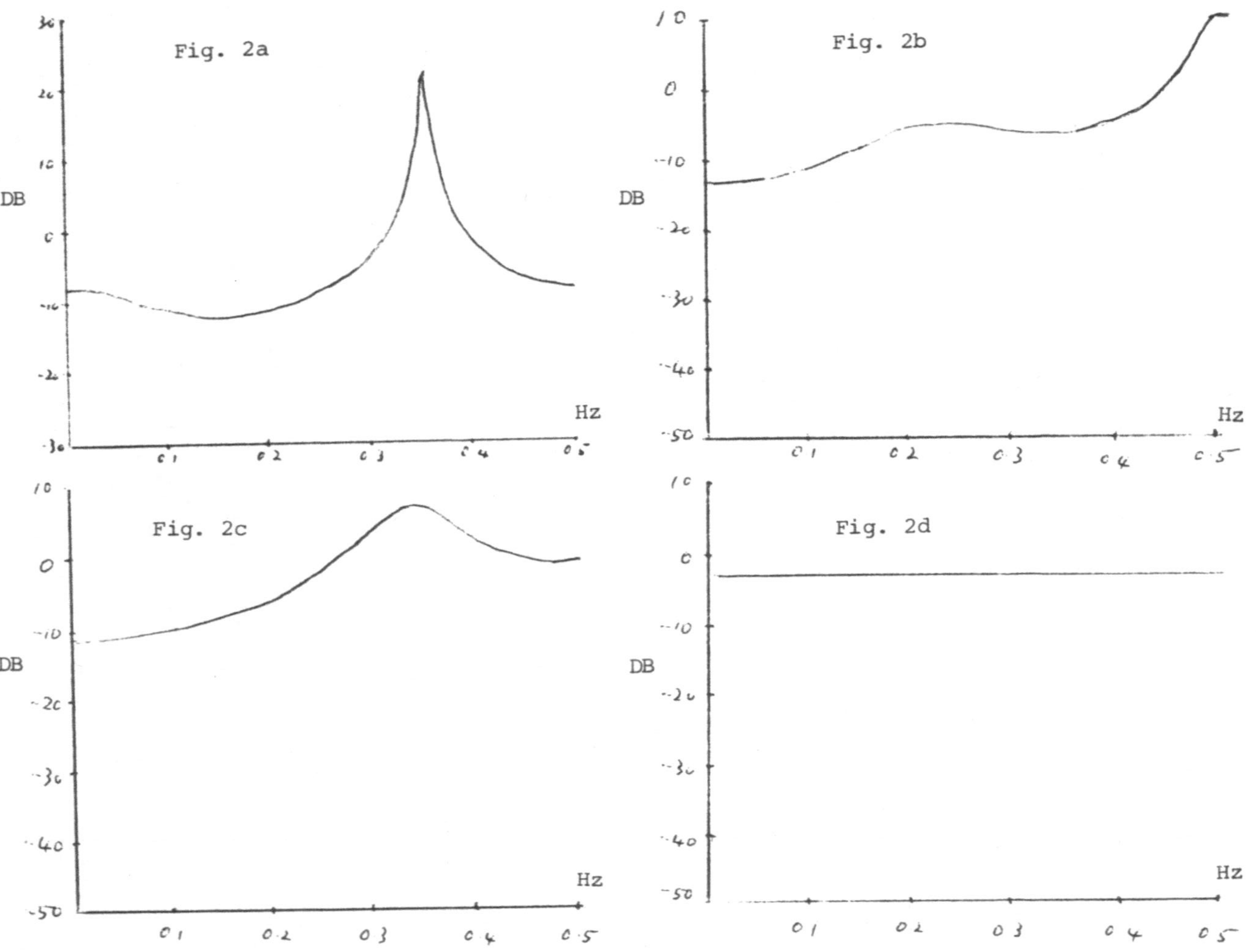
Fig. 2a
Fig. 2b
Fig. 2c
Fig. 2d
DB
Hz

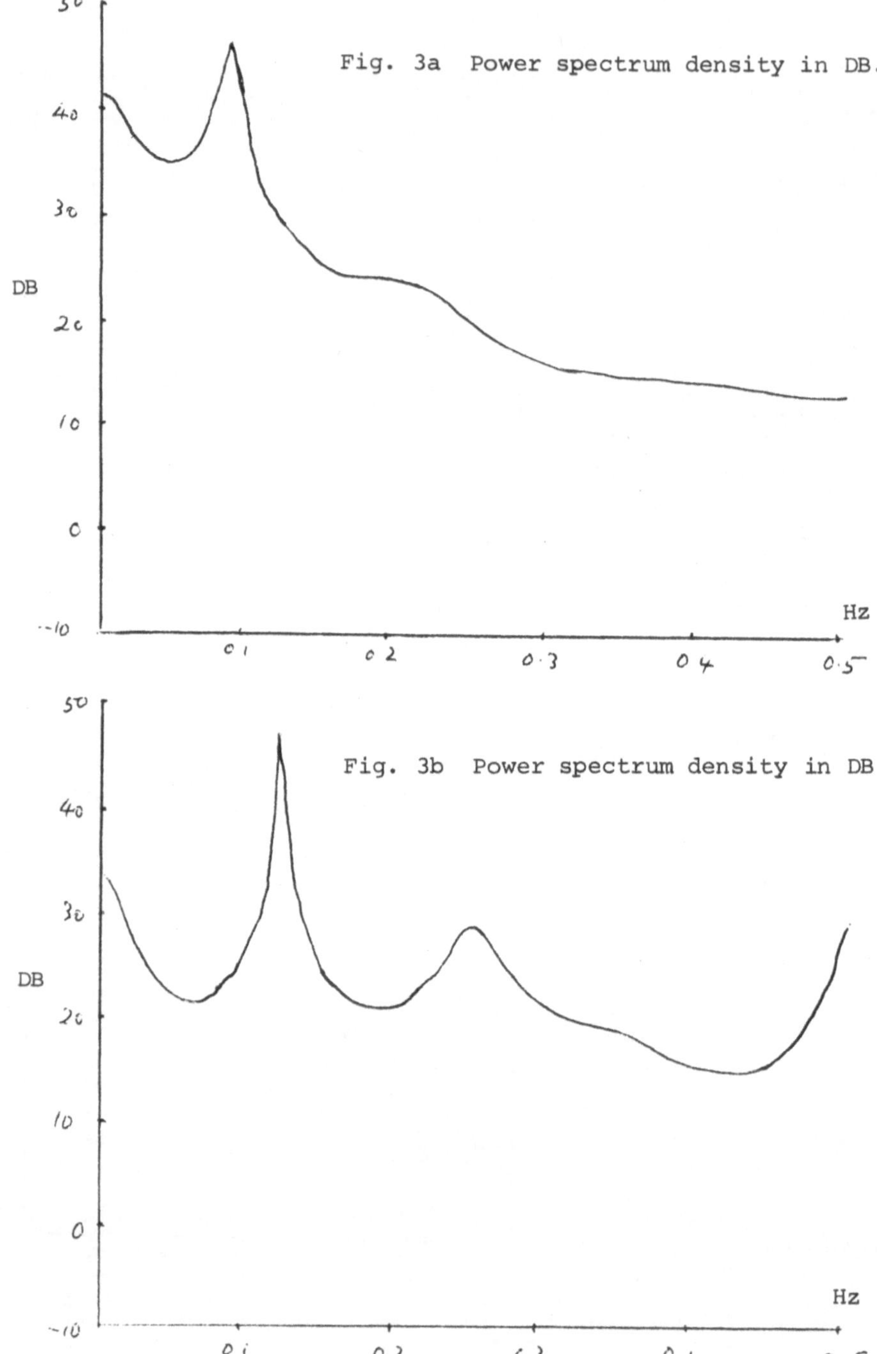

Fig. 3a Power spectrum density in DB.

Fig. 3b Power spectrum density in DB

It is interesting to note the small frequency shift in the AR spectra between the 'high-activity' and the 'low-activity' models. This might perhaps suggest a kind of 'frequency-amplitude dependence', so commonly found in non-linear systems.

It is also found that introducing one more threshold does not give any better fit.

5. DISCUSSION

The introduction of the TAR models as a class of non-linear time series models (signals) as reported here only signals the beginning of a much richer theory and methodology for the analysis of non-linear problems in signal processing. For a number of interesting problems currently under study we refer to Tong (1978). To those, we would also add the most important problem of a structural relation between the TAR and an appropriately enlarged state space model. For applications in other areas, it has been suggested that the information processing within a nerve system is an interesting possibility. Finally, and perhaps most importantly, the extension of a TAR time series to a TAR system of the form

$$X_t = \begin{cases} a_{01} + \sum_{j=1}^{k_1} a_{j1} X_{t-j} + \sum_{j=1}^{k_1'} b_{j1} Y_{t-j} + \varepsilon_t & \text{if } (X_{t-1}, Y_{t-1}) \in R_1, \\ a_{02} + \sum_{j=1}^{k_2} a_{j2} X_{t-j} + \sum_{j=1}^{k_2'} b_{j2} Y_{t-j} + \varepsilon_t & \text{if } (X_{t-1}, Y_{t-1}) \in R_2, \\ \cdots\cdots\cdots\cdots\cdots\cdots & \\ a_{0\ell} + \sum_{j=1}^{k_\ell} a_{j\ell} X_{t-j} + \sum_{j=1}^{k_\ell'} b_{j\ell} Y_{t-j} + \varepsilon_t & \text{if } (X_{t-1}, Y_{t-1}) \in R_\ell. \end{cases}$$

where $R_1, R_2, \ldots, R_\ell$ form a partition of $\mathbb{R}^2$, will provide a starting point of a statistical analysis of a large number of very interesting non-linear systems including, e.g. a saturated system.

ACKNOWLEDGMENT

I sincerely thank Professor C. H. Chen and Mr. T. Ozaki for their interests in the present work. I am also indebted to Mr. K. Komura of the Ministry of Construction, Japan, who has kindly provided me with the Kanna River data.

REFERENCES

1. Akaike, H. (1977a), "On entropy maximization principle", Proc. Symp. Appl. of Statist, Ed. P. R. Krishnaiah, p.p. 27-41, Amsterdam: North Holland.
2. __________ (1977b), "A Baysian extension of the Minimum AIC Procedure of Autoregressive model fitting", Research Memo No. 126, Inst. Statist. Maths., Tokyo. (Nov).
3. Golub, G. (1965), "Numerical methods for solving linear least squares problems", Numerische Mathematik 7, pp. 206-216.
4. Jones, D. A. (1977), "Stationarity of non-linear autoregressive processes", Tech. Rep. Institute of Hydrology, Wallingford, Oxon, U.K.
5. Lawrance, A. J. and Kottegoda, N. T. (1977), "Stochastic modelling of river flow time series (with discussion), Journ. Roy. Statist. Soc. (A), 140, pp. 1-47.
6. Ozaki, T. (1978), "On a model of non-linear dynamic model for river-flow prediction", Tech. Rep. No. 84, Dept. of Maths., UMIST. (U.K.) (April).
7. Sugawara, M. (1961), "On the analysis of run-off structure about several Japenese rivers", Jap. Journ. Geophysics, Vol.2, No. 1, pp. 1-76.
8. Tong, H. (1977), Discussion of Lawrance and Kottegoda's paper (1977).
9. __________(1978), "An approach to non-linear time series modelling", Tech. Rep. No. 85, Dept. of Maths, UMIST. (UK), (April).
10. Waldmeier (1961), "The sunspot activity in the years 1610-1960", Zurich: Swiss Federal Observatory.

NATO RESEARCH STUDY GROUP ON PATTERN RECOGNITION: FINAL REPORT*

David C. Hodge
U. S. Army Human Engineering Laboratory
Aberdeen Proving Ground, Maryland

Bruno Beek
Rome Air Development Center
Griffiss Air Force Base, New York

John S. Dehne
U. S. Army Night Vision Laboratory
Fort Belvoir, Virginia

COL Russell B. Ives, USAF-Ret
Control Data Corporation
Rockville, Maryland

Heywood E. Webb, Jr.
Rome Air Development Center
Griffiss Air Force Base, New York

ABSTRACT. AC/243 (Panel III) RSG-4 on Automatic Pattern Recognition was established in 1971 to promote information exchange on military pattern recognition projects, and to identify topics of mutual concern suitable for cooperative research. RSG-4 developed a topical classification scheme, exchanged project summaries, performed three technology assessments, and initiated two cooperative research programs. This paper describes NATO organization for cooperation on military problems, reviews RSG-4 activities and summarizes its accomplishments, surveys the current cooperative research efforts, notes the group's failures, and touches on the

*The views expressed in this paper are those of the authors, and do not necessarily reflect the official position of the U. S. Department of Defense. Reproduction of this paper for any purpose of the U. S. Government is authorized.

prospects for future NATO activity in this area. Previously unpublished analyses of military problems in automatic pattern recognition are included.

OUTLINE

1. NATO COOPERATION ON DEFENSE RESEARCH

2. RSG-4 ON AUTOMATIC PATTERN RECOGNITION

   2.1 Development of a topical classification scheme

   2.2 Information exchange about national programs

   2.3 Selecting potential topics for cooperation

   2.4 Planning for the technology assessments

   2.5 Technology assessment of image processing

   2.6 Technology assessment of speech recognition

   2.7 Technology assessment of mechanical wave processing

   2.8 Proposed cooperation on image processing

   2.9 Proposed cooperation on speech processing

   2.10 Proposed cooperation on mechanical waves

   2.11 Proposed cooperation on APR in battlefield surveillance

   2.12 Proposed cooperation on APR in sonar signal processing

   2.13 Summary note on proposed cooperative research

   2.14 Search for other military APR topics for study

3. RSG-4 SUBGROUP 1 ON IMAGE PROCESSING (NOW RSG-9)

   3.1 Development of an imagery transfer format

   3.2 Documentation of image processing algorithms

   3.3 Assessing national military priorities

   3.4 Cooperative research on image processing

1. NATO COOPERATION ON DEFENSE RESEARCH

This paper presents a review of the mechanisms of cooperation that exist under the umbrella provided by the North Atlantic Treaty Organization (NATO) and, more specifically, it is a final report on the activities on one NATO group, viz., AC/243 (Panel III) Research Study Group 4 (RSG-4) on Automatic Pattern Recognition. At the outset it will be useful to consider the general structure of NATO and review the various aspects of its activities in scientific cooperation on military problems.

Figure 1 shows an overall, simplified organizational chart of NATO. The main point here is that NATO is divided into two separate parts immediately below the level of the North Atlantic Council. One part consists of the <u>civil</u> side of NATO; the other is the <u>military</u> side. The military side is composed of the Military Committee and its supporting International Military Staff, with three major allied defense commands. The Military Committee advises the North Atlantic Council on military matters.

From this point on, our discussions will be concerned exclusively with activities taking place on the civil side of NATO. The civil side functions through a number of Committees, some of which are shown in Figure 1. There also is an International Staff to provide support for the North Atlantic Council and its Committees.

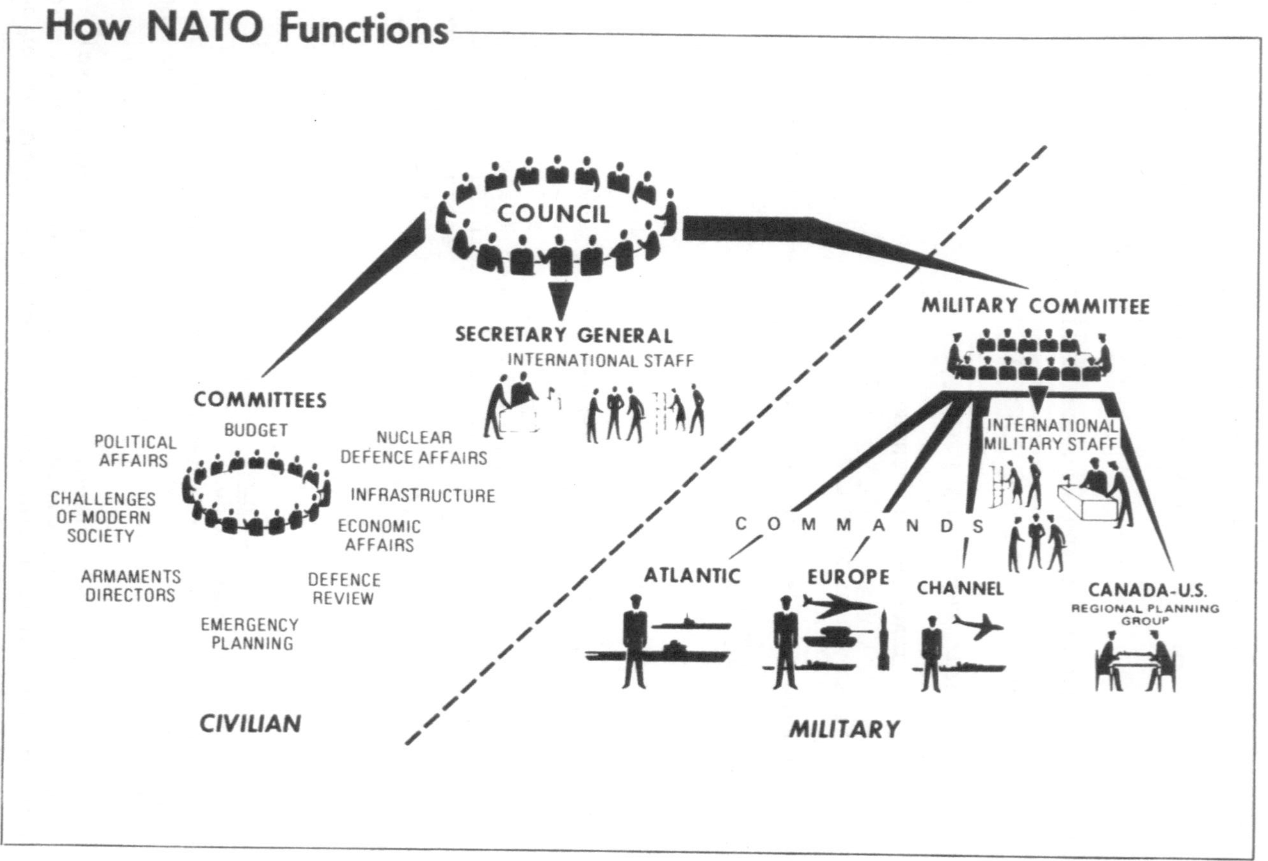

Figure 1. Simplified NATO organizational chart. (From Annex A to Ref. 1.)

The Conference of National Armaments Directors (CNAD), AC/259, was established in 1966 to replace the Armaments Committee. The CNAD and its subordinate bodies have been given Terms of Reference (TOR) laid down by the North Atlantic Council which are directed toward the promotion of cooperation in research, development, and production of future military equipment.

Figure 2 shows how the CNAD is organized to promote cooperation. There are Groups, Panels, Research Study Groups and Subgroups. Directly below the CNAD there are six bodies which are collectively referred to as the "Main Groups". (There also are six other, smaller groups, not shown, having more restricted missions.)

The Defense Research Group (DRG), AD/243, is the body under which the pattern recognition cooperation has been conducted. The functions of the DRG are to [1]:

a. Exchange information on new research and technology which might lead to future equipment.

b. Review the possible military consequences of advances in the field of science and technology.

c. Identify suitable areas or individual projects for bilateral or multilateral cooperation in defense research.

d. Undertake studies, at the request of any of the three service armaments groups, in fields where requirements cannot be met until a breakthrough or a serious advance in technology has been achieved.

e. Cooperate fully and maintain close liaison with the other Main Groups of the CNAD.

The work of the DRG is conducted by a number of Panels which are constituted to address specific topics related to military applications. Note that there is no Panel II or VI. Panels, as well as lesser bodies, are created by the DRG whenever a need arises. They are given a name and assigned a non-recurring number. They function for as long as they are needed, and finally they are disbanded when their work is completed. The pattern recognition activity was organized under Panel III on Physics and Electronics.

The functions of AC/243 Panel III are essentially the same as those of the DRG (see above) except, of course, they are limited to topics in physics and electronics related to military applications. The work of the Panels is conducted by Research Study Groups (RSGs), as well as by Ad Hoc Groups and Exploratory Groups. These latter two types of groups are established on special topics to provide a means of gathering specialists to discuss a subject and

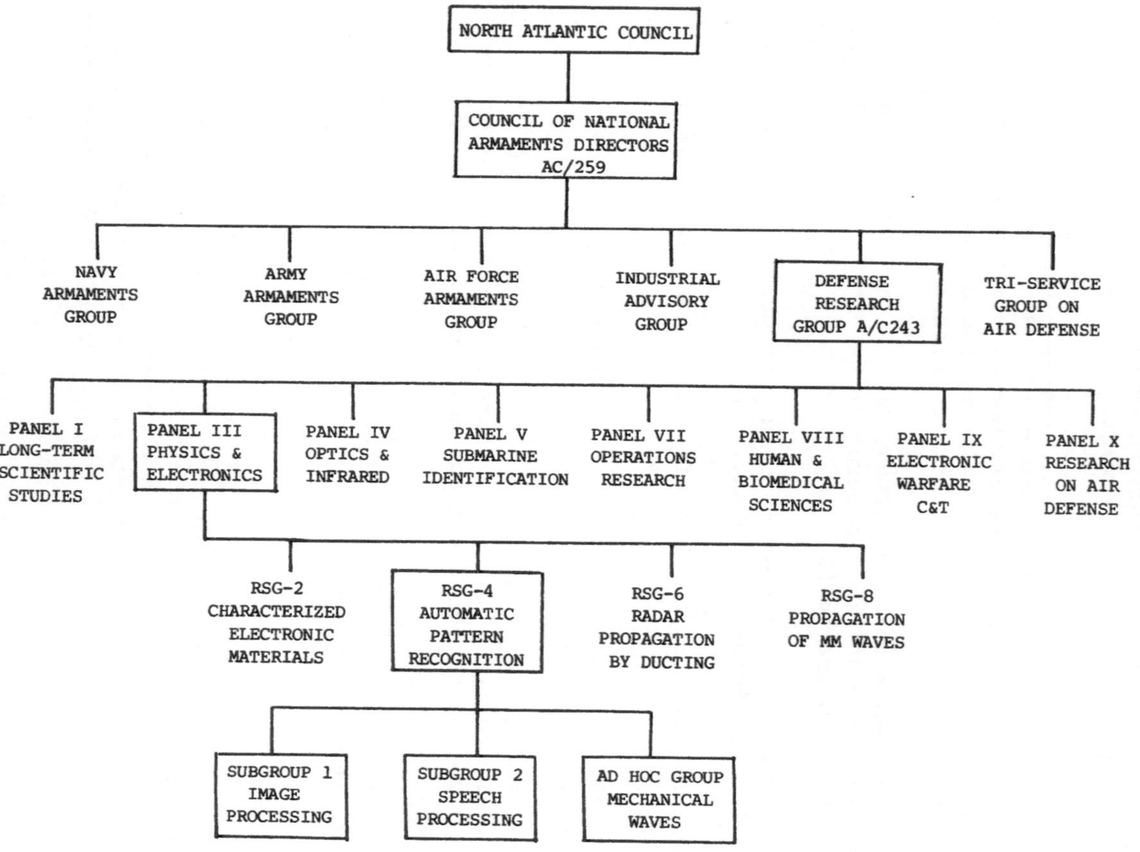

Figure 2. Organizational chart of the CNAD AC/259, DRG AC/243, Panel III, and RSG-4. (As of December 1977.)

recommend for or against establishment of a RSG. Such groups have a very limited life span, often not more than one year.

Research Study Groups (RSGs) are constituted to exchange information on a specific topic of military interest, and to identify projects for potential cooperative research and development. RSGs are usually charged with developing a TOR that is narrow in scope, and RSGs' life spans are normally on the order of five years. Sometimes RSGs are established to pursue cooperative research that has been recommended by another NATO body. In rare circumstances a RSG may be permitted to establish Subgroups and/or Ad Hoc Groups to deal with specific sub-topics.

Referring again to Figure 2, note that as of December 1977 Panel III on Physics and Electronics had four RSGs. There are also several Exploratory Groups not listed in the figure. Also note that RSG-4 on Pattern Recognition is one of the rare RSGs that has been permitted to establish subordinate groups. When RSG-4 was disbanded these Subgroups were elevated to become full RSGs.

In concluding this brief review of NATO organization for cooperation, two points should be made. First, the types of cooperative research and standardization activity conducted under NATO and, in particular, under the DRG, are mainly activities related to _military_ applications; they are also activities for which there is no other appropriate forum or mechanism. The NATO treaty and administrative umbrella provide a means for cooperating on problems for which other (civil) mechanisms are inappropriate. This does not mean that all of the work is classified; in fact, most of it is unclassified and some of it is ultimately published in the open literature. But the NATO umbrella provides a forum for discussing sensitive military topics at a preliminary level without the necessity for completely open circulation of data or results. Conversely, the NATO umbrella is not used to conduct information exchanges or research for which appropriate forums and mechanisms already exist. In other words, if you can do under the IEEE or at the IJCPR you do not need a NATO group.

The second point is that there are _two_ general methods of initiating a NATO study of a scientific topic related to military applications. One way is for one or more nations to petition a Panel directly, through their Delegates, expressing interest in some topic and recommending establishment of a RSG. That is the way in which RSG-4 on Pattern Recognition was established (see below). The other common approach is for another RSG (under Panel III) or some other Panel or body to recommend a topic for study. Frequently, such recommendations result from studies conducted by AC/243 Panel I on Long Term Scientific Studies (see Figure 2). Some time ago, for example, Panel I conducted a study on "Fighting Under Conditions of Limited Visibility". As a result of that study, Panel III established an Ad Hoc Study Group on Military Applications of Millimeter Waves which conducted further discussions, and recommended establishment of a RSG on certain aspects of this topic. As indicated in Figure 2, such an RSG was established and is now functioning [2].

2. RSG-4 ON AUTOMATIC PATTERN RECOGNITION

In 1967 the United Kingdom suggested that the military applications of pattern recognition technology might be suitable as the topic for a RSG, but there was insufficient interest at that time. In 1970 the topic was suggested again, this time by Germany. An exploratory meeting was held in June 1970 at which Canada, Germany, Netherlands, and United Kingdom were represented. It was concluded that there was a basis for further discussions, and that speech recognition should be kept separate from automatic pattern recognition. In July 1970 these conclusions were reported to AC/243 Panel III and, subsequently, the establishment of an RSG on Pattern Recognition was approved by the DRG.

The constitutive meeting of RSG-4 was held in London in November 1971. Countries represented at the first meeting were Canada, France, Germany, Netherlands, Norway, United Kingdom and United States. Germany, as Pilot Nation, was selected to chair the RSG, and Dr.-Ing. Hermut Kazmierczak, from the Forschungsgruppe fur Informationsverarbeitung und Mustererkennung (part of the Gesellschaft zur Forderung der Astrophysikalischen Forschung E. V.), Karlsruhe, was elected Chairman of RSG-4.

2.1 Development of a topical classification scheme

RSG-4's first task was to develop a topical classification scheme for organizing the exchange of information about national pattern recognition programs. Table 1 shows the scheme that was finally adopted. Note that, despite the recommendation of the preliminary 1970 meeting, speech recognition appears in the breakdown. One reason why speech recognition was included undoubtedly was the fact that the Dutch Delegate to RSG-4 was a speech recognition expert! Another may have been that the technical visits held in conjunction with the first RSG-4 meeting included two organizations -- EMI, Ltd., and the National Physical Laboratories -- where automatic speech recognition was being investigated along with a number of other pattern recognition topics, e.g., character recognition, man-machine interaction, adaptive processing, feature extraction, fingerprint recognition, etc.

Another word about the unsuccessful attempt at dichotomizing the pattern recognition topic may be helpful. It now seems perfectly clear (with the benefit of hindsight) that, at the time RSG-4 was constituted, Panel III was of the opinion that the terms "pattern recognition" and "image processing" were synonymous. In fact, comments made to this writer early in the work of RSG-4 made it very clear that Panel III was shocked to discover how broad a charter it has unknowingly given to RSG-4. Whereas RSGs are generally assigned relatively narrow topics and expected to be able to quickly organize them and select relevant military applications and problems for cooperation, in this case the assigned topic was very broad. Indeed, it very soon became apparent that there was interest (in one or more countries) in applying the developing PR technology to virtually all of the complicated military information-handling problems. Thus, whereas Panel III had hoped that RSG-4 would be able to define some topics for cooperation and complete the cooperative projects within five years, in fact it took RSG-4 nearly five years to define the topics and develop the cooperative research proposals. And now there are three RSGs that will each function for five or more years in pursuing cooperative projects!

2.2 Information exchange about national programs

Having agreed on a topical classification scheme (Table 1), the next step was to exchange as much national pattern recognition project information as possible so each participating nation would have as much background information as possible. To that end each participating country was asked to distribute to all the others descriptions of defense-supported programs related to all of the topics listed in Table 1. This first exchange of project information took place at the second RSG-4 meeting (June 1972). Each country prepared a set of project summaries that included information about the technical objectives, approach, recent progress, level of effort, funding, etc. The information exchanged was unclassified and unlimited in distribution.*

Table 2 presents a statistical breakdown of the PR projects that were reported as being pursued by the RSG-4 participants in 1972. Probably the two most informative parts of this table are the extreme right-hand column, and the bottom line. The right column shows the percentage of all the reported projects that relate to the various research and applications topics (Table 1). These data indicate that the largest number of projects was being conducted in the areas related to image processing (Topic 5.0),

*It may be of interest to determine how much information cannot be exchanged under these restrictions. From the U. S. position, less than 10% of PR project summaries were classified or had distribution limitations. Virtually all of the basic and applied research project summaries were exchangeable. Thus, practically no information was lost.

Table 1
Classification Scheme for APR Research Activity

- 1.0 General Methods and Theory
  - 1.1 General PR Theory and Allied Techniques
  - 1.2 Adaptive Processing
  - 1.3 Man-Machine Interaction in APR
  - 1.4 Situation Recognition
  - 1.5 Other
- 2.0 Non-Imaging Signals (Electrical, IR, Video)
- 3.0 Radar Signals
  - 3.1 Target Detection
  - 3.2 Target Tracking
  - 3.3 Target Classification
- 4.0 Acoustic Signals
  - 4.1 Sound, including Seismic and Sonar
    - 4.1.1 Target Detection
    - 4.1.2 Target Identification
  - 4.2 Speech
    - 4.2.1 Speaker Verification
    - 4.2.2 Speaker Identification
    - 4.2.3 Isolated Words
    - 4.2.4 Continuous Speech
- 5.0 Patterns
  - 5.1 Multichannel Signals
  - 5.2 Characters
    - 5.2.1 Multifont
    - 5.2.2 Hand Printed
    - 5.2.3 Hand Written
  - 5.3 Processing of Line Patterns
  - 5.4 Pictures and Scenes
    - 5.4.1 Preprocessing
      - 5.4.1.1 Image Digitalization
      - 5.4.1.2 Image Enhancement
      - 5.4.1.3 Image Filtering
      - 5.4.1.4 Feature Extraction
    - 5.4.2 List Processing
    - 5.4.3 Control and Navigation by Images
    - 5.4.4 Target Detection
    - 5.4.5 Target Classification
    - 5.4.6 Scene Interpretation

Table 2
1972 Project Breakdown by Subject and Country

		CA	FR	GE	NE	NO	UK	US	%
1.0	General Methods & Theory	1	5		1	1		25	19
2.0	Non-Imaging Signals	1	1	1	1	2		9	9
3.0	Radar Signals	2	1	5	1			7	9
4.1	Sound, Seismic, Sonar	1		2				5	
4.2	Speech	1	6	2	1		3	18	18
5.1	Multichannel Signals							3	
5.2	Character Recognition	1	4	3	6		5	8	15
5.3	Line Patterns				2	1	4		
5.4.1	Picture Preprocessing	5	3	2				10	11
5.4.3	Control & Navigation by Images		1	2		1		2	
5.4.4	Target Detection in Images	1						2	
5.4.5	Target Classification							2	
5.4.6	Scene Interpretation			1				1	
	Percentage	7	12	11	7	3	5	55	

the second largest number in general methods and theory (Topic 1.0), and the third largest number in speech processing (Topic 4.2). The bottom line indicates that, as might be expected, over half of the total number of projects was being conducted by the United States, and that France and Germany had the next largest number of pattern recognition projects in progress.

When all of the national project summaries had been circulated, the seven documents were combined into one and distributed to researchers in the participating countries for information and comment [3]. This exchange of PR project information was repeated in 1973 and 1976; the resulting compilations of national summaries are listed as References 4, 5 and 6.

Certain problems were recognized during the 1976 compilation and exchange of project summaries that should be kept in mind in future NATO exercises. These related to the difference between a collection of project (work unit) descriptions and a collection of laboratory or national work programs. (A "project" or work unit is the smallest separately-funded piece of research that is being done; a "program" is a collection of all the projects on a particular military topic.) In general, collections of project descriptions (unclassified and unlimited) can be obtained easily, even in the U. S., and exchanged relatively freely with other NATO countries. (In the U. S., such a collection can be obtained by accessing a computerized data base.) By contrast, the ability to compile a national work program, laboratory by laboratory, seems to vary inversely with the size of the country. Such a compilation is relatively easy for the smaller NATO countries, but nearly impossible for the U. S. (It could only be done by contacting each of the laboratories individually.) Also, while collections of project descriptions may easily be exchanged within the NATO community, a complete national work program description (for any military topic, including basic research) will, undoubtedly, be classified at least NATO RESTRICTED. Early in the work of RSG-4 the collections of project descriptions (upon which the Table 2 statistics were based) were entirely satisfactory. Later, as cooperation got underway and as we searched for additional military topics to assess, these collections became less useful. Managers of future NATO projects should keep this in mind: the broader the topic, the more difficult it will be to compile national program descriptions, and the less likely that they can be freely exchanged, even among friends.

2.3 Selecting potential topics for cooperation

Each NATO RSG has to formulate its own TOR, and to select the most appropriate mechanisms for accomplishing its goals, subject to the approval of its parent panel. Thus, having informed all of the defense PR researchers in the participating countries about the current research projects, RSG-4 began to discuss ways and means of identifying the significant technology gaps that might form the basis for cooperative research. Initially, a survey letter was sent to each researcher and project monitor whose work was listed in the combined PR summary document [3]. Each addressee was asked to indicate: (i) his interests in PR or APR as related to the topics in Table 1; (ii) national projects about which more information was desired; (iii) projects he felt he could make a contribution to; and (iv) projects that seemed to be the most likely candidates to form the basis of cooperative research. The results of this survey were exchanged at the third RSG-4 meeting (November 1972). A very large number of international inquiries about PR projects resulted from this survey. In addition, curiously, at least in the U. S., there were a large number of intranational inquiries. In fact, several U. S. researchers stated that they had not previously been aware of other defense laboratories doing similar work, and one wag even suggested that we should start a cooperative research project among U. S. researchers!

Also discussed at the third meeting was the notion of organizing international workshops for the purpose of identifying topics for cooperation. These would have been conducted by inviting national specialists to talk about military applications of PR, highlighting the unsolved problems and gaps in our knowledge. Moreover, it was proposed that selected discussants could synthesize the presentations and arrive at a series of goals that could be translated into research proposals. Three possible topics for workshops were identified by RSG-4 and a fourth was suggested by Panel III: (i) picture and scene processing; (ii) APR applied to electronic warfare; (iii) speech understanding; and (iv) adaptive and interactive processing.

Interests in these workshop topics were surveyed, and other suggestions were also solicited. Some degree of interest was found for every one of the topics listed in Table 1 [7]. However, when these results were presented at the fourth meeting (May 1973), negative comments from Panel III were also presented. Apparently Panel III had had some bad experiences with workshops--too often they served only as a forum for discussing problems, and no recommended solutions were forthcoming. As a result RSG-4 abandonned the workshop as a means of identifying topics for cooperation.

2.4 Planning for the technology assessments

The alternative procedure adopted by RSG-4 was to devote a portion of each of its regular meetings to discussion of a specific military APR problem. This discussion was structured so it would ultimately result in the development of recommendations to Panel III about directions for cooperative research. Since the national delegates were not expert in all possible military APR problems, these discussions were conducted with the assistance of specialists from countries having ongoing research programs on the particular topics. Specialists were, among other things, chosen because of their better than average understanding of the user requirements for APR applications. Each RSG-4 meeting was to be devoted to one specific topic, and it was anticipated that several meetings might be required to complete the assessment of any one topic. The topics to be selected for discussion were to be limited to those with definite military relevance; topics of interest only to civil researchers were therefore excluded. Each RSG-4 participant could send one or more specialists to discuss any topic, but each would be responsible for presenting an independent assessment of the technology. These independent assessments were to be prepared to answer the following specific questions:

a. What is the present state of the art in automatic processing (name of topic or application)? Specify military application(s) involved in your assessment. To what extent do present systems solve the problem, or part(s) of the problem? What programs are in existence aimed at developing solutions to the problem?

b. What are the unsolved problems? If possible, indicate which of these unsolved problems might form the basis for cooperative research programs.

c. What is the estimated cost (time, money, manyears) of solving the problem? (Assume an orderly program of basic and applied research, but no international cooperation beyond that presently in existence.) Estimate the savings which might result if cooperative research among the NATO countries could be brought to bear on the problem.

d. What are the probable system requirements (hardware, software, speed) necessary to solve the problem, or part(s) of the problem?

Technology assessments were ultimately conducted by RSG-4 on three APR topics of military relevance, as follows:

a. Automatic target detection, identification, and classification in image date (e.g., reconnaissance, picture-aided control, target following, etc.).

b. Automatic speech recognition (e.g., auditory man-machine communication, speaker verification, etc.).

c. Automatic processing of mechanical waves in solids, liquids, and gases (e.g., acoustic, seismic, and sonar target detection and classification).

The results of these three assessments are discussed and presented in the following Sections 2.5, 2.6 and 2.7.

2.5 Technology assessment of image processing

One of the things that had to be decided in conjunction with this, the first of the formal technology assessments, was in what format to present the results. This was a nontrivial point since it would ultimately influence the extent to which the results would be used by the military APR community, as well as the development of proposals for cooperative research. It was decided to attempt to construct four tables:

a. Military tasks that could be automated through the application of APR techniques.

b. Image processing techniques that must be perfected in order to automate the military tasks.

c. State of the art of both applications and processing techniques.

d. Unsolved problems associated with the processing techniques.

After these four tables had been developed, they would be circulated to national researchers for comment and for expressions of interest in cooperative research [8].

The image processing assessment was begun at the fifth RSG-4 meeting (November 1973). Independent assessments addressing the points listed in Section 2.4 were presented by Ian Henderson (Canada), Helmut Kazmierczak (Germany), R. Hoffman and E. J. Simmons (U. S.), B. B. Scheps (U. S.), and V. Shely and B. Schrock (U. S.). Military applications or problems represented in these independent assessments included remote sensing in general, hard target detection, and terrain analysis. The results of the technology assessment were circulated to national researchers for comment, and the formal report of the assessment was finalized at the seventh meeting (August 1974).

It was, of course, concluded by RSG-4 that image processing is a topic of high military relevance. Table 3 lists military tasks that we would like to be able to automate. While most of the participating countries are conducting research on image processing, the results to date indicate that the human operator cannot easily be replaced by automatic procedures. Some semi-automatic solutions might be feasible in the near future, and applying them to the tasks listed in Table 3 would result in improved efficiency of the military systems concerned.

RSG-4 noted in its assessment [9, 10] that present image processing technology is not able to cope with all imagery problems with equal facility. Some relatively straightforward problems, such as IR missile guidance, have been solved by small elegant PR hardware systems. At the other extreme lies the problem of locating targets in noisy backgrounds, such as terrain, in which there are few existing analytic capabilities. This wide range of potential applications is further exemplified by the problems and machine processing capabilities listed in Table 4. It should be noted that the term "target" is here used in a very general sense to mean the focus of interest in a given operational problem. It is inclusive of such objectives as terrain features, cartographic parameters, as well as hostile objects to be tracked and/or destroyed.

The prerequisite for automatic processing, of course, is the discovery of target characteristics which uniquely set it apart from its background and environment. Occasionally, a single target characteristic (such as IR radiance in missile guidance) can be used for automatic processing. Under adverse weather conditions, however, such a single characteristic may not uniquely distinguish a target from its environment, and it becomes necessary to find other, or alternate, characteristics if automatic processing is to be applied. Thusthe target and its environment determine the choice of identifying characteristics.

Table 3
Military Image Processing Tasks for Possible Automation

1 RECONNAISSANCE	Detection & classification of targets & other objects (terrain analysis, establishment of data bases, damage assessment).
2 CARTOGRAPHY	Topographic mapping, thematic mapping, topographic data base, point processing, map revision.
3 WEAPON GUIDANCE	Target tracking & following; guidance to fixed targets.
4 NAVIGATION	Air; ground; sea; space.
5 TECHNICAL INTELLIGANCE	Weapon system assessment; material evaluation.
6 METEOROLOGY	Prediction; real time use.
7 SECURITY	Area defense; personnel identification.
8 REMOTE CONTROL	Robot operations.

Table 4

State of the Art of Some Military (Semi) Automatic Image Processing Applications

ENVIRONMENT	TARGET	PRINCIPAL SENSOR	PURPOSE OF SENSING	STATE OF MACHINE PROCESSING ART
Air/Space	Aircraft; spacecraft; missiles	Radar IR	Tracking; attack	Advanced
Sea surface	Ships	Radar; camera; sonar; scanners	Detection; tracking; classification	Machine processing for tracking possible; requires operator intervention.
Sea sub-surface	Submarines; mines; oceanography	Sonar; magnetic anomoly detection	Detection; tracking; classification; attack	APR methods applied to processing of sensor array data (sensor limited).
Terrain surface	Land vehicles; structures; weapon systems; cartographic & strategic applications	Camera; scanner; IR; radar	Detection; classification; terrain analysis	Preprocessing advanced. APR only applicable in certain limited cases. APR being applied to a limited extent in cartography (sensor limited).

Table 5 lists and defines the image processing techniques that RSG-4 believed needed to be perfected in order to be able to automate the military tasks listed in Tables 3 and 4. Table 6 presents a series of succinct statements about the state of the art of these image processing techniques, and also identifies by number the unsolved problems associated with each technique. The unsolved problems are listed and defined in Table 7. It should be noted that the list in Table 7 is not necessarily all inclusive; in particular there is little or no recognition of that class of problems that arises when several algorithms or procedures are linked together to solve a complex problem. These are basically "systems" problems, and there appear to be three critical areas where much work remains to be done:

a. Digital storage, computing capacity, and hardware architecture.

b. Audio-visual display requirements for semiautomatic interactive pattern recognition systems.

c. Adaptive and/or interactive processing techniques in general, and as design tools for optimizing automatic pattern recognition systems.

Table 5
Image Processing Techniques Requiring Perfection

1 MULTISPECTRAL ANALYSIS. Extraction of information concerning targets in imagery through the use of images made in many parts of the EM spectrum. Typically includes visible, IR, and RF portions of the spectrum.

2 MOTION DETECTION. Recognition of targets through their apparent motion with respect to the background.

3 CHANGE DETECTION. Detecting changes in a scene by subtracting a previous image from a current image of the scene. This highlights any changes in the scene (but one must be careful to ensure that the images are the same scale and orientation).

4 FEATURE EXTRACTION AND TEXTURE ANALYSIS. Feature (characteristic) extraction is the process of converting an image into a set of significant numbers and/or relations, so appropriate automatic data processing techniques can be applied.

5 IMAGE MATCHING. Performing processing such as rectification, rotation, scaling, etc., on different images of the same scene so they may be compared.

(continued)

(Table 5 continued)

6 SCREENING AND PRECONDITIONING. Preprocessing a picture for automatic analysis, and then automatically eliminating those images having no patterns of interest. Also, cueing an analyst as to where in an image a pattern may be found.

7 IMAGE DATA STORAGE, RETRIEVAL, AND MERGING. Storing and locating images so they may be found quickly and merged into reports; or two or more images may be combined into one for analysis.

It should be noted that the third and fourth questions listed in Section 2.4 are not included. It was decided that these questions (regarding costs, and implementation requirements) were very application-specific, and should more properly be answered as part of any cooperative research effort that might be developed. These two questions were also deleted from the remaining technology assessments.

The RSG-4 review of processing capability revealed yet another barrier to the development of improved PR capability: the necessity for processing enormous amounts of data by mostly serial methods. The development of a significant parallel processing capability would overcome many of the present difficulties; a number of architectures are being investigated including optical, parallel digital, associative, pipeline, and multiple processors.

Table 7
Glossary of Unsolved Problems in Image Processing

1 ENVIRONMENTAL INFLUENCES AND NEAR-TIME HISTORY. Effects of current and recent events that affect the response of an object or scene in any frequency band used to form the image.

2 TARGET CHARACTERISTICS. Shape, size, reflectance, emission spectra, color, texture, shadow, etc., enabling identification or detection of targets in a frame of imagery.

3 APPROPRIATE SENSOR RESOLUTION (ACTIVE AND PASSIVE SYSTEMS). Number of separable points contained in any image. May be limited by grain size of photographic film, and memory size of processing system.

4 PRECISE SUPERPOSITIONING OF IMAGERY.

5 COMBINATORIAL ALGORITHMS FOR MULTISPECTRAL DATA ANALYSIS.

(continued)

(Table 7 continued)

6 INTERACTION OF FREQUENCIES. Changes in the response of an object in one frequency band when hit by other frequencies. (Example: Kapton changes its optical reflectivity and color when bombarded with X-rays.)

7 REFERENCE CALIBRATIONS FOR SCANNERS.

8 SCENE NORMALIZATION AND CONTRAST CONTROL FOR SUCH INFLUENCES AS SUN ANGLE, ETC. Processing one or more images so they may be compared. This often includes rectification to normalize look angles, scaling to change altitude, and rotation/translation to correct for differences in orientation/direction of the sensor platform.

9 UNDERSTANDING OF FREQUENCY EFFECTS. Response of an object when illuminated by a given frequency.

10 DATA BASE ORGANIZATION.

11 COMMUNICATION BETWEEN RESEARCHER AND USE.

12 DETECTION OF MOTION AT HIGH AND LOW TARGET SPEEDS.

13 BRIDGING FOR NON-CORRELATION. In image superpositioning, a means for bridging non-correlated areas in automatic processing.

14 DISCRIMINATION BETWEEN BACKGROUND AND TARGET IN NON-DOPPLER MOTION DETECTION.

15 DETECTION OF NON-RADIAL MOTION.

16 CHANGE ANALYSIS AND INTERPRETATION.

17 OPTIMAL FEATURE SELECTION (HEURISTIC/STATISTICAL/ADAPTIVE).

18 NO FEATURE CATALOG EXISTS FOR VARIOUS TYPES OF TARGETS AND BACKGROUNDS.

19 FEATURE EXTRACTION TECHNIQUES ARE LARGELY UNDOCUMENTED.

20 FEATURE EXTRACTION AND CORRELATION.

21 SPEED AND CUEING. Rapidly selecting imagery of interest, and cueing operator to regions of interest.

22 FORMATING.

23 MAKING AND USING MICROGRAPHICS.

Table 6
State of the Art of Image Processing Techniques
(As of August 1974)

TECHNIQUE	PRESENT STATE OF THE ART[1]	UNSOLVED PROBLEMS[2]
1 Multispectral Analysis	B	1-11.
2 Motion Detection	A-for Doppler shift in imaged radar/sonar for middle rates of speed. C-for optical systems.	3, 12, 14, 15. 4-in other than Doppler shift methods.
3 Change Detection	A-C for special cases.	3, 4, 5, 8, 16. 10-for long-term changes.
4 Feature Extraction	B	17, 18, 19.
5 Image Matching	A-for cartography. B-C for reconnaissance.	1, 4, 8, 10, 13, 20.
6 Screening and Preconditioning	B	1-11, 21, 22.
7 Image Data Retrieval and Merging	B	1, 4, 8, 10, 11, 13, 20-23.

[1]Ratings: A=useful now; B=shows promise; C=a long way to go.

[2]See Table 7.

2.6 Technology assessment of speech recognition

The formal assessment of military applications of automatic speech processing and recognition was begun at the sixth RSG-4 meeting (May 1974). However, to obviate some of the difficulties encountered with the image processing assessment, a list of military tasks for possible automation in speech recognition was developed at the fifth meeting. This list is presented in Table 8.

At the sixth meeting, independent assessments were presented by Helmut Mangold (Germany), Louis Pols (Netherlands), Edwin Newman (U. K.), and E. P. Neuberg and D. C. Hodge (U. S.) [11]. A synthesized version [12] of the speech recognition assessment has recently been published in the open literature [13], so only a summary of the results of the assessment will be presented here.

Table 9 lists the speech processing techniques that have to be perfected in order to be able to automate some or all of the military tasks listed in Table 8. Table 10 indicates the state of the art of each technique on a three-point scale, and includes a list of the unsolved problems associated with each of the processing techniques. Table 11 presents a glossary of unsolved problems.

To date the greatest military applications of speech recognition, per se, have been achieved in the areas of speaker verification and voice data entry. Speaker verification is presently being used to control access to secure areas, often in conjunction with other identity verification techniques such as face recognition, signature verification, etc. In the speaker <u>verification</u> problem (as opposed to speaker recognition) the speaker is cooperative, his identity is known to the system, and his spoken data base is in the system's memory. The input communication channel can be a high-quality microphone, or noise-cancelling techniques can be used where necessary. The same communication channel can be used for both the reference and test sets of voice samples. Present techniques in speaker verification produce results on the order of one percent rejection of true speakers, and two percent acceptance of impostors, using only one reference utterance and 120 speakers. By using two reference utterances, the false acceptance rate can be reduced to less than 0.25 percent. Further refinements that are needed for military applications include operation over telephone-bandwidth

Table 8
Military Speech Processing Tasks for Possible Automation

1 SECURITY	Speaker verification (authentication); speaker identification (recognition); determining emotional state of speaker (e.g., stress); recognition of spoken codes; secure access voice identification, whether or not in combination with fingerprints, facial information, identity card, signature, etc.; surveillance of communication channels.
2 COMMAND AND CONTROL	System control (ships, aircraft, fire control, situation displays, etc.); voice-operated computer input/output (each telephone a terminal), data handling and record control; material handling (mail, baggage, publications, industrial applications); remote control (dangerous material); administrative record control.
3 DATA TRANSMISSION. AND COMMUNICATION	Speech synthesis; vocoder systems; bandwidth reduction or, more general, bit-rate reduction; ciphering/coding/scrambling.
4 PROCESSING DISTROTED SPEECH	Diver speech; astronaut communication; underwater telephone; oxygen mask speech; high "G" force speech.

Table 9
Speech Processing Techniques Requiring Perfection

1 SIGNAL CONDITIONING. Some processing of speech signals may be necessary to compensate for different characteristics of input channels, such as overall signal level and differential delay. Also, it may be possible to preprocess to improve speech quality, or S/N ratio, or to remove long silences.

2 DIGITAL SIGNAL TRANSFORMATION. The digitized speech signal is transformed in preparation for the extraction of parameters. Processes used include Fourier and Walsh transforms, correlation, linear predictive coding and digital filtering.

3 ANALOG SIGNAL TRANSFORMATION AND FEATURE EXTRACTION. The signal can be transformed by hardware, such as filter banks and correlation devices. Transforms can be digitized for further processing, or parameters and features can be extracted in a continuous manner for presentation to decision networks or algorithms.

4 DIGITAL PARAMETER AND FEATURE EXTRACTION. Calculations are done on the transformed signal to extract relevant parameters, such as formant tracking, pitch extraction and principle components analysis.

5A RESYNTHESIS. Speech parameters extracted, as mentioned above, in speech compression systems or stored in voice playback systems, may be transformed into acceptable acoustic speech signals.

5B ORTHOGRAPHIC SYNTHESIS. In the translation of written materials to speech, a number of techniques must be developed. Some of these techniques are similar to those cited in the paragraphs above. One of the most important is the development of speech morphology.

6 SPEAKER NORMALIZATION, SPEAKER ADAPTATION, SITUATION ADAPTATION. The effectiveness of parameters in carrying relevant speech information depends on characteristics of individual speakers and on operational situations. This could mean that systems must be trained or must adapt to optimize parameters.

7 TIME NORMALIZATION. In recognition of isolated utterances, normalization is imposed to compensate for local and global differences in speech rate. Both linear and nonlinear schemes can be used.

(continued)

(Table 9 continued)

8 SEGMENTATION AND LABELING. Segment boundaries are set, e.g., at points of rapid change, formant positions, voicing, spectral shape or other parameters. Segments may be labeled probabalistically to acoustic-phonetic classes. Prestored knowledge of features and parameters for the various classes of segments are used in the decision.

9A LANGUAGE STATISTICS. Language statistics and partial recognition are used to predict and evaluate words at specific points in an utterance.

9B SYNTAX. The grammar of the task is used to predict and evaluate word categories at specific points in an utterance.

9C SEMANTICS. Knowledge of the task domain is used to predict and evaluate subject matter at specific points in an utterance.

9D SPEAKER AND SITUATION PRAGMATICS. In determining the semantics of speech, certain aspects of the utterances are related to an underlaying assumption about what the speaker would generally consider an appropriate response. The development of this type of knowledge is required for speech understanding systems. Knowledge of the situation that gave rise to the utterances is also required for reliable interpretation and execution of the task to be performed in response to the utterance.

10 LEXICAL MATCHING. Strings of linguistic-phonetic elements hypothesized by the linguistic part of the system are compared with strings of acoustic-phonetic elements derived from an utterance. A quantitative goodness of match is calculated.

11 SPEECH UNDERSTANDING. All sources of knowledge (acoustic, phonetic, pragmatic, semantic, syntactic) are used in combination to reconstruct the utterance and/or determine its meaning.

12 SPEAKER RECOGNITION. Speaker-specific parameters are extracted and compared with stored parameter sets from known speakers.

13 SYSTEM ORGANIZATION AND REALIZATION. Systems must be developed keeping in mind use by humans and cost-effectiveness factors.

14 PERFORMANCE EVALUATION. Present development of all speech systems requires the determination of the quantitative value of each possible technique studied. Only by the use of stored speech samples is this performance evaluation possible.

Table 10
State of the Art of Speech Processing Techniques

PROCESSING TECHNIQUE	STATE OF THE ART[1]	UNSOLVED PROBLEMS[2]
1 Signal Conditioning	A, except speech enhancement (C)	1,15,20,23
2 Digital Signal Transformation	A	1,15,20
3 Analog Signal Transformation & Feature Extraction	A, except feature extraction (C)	1,2,6,14-16,20,24,25
4 Digital Parameter & Feature Extraction	B	1,2,6,14,16,20,24,25
5A Resynthesis	A	4,7,20,25
5B Orthographic Synthesis	C	4,6-8,19,26-28,29
6 Speaker Normalization & Adaptation; Situation Adaptation	C	15-17,19,20,23-25,29
7 Time Normalization	B	3,16,20,25,29
8 Segmentation & Labeling	B	1,4-9,11,13,16 18-20,24,25
9A Language Statistics	C	5,8,9,11,12,14,20,24,25
9B Syntax	B	6,7,9,12,14,20,25
9C Semantics	C	6,7,9,10,12,14,20,25
9D Speaker & Situation Pragmatics	C	3,6,12,14,16,18,19,23
10 Lexical Matching	C	7-9,12-14,20,25

(continued)

[1]Ratings: A=useful now; B=shows promise; C=a long way to go.

[2]See Table 11.

(Table 10 continued)

PROCESSING TECHNIQUE	STATE OF THE ART[1]	UNSOLVED PROBLEMS[2]
11 Speech Understanding	B-C	5,9,12,14,16,18, 20,23,25
12 Speaker Recognition	A for speaker verification; C for all others	14,16,17,19,20,24, 25
13 System Organization & Realization	A-C	21,22
14 Performance Evaluation	C	1,6-11,18-20,24-28

[1]Ratings: A=useful now; B=shows promise; C=a long way to go.

[2]See Table 11.

channels, and over radio links which involve peculiar S/N and level fluctuation problems, plus developing techniques that will permit context-free speaker verification.

Voice data entry is the other military application area in which speech recognition, per se, is presently being applied. Voice data entry is being used in a variety of hands-busy situations as well as to reduce data transmission errors by eliminating hand transcription (as for computer entry). Present voice data entry systems are based on isolated word recognition techniques, and these types of systems generally involve the following limitations: (i) vocabulary limited to about 75 words; (ii) no speaker normalization is used; (iii) every speaker must train the system on every word in the vocabulary; and (iv) the words must be spoken in a discrete manner.

Further developments in voice data entry in the immediate future are likely to be in two directions. One will be improving systems so they will be speaker-independent, but have other current limitations remain the same. The other direction is to eliminate the requirement for discrete word input by developing techniques that will allow short phrase recognition and connected word or connected digit recognition.

If we broaden this topic to include speech <u>processing</u> as well as recognition (which, of course, was done in the technology

Table 11
Glossary of Unsolved Problems in Speech Processing

1 Detect speech in noise; speech/nonspeech.

2 Extract relevant acoustic parameters.

3 Dynamic programming (nonlinear time normalization).

4 Detect smaller units in continuous speech.

5 Establish anchor point; scan utterance left to right.

6 Stressed/unstressed.

7 Phonological rules.

8 Missing or extra added ("uh") speech sounds.

9 Limited vocabulary; possibliity of adding new words.

10 Semantics of (limited) tasks.

11 Limits of acoustic information only.

12 Combine acoustic, syntax and semantic information.

13 Recognition algorithms.

14 Hypothesize and test; backtrack; feed forward.

15 Effect of distortions (physiological, physical, mechanical).

16 Adaptive and interactive quick learning.

17 Mimicking; uncooperative speaker(s).

18 Necessity of visual feedback, error control, rejection level.

19 Consistency of references.

20 Real-time processing.

21 Human factors engineering problems.

22 Cost effectiveness.

23 Detect speech in presence of competing speech.

(continued)

(Table 11 continued)

24 Economical ways of adding new speakers to system.

25 Use of prosodic information.

26 Coarticulation rules.

27 Morphology rules.

28 Syntax rules.

29 Vocal tract modeling.

assessment--see Table 8) then other current military applications include channel vocoders and helium speech unscramblers.

Despite the large amount of money which has recently been devoted (particularly in the U. S.) to research and development in the area of continuous speech understanding, there is no stated military requirement for such a capability at present. On the other hand, once the next generation of such systems comes along, and some of the present limitations have been overcome [14], we will probably begin to see some of this developing technology applied to military problems.

In addition to the specific applications mentioned above, the following are additional examples of military problems to which speech recognition and processing technology is presently being applied:

a. Digital narrowband communication system: A massive effort is underway to develop and implement an all-digital communication system.

b. Training systems: A limited speech understanding system is under study for use in a military training system.

c. On-line cartographic processing system: Studies are underway to use speech recognition and voice response techniques with cartographic point and trace processing systems.

d. Word recognition for militarized tactical data systems : Word recognition, speaker verification, and voice response will be used for message entry to a tactical data system.

e. Voice recognition and synthesis for aircraft cockpit: Existing word recognition systems are being tested under simulated cockpit environments.

2.7 Technology assessment of mechanical wave processing

The first discussion of the application of APR techniques to acoustic, seismic, and sonar signals took place at the fifth RSG-4 meeting (November 1973). (For convenience, this topic was referred to as "sound/seismic/sonar.") Surveys of national MOD interests in an assessment and/or cooperative research were discussed at the next two meetings; in general, the consensus was that the results of a technology assessment would be very useful, but there was considerable doubt about cooperative research due to the sensitivity of all the data.

Largely to avoid the continued reference to "sonar", the name of this topic was changed at the seventh meeting. All of the signals of interest represent mechanical waves traveling in various media, so it was decided to re-title this topic "Automatic Processing of Mechanical Waves in Solids, Liquids, and Gases, for Military Purposes Such As Acoustic, Seismic, and Sonar Target Detection and Classification". Also at the seventh meeting a list of possible military problems and applications was developed to guide further discussion; that list is shown in Table 12.

Table 12
Possible Military Problems and Applications in the Automatic Processing of Mechanical Waves in Solids, Liquids and Gases*

1 SURVEILLANCE (PASSIVE AND ACTIVE)	Detection and classification of military vehicles (air/land); fixed objects (mines); explosions/implosions (type/location); underwater and surface vehicles; personnel.
2 NAVIGATION	Position checking; confined and hazardous sea areas; underwater mapping.
3 COMMUNICATION	Underwater communications (underwater telephone); IFF.
4 INTERCEPT	Detection and classification of hostile transmissions.
5 DAMAGE DETECTION	Anticipating equipment failures (engines; gun tubes); quality control (nondestructive testing).
6 SIMULATORS	Realistic signals for operator and command training systems.

*Revised from Annex E to Ref. 15.

At the eighth meeting (July 1975) RSG-4 constituted an Ad Hoc Group on Mechanical Waves, composed of the national specialists present at the meeting. This group was charged with the responsibility of deciding how to structure a technology assessment, and with actually carrying out the assessment. COL Russell B. Ives of the U. S. was appointed Chairman of this group, a position that he held until the conclusion and publication of the assessment.

COL Ives presented a proposed outline for the mechanical waves assessment for discussion at the eighth meeting -- see Table 13. This "straw man" was used to stimulate discussions of the state of the art, directions of current national research programs, and potential areas for future cooperation. At that time the greatest interest was shown in passive sonar, with active a close second; however it was recognized that any exchange of passive sonar signatures would be impossible, and active sonar data exchange would be difficult. It was agreed, however, that an attempt would be made between the eigth and ninth RSG-4 meetings to draft a technology assessment of the topic for discussion at the ninth meeting.

At the ninth meeting the proposed outline (Table 13) was found to be unworkable by some of the specialists. As a result the technology assessment was formulated as shown in Tables 14 and 15.

Table 14 lists the main applications of military interest. These have been divided into three areas: passive listening systems, active echo analysis systems, and communication systems. Against each of these major headings, various systems operating in the three different media are considered. For each particular system, the type of sensor platform used is listed together with the targets or objects which provide the focus of interest. In addition, the purpose of the system in terms of its ability to detect, classify, localize, track, etc., is identified.

(Keep in mind that some systems exist which use active non-mechanical waves to detect and analyze mechanical waves, e.g., EM, radar, and lasers. These systems are not considered to be active mechanical wave systems and are, therefore, excluded.)

The brief statements made in Table 14 concerning the state of the art against the various applications are included to indicate the degree to which operational systems, and those in advanced stages of development, make use of APR techniques. Although certain fully automatic systems are identified (e.g., homing torpedoes, and mine detection systems), these are almost invariably based on operating concepts which are too simple to be properly considered as APR techniques.

Table 13

Proposed Outline for a Technology Assessment of the Automatic Processing of Mechanical Waves in Solids, Liquids, and Gases

I. Current Military Requirements
   A. Successful Applications
      1. Development time
      2. Useful life span
      3. Anticipated obsolescence
   B. Unsuccessful Applications
      1. Development time
      2. Deficiencies noted
      3. Proposed vs. actual performance
      4. Identified corrective research

II. Future Military Requirements
   A. Identified Applications
   B. Projected Technological Deficiencies
      1. Requisite basic research (6.1)
      2. New engineering techniques (6.2)
      3. Essential advances in technology (6.3+)
   C. Anticipated Resource Allocations
      1. Time
      2. Cost
      3. Manpower

III. Technical Tools
   A. Signal Conditioning Techniques
      1. Active and passive filters
      2. Compression and expansion methods
      3. Multiple channel superposition
   B. Feature Extraction
      1. Physical features
      2. Structured features
      3. Mathematical features
      4. Transform domains
         a. Fourier
         b. Hadamard
         c. Binary
   C. Clustering algorithms
      1. Statistical
         a. Means
         b. Covariance
      2. Clustering
         a. Isodata
         b. Hierarchical

(continued)

(Table 13 continued)

III. Technical Tools
- C. Clustering Algorithms
  - 3. Distance measures
    - a. Geometric
    - b. Mahalanobis
    - c. Bhattacharya
  - 4. Mappings
    - a. Linear (Karhunen-Loeve)
    - b. Nonlinear
      - (1) Sammon
      - (2) Fukunaga-Olsen
  - 5. Decision functions
    - a. Bayes linear and quadratic
    - b. Piecewise linear
    - c. Fisher discriminants
    - d. Language strings
- D. Hardware
  - 1. Transducers
  - 2. Data channels
  - 3. Remote vs. central processing
  - 4. Processors
    - a. Minicomputers (single and clustered)
    - b. Large scale processors
    - c. Peripherals
  - 5. Displays
    - a. Environmental restrictions
    - b. Media
    - c. Complexity required

IV. International Areas of Cooperation
- A. Areas Which Would Benefit
  - 1. Time
  - 2. Cost
  - 3. Accuracy
  - 4. Reliability
- B. Areas Which Preclude Cooperation
  - 1. Sensitive due to expressed interest
  - 2. Sensitive due to data or sources of data
  - 3. Lack of interest

V. Recommendations
- A. Resource Allocations Within NATO
- B. Realistic End Products Within Specified Time Frame
- C. Areas of Responsibility
- D. Reporting Procedures and Dissemination of Results

Table 14
Military Applications of Mechanical Waves and State of the Art

METHOD	MEDIUM	TARGETS	PURPOSE	SENSOR CARRIER	STATE OF THE ART
			A. Passive Listening Systems		
Sonar	Water, water-air	Ships, submarines, aircraft, decoys, personnel, torpedoes, explosions, implosions	Detection, classification, localization, tracking	Bottom arrays, towed arrays, helicopters, surface ships, submarines, sonobuoys, torpedoes, mines	Limited automatic tracking. Most tasks require operator intervention.
Doppler Analysis	Water air	Submarines, ships, torpedoes, aircraft, decoys	Classification, tracking, localization	(same as above)	(same as above)
Intercept	Water	Active range measurement, communication systems	Detection, classification, localization, tracking	Submarines, torpedoes, ships, fixed installations, mines	No existing systems using APR techniques are known.
Airborne Sound	Air	Detonations, aircraft, personnel, vehicles, stationary engines	(same as above)		Some elementary APR systems exist. Greater success has been achieved with detonations.

(continued)

(Table 14 continued)

METHOD	MEDIUM	TARGETS	PURPOSE	SENSOR CARRIER	STATE OF THE ART
Vibration Analysis	Mechanical structures, air	Vehicles, stationary engines	Detection, classification, failure prediction, quality control	N/A	Fully automatic systems are not sufficiently reliable.
Seismic	Earth (surface & body waves)	Personnel aircraft, vehicles, fixed	Detection, classification, localization, tracking	Surface of earth	Some APR techniques exist for a limited number of single targets.
Sea Surface Waves (hydrostatic pressure)	Water	Ships, submarines, decoys	Detection, classification	N/A	Some automatic detection systems exist.

(continued)

(Table 14 continued)

METHOD	MEDIUM	TARGETS	PURPOSE	SENSOR CARRIER	STATE OF THE ART
		B. Active Echo Analysis Systems			
Echo Sonar	Water	Surface ships, submarines, mines, torpedoes, wrecks, sea bottom	Detection, classification, localization, tracking	Surface ships, submarines, helicopters, buoys, torpedoes, divers	Some semiautomatic methods giving limited detection, tracking, & classification have been developed.
Echo Sounding	Water	Bottom	Depth measurement, reflection coefficients, profiling	Surface ships, submarines	No existing systems using APR techniques are known.
Doppler Analysis	1) Water	Bottom	Navigation	Surface ships, submarines	Some systems using elementary APR techniques have appeared.
	2) Air	Personnel, waves, vehicles	Detection, classification, localization, tracking	Fixed installations	(same as above)

(continued)

(Table 14 continued)

METHOD	MEDIUM	TARGETS	PURPOSE	SENSOR CARRIER	STATE OF THE ART
Echo Seismic	1) Earth	Profiles, discontinuities	Resource exploration, tunnel location	Surface of the earth	Some APR systems exist for ranges to 100 M.
	2) Water	Submarines; bottom	Resource and profile exploration	Surface of the earth	No known APR systems exist.
Structure Analysis	Mechanical structures	Material discontinuities	Nondestructive testing	N/A	Some elementary APR systems beginning to appear.
			C. Communication Systems		
Data Transmission	Water	Surface ships, submarines, buoys	Reliable machine to machine communication; IFF	Surface ships, submarines, buoys, fixed installations	No known APR systems. Some sophisticated error correcting codes developed.
Speech Transmission	1) Water	Ships, divers, submarines	Person to person communication	Ships, submarines, divers, fixed installations	See Ref. 13.
	2) Artificial Atmospheres	Personnel	(same)	Underwater chambers	See Ref. 13.

Table 15
Unsolved Problems in Mechanical Wave Processing

1 ENVIRONMENTAL EFFECTS. Imperfect understanding of the physics of the medium and its boundaries gives rise to uncertainty in the time and frequency response observed from both the target of interest and other sources (i.e., nonstationary and multipath effects).

2 TARGET CHARACTERISTICS. Complex target structure causes uncertainties in their reflection and/or emission characteristics. This problem is compounded when multiple targets are observed.

3 BACKGROUND CHARACTERISTICS. Insufficient knowledge about noise generating and back scattering mechanisms in the propagation media make prediction of the performance of processing systems difficult.

4 SENSOR RESOLUTION. The choice of resolution affects the performance of processing systems. An optimum choice is complicated by distorting effects of the medium and uncertainties about target and background characteristics. In many practical cases, sensor size, resolving power, and time constraints may be limited by considerations relating to the platform on which it is to be mounted as well as the physical characteristics of the sensor itself.

5 OPTIMAL FEATURE SELECTION. A variety of theoretical and empirical techniques have been developed for the generation and selection of feature sets to describe patterns. Little is known about techniques for optimizing the generation of feature sets and feature generation is highly data dependent.

6 COMBINATORIAL ALGORITHMS. The best methods for combining features, or reducing the feature set, to obtain maximum discrimination between pattern classes have not been formulated.

7 NORMALIZATION. Conflicting requirements of both targets and background give rise to compromise in the choice of a strategy for normalizing data.

8 MOTION DETECTION BY NON-DOPPLER MEANS. The detection of nonradial motion and radial motion is complicated by problems associated with sensor resolution and the difficulty of correctly associating background and target data.

(continued)

(Table 15 continued)

9	DOPPLER DETECTION. Although a well established technique, problems still exist in discriminating slow speed radial targets from a background.
10	DATA BASES. Frequency methods and results are based on a statistically insignificant number of samples collected in a variety of uncontrolled conditions. This gives rise to uncertainty as to how representative are the results obtained as well as the performance which might be observed in other conditions. Additional work is needed to optimize the size of a data base. Specifically, to ascertain under what conditions a representative target would be added to a data base or replace an existing entry within the data base.
11	PRECONDITIONING. Like feature generation and selection, a variety of techniques have been studied. Little, however, is known about optimum methods to preprocess data in order to reduce processor loading without undue loss of performance.
12	FEATURE TRACKING. Features may not be time invariant. Superior algorithms must be developed to cope with rapidly changing features and widely varying S/N ratios.
13	DATA PROCESSING. Due to the fact that large time-bandwidth product signals are to be processed, there is a need for high capacity, on-line processors.
14	CLASSIFICATION LOGIC RESTRUCTURING. For each incorrect classification, a technique is required to identify the responsible features and associated decision function. Some interactive process would be desireable to permit an immediate upgrading of the classification logic.

The problems associated with the successful applications of fully automatic and semi-automatic pattern recognition techniques to the majority of the systems mentioned are considered to be too numerous and varied to permit a comprehensive enumeration. An attempt has been made, however, to cite the major problem areas which require a fuller understanding. Research in these areas should provide the information necessary to make more extensive applications of these APR techniques possible.

The requisite research must provide a better understanding of all of the parameters which characterize the desired target, and the identification of those features which uniquely separate the target from the other distracting data. More insight must also be gained with respect to the complex effects of the dynamic, non-homogeneous propagation media in which the target and sensor platform are usually immersed. This is reflected in the first three problem areas identified in Table 15. Sensor resolution has been mentioned as a problem, because it has a significant effect on the received data. It should be noted that this point was of no direct concern to RSG-4, since development of sensors is the responsibility of other NATO groups (see Table 2). Also, in the majority of cases, the interests of the participating nations was in applying APR techniques to existing sensor systems.

The mechanical wave assessment [16] concluded by noting that until a much better understanding of the various problem areas has been reached, a great deal of system design will continue to rely heavily on the experience and intuition of the individual designer. Also, progress by individual designers may be limited by the quality and quantity of suitable data bases as well as insufficient funds to explore a meaningful variety of analytical techniques. These difficulties, among others, could probably be largely overcome through the adoption of unified approaches by the RSG-4 countries including structured cooperative research programs.

2.8 Proposed cooperation on image processing

The first draft of the image processing technology assessment [8] was developed at the November 1973 meeting of RSG-4 (see Section 2.5, esp. Tables 4 - 8). Thereafter this draft was circulated to defense researchers in the participating countries for comment, and for expression of interest in cooperative research on the various unperfected techniques and unsolved problems. The survey results were first discussed at the May 1974 meeting of RSG-4, but no conclusions were reached. However, it became clear that we would have a better chance of obtaining substantive comments if we had a "strawman" proposal to circulate for comment.

The first complete proposal for cooperation was developed at the August 1974 meeting [Annex G to Ref. 15]. This proposal was based on a recognition of the fact that in order to obtain national support of specific projects it would first be necessary to demonstrate RSG-4's ability to undertake cooperation. This situation (which may be obvious to some readers) derives from a number of considerations, including the following:

a. All of the participating nations had on-going research programs planned for two or more years in advance, and would be unlikely to change those plans.

b. Funding levels are likewise programmed for two or more years in advance; therefore, no major sources of funding are immediately available.

c. Parallel approaches to common image processing problems are being pursued in the various countries; the managers of such programs would be unlikely to abandon the hardware systems involved, nor the researchers the theoretical models nor priorities involved.

RSG-4 thus concluded that the immediate goal of a cooperative research plan should have the following properties [15]:

a. It should recognize that ongoing research in the participating countries cannot be changed (at all) without allowing a period of one to two years of adjustment.

b. It should be based on current efforts and exploit current capabilities to the extent possible.

c. Initially, it should involve the minimum amount of additional effort on the part of the national scientists.

d. It should be based on common problems, i.e., problems common to every laboratory in every country.

e. Above all, it should be possible of achievement in a realistic time frame; and it should be a phased plan so some useful capabilities emerge relatively quickly.

After reaching these conclusions an ad hoc working group (headed by Warren E. Grabau, U. S. Army Engineer Waterways Experiment Station) spent a weekend writing the first plan for cooperation on image processing. This plan consisted of five phases:

a. Phase 1: Development of software that would make it possible to freely transmit data bases among laboratories in the participating countries.

b. Phase 2: Adapt existing computer software available in the various laboratories to the hardware of such other laboratories as need the capability.

c. Phase 3: Formulate critical military image processing requirements; document national priorities.

d. Phase 4: Prepare engineering specifications for parallel processing hardware and related software.

e. Phase 5: Conduct research on location and identification of tactical targets in complex scenes.

Phase 1 of this plan was derived from the fact that, at that time, it was commonly impossible or impractical for image processing laboratories to exchange data bases. (But, obviously, utilization of common data bases would be an essential part of any international cooperative program that might result in standardized military procedures.) For a variety of technical and administrative reasons each laboratory has developed its own unique format for storing its digitized image data base. Even if we ignore the 7- vs. 9-track tape incompatibility, there were (and probably are) both hardware and software differences that make data base exchange difficult or impossible. The problem had reached such ridiculous dimensions that we found, for example, that two research groups in the same building at a U. S. defense installation could not exchange data bases!

Not only had each laboratory developed its own storage format for digitized images, but each such format tended to perpetuate itself. Thus it was not appropriate to try to define a "standard" data format to which all participating laboratories could convert their data. Instead, what was proposed was the specification and standardization (among the participating laboratories) of a <u>transfer format</u> into which data bases could be transformed, and from which each laboratory could transform data in its own storage format. In other words, to utilize such a transfer format, each participating laboratory would have to write only two new programs: one to transform its data into the transfer format, and another to transform data out of the transfer format.

Apparently, there had been one or more earlier efforts in the image processing community to accomplish this same thing. I can recall discussion of this topic at a symposium organized by the EIA Committee on Imagery Pattern Recognition [17]. However, that effort had gotten hung up on "standardization" in the NBS or IEC sense, whereas RSG-4 was proposing standardization in the sense of agreement among specific parties who want to do something collectively.

The steps involved in accomplishing the goal of developing a transfer format were envisioned by RSG-4 as consisting of the following:

a. Determine the precise data storage formats currently used by the participating laboratories.

b. Formulate transfer formats. It was assumed that two such formats would be required, one for 7-track (556 BPI) and one for 9-track (800 BPI) hardware.

c. Disseminate formats to labs, who would attempt to write the necessary transfer programs.

d. Evaluate all proposals for modifications to the transfer format(s) that arise from difficulties encountered in the trials.

e. Finalize formats, and run systems checks consisting of trial data exchanges among the participating labs. Prepare report describing the procedure and software.

Phase 2 of the proposed cooperative research plan, dealing with documentation and exchange of existing algorithms, was based on the common problem that many (if not most) image processing algorithms are not sufficiently documented. Each image processing laboratory usually has developed its own collection of algorithms (sometimes referred to as a "bag of tricks") such that each lab can do about the same kinds of processing of image data. For example, most laboratories have developed one or more algorithms for separating an image into fields on the basis of density classes. However, close examination often reveals that each laboratory has a unique procedure for achieving the desired product. As a result of the inadequate documentation of what an algorithm does, as well as how it does it, laboratories find it difficult, if not downright dangerous, to expand image processing capabilities by using borrowed algorithms. This problem has also been the subject of some investigation by other bodies [18].

As originally envisioned by RSG-4 this phase of the cooperative program involved the following steps:

a. Each laboratory prepares a list of its available algorithms, and decides which ones to develop documentation for.

b. A subcommittee collects the documented algorithms, and reviews them for needed amplification.

c. Laboratories revise the documentation as required.

d. The subcommittee collects the final versions, and organizes them for dissemination.

Phases 3 - 5 were not drafted in detail at the outset; it was assumed that these phases might change (they did) and that they could better be drafted in detail by the national image processing specialists who were to be appointed by the participants to conduct the cooperative programs.

The draft proposals were circulated to national researchers for comment, and Panel III was requested to approve establishment of a subgroup under RSG-4 to conduct the program. The remainder of the activities in image processing cooperation is covered in section 3.

2.9 Proposed cooperation on speech processing

The technology assessment of military applications of automatic speech recognition was initiated at the sixth RSG-4 meeting in May 1974 [11]; after it had been circulated to national speech researchers for comment it was published as a NATO report [12] in April 1976.

When the list of techniques to be perfected, and the list of unsolved problems in speech processing, were circulated for comment and for expressions of interest in cooperation, the initial returns were disappointing. At the August 1974 meeting only two countries were interested in cooperating on speech recognition, per se; by broadening the topic to speech processing, a total of three expressed interest. At that point it was decided to shelve the topic of cooperation and put our energy primarily into writing the combined NATO report [12].

At the July 1975 (8th) RSG-4 meeting there was further discussion of cooperation on speech processing, which may be summarized as follows:

a. There is already a great deal of international cooperation on speech processing, via national technical societies and regular international conferences. This cooperation includes countries outside NATO (e.g., Japan and USSR) where there is much interest in the topic. As a result, little interest has been expressed in RSG-4 for cooperation on unclassified problems.

b. National language differences cause processing problems, i.e., processing techniques that work with one language may not work with another. There are, however, a number of problems in common, such as speaker normalization, that might form the basis for cooperation.

c. The availability of the speech technology assessment report [12], with its emphasis on current and near term military applications, may stimulate interest in cooperation, particularly on classified problems.

Thus it was decided to continue internal discussions of possible topics for cooperation, and to continue dialogue with national researchers about their interests in cooperation.

At the 9th meeting (February 1976) four countries expressed some degree of interest in speech processing topics. As a result of this increase in interest, a very general proposal was developed for circulation among speech researchers in the participating countries [19]. This proposal consisted of a brief discussion of three speech topics to be further explored for degree of interest: (i) speech feature extraction in vocoder systems; (ii) automatic speaker verification; and (iii) voice control (limited vocabulary--100 words; many speakers). The survey results reported to the 10th meeting (November 1976) [20] indicated that four countries were interested in cooperating on vocoders, three on speaker verification, and four on voice control. For three of these respondents, voice control represented their area of greatest interest. These data were taken by RSG-4 as a clear indication of sufficient interest to warrant preparation of more detailed proposals for cooperation.

The speech processing specialists who were present at the 10th RSG-4 meeting proceeded to discuss the prospects for cooperation, and to draft a Terms of Reference and a research plan for a subgroup on speech processing. The main thrust of this research plan was in the area of limited vocabulary voice input systems. This topic is one in which all the potential participants could conceptualize military tasks to which such systems would be applicable; this fact also meant that each country could list some (possibly unique) requirements that voice input systems would have to meet. For example, one application of considerable interest is voice input in aircraft cockpits. Such an application requires that the speech recognition system be able to overcome problems such as high ambient noise, mask breathing, high "G" forces, etc. Another country was interested in using voice input over a radio link in tactical situations; this use implies insensitivity to atmospheric interference and varying signal levels, as commonly encountered in radio communications.

The implementation plan that was proposed was to consist of several tasks (phases), but only the first one was developed in any detail at the 10th meeting. Task 1 was at that point envisioned as consisting of the following elements:

a. Developing a list of the possible military applications of voice input, i.e., as related to specific national interests.

b. Itemizing the specific scientific problems associated with each of the applications. Then listing national interests and existing research programs aimed at the various problems.

c. Discuss, and reach consensus, on the specific cooperative projects to be undertaken.

The draft Terms of Reference and research plan [Annex G to Ref. 20] were submitted to AC/243 Panel III with a recommendation that a subgroup on speech processing be established. The remainder of the cooperative activities on this topic are covered in Section 4.

2.10 Proposed cooperation on mechanical waves

Simultaneous with the preparation of the mechanical waves technology assessment [Annex G to Ref. 19], a list of possible areas of cooperation was developed, including: (i) general purpose algorithms, theory, and computer programs for APR; (ii) vibration analysis with respect to fault detection, quality control, and nondestructive testing; (iii) mathematical models of propagation; (iv) detection and classification of explosions by seismic waves; (v) theory and algorithms governing the use of multi-microprocessors in a single environment; and (vi) seismic and acoustic detection and classification of military vehicles and personnel (this would have included the development and exchange of data bases).

At the 10th meeting the Ad Hoc Group on Mechanical Waves concluded that there were four military applications areas in which there was interest in cooperation on processing techniques: (i) active sonar; (ii) passive sonar; (iii) bettlefield surveillance; and (iv) fault detection and maintenance in structures. Three research tasks were agreed on, viz.: (i) pooling of data bases; (ii) feature generation, reduction, and selection techniques; and (iii) classification methods. Each of these research tasks was viewed as intersecting with each of the four military applications areas [Annex I to Ref. 20]. A Terms of Reference was drafted, but completion of the research plan was deferred to the 11th meeting.

At the 11th meeting the Ad Hoc Group drafted a research plan encompassing all four of the military applications areas listed above. A set of nine phases were identified which applied to each of the applications areas:

a. Establish a standard data base format to facilitate an international exchange of data.

b. Exchange the data bases and insure that they are valid on the target computer.

c. Identify specific personnel to serve on the projects of interest.

d. Identify, develop, and exchange applicable APR algorithms.

e. Propose and examine any systems engineering required to implement the APR algorithms in a useful fashion.

f. Compile a comprehensive catalog of all of the APR algorithms, identify the purpose for which they were created, and cite any special considerations for their use.

g. Submit the algorithms to a variety of tests and record the testing environment and results.

h. Conduct a careful evaluation of the information collected above.

i. Prepare a final report.

This proposal [Annex F to Ref. 21] was submitted to Panel III for approval, along with a recommendation that a subgroup on mechanical waves be established under RSG-4 to conduct the cooperative research. However, at its subsequent meeting, Panel III rejected the Terms of Reference and research plan as being too general; instead, Panel III requested that a TOR be prepared for a subgroup on APR applied to battlefield surveillance. Also, RSG-4 was asked to prepare a statement as to the feasibility of cooperation on the sonar signal processing aspects of mechanical waves.

2.11 Proposed cooperation on APR in battlefield surveillance

RSG-4 participants having ongoing interests in the acoustic and seismic aspects of APR in battlefield surveillance sent specialists to the 12th RSG-4 meeting. A TOR and research plan were prepared as requested by Panel III.

In battlefield surveillance the objective is to detect, locate, and identify enemy forces on the battlefield by remotely-emplaced passive devices which sense mechanical waves. The problem is complex. The potential targets (men and machines) generate widely varying acoustic and/or seismic signals, depending on type of activity, rate of speed, conditions of the ground, meteorological conditions, etc. The media (earth, water, and air) through which mechanical waves propagate are far from being uniform, and thus the wave trains may be significantly distorted within relatively short distances. Battlefields are notoriously noisy, both acoustically and seismically, and this may add greatly to the problem of identifying the signal generated by a specific target or category of targets.

There are several major impediments to the development of greatly improved sensor systems, including the following:

a. The design data bases are inadequate. A design data base is a library of records of the signals generated by known targets (chiefly men, ground-contact vehicles, and low-flying aircraft) operating under carefully documented environmental conditions. The existing design data bases are inadequate in that: (i) too few target types are represented; (ii) records do not include an adequate sample of environments; and (iii) the site characteristics of many records may be inadequate.

b. The available algorithms for detecting target signals and classifying them according to position and target type have not been adequately validated.

Thus, the cooperative research proposal drafted at the 12th RSG-4 meeting [Annex A to Ref. 22] was aimed at mitigating these impediments to improved sensor systems. In brief, the proposed program involves combining design data bases, and transmitting and testing classification algorithms.

The proposed cooperative program on APR applied to battlefield surveillance consists of three primary tasks, each of which is briefly described below:

a. Task 1: Development of data transfer format. The objective of this task is to develop and disseminate to the participating nations a standardized format for recording mechanical wave data, plus all necessary ancillary data, in digital form on magnetic tape. This step is required to overcome the problems resulting from unique laboratory and national data formats. A standard data transfer format would largely solve this problem. It is planned to follow the transfer format developed by the image processing subgroup to the extent possible.

b. Task 2: Development of an algorithm transfer format. Such a format would make it possible for all interested laboratories to use all relevant computer programs available within the participating nations with a minimum of cost and time delay; it is also essential to the conduct of Task 3, below. It is planned to follow the existing algorithm transfer format developed by the image processing subgroup to the extent possible.

c. Task 3: Development of improved APR algorithms. Many algorithms and procedures already exist, including a variety of APR procedures, feature extraction algorithms, digital models of sensor logics, digital filtering techniques, as well as mathematical models that purport to simulate propagation of mechanical waves through layered and nonhomogeneous media. Many of these are of recent development, and have not been adequately validated. The objective of this task is thus to develop an improved set of algorithms and mathetical models of such reliability that they can be used with confidence by the sensor design and military planning communities to predict the performances of existing and conceptual sensors on the battlefield.

Panel III has recently approved this TOR and research plan, and established RSG-11 on APR in Battlefield Surveillance. Three countries are actively participating at present and several more are planning to initiate programs in this area.

2.12 Proposed cooperation on APR in sonar signal processing

As noted above (Section 2.10), Panel III rejected the notion of general cooperation on a mechanical waves program that included both active and passive sonar, and requested a statement from RSG-4 concerning the feasibility of cooperation on APR applied to sonar signal processing. The Ad Hoc Group on Mechanical Waves, which consisted mainly of sonar specialists, discussed this matter in detail at the 12th meeting and again agreed that cooperation is feasible. Moreover, a plan of cooperative research on APR in sonar signal processing was drafted [Annex B to Ref. 22], which consisted of the following elements:

a. Step 1: Creation and exchange of a common data base of sonar signatures. This would provide a uniform standard against which to test various sonar signal processing algorithms. It would also aid in the development of new algorithms.

b. Step 2: Identification, development and testing of appropriate APR algorithms. Participants will individually compile a set of APR algorithms, then share them to form a central listing. Each participant will then be assigned a subset of APR algorithms to apply to the common data base (Step 1, above).

c. Step 3: Evaluation and final report. At the completion of the testing phase, the results will be evaluated by a panel appointed by the participants. The findings of this panel will be presented to Panel III and, it is hoped, would form the basis for the development of more fully automatic APR techniques for sonar signal processing.

This proposal was subsequently presented to Panel III, which shelved the topic pending further discussion. Apparently there is concern at the Panel III level that all of the NATO bodies that are involved with any aspect of sonar should be fully informed about this proposed cooperation prior to taking any action.

2.13 Summary note on proposed cooperative research

The reader will undoubtedly have noted the high degree of communality among the cooperative research proposals developed in the areas of images, speech, battlefield surveillance, and sonar signal processing (Sections 2.8, 2.9, 2.10, 2.11, and 2.12). In effect, the APR specialists in each of these applications areas have said the same thing: before we can perform substantial cooperation in these areas we must (i) have a common data base, or be able to freely transmit data bases among the participating laboratories; and (ii) be able to share existing algorithms and other APR procedures, preferably in digital form using a suitable transfer format.

Stated another way, these specialists are saying: "There is a communication gap in many areas of APR, and international co-operation depends on eliminating this communications gap".

RSG-4 can hardly take credit for this discovery (although, certainly RSG-4 should take credit for awakening the military APR community to the gravity of the situation). Various other groups, including the IEEE, National Bureau of Standards, and Electronic Industries Association AIPR Committee have tried to make the same points for years. Also, there were two panel sessions at the Second International Joint Conference on Pattern Recognition [23,24] which addressed some of these same issues.

The situation is mentioned here primarily because RSG-4 has been subjected to some criticism for having "re-invented the wheel" by listing the same, or a very similar, problem at the outset of proposals for cooperative research in five military applications areas. In conclusion, it is hoped that, by now, everyone is aware of the existing communications gap. International cooperation is only feasible if the gaps can be closed.

2.14 Search for other military APR topics for study

When RSG-4 had scheduled assessments for all the military APR applications originally selected for study (vix., image, speech, and mechanical wave processing), it began to look for "other topics" that should be considered for assessment and/or cooperation. This search for other topics was initiated at the 6th meeting held in August 1974 [11]. Three candidate topics were brought up at the 8th meeting (July 1974): (i) signal processing in general; (ii) robotics; and (iii) interactive processing from the view of man-computer task optimization.

At the 9th meeting (February 1976) [19], five possible topics were discussed:

a. Interactive methods of semi-automatic pattern recognition Earlier, RSG-4 had accepted, but then rejected, this topic for analysis; the final consensus was that most aspects of the topic would more appropriately be covered under the various technology assessments [9, 12, 16]. However, a new aspect of the topic was raised at this meeting: man-machine task optimization.

b. Interactive methods for APR system design. The OLPARS system at Rome Air Development Center [25, 26] had already proven to be a valuable tool for optimizing APR techniques and system design. It was thought that possibly an OLPARS-like facility should be established somewhere in Europe.

c. APR facilities of research laboratories. This would be a catalog of facilities in both participating and nonparticipating countries and laboratories. Another aspect of the topic was the notion of developing a NATO-supported APR research facility somewhere in Europe.

d. Fault detection (all sensors). This involves the general problem of fault detection and anticipation of equipment failures, using all possible sensors (not just acoustic and seismic, as proposed by the Mechanical Waves Group).

e. Microwave processing and signature analysis. This topic included (but was not necessarily limited to) radar signal processing and signature analysis. The topic could have included ECM, although it was generally believed that other NATO bodies were more appropriate for discussions of ECM.

(A word should be injected at this point concerning this search for "other topics" and what would have happened had consensus been reached. One of the reasons why this was such a difficult question is that some of the participating countries did <u>not</u> want to discover <u>any</u> new topics for assessment and/or cooperation. And the reason for that position is really quite simple: they could not afford it. In other words, by this time we had already recommended establishment of a subgroup to prosecute the image processing project, and had two other topics scheduled for assessment for which subgroups would eventually be required. Some of the smaller NATO countries are simply not able to send representatives to an infinite number of NATO groups, regardless of the military significance of a topic.)

At the 10th meeting (November 1976) [20] the responses to the five topics listed above was predictable. Four countries wanted practically all of the new topics incorporated into the work of the three topical subgroups. Two countries were interested in interactive processing, one in APR system design, one in a catalog of facilities, three in fault detection, and two in microwaves. However, there were, in addition, several responses in an area that could be characterized as "multi-sensor pattern recognition" which was, through discussion, elaborated as consisting of several aspects: (i) battlefield surveillance and reconnaissance; (ii) fault detection, quality control, and nondestructive testing; and (iii) medical diagnosis. It was agreed that additional national interests should be surveyed, and that, in addition to these topics, the surveys should also include "artificial intelligence and/or robotics."

Some of the participating countries presented a great deal of information about possible other topics at the 11th meeting [21], so much in fact that it was not possible to reach any consensus at the meeting; however, some participants were still adamantly opposed to the selection of any new topic for study (see above). There was some discussion about the possibility of organizing a NATO workshop to identify military problems to which APR techniques could be applied. There was also consideration of requesting AC/243 Panel I on Long-Term Scientific Studies to initiate a study. However, these discussions were terminated by Panel III's decision to disband RSG-4.

3. RSG-4 SUBGROUP 1 ON IMAGE PROCESSING (NOW RSG-9)

As previously noted (Section 2.8) the initial proposal for cooperation on image processing was drafted at the August 1974 meeting. Thereafter, the proposal was circulated within the participating countries for comment. The initial meeting of the subgroup on image processing was held in July 1975, at which time the image processing specialists began to plan the conduct of the cooperative tasks. Meanwhile, RSG-4 wrestled with questions about how best to manage the cooperation.

The alternatives considered were: (i) establishment of another RSG; (ii) establishment of an internal subgroup; and (iii) operation under a signed Memorandum of Understanding (MOU). The MOU approach was at first favored by the U. S. because of its visibility and because it would have permitted participation by non-NATO nations. The disadvantages of the MOU approach include the fact that it divorces the project from the NATO group that proposed it, and it requires creation of a separate project management group and structure of the type that already exists under the NATO umbrella (see Section 1). While RSG-4 at first concurred with the MOU approach, Panel III recommended that the cooperative work be conducted by a subgroup established under RSG-4.

3.1 Development of an imagery transfer format

Phase 1 of the cooperative research program on image processing involved reaching agreement on a magnetic tape format to be used in exchanging digitized image data bases among the participating laboratories. Such a format would enable international exchange of image data without necessarily imposing any changeover in image storage format on the laboratories. To exchange data, each laboratory would only have to write two new computer programs, viz., one to transform its data <u>into</u> the transfer format, and another to transform data <u>out of</u> the format.

Preliminary discussions about the transfer format were held at the formative meeting of the image processing subgroup (hereinafter referred to as SG-1) in July 1975. Prior to the 2nd meeting of SG-1

(February 1976) the first draft of a transfer format was circulated [Annex G to Ref. 26]; the discussions at the 2nd meeting then centered around various laboratory hardware and software considerations requiring modification(s) of the proposed format. Minor changes were made as required, and the transfer format was finalized at the 3rd SG-1 meeting in November 1976. This format was subsequently received considerable attention [28, 29] from the pattern recognition community, and is being considered for use as an Electronic Industries Association standard [29].

Table 16 summarizes the characteristics of the SG-1 imagery transfer format. A paper is presently being prepared in which all the details will be presented [30].

Table 16
Characteristics of the SG-1 Imagery Transfer Format

General:

a. Data is recorded on 1/2 inch wide magnetic computer tape.
b. Format allows either 7 or 9 tracks at 800 BPI, NRZI.
c. Maximum record length allowed is 4K tape characters (6 or 8 bits).
d. A tape contains one or more files.
e. Files are separated by single EOF marks.
f. Last file on tape is followed by at least two EOF marks.
g. A program unit is a program or subroutine or function or card input or output decks.
h. Each file contains a whole image or one or more program units. Images too large for one tape must be divided into sub-images which are then handled as individual images.
i. Each file consists of two header records followed by N image data records and N card image records.

Header 1:

a. The first record of any file (Header 1) is 128 bytes long.
b. Header 1 is coded in BCD for 7 track tapes, even parity; and in ASC II (Modified) for 9 track tapes, odd parity.
c. A symbol is recorded in one byte on the tape.

Header 2:

a. The second record of any file contains a free form description of the image. This may include specification of the physical significance of the various channels, description of scanner, meteorological conditions when image was taken, location of objects of interest in image, etc.
b. Header 2 has variable length as specified in Header 1(L2). ($0<L2\leq 4096$ tape characters.)
c. Coding is same as Header 1.

(continued)

(Table 16 continued)

d. For source code exchange, Header 2 should contain the name, telephone number and address of the individual responsible for each program unit.

Image Data:

a. Image data is coded in binary, odd parity, one's complement.

b. Integer format is basic to all others.

c. Integer format:

(1) Most significant part is recorded as first tape character.

(2) Most significant bit of first tape character is sign bit (0→positive, 1→negative).

(3) Most significant bit is the left most bit.

d. Real format:

(1) The exponent and mantissa (base 2) are reduced to integers. Both of these are then stored as separate integers on the tape for each sample.

(2) First E_E tape characters for the exponent, then E_m characters for the mantissa.

e. Complex format:

(1) Integer values (T=1) stored as integers.

(2) Real integers (T=3) stored as real values. Each channel consists of two values. The first value is the real part, the second is the imaginary part.

Multichannel Data:

a. Nonregistered data stored as separate images in separate files.

b. Registered data stored on pixel interleaved basis. (For one sample the values of the different channels are recorded adjacently.)

3.2 Documentation of image processing algorithms

Phase 2 of the cooperative program, as originally proposed, consisted of developing a standardized format for documenting existing image processing algorithms, compiling a list of available algorithms, and disseminating the list of standardized documentation among the participating laboratories. The following activities have actually been accomplished under this task:

a. All the participants exchanged representative lists of their computer programs and algorithms.

b. It was determined that a wholesale exchange of image processing algorithms would not be useful (and would be very expensive), because many of the existing algorithms had been written in assembly language and were very machine dependent. So it was decided not to develop a standardized algorithm description for existing programs. However, a descriptive format may be developed for new programs.

c. Minor modifications were made to the imagery transfer format [29], making it possible to exchange FORTRAN IV source codes. This modified format is described in Ref. 30.

3.3 Assessing national military priorities

Phase 3 of the cooperative program was aimed at conducting such discussions as were required to explore national priorities for achieving military image processing applications and, from the resulting consensus, to derive a prioritized list of topics that could form the basis for substantive cooperative research. Such discussions were conducted at the first three meetings of SG-1. (The national positions are not included in this report.) Four cooperative research projects were drafted as a direct result of these discussions; the first two projects have been approved by Panel III.

3.4 Cooperative research on image processing

3.4.1 Project 1: Discrimination and classification of operating military targets in natural scenes from multispectral data [Annex L to Ref. 20].

a. Data consists of images of operating military targets such as tanks, trucks, ships, planes, etc., received by active reflection or passive emission of visible, IR or microwave radiation. Data to be available in the transfer format [29, 30].

b. Preprocessing includes digitizing, rectification, registration of the images in the several spectral bands and transferring to the transfer format.

c. Goal 1: Discrimination of possible targets from the background. The result will be presented to the operator as a processed image. Possible approaches include multispectral analysis, spatial analysis (e.g., texture analysis).

d. Goal 2: Classification of operating military targets. The results of the processing will be presented as a list of targets.

e. Subgoals to be reached under Goals 1 and 2:

(1) Generation and evaluation of multispectral features for the description of targets and backgrounds.

(2) Generation and evaluation of spatial features (texture, local shape) for the description of targets and backgrounds.

(3) Description of targets and backgrounds based on (1) and (2).

(4) Classification or cueing of targets based on (1), (2) and (3).

(5) Comparison and evaluation of techniques.

(6) Evaluation of the hardware requirements for implementation of the methods.

3.4.2 <u>Project 2</u>: <u>Detection of geographical targets of military relevance from aerial multispectral imagery</u> [<u>Annex M to Ref. 20</u>]

a. Data consists of multispectral images from reconnaissance satellites, airborne sensors, etc.

b. Preprocessing includes digitizing, rectification, registration of the images in the several spectral bands and transferring to the transfer format.

c. Goal 1 is to detect planimetric patterns on the analyzed scene, e.g., roads, railroads, towns, boundaries of lots, industrial plants, air fields, missile sites, and terrain types. Possible approaches include multispectral analysis, texture and shape analysis.

d. Subgoals to be reached:

(1) and (2) Same as Project 1, for ground targets.

(3) Automatic image segmentation based on (1) and (2).

(4) Description of patterns based on (1), (2) and (3) (shape of the pattern, texture, spectral criterion of the surface of the pattern).

(5) Classification of the patterns.

(6) and (7) Same as Project 1 (5) and (6).

3.4.3 Project 3: Tracking of operating military targets in natural scenes from multispectral data [Annex E to Ref. 27].

a. Data consists of images of operating military targets such as tanks, trucks, ships, planes, etc., received by active reflection or passive emission of visible and/or IR radiation. Sequences of frames showing target and/or sensor motion are required. Data should be in the transfer format.

b. Preprocessing is essentially the same as Projects 1 and 2.

c. Goal 1: Target tracking for fire control. Tracking of stationary and/or moving targets from stationary and/or moving platforms by use of an imaging sensor for fire control purposes. Possible approaches include correlation tracking, color tracking, and edge-centroid tracking.

d. Goal 2: Terminal homing for smart ordnance. Tracking of stationary and/or moving targets from a moving missile for the purpose of directing the missile into the target. Possible approaches include same as above plus reticle tracking.

3.4.4 Project 4: Automatic navigation using real time image data [Annex F to Ref. 27].

a. Method: Radar or other EM radiation reflection correlated in very near real time with suitably accessible reference imagery.

b. Data: Real or synthetic images of geographical terrain taken in applicable EM wave band or detailed map data from which such imagery can be synthesized.

c. Preprocessing: Image rectification and scaling, selection of catalog of most readily correlated features, development of map formats (if these are used as reference) in order to convert cartographic data into reference imagery for the purpose of correlation, conversion of polar display to rectilinear format.

d. Goal 1: To achieve success in the preprocessing steps noted above.

e. Goal 2: To generate parameters for auto pilot update.

Projects 3 and 4 (described above in Sections 3.4.3 and 3.4.4) have not yet been approved by Panel III. It should also be noted that, with the termination of RSG-4, Projects 1 and 2 (Sections 3.4.1 and 3.4.2) are being continued by AC/243 (Panel III) RSG-9 on Image Processing.

4. RSG-4 SUBGROUP 2 ON SPEECH PROCESSING (NOW RSG-10)

The first cooperative task proposed by SG-2 on Speech Processing was in the area of voice data entry for command and control. It was agreed that the first two goals of that task should be: (1) to develop specifications and formats for the exchange of speech data bases; and (ii) compilation of a complex matrix relating military applications of voice data entry to the scientific problems involved and the state of the art [31]. The exact nature of the cooperative research on voice data entry would be derived from the conclusions reached in evaluating the matrix.

4.1 Specifications for analog speech tapes

Table 17 summarizes the specifications agreed upon for the exchange of data bases in the form of analog tape recordings [32]. (Formats for digital recordings, algorithm exchange, etc., are to be decided in the future.) This format is being tested by having each country generate a tape recording to be distributed to each other country. The tapes contain the digits, spoken in the project officers' native languages. Evaluation of the tape format is in progress.

Table 17
Specifications for Analog Speech Tapes

1. Reel to reel tapes only; two track.
2. Minimum reel size 5 in (12 cm); maximum 10 in (24.5 cm).
3. No Chromium dioxide tape to be used.
4. Correction or equalization network to be specified.
5. Specify recorder and recorder correction network.
6. Tape speed 7.5 in/sec (19 cm/sec) or 3.75 in/sec (9.5 cm/sec).
7. Spoken header and written specification (date, organization, time, etc.).
8. Reference tone (1000 Hz) after header.
9. Only one track to be used; other track blank.

4.2 Assessment of voice data entry status and priorities

At the second SG-2 meeting (December 1977) a mini-technology assessment was conducted for the specific topic of voice data entry [32]. Table 18 shows the current (i.e., December 1977) ratings of the state of the art in voice data entry in relation to military applications. Note that <u>none</u> of these applications is presently regarded as being solved from a military applications standpoint.

Table 18
Current State of the Art in Voice Data Entry in Relation to Potential Military Applications

Potential Applications	Rating*
A. Avionic systems	2
1. Query systems	
a. Weather information via voice response	
b. Information retrieval from aircraft instruments	
2. Data entry	
a. Voice input to flight information center	
b. Voice control over instrumentation	
c. Direct voice input to avionic systems	
d. Preparing and arming weapon systems	
e. Changing radio channel frequencies	
f. Preparing flight plans	
B. Interactive Data Handling for Command, Control, Commo	
1. Query systems	3-4
a. Data base query over telephone channels	
b. Inventory control (logistics)	
2. Data entry	3-4
a. Computer input/output	
b. Hands/eyes busy & complete mobility situations	
c. Machine control (semiautomatic assembly)	
d. Remote control (dangerous materials)	
3. Battlefield/tactical data entry	2
a. Information from battlefield to command post	
b. Final control information by front line observer	
4. Maps, charts, photo data entry	3-4
a. Source data capture for generating maps & charts	
b. Aids to photo interpreters & analysts	
5. Simulation for training	2-3
a. Air to air intercept	
b. Cockpit situation simulation	
c. Human operator simulations	
d. Ships maneuvering simulation	

*Ratings: 1 = Early stages of basic investigation
2 = Laboratory investigation
3 = Operational testing
4 = Well in hand
5 = Solved

Table 19 shows the state of the art in voice data entry in relation to desired system capabilities. For this analysis the ratings apply to the general topic, i.e., composed of all of the applications listed in Table 18. Note that no capabilities have been assigned a rating higher than "3" (denoting operational testing).

Table 20 is a matrix which shows the degree to which the capabilities of voice data entry systems are important to the various potential military applications.

Table 21 shows the preliminary consensus about priorities for achieving voice data entry capabilities [32]. These rankings were generated by collapsing across all the national interests expressed at the second SG-2 meeting.

4.3 Cooperative research on voice data entry

Based on an evaluation of the information contained in Tables 18-21, and other national inputs, SG-2 proposed its first substantive cooperative research project which is titled: "Connected-Word Voice Data Entry for Military Command and Control." In Table 20, for example, the connected digit/word capability is rated either 4 or 5 in four of the military applications areas. In Table 21, connected digit recognition was ranked as having the highest priority among the participating countries.

This project aims to overcome the restriction of many present military applications which is that the words must be spoken with pauses between them. It aims to perfect techniques that would permit the digits and a restricted set of command words to be spoken in a connected fashion.

As proposed, the SG-2 participants would synchronize and, in some cases, redirect national efforts toward development of techniques for recognizing connected words. Initial attempts will concentrate on connected digit recognition under laboratory conditions, i.e., using high-quality speech input, etc. After the capability to recognize connected words (digits) has been demonstrated in the laboratory, the project will be broadened to include realistic operational military conditions (telephone bandwidth, etc.). This approach implies that experts think it will be easier to overcome the segmentation problem than the signal quality problem.

At the same time the various ergonomic aspects of military applications will be studied to estimate the benefits accruing from the use of automatic speech recognition, the extent of the problem areas such as poor signals, and the exploitable structures of the tasks leading to techniques such as dynamic sub-vocabulary

Table 19
Current State of the Art in Voice Data Entry in Relation to System Capabilities

Desired System Capabilities	Rating*
1. Use of task syntax, i.e., dynamic sub-vocabulary selection	3
2. Operation over communication networks, e.g., commercial telephone	2-3
3. Operation in very bad conditions, e.g., tactical radio, secure voice	1
4. Complete spoken numbers connected with a few other words	2-3
5. More general connected sequences of words; limited vocabulary; simple syntax	2
6. Ability to set up for new user without him uttering all the vocabulary words	2
7. Adaptation of speaker to the situation while he is using the system; use of dialogue utterances to improve subsequent recognition	1-2
8. Small size and weight	3
9. Ability to do verification in conjunction with speech recognition	3
10. Prior study of the task (ergonomics, systems analysis, operational)	2

*Ratings: 1 = Early stages of basic invertigation
2 = Laboratory investigation
3 = Operational testing
4 = Well in hand
5 = Solved

Table 20
Degree to Which the Desired Capabilities of Voice Data Entry Systems Are Important to Potential Military Applications

APPLICATIONS \ CAPABILITIES	Task syntax; dynamic selection	Operation over communication networks	Operation in very bad conditions	Spoken digits; a few other words	More general connected words	Ability to train new users	Adaptation to the situation	Small size and weight	Verification with recognition	Ergonomics; systems analysis
Avionic query/entry	4	1	5	5	2-4	2-4	3-4	5	2-3	5
Interactive C^3	4-5	5	2	4	3-5	3-5	3-4	2	4	4
Tactical data entry	4	4-5	5	5	1-5	4	3-5	5	4	5
Graphics data entry	3	2	1	4	3	3-4	3	2	1	3
Simulation	4-5	2	1	3-5	1-5	3-5	3-5	2	1	4-5

Ratings: 1 = not required; 2 = advantageous; 3 = important; 4 = very important; 5 = necessary.

Table 21
Preliminary Consensus About Priorities for Achieving Voice Data Entry Capabilities

Desired Capabilities	Rank Order*
1. Use of task syntax	3
2. Operation over communication networks	5
3. Operation in very bad conditions	9
4. Recognize connected digits, few other words	1
5. More general connected words	8
6. Ability to train new users	4
7. Situation adaptation	7
8. Small size and weight	6
9. Simultaneous verification & recognition	10
10. Ergonomics; systems analysis	2

*1 = highest priority.

selection will be determined. The applications which will receive the most attention (see Table 18) are voice control of avionic systems and interactive voice data entry for command and control.

In January 1978 Panel III elevated RSG-4 SG-2 to be AC/243 (Panel III) RSG-10 on Speech Processing. The cooperative projects described in this Section are continuing.

5. COOPERATIVE RESEARCH ON APR IN MECHANICAL WAVES

5.1 Battlefield surveillance (RSG-11)

In January 1978, Panel III approved the Terms of Reference of RSG-11 on APR in Battlefield Surveillance. At this writing, however, the national project officers have not been appointed, the number of participating countries is unknown, and no information is available about possible cooperation beyond that presented in Section 2.11.

5.2 Sonar signal processing

The possibility of cooperation on this topic is still being discussed in Panel III; the prospects for establishment of RSG are uncertain. In addition to the problems of data sensitivity, there are internal NATO coordination considerations involved.

6. SUMMARY OF RSG-4'S ACCOMPLISHMENTS

RSG-4 was the first (but, hopefully, not the last) NATO body to undertake any sort of long range study of pattern recognition applications to military problems. As such, it established precedents which will probably guide future activity in this area, particularly if conducted under the auspices of AC/243 Panel III on Physics and Electronics.

This Section merely summarizes the most important things we can point to as accomplishments (products, if you will) of RSG-4's more than six years of existence:

a. A classification scheme for organizing the topic of military applications of APR technology was developed. In 1972 when this was done no other such scheme was in existence.

b. Three iterations of a comprehensive information exchange on PR research project summaries (military and civil) were conducted. The combined summaries were distributed to all identifiable defense researchers and contractors in all the participating countries. This was the first effort of this sort (at least in the U. S.), and it prompted one U. S. researcher to suggest that we should start a cooperative research project among U. S. researchers!

c. Technology assessments of military applications of PR technology to image, speech, and mechanical wave processing were conducted and published as NATO reports. (The speech processing assessment was also published in the open literature.) Each of these assessments summarized the state of the art with respect to military applications, identified the PR techniques that needed to be perfected, and listed the unsolved problems associated with each PR technique. Judging from comments already received from various national defense authorities, these assessments have already been used to determine (in some instances) directions for future military-sponsored research.

d. Cooperative research programs have been established in three applications areas (images, speech, and mechanical waves applied to battlefield surveillance). These programs were at first prosecuted by subgroups established under RSG-4 (in itself, a rarity) and now, with the conclusion of RSG-4's work, three new RSGs have been constituted to carry out these programs (viz., RSG-9, 10 and 11 under AC/243 Panel III on Physics and Electronics).

e. A transfer format for digitized imagery has been developed and standardized by RSG-4 Subgroup 1 (now RSG-9). This format makes it possible for laboratories to exchange image data bases by writing only two new computer programs: one to transform their data into the transfer format, and another to transform data out of the transfer format. This transfer format has already met with international acclaim, and is being considered for use as a national standard data storage and transfer format.

f. A transfer format has likewise been developed for the exchange of image processing and APR algorithms and procedures.

g. Although relatively intangible, there has been a great increase in the amount of communication among PR researchers, particularly international communication among military PR researchers, as a direct result of the information exchanges and proposals for cooperation developed by RSG-4.

h. An OLPARS-like system [25, 26] was established in the U. K. as a direct result of information exchanged, and technical visits conducted, under RSG-4. U. K. representatives witnessed a demonstration of the OLPARS capabilities during the 4th meeting (November 1973); they requested the software in May 1974, and the final transfer of system software took place at the 12th meeting in November 1977.

7. WHAT RSG-4 FAILED TO ACCOMPLISH

RSG-4 failed to accomplish any sort of information exchange, assessment or cooperation directly related to the human factors aspects of military APR applications. Despite the fact that all APR applications to complex problems in both military and civil situations presently involve man-machine interaction, and despite the fact that such is likely to be the case for the foreseeable future, no sustained interest could be generated among the majority of the RSG-4 delegates to consider the human factors aspects of APR applications in any separate and systematic fashion. This fact seems particularly ironic in view of the fact that two countries (Netherlands, and U. S.) appointed delegates to RSG-4 who had backgrounds in computer science and human factors engineering, for the express purpose of injecting some human factors consideration into the work of RSG-4.

This unwillingness to consider the human factors aspect may be illustrated by an incident that occurred early in the deliberations of SG-1. The first proposal for cooperative research on detection of operating military targets gave as one of the purposes the "cueing of operators". But the proposal included no procedure for determining that the operator had, in fact, been cued! This was pointed out, and the proposal's author stated that he was interested only in image processing, not in operator cueing. As a result, the final version of the project statement (Section 3.4.1) makes only an oblique reference to operator cueing and does not cite it as one of the goals of the project.

It would, however, only be fair to make two additional points in this regard. First, the other five national delegates to RSG-4 had no background in human factors and little interest in those aspects of APR applications; thus, they would probably have felt out of place trying to deal with such problems. Secondly, some sentiment was expressed to the effect that the human factors aspects of APR applications should be treated by RSG-5 on Human Engineering (which was later elevated to become AC/243 Panel VIII on Human and Biomedical Sciences [33]). In fact, due to the communality of interest in pattern recognition problems between RSG-4 and RSG-5, liaison was established at the national level to keep RSG-5 informed about RSG-4 activities. However, RSG-5 did not then have any working groups addressing topics relevant to APR, nor does Panel VIII have any such groups at this time.

The closest that RSG-4 came to addressing any of the human factors aspects of APR applications was in its search for other topics for study (see Section 2.14). One of the possible topics that was discussed was "Interactive Methods of Semi-Automatic Pattern Recognition", which was elaborated to include man-machine task optimization. Task optimization has recently received considerable attention [34-37]. In order to optimize the design of a pattern recognition system, one has to be able to define the capabilities of both the human and the machine components; then, presumably, one can configure a system for a particular application that will make the best use of the components' capabilities and limitations. However, consideration of this topic was cut short by Panel III's decision to disband RSG-4.

8. PROSPECTS FOR FUTURE NATO PATTERN RECOGNITION ACTIVITY

AC/243 (Panel III) RSG-4 on Automatic Pattern Recognition was disbanded by Panel III at its January 1978 meeting [38]. This major scientific information exchange and technology assessment program, whose activities spanned more than six years, has ended. But in its place, three new RSGs continue to function: (i) RSG-9 on Image Processing; (ii) RSG-10 on Speech Processing; and (iii) RSG-11 on Automatic Pattern Recognition in Battlefield Surveillance with

Mechanical Waves. In one sense, then, future NATO activity in the area of pattern recognition is virtually assured for the next five years.

The main question, however, is: "What are the prospects for NATO getting involved in other pattern recognition applications in the future?"

One of the factors that will have to be considered is the degree of support that the smaller nations may be willing or able to provide. The smaller NATO nations cannot afford to participate in an infinite variety of groups. Thus, one requirement for further NATO pattern recognition activity will be identification of an application that is so important it will cause nations to reorder their priorities.

Another factor governing future activity will be a determination of the most appropriate NATO body to serve as a forum. At present, Panels III and VIII appear to be the most likely candidates; but other choices should probably be considered. The matters discussed in Section 7 may also be relevant here.

For the immediate future (next five years) NATO pattern recognition activity will probably be confined to RSGs 9, 10 and 11. After that Panel I should probably be petitioned to initiate one of its long-term study programs.

REFERENCES

1. U. S. Army Regulation 34-1. Standardization: United States Participation in NATO Military Standardization Research, Development, Production, and Logistic Support of Military Equipment. Headquarters, Department of the Army, Washington, D. C., 5 April 1974.
2. AC/243 (Panel III) Document 115. Report of the Ad Hoc Group on the Military Applications of Millimeter Waves. NATO Headquarters, Brussels, Belgium, 28 June 1974.
3. D. C. Hodge. Automatic Pattern Recognition: Defense Research Projects in Seven NATO Countries. Report RSG-4-1A, July 1972. (Available only from U. S. Defense Documentation Center, Alexandria, Va.: AD905384L.)
4. D. C. Hodge & E. J. Simmons. Automatic Pattern Recognition. Report RSG-4-2A, June 1973. (Available only from U. S. Defense Documentation Center, Alexandria, Va.: AD914008L.)
5. D. C. Hodge & H. E. Webb. Automatic Pattern Recognition Research in the United States. Part 1: Department of Defense Supported Projects (Unclassified; Unlimited). Part 2: Projects Supported by Other Sources. Report RSG-4-3A, September 1976.

6. D. C. Hodge & H. E. Webb. Automatic Pattern Recognition. Part 3: Projects Conducted by Countries Other Than the United States. Report RSG-4-3A, December 1976.
7. E. J. Simmons, Jr. A Summary of Expressions of Interest in International Workshops in the Area of Automatic Pattern Recognition. Presented to NATO AC/243 (Panel III) RSG-4, May 1973.
8. Unpublished Minutes of 5th RSG-4 meeting, November 1973.
9. An Assessment of the Technology of Automatic Image Processing for Military Purposes. Annex G to Minutes of 7th RSG-4 meeting, August 1974.
10. D. C. Hodge. NATO RSG-4 on Pattern Recognition: Proposed Cooperative Research on Image Processing. In L. N. Kanal (Ed.), Proceedings of EIA-NSF Workshop on Near Future Prospects in Image Pattern Recognition. Silver Spring, Md., 1975, pp. 157-163.
11. Unpublished Minutes of 6th RSG-4 meeting, May 1974.
12. AC/243 (Panel III/RSG.4) Document 13. An Assessment of the Technology of Automatic Speech Recognition for Military Applications. NATO Headquarters, Brussels, Belgium, 13 April 1976.
13. B. Beek, E. P. Neuberg, & D. C. Hodge. An Assessment of the Technology of Automatic Speech Recognition for Military Applications. IEEE TRANSACTIONS on Acoustics, Speech, & Signal Processing, 1977, vol. ASSP-25, 310-322.
14. D. H. Klatt. Review of the ARPA Speech Understanding Project. J. Acoust. Soc. Amer., 1977, 72, 1345-1366.
15. Unpublished Minutes of 7th RSG-4 meeting, August 1974.
16. AC/243 (Panel III/RSG.4) Document 15. An Assessment of the Technology of Automatic Processing of Signals Due to Mechanical Waves in Solids, Liquids and Gases for Military Purposes. NATO Headquarters, Brussels, Belgium, 15 March 1977.
17. EIA Committee on Automatic Imagery Pattern Recognition. 5th Annual Symposium, College Park, Md., April 1975.
18. R. Franke. The Specification of Algorithms. Technical Report, U. S. Naval Postgraduate School, Monterey, Calif., July 1975. AD A013564.
19. Unpublished Minutes of 9th RSG-4 meeting, February 1976.
20. Unpublished Minutes of 10th RSG-4 meeting, November 1976.
21. Unpublished Minutes of 11th RSG-4 meeting, June 1977.
22. AC/243 (Panel III/RSG.4). Final Report of RSG-4 to Panel III. NATO Headquarters, Brussels, Belgium, in press. (Submitted December 1977.)
23. L. N. Kanal (Chm.). Panel on General and Dominant Problems in Pattern Recognition. 2nd International Joint Conference on Pattern Recognition, Lyngby, Denmark, August 1974.
24. D. H. Foley & H. L. Oestreicher (Chm.). Panel on Bridging the Gap Between Pattern Recognition Theory and Practice, With Particular Emphasis on Representative Data Bases. 2nd International Joint Conference on Pattern Recognition, Lyngby, Denmark. August 1974.

25. J. Faust, H. Webb, & L. Gerhardt. RADC Interactive Laboratory for Design of Pattern Recognition, and Its Function. Proceedings of the Conference on Computer Graphics, Pattern Recognition and Data Structures, May 1975. IEEE Catalog No. 75CH0981-1C.
26. Pattern Analysis & Recognition Corp. Multics OLPARS Operations System, Vol. 1. Technical Report 76-271, Rome Air Development Center, Griffiss Air Force Base, N. Y., 1976.
27. Unpublished Minutes of 2nd RSG-4/SG-1 meeting, February 1976.
28. J. M. Evans, Jr., R. Kirsch, & R. N. Nagel (Eds.). Proceedings of Workshop on Standards for Image Pattern Recognition, held in Gaithersburg, Md., 3-4 June 1976. National Bureau of Standards Special Publication 500-8, May 1977.
29. J. S. Dehne. Proposed EIA Standards for Digital Imagery on Magnetic Tape. Paper presented to 7th Annual Symposium of EIA Committee on Automatic Imagery Pattern Recognition, College Park, Md., May 1977.
30. J. S. Dehne. NATO Transfer Format for Digitized Imagery and Source Codes. In preparation for submission to Pattern Recognition, 1978.
31. Unpublished Minutes of 1st RSG-4/SG-2 meeting, June 1977.
32. Unpublished Minutes of 2nd RSG-4/SG-2 meeting, December 1977.
33. AC/243 (Panel VIII) Document 1. Report to the DRG by the Chairman of Panel VIII. NATO Headquarters, Brussels, Belgium, 9 October 1974.
34. E. C. Poulton. Experiments on Detection and Classification with the Assistance of a Computer. Report No. DRIC-BR-48925, Royal Naval Personnel Research Committee, March 1974. AD A019314.
35. C. D. Wylie, R. A. Dick, & R. R. Mackie. Toward a Methodology for Man-Machine Function Allocation in the Automation of Surveillance Systems. Vol. I: Summary. Human Factors Research, Inc., Goleta, Calif., 1975. AD A017013.
36. D. A. Topmiller. Man-Machine Command-Control-Communication Simulation Studies in the Air Force. Technical Report 76-122, Aerospace Medical Research Laboratory, Wright-Patterson Air Force Base, Ohio, 1976. AD A042148.
37. B. M. Crawford, D. A. Topmiller, & G. A. Kuck. Man-Machine Design Considerations in Satellite Data Management. Technical Report 77-13, Aerospace Medical Research Laboratory, Wright-Patterson Air Force Base, Ohio, 1977. AD A041287.
38. AC/243 (Panel III) Document 159. Draft Terms of Reference for New Research Study Groups. NATO Headquarters, Brussels, Belgium, 6 February 1978.